Analytical Approaches for Reinforced Concrete

Woodhead Publishing Series in Civil and Structural Engineering

Analytical Approaches for Reinforced Concrete

Yufei Wu

WP
WOODHEAD PUBLISHING
ELSEVIER An imprint of Elsevier

Woodhead Publishing is an imprint of Elsevier
50 Hampshire Street, 5th Floor, Cambridge, MA 02139, United States
The Boulevard, Langford Lane, Kidlington, OX5 1GB, United Kingdom

ISBN: 978-0-12-821164-9 (print)
ISBN: 978-0-12-823201-9 (online)

For information on all Woodhead publications
visit our website at https://www.elsevier.com/books-and-journals

Publisher: Matthew Deans
Acquisitions Editor: Glyn Jones
Editorial Project Manager: Howi M. De Ramos
Production Project Manager: Nirmala Arumugam
Cover Designer: Christian J. Bilbow

Typeset by STRAIVE, India

To my parents

Contents

Preface

This book is intended to facilitate a better understanding of the fundamentals, principles, mechanisms, modeling, and difficult concepts that readers may encounter when learning about reinforced concrete (RC). It is suitable for readers of all levels, including undergraduate students, postgraduate students, practicing engineers, researchers, academics, and professionals in the field, as a reference book for further reading, study, teaching, research, and development of professional guidelines.

This text does not cover all the common elements of a reinforced concrete course. Instead, it endeavors to give the reader a better understanding and training in the fundamental concepts, including the failure modes, failure mechanisms and key factors, critical design considerations, general design requirements and design principles, and development of design and modeling approaches. Rather than introducing detailed design procedures for RC members such as beams, slabs, columns, and walls, the book covers the general design requirements and considerations for flexure, shear, bond, serviceability (cracking and deflection), ductility, composite action, and confinement effects applicable to all RC members. Particular emphasis is placed on the rationales and explanations behind the design approaches and models. These general considerations and principles are a common basis for the design codes in different countries and regions. Therefore the book is suitable for all readers regardless of their background in design codes of practice.

Both beginners and experienced professionals will find the book useful, as the fundamentals are the same and are important to all, and there are some elements in the book that readers will not have encountered elsewhere. The text attempts to explain difficult concepts and models in simple language, using analogies; hence, it suits readers of all levels. Some sections are more advanced, or deal with a particular element at a higher level. The reader can first browse through the book and skip those parts that are more difficult or that involve more complicated details than are needed in a first reading. Readers can then go back and study the areas of particular interest in more detail. The mathematical level of most sections of the book should pose little difficulty for engineering students and professionals. Some sections do involve substantial mathematical derivations and details, but these can be skipped without affecting basic understanding.

The text is an accumulation of works of the author from his three decades of experience in the industry as a professional civil/structural engineer in consultancies in China, Singapore, Australia, and Hong Kong SAR, and as an academic in Shanghai Jiao Tong University, City University of Hong Kong, RMIT University, and Shenzhen University. Much of the content is based on the lecture notes used by the author for teaching university-level reinforced concrete and other structural design courses, as well as the author's notes from invited lectures and seminars at many different universities. The content is also the result of the long-term research of the author and his co-workers. Some elements of the book are new research findings and ideas that have not been published elsewhere.

The author believes that the prime needs of a tertiary student at a good university are a deep understanding of fundamental concepts, basic skills training, and development of problem-solving ability, all of which provide "a solid foundation to construct a tall building." Students' level of understanding of difficult concepts depends very much on the level of understanding of their teacher (at least for most students, and particularly in the beginning). To better understand a problem, the teacher needs not only extensive knowledge but also good sense, insight, and the necessary practical experience in the relevant area. A broad engineering background and practical experience of the teacher provides fertile ground for a thriving crop. It is said that the most profound takes the simplest form, and it is important for teachers to be able to explain the profound in simple terms. This book tries to provide teachers with some extra tools for accomplishing this.

Adequate knowledge acquisition by students is necessary, but comprehensive or even extensive knowledge is not essential, as current knowledge can be obsolete in a few years' time. However, the fundamentals are not subject to obsolescence. Students trained in solid fundamentals will find themselves in a better position in their future careers, in terms of adaptability to the changing and challenging working environment and achievability of higher career goals. I quote here a statement from the former cochairman of Ove Arup and Partners, G.J. Zunz, as the purpose and theme of this book: "As my experience and that of my colleagues develops, I find more and more that it is the fundamentals that matter and those without a sound training in them suffer for the rest of their careers."

Since this book is somewhat different from other texts on reinforced concrete, and since it is the first edition of the book, errors, omissions, misinterpretations, and subjectivity will be inevitable. I would be grateful to receive any error notifications, constructive comments, and criticisms that readers may have.

I am grateful to my colleagues and friends in my former and current universities and to former students who provided me with strong encouragement and support over the course of writing this book. Special thanks goes to the following people:

- Fang Yuan, who produced Figs. 5.1C, 7.40, 9.46–9.48, 11.4, and helped to find Fig. 9.42.
- Biao Hu, who produced Figs. 2.18, 6.11, 8.5, 9.36, and proofread Sections 10.1–10.5.
- Lei Gao, who produced Fig. 5.17.
- Cheng Jiang, who produced Fig. 7.32.
- Pengda Li, who produced Figs. 8.39, 9.53–9.54.
- Rongqiao Xu, who proofread Section 6.2.7(1).

Finally, a heartfelt thanks goes to my wife Chunmei for her full support. This book could never have been completed without her patience, encouragement, and understanding.

Yufei Wu
Melbourne, Australia; Shenzhen, China

Notations

A	area
$A(\varepsilon_c)$	area enclosed by stress-strain curve from zero strain up to strain ε_c
A_b	area of balanced reinforcement
A_c	area of concrete
A_f	area of FRP
A_g	gross cross-sectional area
A_s	$A_{sc} + A_{st}$
A_{sc}	area of compression reinforcement
A_{st}	area of tension reinforcement
$A_{st,e}$	maximum tension steel area of cross section that can be loaded to yield under flexure
A_{sv}	area of transverse reinforcement
a	crack spacing, or length of shear span
a_1	experimentally determined material constant of steel for Menegotto-Pinto model
a_2	experimentally determined material constant of steel for Menegotto-Pinto model
a_{min}	minimum crack spacing
b	breadth of section
b_c	breadth of concrete member
b_f	breadth of FRP plate
C	compression force
c	depth of concrete compression zone
D	diameter of circular member, or damage variable
d	effective depth
d'	concrete cover thickness from compression face to centroid of compression bar
d_b	diameter of bar
E	elastic modulus
E_1	$A_{sc}E_s/bd'$, or slope of elastic part of curve
E_2	modulus/slope of curve after transition point
E_b	elastic modulus of CY block
E_c	elastic modulus of concrete
E_{cT}	elastic modulus of concrete under elevated temperature
E_{eff}	long-term effective modulus of concrete
E_f	elastic modulus of FRP
E_h	hardening modulus of reinforcement
E_l	confinement rigidity
E_s	elastic modulus of steel
E_{sec}	secant modulus of concrete

E_{sT}	elastic modulus of reinforcement at elevated temperature
$(EA)_i$	EA of element i
$\overline{EA}$	axial rigidity defined as $\frac{1}{\overline{EA}}=\frac{1}{(EA)_1}+\frac{1}{(EA)_2}$
EI	flexural rigidity of section
$\overline{EI}$	flexural rigidity of composite member under full interaction
F	force
F_b	shear force of a shear connector
F_c	compression force of concrete
F_{sc}	force of compression reinforcement
F_{shr}	interfacial shear force
F_{st}	force of tension reinforcement
$f(\varepsilon)$	stress of concrete at strain ε
f_b	yield strength of CY block
f_{bu}	ultimate bond strength, or ultimate strength of CY block
f_c	stress of concrete, or concrete stress at extreme compression fiber
f_c'	cylinder compressive strength of concrete
f_{cc}	strength of confined concrete
f_{cm}	stress of equivalent rectangular stress block
f_{co}	compressive strength of concrete under flexure, sometimes used as f_c'
f_{ct}	concrete tensile strength
f_{cT}	concrete stress under elevated temperature
f_{cT}'	concrete strength under elevated temperature
f_{cu}	concrete stress at ultimate failure point, or cube strength of concrete
f_f	stress or strength of FRP
f_{fu}	strength of FRP at rupture
f_j	strength of FRP jacket
f_l	lateral confining pressure
f_o	concrete stress at transition point, or at intersect with y-axis
f_s	stress of reinforcement
f_{sc}	stress of compression reinforcement
f_{st}	stress of tension reinforcement
f_{su}	ultimate strength of steel
f_t	tensile strength of concrete
f_{to}	tensile strength of concrete in flexure
f_y	yield stress of reinforcement
f_{yT}	yield stress of reinforcement at elevated temperature
f_{yv}	yield stress of transverse reinforcement
G_{fc}	postpeak fracture energy
g_{fc}	prepeak fracture energy
h	depth of cross section
I	moment of inertia of section
I_c	moment of inertia of concrete section
I_{cr}	moment of inertia of cracked concrete section
I_e	effective moment of inertia
I_g	moment of inertia of uncracked gross section

I_s	moment of inertia of steel bars
I_t	moment of inertia of transformed section
K_b	rigidity of bolt or shear connector
k_0	confinement effectiveness coefficient
k_s	shape factor of confinement effectiveness
k_u	c/d, neutral axis parameter
k_{ub}	neutral axis parameter for balanced failure
L	length, or span length
L_b	bond length
L_{cc}	length of compression zone of member where compressive strain is larger than ε_{cu}
L_{cs}	length of compression zone of member where compressive strain is larger than ε_{co}
L_d	development length of bond, or length of failure localization zone
L_e	effective bond length
L_e'	effective bond length of HB system
L_f	mechanically fastened bond length
L_p	equivalent plastic hinge length
L_{p0}	plastic hinge length of member without confinement
L_{pb}	additional plastic hinge length due to yield penetration into column base
L_{pc}	length of significant curvature localization zone
$L_{p\text{-}c}$	equivalent plastic hinge length under cyclic load
L_{pf}	additional plastic hinge length due to confinement
L_{ph}	physical plastic hinge length
$L_{p\text{-}m}$	equivalent plastic hinge length under monotonic load
L_s	longitudinal spacing of shear connectors
L_{sy}	length of rebar yielding zone
M	bending moment
M_{ACI}	flexural strength calculated from ACI 318 rectangular stress block
M_{co}	bending moment at the onset of ε_{co}
M_{cr}	bending moment at concrete cracking
M_{cu}	bending moment at the onset of ε_{cu}
M_e	elastic limit-resisting moment
M_m	maximum moment in the span of member, or moment at the onset of allowable FRP strain
M_{max}	maximum moment resistance of RC section
M_p	peak moment resistance of RC section
M_{Pop}	flexural strength calculated from rectangular stress block based on Popovics model
M_y	yield moment
M_u	ultimate (peak) moment resistance of RC section
N	axial or normal force
N_a	normal compression force to plate from one fastener
N_c	$f_{co} \cdot b \cdot h$, crushing axial load
N_u	ultimate (peak) axial resistance
n	$(N-A_{sc}f_y)/bd$, net axial stress, or $n=E_s/E_c$, or axial load level P/P_u
n'	$(N-A_{sc}f_y)/bdf_{co}$, net axial force ratio
n_1	$(N+A_{st}f_y)/bd'$

P	force, or axial load
P_a	adhesive force
P_d	dowel force
P_f	frictional force
P_u	ultimate axial load-carrying capacity
p	pressure
Q	first moment of cross section
q	bond/shear force per unit length
q_t	tensile stress ratio, $A_{st}f_y/(bdf_{co})$
R	resultant force, or radius of circular section
R_0	experimentally determined material constant of steel for Menegotto-Pinto model
R_{st}	rigidity ratio of tension reinforcement, $A_{st}E_s/(bd)$
r	radius of circle, or corner radius
s	interfacial slip, or spacing of anchors, or spacing of transverse reinforcement
s_p	boundary interfacial slip
s_u	ultimate slip where interfacial bond stress drops to zero
T	tensile force, or temperature
t	thickness
t_f	thickness of FRP laminate/plate
u_c	displacement of concrete
u_s	displacement of steel
V	shear force
V_c	shear strength contributed by concrete
V_{cr}	shear strength at the onset of critical diagonal crack, defined as the shear strength of RC members without web reinforcement
V_s	shear strength contributed by transverse reinforcement
V_z	shear adjacent to a concentrated load or reaction at a section of maximum moment
v	shear stress
v_c	shear stress due to V_c
v_{cr}	shear stress at the onset of critical diagonal crack
v_s	shear stress due to V_s
w	distributed load, or crack width
w_{max}	maximum crack width
x	distance from a compression fiber to neutral axis
$y(\varepsilon_c)$	$(\varepsilon_c \int_0^{\varepsilon_c} f(\xi)d\xi - \int_0^{\varepsilon_c} f(\xi)\xi d\xi)/(\int_0^{\varepsilon_c} f(\xi)d\xi)^2$, a function of ε_c
y_e	elastic displacement
y_p	plastic displacement
y_u	ultimate displacement
z	lever arm, or shear span length
α	ratio of average compressive stress in the compression zone to f_{co}, or plastic hinge length ratio L_p/L
α_{01}	zero-one variable for absence or presence of strain penetration effect
α_1	ratio of compressive stress of equivalent rectangular stress block to f_{co}
α_m	α value at $\varepsilon_c = \varepsilon_{cm,u}$
β	ratio of distance from compression centroid to compression face to c, or reliability index

β_1	ratio of depth of equivalent stress block to c
β_m	β value at $\varepsilon_c = \varepsilon_{cm,u}$
β_T	target reliability index
Δ	displacement
Δ_y	yield displacement
Δ_u	ultimate displacement
δ	damage degree
ε	strain
ε_{by}	yield strain of CY block
ε_{bu}	ultimate strain of CY block
ε_c	strain of concrete, or concrete strain at extreme compression fiber
ε_{cc}	strain at the peak of compressive stress-strain curve of confined concrete
ε_{cm}	flexural design strain, i.e., strain at extreme compression fiber at peak moment, or mean strain of concrete
$\varepsilon_{cm,c}$	flexural design strain for sections with elastic compression reinforcement
$\varepsilon_{cm,e}$	flexural design strain for sections with elastic tension reinforcement
$\varepsilon_{cm,t}$	flexural design strain for transition stage
$\varepsilon_{cm,u}$	flexural design strain for underreinforced sections
ε_{co}	strain at the peak of concrete compressive stress-strain curve
ε_{cp}	extreme concrete strain at the yield of tension reinforcement
ε_{cr}	creep strain of concrete
ε_{cs}	shrinkage strain of concrete
ε_{cT}	concrete strain at elevated temperature
ε_{cy}	solution of Eq. (3.47), extreme concrete strain at yielding of maximum possible tension reinforcement
ε_{cu}	ultimate strain of concrete
ε_{cue}	elastic component of concrete compressive strain at the ultimate stage (0.001–0.002)
ε_{cuo}	basic maximum compressive strain in concrete
ε_e	strain due to elastic deformation
ε_f	strain of FRP
ε_{fu}	ultimate strain of FRP material
ε_j	ultimate strain of the FRP jacket
ε_o	concrete strain at transition point
ε_{oT}	strain at peak concrete stress under elevated temperature
ε_{pd}	debonding strain of FRP plate
ε_s	strain of steel
ε_{sc}	strain of compression reinforcement
ε_{slp}	slip strain
ε_{sm}	mean strain of steel
ε_{sp}	spalling strain of cover concrete
ε_{st}	strain of tension reinforcement
ε_t	tensile strain
ε_{to}	strain at the peak of concrete tensile stress-strain curve
ε_u	strain of concrete at the onset of M_u
ε_{un}	unloading strain of concrete
ε_y	yield strain of reinforcement

ε_z	longitudinal strain
ϕ	diameter of bar, or resistance reduction factor
ϕ_c	creep coefficient
η	relative height of CY block
κ	curvature
κ_{co}	curvature at the onset of ε_{co}
κ_{cr}	additional creep curvature
κ_{cs}	shrinkage curvature
κ_{cu}	curvature at the onset of ε_{cu}
κ_e	elastic curvature
κ_p	plastic curvature
κ_y	yield curvature
κ_u	curvature at the onset of ultimate moment
λ_f	confinement ratio
μ	frictional coefficient, or Poisson's ratio
μ_Δ	Δ_u/Δ_y, ductility factor
θ	rotation angle, or drift ratio
θ_u	ultimate drift ratio (rotation angle)
ρ	reinforcement ratio, or effective reinforcement ratio, or corner radius ratio
ρ'	$\rho_{st}-\rho_{sc}+N/f_ybd$, equivalent reinforcement ratio
ρ_c	tension reinforcement ratio at the intersection of elastic stage and transition stage
ρ_e	minimum elastic reinforcement ratio
ρ_{eff}	effective confinement stiffness ratio
ρ_f	reinforcement ratio of FRP
ρ_m	maximum tension reinforcement ratio of underreinforced section
ρ_s	$\rho_{st}+\rho_{sc}$
ρ_{sc}	compression reinforcement ratio
ρ_{st}	tension reinforcement ratio
ρ_{sv}	volumetric ratio of transverse steel
σ	stress
σ_c	stress of concrete
σ_l	lateral stress
σ_r	radial stress
σ_s	stress of steel
σ_{sc}	stress of compression steel
σ_{sn}	$(N-A_{sc}f_y+A_{st}f_y)/bd$, net axial stress
σ_{st}	stress of tension steel
σ_t	tensile stress
σ_z	longitudinal stress
τ	shear stress
τ_a	adhesive bond stress
τ_b	bond stress
τ_f	frictional shear stress
τ_m	bond stress due to mechanical fastener
τ_{max}	peak bond stress
ξ	strain of concrete at location x, or strain hardening/softening modulus ratio of CY block

Failure of reinforced concrete members

1

1.1 Introduction

The purpose of structural design is to provide a structure that can serve its function and avoid failure during its intended lifetime, with reasonably low costs, construction efficiency, and esthetical acceptance. Among all the factors considered, safety is the prime concern in structural designs. Until today, structural failure is evaluated through its various loading capacity. As a result, the prime tasks of structural research and design are to

(1) find out how a structure fails, or identify failure modes; and
(2) quantify or evaluate the conditions at which a particular mode of failure occurs.

Structures are made up of structural elements (or members or components), such as beams, slabs, columns, walls, foundations, etc. The assembly of elements forms a structure, such as a building, a dam, a tunnel, an airplane, etc. As a result, failure of a structure can occur under two circumstances:

(1) failure of a structural system, which occurs if the elements are not assembled properly; and/or
(2) failure of structural elements, which occurs if a structural element is not designed properly or overloaded.

Development of adequate understanding and knowledge of structural systems requires a systematic training through many courses, such as statics, structural mechanics, structural analysis and design, earthquake engineering and design, design of tall buildings, advanced steel and concrete design courses, and more importantly, practical experience. This book focuses on structural elements and introduces different failure modes, their failure mechanisms, fundamental design principles, and theoretical modeling of reinforced concrete elements/members.

Concrete is the most widely used structural material, but it has its disadvantages. The most critical ones are its very low tensile strength and high brittleness. Engineers have long ago resolved the problems by "reinforcing" concrete with steel, hence reinforced concrete (RC), through two principles:

(1) enhancing the tensile strength of a member by inserting steel bars in regions where tensile stress is high; and
(2) making use of the high deformability of steel to increase the deformability of RC members by allowing tensile reinforcement to deform sufficiently before failure of the concrete.

Therefore, reinforced concrete is an ideal combination of concrete and steel, which offsets the disadvantages of the two constituent materials and makes use of the best properties of them. As the two constituent materials are compositely integrated

Analytical Approaches for Reinforced Concrete. https://doi.org/10.1016/B978-0-12-821164-9.00001-2

together, RC members are composite members. As a result, failure of an RC member can be caused by failure of a constituent material or the loss of the composite action. In fact, an RC member such as a beam can be visualized as a global building structure, where the constituent materials are the elements, and the way the constituents are integrated together or the assembly of them is the structural system. Therefore, constituent material failure can be considered as a structural element failure and the failure of composite action is considered as a structural system failure. Consequently, the principles for the design of global structures are similarly applicable to the design of RC members and vice versa, particularly from a design philosophy point of view.

1.2 Typical structure of RC beam

A typical RC beam is illustrated in Fig. 1.1. Loading causes various effects or actions in the structure, such as bending moment M, shear force V, and deflection Δ (Fig. 1.1A). Structural design is to ensure that these effects do not affect functioning and safety of the structure. The structure of a typical RC beam is illustrated by the simply supported beam in Fig. 1.1B. Under a bending moment, the top of the beam is under compression and the bottom of the beam is under tension. As tensile resistance of concrete material is small, the tension zone will crack under significant

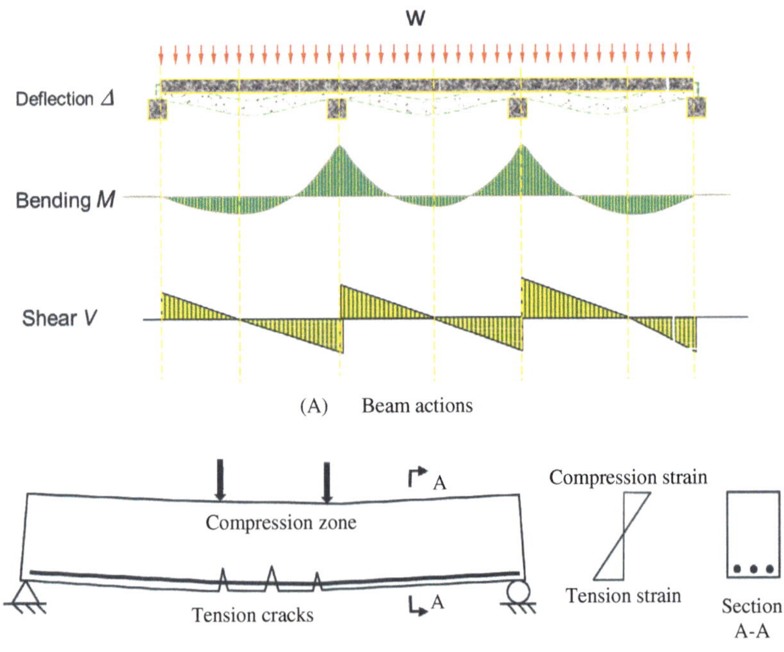

(B) Concrete beam reinforced in tension zone

Fig. 1.1 Typical beams.

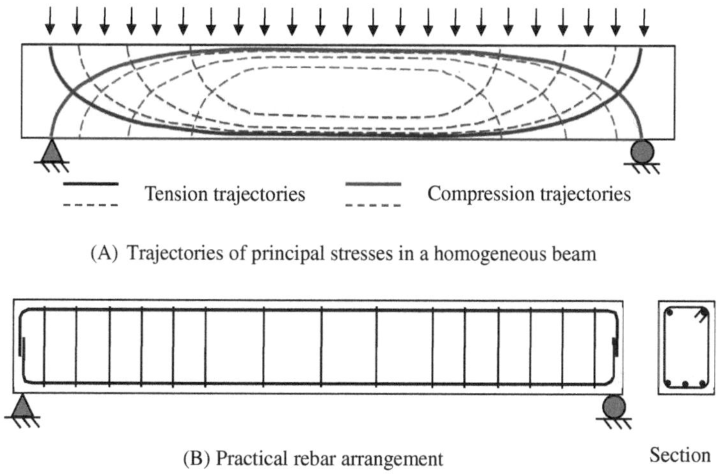

Fig. 1.2 Placement of reinforcing bars.

loading. Therefore, tension reinforcing bars (rebars) are generally placed at the tension side of the beam to take tensile stress and restrain opening of cracks. As rebars are used to take tensile stress, theoretically speaking the optimal positioning of rebars should be along the tension trajectories as shown in Fig. 1.2A so that the rebars are used most efficiently. However, such an arrangement is not practical as it is difficult to form and place the rebars in this way from a construction point of view. Furthermore, the tension trajectories change when loading changes.

Therefore, a typical rebar arrangement for RC beams is shown in Fig. 1.2B, where a reinforcement cage is used that is most convenient for construction. The longitudinal bars in the tension zone are used to resist the longitudinal component of the tension stress. The lateral bars, also known as transverse reinforcement, fitments, ligatures, links, or stirrups, are used to take the transverse component of the tensile stress. The longitudinal bars in the compression region are used for two main purposes: (1) as hanger bars to fix stirrups in position during construction, and (2) to share compressive stress together with concrete when concrete strength is insufficient to resist compression force alone. There are other purposes of using bars, such as serving as a space truss to resist torsion, confining core concrete, restraining shrinkage cracking, reducing long-term deformation, etc., which can be found in all reinforced concrete textbooks.

1.3 Experimental observations—An RC practical

A simple RC beam test program that was used as a practical for the course of reinforced concrete structures conducted at the University of Adelaide and further developed by the author at City University of Hong Kong is used in this section to facilitate understanding of the structural failure mechanisms of RC beams. The beams were tested under four-point bending. The test scheme is illustrated in Fig. 1.3. Six different designs of beams were experimentally tested to observe their behavior. The casting of test beams and test setup is shown in Fig. 1.4. A load *P* was applied at the mid-span of the beam from a loading jack (Fig. 1.4C). The applied load *P* was spread into two point-loads *F* by a distribution frame ($P = 2F$). The relevant bending moment diagram (BMD) and shear force diagram (SFD) are shown in Fig. 1.3.

Two identical beams were tested for each beam design. The tests were conducted under displacement control, that is, pushing down by the loading jack to have a continuous increase in the mid-span displacement. Details of the test beams are given here:

Dimensions: 1000 mm (span length) × 125 mm (depth) × 100 mm (width).
Concrete cover thickness to the bottom of the longitudinal reinforcement bars: 15 mm.
Concrete strength: 40 MPa.
Beam 1: Unreinforced beam—plain concrete beam without reinforcement.
Beam 2: Longitudinal reinforcement—2 No. of 6 mm diameter mild steel round bars (smooth surface without ribs) with a yield strength of 384 MPa (R6 bar), no stirrup.
Beam 3: Longitudinal reinforcement—2 No. of 10 mm mild steel round bars with a yield strength of 361 MPa (R10 bar), no stirrup.
Beam 4: Longitudinal reinforcement—2 No. of 12 mm ribbed (deformed) high yield strength bars with a yield strength of 527 MPa (T12 bar), no stirrup.

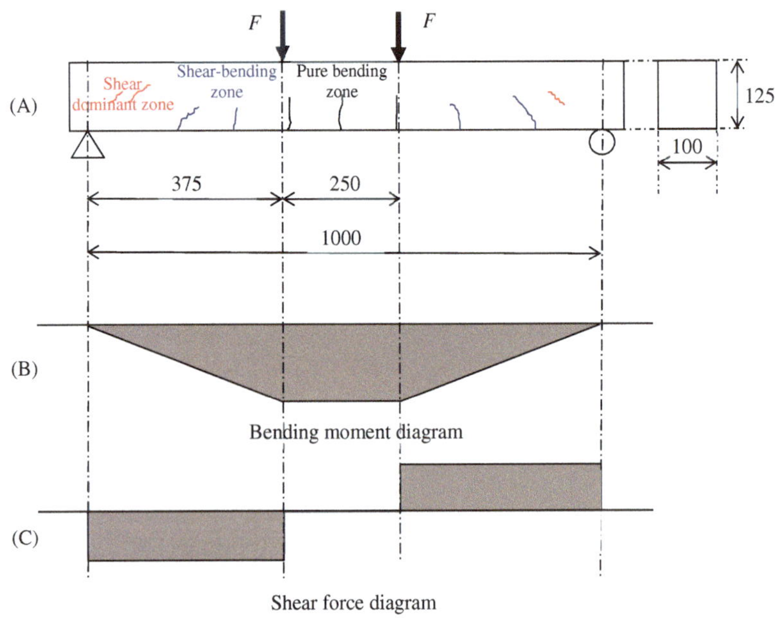

Fig. 1.3 Four-point bending test of beam.

(A) Beam casting (beams 1 and 4)

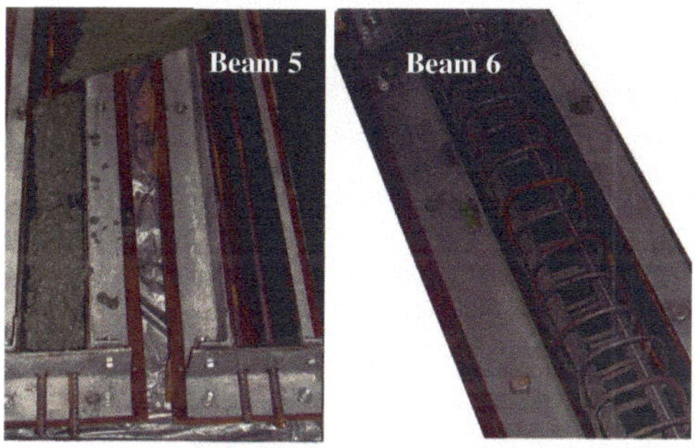

(B) Beam casting (beams 5 and 6)

(C) Test setup

Fig. 1.4 The RC practical.

Beam 5: Longitudinal reinforcement—2 No. of 16 mm ribbed high yield strength bars with a yield strength of 562 MPa (T16 bar), no stirrup.
Beam 6: Longitudinal reinforcement—2 No. of T16 bars, stirrups—6 mm mild steel bars at 50 mm spacing with a yield strength of 384 MPa (Fig. 1.4B).

The applied load vs the mid-span displacement response curves obtained from the RC practical are shown in Fig. 1.5. The failure modes of the beams are shown in Fig. 1.6.

The following observations were from the tests.

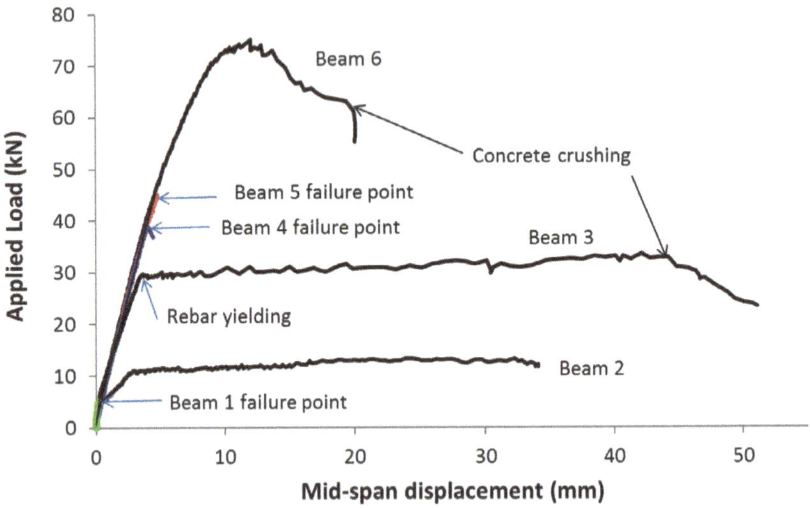

Fig. 1.5 Response curves.

1.3.1 Beam 1

The beam failed due to tension cracking initiated at the bottom of the beam that quickly extended to the top of the beam, and the beam broke into two pieces suddenly. The maximum applied load on the beam was 5.3 kN. The response curve was basically a straight line up to the failure point (Fig. 1.5). There was only one major crack that occurred within the maximum moment region between the two point-loads (Fig 1.6A). Theoretically speaking, the position of first cracking can be any place within the region between the two point-loads, because the bending moment in the region is constant. In reality, concrete is not a uniform material and the first crack occurs at the weakest cross-section within the maximum moment region.

1.3.2 Beam 2

Cracking initiated at the bottom of the beam at a similar load as Beam 1 (about 5 kN). After cracking, the stiffness of the beam was significantly reduced, as shown by the reduced slope of the response curve after the turning point where the first crack occurs (Fig. 1.5). When the load increased to about 10.5 kN, another turning point occurred on the response curve. A long plateau of the response curve was then observed after this turning point. The strain of the rebar was found to be smaller than the yield strain, and therefore the plateau of the response curve was not caused by yielding of the rebar. Careful observation during the test found that the rebar was gradually pulled out from the major crack (can be more than one major crack), as shown in Fig. 1.6B between the two crack faces. The plateau of the response curve was a result of rebar slipping or debonding of rebar from the concrete substrate. As the load-carrying capacity of the beam was determined by the bonding resistance between the rebar and substrate concrete, the failure mode is called debonding failure.

(A) Beam 1

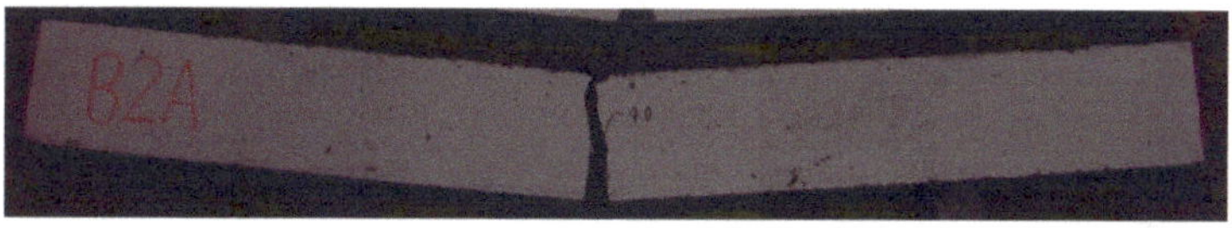

(B) Beam 2

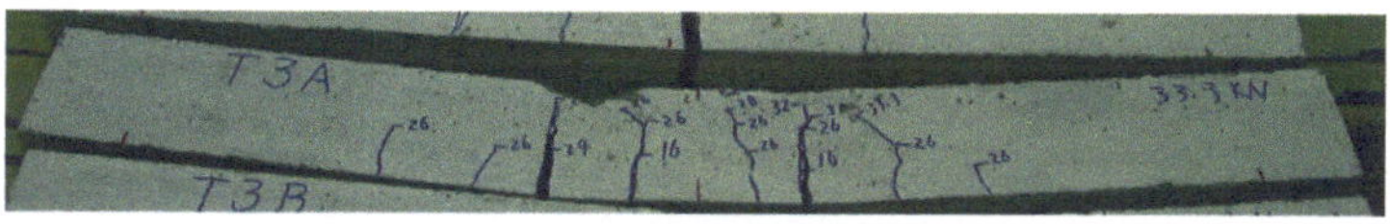

(C) Beam 3

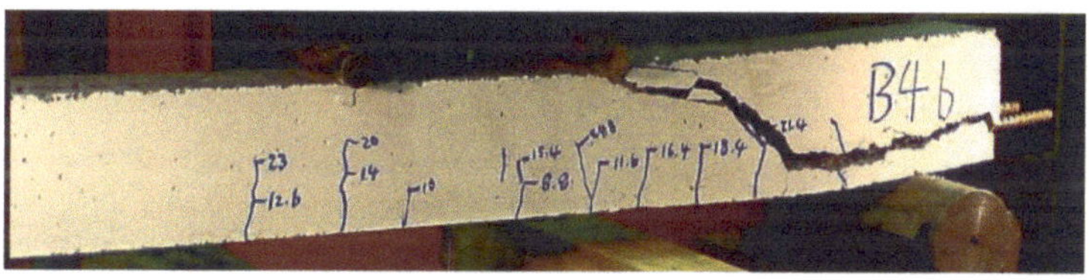

(D) Beam 4

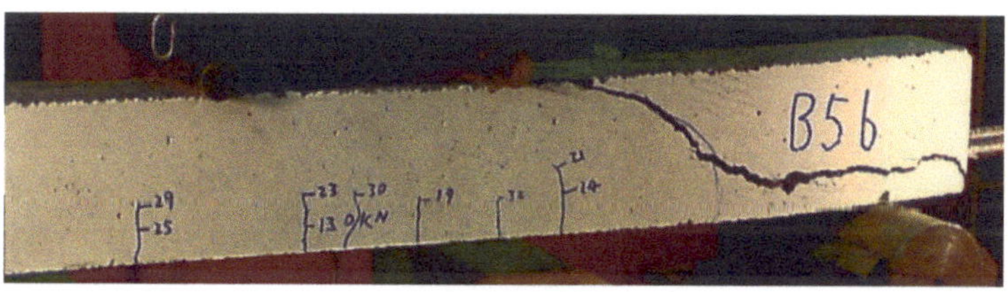

(E) Beam 5

(F) Beam 6

Fig. 1.6 Failure modes.

The applied load was maintained (slightly increasing) during the slipping of the rebars when the displacement of the beam increased continuously. This phenomenon showed that the resistance to the rebar slipping was nearly constant. In other words, the tensile resistance of the beam was maintained until the end of the test. When the displacement increased, the major crack kept opening. In the meantime, the crack tip extended toward the top of the beam, causing a gradual reduction of the compression zone of the beam. As the compression zone reduced gradually, the compressive stress of the concrete in the compression zone kept increasing, until the moment that crushing of concrete occurred at the top of the beam. From this moment, the beam resistance started to drop. Therefore, the onset of concrete crushing is considered to be the ultimate failure point of the beam.

1.3.3 Beam 3

Initial cracking and reduction of stiffness after cracking of the beam were similar to those of Beam 2. However, the slope of the response curve after cracking was larger than that of Beam 2 (Fig. 1.5). This was because the rebars in Beam 3 were larger than those in Beam 2, leading to a greater stiffness of the beam after cracking. The response curve after cracking was essentially a straight line, which indicated that the materials in the beam were still elastic. The elastic response stage ended at the turning point after which a long plateau of the response curve occurred. Yielding of the rebars occurred exactly at this turning point. After yielding of the rebars, the tensile resistance kept constant while rebar strain continued to increase. Therefore, it was the yielding of the rebars that produced the plateau of the response curve.

Further yielding of the rebar caused opening up of the cracks and extension of the crack tips toward the top of the beam, similar to the process described for Beam 2. This process continued until the moment that the concrete in the compression zone was insufficient to sustain the demand in compression force and crushed at the top of the beam (Fig. 1.6C), by which moment the load started to drop (Fig. 1.5). It will be seen in Chapter 2 that this failure was caused by bending moment, and hence is called flexural failure. Furthermore, the load-carrying capacity of this beam was determined by the yielding of the rebars and, hence, the failure mode is called underreinforced failure.

1.3.4 Beam 4

The beam cracked at a similar load as previous ones, followed by the second elastic deformation stage with a reduced stiffness that was slightly greater than that of Beam 3 (Fig. 1.5) due to the larger longitudinal bars. The beam suddenly failed at the load of 39 kN (Fig. 1.5). The failure was so sudden and brittle that the whole chunk of concrete near the support of the beam flew away explosively. The failure process was video recorded, and the snapshots are shown in Fig. 1.7. The fly-off concrete chunk was placed back in Fig. 1.6D after the test.

The load-displacement curve terminated suddenly at the failure point without any softening of the curve (Fig. 1.5). The displacement and crack width of the beam were all small at the time of failure, giving little sign of warning before failure. The cracks in Figs. 1.6D and 1.7 were highlighted by a marker during the test to facilitate

(A)

(B)

(C)

(D)

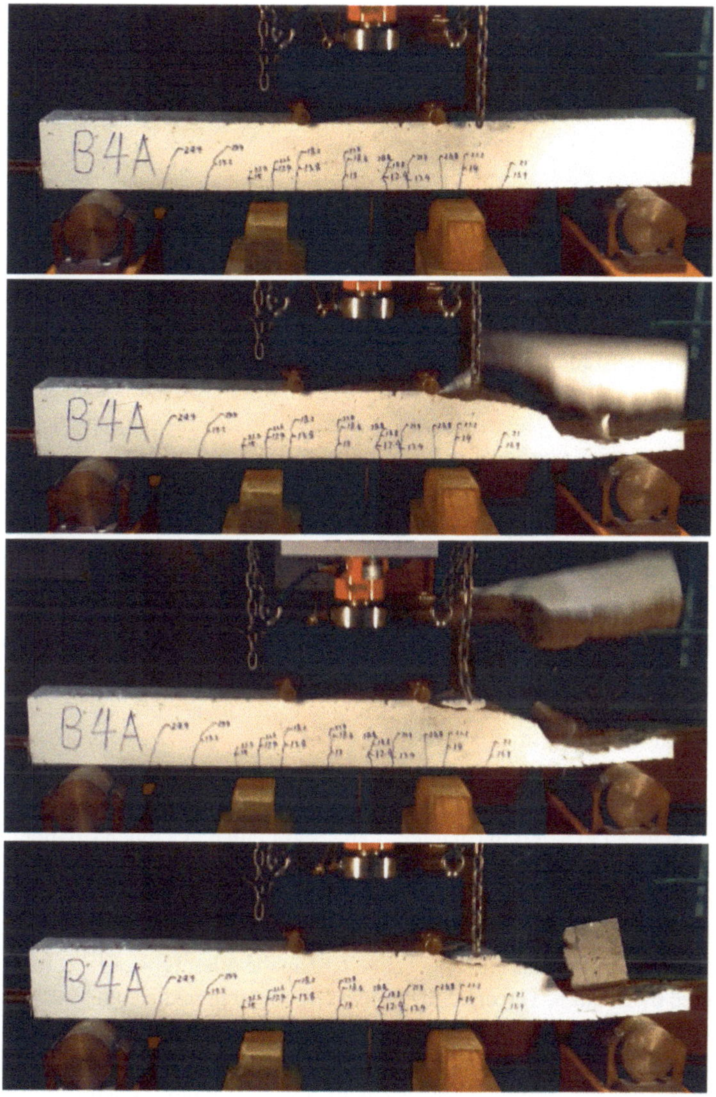

Fig. 1.7 Explosive shear failure.

visibility of the cracks. The cracks were not easily seen from a distance, and thus were difficult to be observed in reality. Chapter 10 will show that this beam failed due to excessive shear force. Therefore, the failure mode is called shear failure.

1.3.5 Beam 5

The beam had a behavior very similar to that of Beam 4 and also failed suddenly without warning. This beam had the same failure mode as Beam 4—shear failure (Fig. 1.6E). As the diameter of the longitudinal bars of this beam were increased from

12 mm for Beam 4 to 16 mm, the slope of the response curve after cracking was slightly steeper than that of Beam 4. The failure load was also increased from 39 to 44 kN. It will be seen in Chapter 10 that the increase of the area of longitudinal bars increases the shear failure strength of RC beams (Eq. 10.11).

1.3.6 Beam 6

The behavior of Beam 6 was nearly identical to that of Beam 5 up to the failure point of Beam 5, because the longitudinal rebars of this beam were identical to those of Beam 5. The provision of stirrups in this beam restrained diagonal cracking and prevented shear failure. In fact, the diagonal shear cracks (the inclined cracks at both sides of the beam in Fig. 1.6F) occurred at a similar load as the failure load of Beam 5. However, further development of the diagonal cracking was restrained by the stirrups; and hence shear failure did not occur. The beam kept its nearly elastic behavior up to the load of about 70 kN, after which nonlinear behavior occurred. Softening of the response curve was observed after the peak load of 76 kN (Fig. 1.5). Further increase in displacement of the beam caused faster softening, and finally a steep drop of the load.

It was observed in the test that small and local crushing of concrete occurred at the top of the beam within the maximum moment region when softening of the response curve occurred. The size of the concrete crushing zone and degree of crushing developed quickly as the beam deformation continued to increase in the descending part of the curve, which was consistent with the faster drop of loading after the peak load in Fig. 1.5. The abrupt drop of load occurred at the sudden crushing of a large extent of concrete as shown in Fig. 1.6F.

The rebars did not yield throughout the test of Beam 6. As a result, the yield plateau of Beam 3 (Fig. 1.5) did not occur. This failure mode is called overreinforced failure, and is discussed in detail in Chapter 2. The beam failed at a much smaller deformation compared with Beam 3 (Fig. 1.5) and with much smaller crack width before failure (the concrete crushing on top of the beam in Fig. 1.6F occurred when it failed). These signals issued fewer warnings before failure. In the meantime, it failed more suddenly with a faster drop in load-carrying capacity than the other type of flexural failure mode exhibited by Beam 3. The crushing process of concrete in the compression zone determines the suddenness or brittleness of the beam failure. The crushing of higher strength concrete is more brittle than that of lower strength concrete, and hence the overreinforced failure of RC beams with higher strength concrete is more sudden and dangerous.

1.4 Failure modes

1.4.1 Evolution of failure modes

Beam 1 failed in tension cracking of concrete with a very small load-carrying capacity. To prevent the brittle tension failure mode, two small R6 bars were used in Beam 2, which significantly increased the load-carrying capacity and changed the failure mode to rebar debonding. To avoid rebar debonding, larger bars (R10) were used in Beam 3, which increased the bond surface area, and hence the bond strength. The use of larger rebars (2R10) in Beam 3 successfully changed the failure mode from debonding to yielding of the longitudinal bars, and further increased the load-carrying capacity. Further increase in the reinforcement of 2T12 in Beam 4 provided a higher rebar strength, which changed the failure mode from yielding of the longitudinal bars to the shear failure, with a further increase in the load-carrying capacity. When the reinforcement was increased to 2T16 in Beam 5, the load-carrying capacity continued to increase, however, with the same failure mode this time as that of Beam 4. The use of stirrups in Beam 6 prevented shear failure and changed the failure mode to concrete crushing without yielding the tension bars, and, similarly, further increased the load-carrying capacity.

These discussions show that increasing reinforcement in the failure zone can generally increase the load-carrying capacity of RC members, but at the same time, changes the failure mode from one to another. When one failure mode is suppressed by reinforcing the failure zone, the failure moves to another place, in another failure mode, and with a higher load-carrying capacity.

1.4.2 Safe failure mode and control of failure mode

The RC practical shows the following failure modes of RC beams:

(1) Brittle concrete tension failure

The plain concrete beam (Beam 1) fails suddenly with little sign (warning) and a very small load-carrying capacity. The failure is sudden and nonductile. Therefore, flexural failure caused by tension failure of concrete without reinforcement in the tension zone is prohibited in practice.

(2) Debonding failure

The rebar debonding failure that occurs before rebar yielding cannot make full use of the rebar material strength, and hence is uneconomical and usually undesirable. In RC design, engineers try to avoid debonding failure by using bars with a larger surface (bond) area or ribbed bars that have sufficient bond resistance to avoid the debonding failure mode (see Chapter 5).

(3) Shear failure

Shear failure is sudden and brittle with little ductility. This failure mode is very dangerous and must be avoided in practice. Details of shear failure are discussed in Chapter 10.

(4) Flexural failure—underreinforced beams

The failure of Beam 3 starts with yielding of the rebars. When yielding develops and cracks continuously open and extend, the compression zone of concrete will reduce gradually and finally reach a sufficiently small size so that the concrete stress is higher than the concrete material strength that crushes the concrete in the end. The crushing of concrete causes significant loss of concrete compressive resistance, and consequently a fast drop in load-carrying capacity. Therefore, the final concrete crushing is the cause of the beam failure. As this failure mode is ductile with ample warning before it loses its load-carrying capacity, it is relatively safe, and hence the most desirable failure mode in RC design.

(5) Flexural failure—overreinforced beams

Beam 6 also fails in concrete crushing. But the longitudinal reinforcement does not yield before the concrete crushes. The reinforcing material is not sufficiently used or oversupplied in such beams. For this reason, the failure mode of Beam 6 is called overreinforced failure. In contrast to Beam 6, the failure mode of Beam 3 is called underreinforced failure. An overreinforced beam fails at a small deformation and barely visible cracking of the beam before it fails. This failure mode gives insufficient warning before final failure. The loss of load-carrying capacity is also faster and the concrete crushing is more severe (Fig. 1.6F) compared with that in Beam 3. Therefore, this failure mode is also nonductile, sudden, and dangerous, hence undesirable. Although the provision of more longitudinal reinforcement in Beam 6 provides a much higher load-carrying capacity than Beam 3, this kind of overreinforced failure is considered unsafe and is not allowed in practice. The reasons are discussed in Section 1.5.

The consequence of failure of an RC member is different when it fails at different failure modes. Sudden failure modes cause more damage and losses than ductile ones because sufficient signs of failure can warn people to run away or even take steps to avoid failure. An important aim of structural design is to provide a structure that has a minimum loss when its design loading is exceeded. This requires structures to fail under desirable failure modes. Structures failing in this way are said to have a safe failure mode. Failure mode can be controlled by reinforcing concrete members in correct places, ways, and with adequate quantity. Sufficient yielding of rebar is critical to facilitate safe failure modes. The elongation of rebar under sustained resistance, i.e., yielding, leads to large deformation of RC members while maintaining the load-carrying capacity of the structure. Therefore, rebar is usually required to be ductile in RC structures.

The stress-strain curves of a few typical reinforcing bars are shown in Fig. 1.8. The reinforcing bars made of fiber-reinforced polymer (FRP) are linear elastic and fail abruptly without yielding. Therefore, concrete beams reinforced with FRP bars have little ductility if FRP bars are used in the same way as steel bars. Consequently, FRP

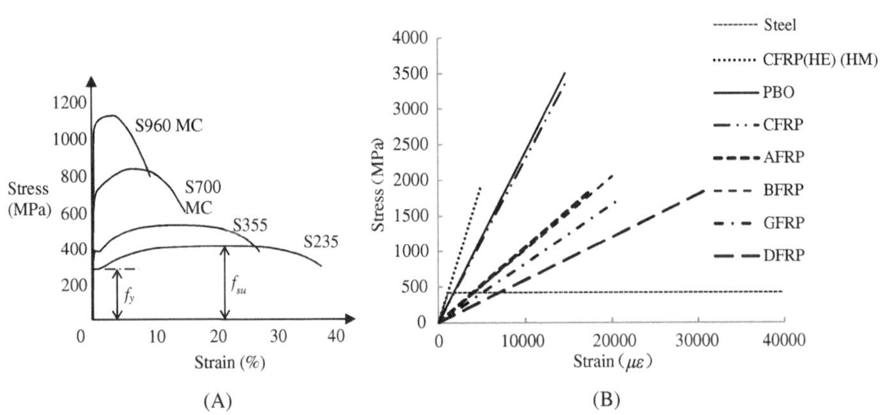

Fig. 1.8 Stress-strain curves of typical reinforcements: (A) structural steel; and (B) FRP. From Wu, Z. S., Y.F. Wu, and M. F. Fahmy. 2020. *Structures strengthened with bonded composites*. Woodhead Publishing, Elsevier.

bar reinforced concrete beams must be designed differently to provide sufficient ductility for the members. Nonductile high yield steel bars may also lead to insufficient ductility of structures due to the lack of sufficient yield plateau of the reinforcement (Fig. 1.8A). Therefore, care should be exercised in selecting the appropriate type of longitudinal bars to ensure a safe failure mode. When material yielding is unavailable to deliver sufficient ductility of the structure, other structural approaches should be adopted in the design (see Chapter 9).

1.4.3 The principle of weakest link

It is shown in Section 1.4.1 that when one failure mode is suppressed by reinforcing one part, another one will occur in another place. What happens if additional steel bars are added in the compression zone of Beam 6 to increase the compressive resistance so that concrete crushing in Beam 6 is prevented? What will be the next failure mode? This question can be answered using the simple chain in Fig. 1.9.

When the force F applied at the loaded end of the chain is increased continuously, there must be a moment that the chain breaks. Clearly, it will break at the weakest link. If we strengthen the weakest link so that it is stronger than others and the force F continues to increase, breaking will take place at the next weakest link or the weakest link after strengthening. This strengthening process can continue, but failure will still occur sooner or later, and it always occurs at the current weakest link.

If the beam has n failure modes, the load-carrying capacity corresponding to the 1st, 2nd, …, ith … nth failure mode is R_1, R_2, …, R_i…, and R_n, respectively. If R_i is the smallest value among all, clearly the beam will fail at the ith failure mode. After a strengthening work, the values of R_1, R_2, … R_i…, and R_n will change. If R_k is the smallest value after strengthening, the next failure will occur under the kth failure mode.

This analysis shows that the correct judgment of structural failure requires the knowledge of

(1) all possible failure modes of a given structure, and
(2) the model or formulas to calculate the failure load corresponding to each failure mode.

Therefore, a major objective of structural engineering research is to discover or identify possible failure modes of different structures and develop models (equations, criteria, etc.) to quantify or evaluate the failure load or condition corresponding to

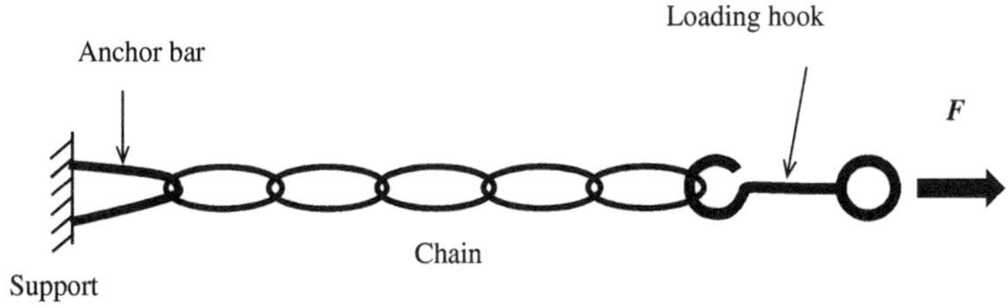

Fig. 1.9 Weakest link principle.

each failure mode. Only with the knowledge of these two can a structure be designed adequately to satisfy the following two aims:

(1) Avoiding failure under design loading and conditions

This involves calculations or evaluations of all the links in the chain of Fig. 1.9 to ensure that R_i of each link is greater than the actions caused by design loading. However, the number n of possible failure modes may be very large. Evaluation of all links will involve too much design workload, long design period, and high design costs. Therefore, an important job of a design code of practice is to specify the scope of design works for certain types of structures, i.e., specify the links that need to be evaluated. For more details refer to Section 1.7.3.

(2) Failing in a safe mode

As overloading (higher than design loading) does happen in reality, it is critical for engineers to design a structure so that the links in Fig. 1.9 that would fail in unsafe modes are stronger than those that would fail in safe modes. More discussions are given in Section 1.5.

1.4.4 Other failure modes

RC members have many more possible failure modes than those discussed in Section 1.4.2. Common failure modes of RC structures are introduced in textbooks, design codes of practice, and other professional publications. However, readers cannot assume that they are exhaustive. For the chain in Fig. 1.9, apart from the failures at one of the five links, there could be failure at the anchor bar or loading hook, or even pulling out failure at the fixed support, and so on. For the beam in Fig. 1.10, apart from the common failure modes and design works described in textbooks including bending at the mid-span and shear at the support, it is also possible that concrete may crush locally under the point load or support positions, as shown. The pipe penetration weakens section A-A and the local flexural and shear strength at the penetration region may be significantly reduced, which cannot be designed using design models that are developed for ordinary RC sections. If the beam is cast in two batches of concrete at different times and the joint between the old and new concrete was not treated properly, the joint may split (longitudinal shear failure) under loading, which cannot be designed based on a typical beam cross-section for normal failure modes. On the other hand, if the new concrete material in Fig. 1.10 has a significantly different stress-strain behavior from that of normal concrete, the flexural and shear strength of the beam can

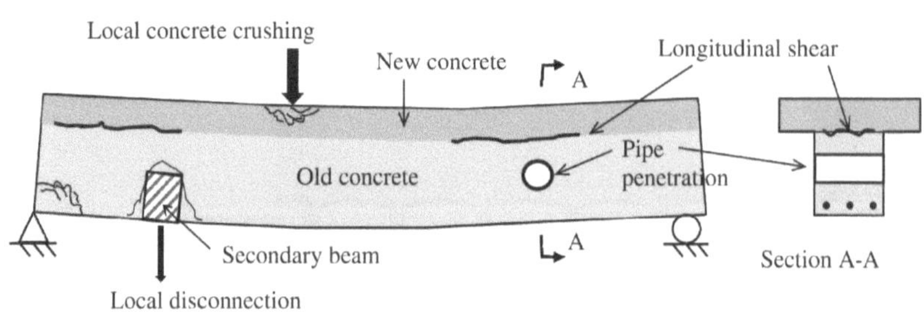

Fig. 1.10 Uncommon failure modes.

be different from those of ordinary RC members, and hence the design models/equations provided in design codes for normal RC members may not be applicable.

It is also important to note that the typical failure modes discussed in previous sections are observed from beams where loading is applied on top of a beam or in the compression zone of concrete. The failure modes could be quite different if loading is applied at the bottom or in a tension zone of a beam. On one hand, the point load applied at the bottom of a beam (such as by a secondary beam) may pull apart the concrete locally as shown in Fig. 1.10. On the other hand, even the common failure modes and failure loads may be significantly affected by the local point load. For example, the secondary beam in Fig. 1.10 would increase the local tensile and shear stress in the region around the secondary beam and the shear strength of the main beam could be significantly reduced. In all these cases, the normal design approach and models may not be applicable. For loads applied to the tension zone of a concrete beam, a simple solution is to provide "hanger" reinforcement of stirrups that are anchored to the compression side of the main beam so that a point load applied at the tension zone is transmitted to the compression zone by the additional reinforcement and the beam can be designed as normal RC beams.

Although the failure modes discussed previously are not typical of the design of ordinary RC members, many of these failure modes have been observed in engineering practice and investigated by researchers. Design guidelines can be found in specialist literatures for these special cases. However, there is always the possibility that a failure mode or design condition has never been encountered. Furthermore, special and new kinds of structures or abnormal service conditions of normal structures are encountered in engineering practice from time to time. In such cases, engineers are required to do research during engineering practice to identify new failure modes and develop corresponding models for evaluation of relevant failure load. Section 1.8 provides more discussion of the cases.

1.5 Failure consequence, safety factor, and ductility

The RC practical shows that all structures will fail in one mode or another when loading is sufficiently large. It is uneconomical and impractical to design a structure that does not fail under practically possible loadings (more precisely, the loads with such a small probability that they are unlikely to occur). Structural design is a balance between failure risk and structural costs, which aims at a reasonable cost for acceptable level of failure chance. However, for the same structural costs, the design of a structure can be different with different failure consequences. Designing structures so that they fail in a safer failure mode and a lower failure consequence when excessive loading occurs is the prime task of engineers.

All structures are designed to have a certain safety margin or safety factor under design loading. The value of the safety factor should be closely related to failure consequence or failure mode. A more dangerous failure mode, such as the shear failure of RC members, requires a higher safety factor than a safer failure mode, such as under-reinforced flexural failure. This is to ensure a safer failure mode occurring before a more dangerous one. Consequently, it results in the links in Fig. 1.9 that, in a more dangerous failure mode, have to fail at a higher loading than those that fail in a safer mode. As a result, a more dangerous failure mode will not occur in structures due to the use of a higher safety factor in design for the relevant failure mode.

Ductility is critical for a safe failure mode. The ultimate deformation of a structure at failure Δ_u relative to its deformation at yielding Δ_y is defined as the ductility, or

$$\mu_\Delta = \frac{\Delta_u}{\Delta_y} \tag{1.1}$$

where μ_Δ is referred to as the ductility factor (or ductility ratio); Δ_y and Δ_u are illustrated in Fig. 1.11A. A higher value of ductility usually means a safer and more desirable failure mode. The response curve of a real structure is different from the ideal case in Fig. 1.11A. For a real response curve, different definitions of ultimate deformation and yield deformation are used by different people and in different literature. One popular definition of ductility is shown in Fig. 1.11B. Even with the definition given in Fig. 1.11B, the ductility value can be different when different p values are adopted. Therefore, the calculated value of ductility of a certain structure may vary significantly when different definitions are used. As a result, it is important to know the definition of ductility factor before evaluating and comparing the ductility of structures.

Appropriate ductility of a structure provides adequate safety of the structure for the following reasons:

(1) Sufficient warning before failure

Higher ductility gives larger structural deformation, and hence a clearer sign of structural distress before failure. When large deformation is exhibited before a structural failure, people are alerted and can escape. For people to evacuate safely, the yield plateau from Δ_y to Δ_u must possess the following two basic features: (i) it is sufficiently long so that sufficient sign of structural distress is exhibited to provide a warning; (ii) sufficient load-carrying capacity in the yield plateau is sustained.

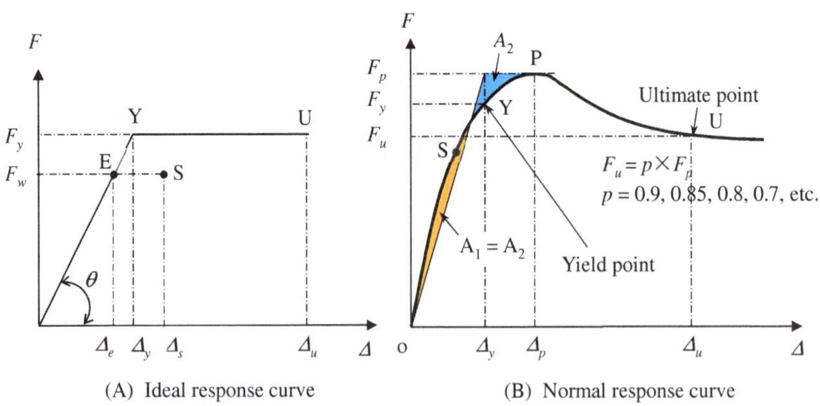

Fig. 1.11 Definition of ductility.

(2) Limiting loading and internal stress of a structure

A structure with a higher ductility attracts a smaller load in an earthquake. Life experience shows that, if a person stands firm on a bus, the acceleration of the bus produces a large force on his/her body. If his/her body allows for certain flexibility or deformability, the force produced on the body will be significantly reduced. Structures cannot be too flexible, i.e., the angle θ in Fig. 1.11A is required to have a lower bound of value, due to serviceability requirements. On the other hand, satisfactory seismic performance of a structure usually requires good energy dissipation instrument. Deformability due to elastic deformation does not dissipate energy but a plastic one does. Therefore, higher demand on deformability of a structure often requires a higher ductility (higher value of μ_Δ) rather than smaller rigidity (smaller value of θ).

To deal with exceptionally large but rare loading such as in a major earthquake, a wise and practical strategy is to give in, or allow structures to yield, when such an event occurs. When a structure yields in an earthquake, a further or stronger earthquake excitation (larger base displacement) only causes the structure to deform further along its yield plateau without inducing a larger force to the structure (see the yield plateau in Fig. 1.11A, and Beam 3 in Fig. 1.5). Therefore, yielding caps the loading applicable to structures.

The structural design strategy of providing sufficient ductility to curtail force induced by earthquakes is a general structural design tactic that is applicable not only to seismic loading but also to many other loadings, such as impact loading. Ductility or yielding is also essential for economic design of structures by limiting internal stress in certain parts of a structure while distributing additional loading effects to other less stressed parts through moment redistribution (see Section 2.6.2) so that a structure is more uniformly stressed.

(3) The last line of defense

Ensuring that a structure is safe at design loading cannot really avoid failure of the structure. Design loads are based on probability and must not be too large for reasonable structural costs. Therefore, overloading (larger than design load) is possible in reality. In fact, users of a structure usually do not know the design load of the structure. It is common for people to continue to add loading to the structure until a distress is observed. Therefore, the distress is the real brake that stops people from further loading the structure. When overloading occurs and the structure has a sufficient ductility, structural distress will be clearly observed before its final failure. In this case, measures can be taken to reduce the load, thereby avoiding final structural collapse. Designing structures with a higher safety factor or larger load-carrying capacity actually cannot avoid structural failure, but reduces the failure probability only. Instead, it may cause an even more drastic failure if the failure is not ductile. Therefore, providing sufficient ductility is the last, and a real effective, line of defense when overloading occurs.

It is noted that the ductility definition of Eq. (1.1) may be misleading sometimes for structures with significant P-Δ effect. This is explained in Section 7.3.5.1.

1.6 Limit state design approach

Structural designs typically use safety factors to set the margin of safety. Historically, three basic methods have been developed to achieve safe and feasible structures using safety factors:

(1) The permissible stress method (also known as Allowable Stress Design (ASD) or elastic design). The ultimate strengths of the materials are divided by a factor of safety γ_m to provide allowable design stresses which are usually within the elastic range of material. No factor of safety is applied to loading.

(2) The load factor method (also known as collapse or plastic design). The load that causes the collapse of a structure is divided by a factor of safety γ_f to give the working loads.

(3) The limit state design method. The working loads are multiplied by the partial factors of safety for loading γ_f (also known as load factors), and in the meantime, the ultimate strengths of the materials are divided by the partial factors of safety for material γ_m (also known as resistance factors). The global safety factor is equal to $\gamma_m \times \gamma_f$.

The first two methods were widely used in structural designs in the past. The method of limit state design (LSD) was first developed in the USSR based on research led by Professor N.S. Streletski. It was introduced in the USSR building regulations in 1955. The LSD has been enforced in Eurocodes and is commonly adopted in most building codes all over the world nowadays, including Australia, Canada, China, France, New Zealand, the United States, among many others. The United States has been particularly slow in adopting the LSD, and the ASD is still used currently in some US codes. LSD is also known as the Load and Resistance Factor Design (LRFD), particularly in the United States.

The advantage of LSD can be seen in Table 1.1. The material partial safety factor γ_m is used to set the safety margins for different failure modes so that a more dangerous failure mode can be allocated a higher safety factor. The partial safety factor for loading γ_f is used for the safety margin related to the uncertainty of loading so that a higher safety factor can be used for loads with higher uncertainty. By using the two types of safety factors simultaneously, a more dangerous failure mode caused by more uncertain loads can have a higher overall factor of safety. In this way, the margin of safety can be set more reasonably, as shown in Table 1.1.

It is noted that different partial material safety factors γ_m are used as the gauge of safety margin for different failure modes in the former BS (British Standards) codes and Eurocodes. Another alternative to the material partial safety factor γ_m for the

Table 1.1 Factors of safety for limit state design method.

Example of failure mode	Material safety factor γ_m	Load safety factor γ_f	Global safety factor $\gamma_m \times \gamma_f$
Flexural failure due to yielding of rebar ($\gamma_m = 1.15$) under dead load ($\gamma_f = 1.35$)	1.15	1.35	1.55
Flexural failure due to yielding of rebar ($\gamma_m = 1.15$) under live load ($\gamma_f = 1.5$)	1.15	1.5	1.73
Flexural failure due to concrete crushing ($\gamma_m = 1.5$) under live load ($\gamma_f = 1.5$)	1.5	1.5	2.25

Note: The partial safety factors are based on Eurocode 2 (BSI 2004).

control of safety margin for different failure modes is to use the capacity reduction factor ϕ for different failure modes, which is adopted widely in other western countries such as the United States and Australia. The capacity reduction factor ϕ is multiplied by the calculated value from a model for a specific failure mode to give the design strength. Therefore, it is directly related to a failure mode, and, furthermore, different values of ϕ can be assigned to different models for the same failure mode to consider their difference in accuracy. Therefore, the use of ϕ is more rational than γ_m from this point of view. For more discussions on safety margin, see Section 11.8.

Designers are usually interested only in designing a structure that satisfies design requirements rather than in knowing its complete behavior in all scenarios. Designing structures to satisfy a limited number of requirements is the limit state design approach. A limit state is defined as the state of a structure that represents the acceptable limit of an aspect of structural behavior. A limit state involves two important aspects: (1) the consequence being exceeded, and (2) the cause of the consequence. Margins of safety for these two parts are set separately in the limit state design method, and therefore it is more rational than the permissible stress method and the load factor method.

Limit states are generally divided into two classes:

(1) Ultimate limit states (ULS)

An ultimate limit state concerns the safety of a structure. When a ULS is exceeded, it results in the loss of its load-carrying capacity or structural integrity, such as partial or total collapse of the structure, buckling, overturning, material overstressing such as crushing of concrete and yielding of steel, disconnecting, etc. The difference between ultimate state (US) and ULS should be noted. The US is a physical state in which the integrity of a structure is lost, whereas the ULS is a design limit to comply with the structural safety requirement. For example, one can say that the US of the member is reached in an experimental test, while in design one should use the words "to design the member for a ULS." When a structure reaches the design ULS, it may not reach its US, due to safety margins and inaccuracy of calculations.

(2) Serviceability limit states (SLS)

A serviceability limit state concerns the functionality or fitness of a structure for use. When an SLS is exceeded, it leaves the structure in an unserviceable condition, such as excessive deformation (or settlement, cracking, stress), unacceptable vibration (or dynamic condition, occupant discomfort), low durability, inadequate acoustics and heat transmission conditions, etc.

In layman's language, a limit state is a specific condition or requirement on the structure, and when exceeded, the structure is considered no longer fulfilling the relevant design criteria, i.e., considered as failed. When the failure is related to its functioning, it is an SLS, and when it affects the safety of a structure, it is a ULS. Design for serviceability limit states is to ensure a structure functioning properly at all working conditions. Ensuring safety of a structure under all design conditions is called designing for ultimate limit states. As there can be many different requirements on the functioning of a structure and many possible failure modes, the number of limit states to be designed can be large. The task of designers is to ensure that the structure does not fail in any of the limit states.

All structures are required to function properly under service loads and be safe under ultimate loads. Service loads are characteristic loads provided in Building Regulations that regulate the design loading requirements on specific types of structure under normal working conditions. The characteristic load values given in Building Regulations are usually mandatory. Ultimate loads are maximum loads considered in the design of a structure. It is usually given by characteristic loads multiplied by the partial safety factor of loading γ_f. It reflects an accidental loading condition for which the probability of being exceeded during the intended life of the structure is acceptably low.

1.7 Structural design procedure

All limit state requirements, including serviceability limit states and ultimate limit states, should be met in a structural design. The objectives of a structural design can be explained through Fig. 1.11.

1.7.1 Design for serviceability limit states

Design for serviceability limit states concerns the structural behavior in its elastic deformation stage, i.e., the response curve before the yielding point Y in Fig. 1.11. The design is to ensure that, at normal working conditions, i.e., under working loads including minor earthquake loading, the responses of the structure are smaller than allowable limit values. For example, the deformation of the structure Δ_e at working load F_w is less than the code-specified maximum allowable displacement Δ_s. Using elastic structural analysis, the structural responses under working loads can be determined. If a limit value is exceeded, the design should be altered until it is within the limit. This step of design is largely related to the determination of the minimum structural rigidity (minimum angle θ in Fig. 1.11A). No significant structural damage is allowed under working conditions.

1.7.2 Design for ultimate limit states

This step of design is to ensure that, when expected accidental loading or ultimate loading occurs, the structural responses do not exceed their peak loading capacity. The design is different for structures under a displacement-controlled loading and a load-controlled loading (see Section 2.2.1). Under a displacement-controlled loading, such as seismic loading, the design is to ensure the deformation capacity or ductility is not less than a code specified limit value. For a structure with the response curve shown in Fig. 1.11, it is to ensure that Δ_u is greater than the code required minimum value. The requirements on the deformation capacity or ductility given in design codes depend on loading demand, such as earthquake intensity. A stronger earthquake needs a higher deformation capacity and ductility.

For a structure under a load-controlled loading condition where the load cannot be reduced once applied, such as gravitational loads, the design needs to satisfy two conditions:

(1) Under ultimate loads, the peak resistance of the structure is not less than the effect caused by the ultimate loads, i.e., $F \leq F_p$ in Fig. 1.11B. This is to ensure that the structural response does not exceed point P at ultimate loading because the structure collapses immediately when point P is passed under load control.
(2) As discussed in Section 1.5, the first condition here may not provide a real protection to structures, because there is a possibility that the design maximum loading is exceeded. When overloading happens, Δ_p should be large enough to give sufficient warning. To this end, the current RC codes usually provide additional ductility requirements at onset of the peak resistance so that Δ_p is sufficiently large to give warning before failure. This is usually achieved by limiting the compression zone of concrete to be less than a limit value so that the curvature at ultimate failure is greater than a certain value (see the value of k_{u0} in Section 2.5).

1.7.3 Design procedure

As discussed in Section 1.6, an acceptable structural design needs to satisfy many limit states. One limit state can be considered as one link of the chain in Fig. 1.9. The structural design is to ensure that all links do not break in their intended lifetime. If all the links are designed or checked, the workloads of designers will be excessive. However, there are only a few limit states among all the possible limit states that are critical and govern the design of a structure. Other limit states are not critical and can be automatically satisfied by following the deemed-to-satisfy rules recommended in design codes, for example, by using minimum concrete strength and concrete cover thickness, and adopting recommended rebar detailing, etc. Therefore, designers only calculate and design for a small number of limit states in practice. More discussions on "noncritical" limit states are given in Section 11.9.

As critical limit states usually govern a design, a proper design procedure that starts with the possible weakest links can save time and reduce the workload of designers. For example, if the deflection of a beam is calculated first and found satisfactory and the flexural strength is designed after that, it is possible that the flexural design requires a change of the cross-section of the beam. In this case, the deflection calculation needs to be repeated after the flexural design. This is because ultimate limit states are usually more critical than serviceability limit states in the design of reinforced concrete members.

Structures are different. The same structure can also have different working conditions. All these may result in different weakest links or critical limit states. Change of design details may also shift the weakest link from one place to another, as discussed in Section 1.4.1. Therefore, the possible weakest links can be more than one in a structure. This is the reason that engineers need to design for a few limit states in a structural design rather than one. Consequently, different structural designs may require different design procedures. For example, for the design of conventional reinforced concrete beams, the normal design procedure is to design for the flexural ultimate limit state first that determines the flexural reinforcement, and then design for the ultimate limit state of shear that determines required stirrups. Afterward, the serviceability limit states of deflection and crack width are checked, which could have been automatically satisfied, and if not, small adjustments in the design can be made to satisfy the serviceability limit states. This design procedure ensures a minimum design workload to produce a structure that satisfies all required limit states. On the contrary, the design of prestressed concrete beams should be done by designing for the serviceability limit states on working stresses and deflection first, and after that, checking for the ultimate limit states. This is because the serviceability limit states are usually more critical than the ultimate limit states for prestressed concrete members. If a prestressed beam is designed following the procedure for normal reinforced concrete, it is likely that the consequent design does not satisfy all the limit states and needs to be repeated.

A good design procedure ensures that all limit states are satisfied while minimizing the workload to provide a most efficient design. The normal design procedures and approaches provided in textbooks, design manuals, and design codes of practice are developed from practical experience that facilitates the most efficient design practice. However, there are always special cases where critical links change due to special circumstances. Good designers are able to make appropriate adjustments based on their own experience for a particular type of design work.

1.8 Design codes of practice

Due to the deadly consequences of structural failure, structural designs are regulated and carried out by qualified personnel in most modern societies. The safety control framework of structural design is largely similar in many countries, although differences do exist. At the top of the framework are building regulations that are published and enforced by the government as part of the law of a country. Building regulations set out basic requirements for buildings, particularly those related to safety. The next level of the framework is the codes of practice that give guidance to professionals on methods of design to meet the building regulations and provide an effective and efficient design. The third level of the control is various standards that specify the quality and properties of construction materials.

It is noted that codes of practice are intended to provide guidance to designers rather than a set of rules that must be obeyed. Although codes are not mandatory, most designers always follow them rigidly. While design codes do provide the users with some protection against accusations of negligence, this protection is not as solid as some would like to think. It remains the responsibility of designers to ensure the adequacy and safety of a design. Although mankind has made great progress in science, technology, and culture, which is reflected in the progress in design codes of practice, some basic elements have been preserved in practice. The earliest written attempt to set out a uniform system of law was in about 1754 BC by King Hammurabi in Mesopotamia, in which one provision that dealt with building stated: "If a builder builds a house for a man and does not make its construction sound, and the house which he has built collapses and causes the death of the owner of the house, that builder shall be put to death." (Harper 1904). This was the earliest extant "building regulation." Although the penalty of "eye for eye, tooth for tooth" has changed in modern building regulations, this statement reflects the designer's responsibility to some extent.

All design codes of practice are developed from past research findings and practical experience. Furthermore, design codes normally cover ordinary structures and current structural systems. There are always cases that existing design codes of practice do not cover or are inapplicable, for example, when new materials are used in a conventional structure, when a new type of structure is designed, when a normal structure is under new/special loading conditions or in a new/special service environment, or when new/special requirements are imposed on a structure. As technologies are developing all the time and the environment is also changing, such cases are encountered from time to time. In practice such a problem is normally dealt with in the following ways:

(1) Waiting for the professional bodies to develop relevant new design codes/guidelines;
(2) By proof tests, i.e., undertaking adequate experimental tests to ensure satisfaction of the design. Existing codes of practice may allow for such proof tests as an acceptable design approach, as an alternative to the design calculations specified in the design codes; and
(3) Special approval from qualified personnel and authorities after detailed studies of the structural design scheme.

All of these alternatives should, and can only, come from and be based on the first principles of mechanics of structures and materials as discussed in this book. Therefore, good understanding of the fundamentals is essential and critical to be a competent engineer and building professional. The world-renowned structural engineer, former co-chairman of Ove Arup and Partners Sir G. J. Zunz, pointed out

> *As my experience and that of my colleagues develops, I find more and more that it is the fundamentals that matter and those without a sound training in them suffer for the rest of their careers.*

This book focuses on the fundamentals for applications and developments of the theories and models in reinforced concrete structures.

References

British Standards Institution (BSI), 2004. *Design of concrete structures—general rules and rules for buildings*. BS EN 1992-1-1, Eurocode 2—Part 1-1, London.

Harper, R. F. 1904. *The code of Hammurabi*. Chicago: University of Chicago Press.

Flexural failure and design theory 2

2.1 Introduction

As shown in the RC practical in Chapter 1, flexural or bending failure is a major failure mode of RC members. This chapter studies the failure mechanism and relevant moment resistance of RC members under flexure. Bending moment and shear force often occur simultaneously at one cross-section. However, normal stress caused by a bending moment generally localizes on the top and bottom of a cross-section while shear stress caused by a shear force concentrates in the middle of the cross-section. Therefore, shear force usually has an insignificant effect on flexural resistance of RC members, except in some special cases such as deep beams. Only pure bending is considered in this chapter.

2.2 Failure process

The typical flexural failure process of an RC member (such as a beam and column) is illustrated in Fig. 2.1. When the RC member is subjected to a continuously increasing mid-span displacement Δ, it experiences the following four phases divided by points A to D successively:

- Elastic deformation until point A at which concrete cracks at the bottom of the member at the load of P_A;
- Nearly elastic stage with a reduced rigidity until point B where tension bars start to yield at the load of P_B;
- Hardening of member until point C where peak resistance is reached at the load of P_C; and
- Softening of member until point D where significant concrete crushing occurs at the compression face and the member resistance drops quickly after the load of P_D.

Cracking is unavoidable and allowable in concrete structures. Although cracking causes a reduction in the effective cross-sectional area and hence the rigidity of an RC member, it generally does not affect the functioning of the RC member provided the serviceability limit states are satisfied, e.g., the maximum allowable deflection and crack width are not exceeded. RC flexural members usually cannot be loaded to pass point B at working loads because yielding of rebar will cause not only extensive cracking and displacement but also significant residual or permanent deformation of the structure. Point D is considered as the failure point because the load-carrying capacity drops quickly due to significant concrete crushing after this point. The beam is extensively damaged physically after concrete crushing and is no longer serviceable. The load-carrying capacity at point C is used as the flexural strength as it is the maximum resistance of the RC member.

Point B may not exist when excessive reinforcement is provided, so that the rebar does not yield throughout the failure process. When a large axial force is applied to an

Analytical Approaches for Reinforced Concrete. https://doi.org/10.1016/B978-0-12-821164-9.00002-4

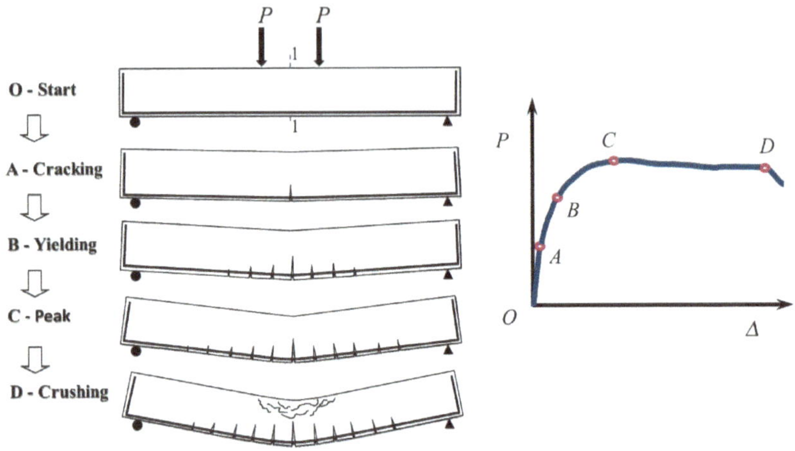

Fig. 2.1 Flexural failure process.

RC member, point B may also disappear even with a low reinforcement ratio, as the axial force exerts additional compression force on the rebar so that it does not yield on its tension face under bending. Occasionally, point A may not present in a loading process, e.g., when an RC member is loaded under a very large axial force or under large prestressing.

2.2.1 Effect of loading type on failure point

The displacement at point D is usually considered as the ultimate displacement (see flexural Theorem 3.3 in Chapter 3) of a member and is often used as the deformation capacity or for the calculation of ductility of RC members. However, the displacement at point D can only be achieved under a deformation (or displacement)-controlled loading, e.g., earthquake loading. Under earthquake loading, ground movement causes displacement of a structure rather than directly applying a certain force to the structure. Therefore, the structure can actually achieve the ultimate deformation at point D before it collapses. However, in a load-controlled system where loading increases gradually, such as a beam under gravitational loads, the structure will collapse immediately after the loading increases to point C, because the beam cannot resist a load that is greater than that at point C and a gravitational load usually does not reduce once it increases to a certain value. As the structure immediately collapses after point C is passed under load control, the displacement at point C must be considered as the ultimate deformation capacity in this case, rather than that at point D.

Based on the preceding discussions, the general structural design considerations can be summarized as follows:

(a) For serviceability limit states, a point lower than point B (a point before point C if point B does not exist) should be used as the structural design limit. The actual location of a design limit point on the curve depends on specific design requirements such as allowable displacement and crack width, which are different for different structures based on their functional requirements.

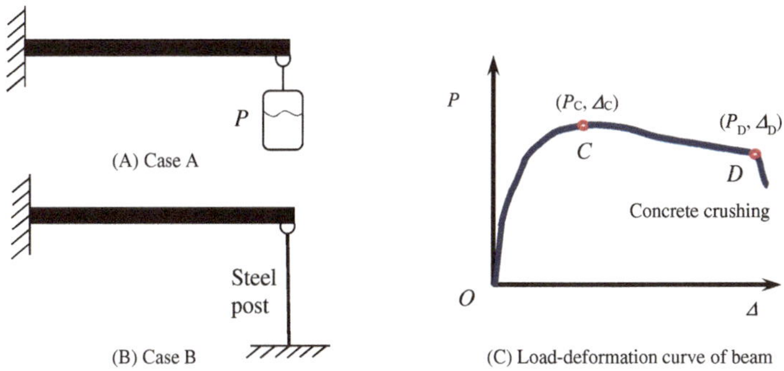

Fig. 2.2 Load types and failure points.

(b) For the design of ultimate limit states of structures subject to applied force (e.g., gravitational force, wind force, and water pressure), the load resistance and deformation at point C is the load-carrying capacity and deformation capacity, respectively.

(c) For the design of ultimate limit states of structures subject to applied deformation (e.g., earthquake, impact, movement/settlement due to temperature/shrinkage, and support movement), the deformation and resistance at point D should be considered as the deformation capacity and corresponding load-carrying capacity, respectively, at failure of the structure. This is because the loading is deformation, and the concern of the design is how much a structure can move before it fails.

Clear understanding of design requirements is essential to ensure a safe design. Some people use the resistance at point C, P_C, and the displacement at point D, Δ_D, as the load-carrying capacity and deformation capacity of a structure, respectively. However, P_C and Δ_D are not the same point and do not occur at the same time. Under a certain type of loading, you either have P_C or Δ_D but not both P_C and Δ_D when failure occurs. For example, the cantilever beam in Fig. 2.2A supports a water tank. The load deformation curve of the cantilever beam is shown in Fig. 2.2C. When the weight of water in the tank increases to P_C, the structure fails immediately. The ultimate load and displacement in this case are $P_u = P_C$ and $\Delta_u = \Delta_C$, respectively. The failure point at D is irrelevant to the design. If the loading is changed to the movement of the cantilever tip caused by the drop of temperature in the steel post (or the movement of the post base), as shown in Fig. 2.2B, the allowable drop in temperature (or the movement of the post base) is determined by the displacement of Δ_D. In this case, the peak point C is irrelevant to the design.

2.3 Moment-curvature relationship

The flexural behavior of RC members is often studied through the moment vs. curvature relationship that can be produced by adopting the assumption that the plane section remains plane. The validity of the plane section assumption is discussed in Section 2.8.2. For flexural members, the curvature of a given cross-section is equal to the angle between the strain profile and the vertical line (κ in Fig. 2.3). The reasons and mathematical derivation of the curvature of the member and strain profile are

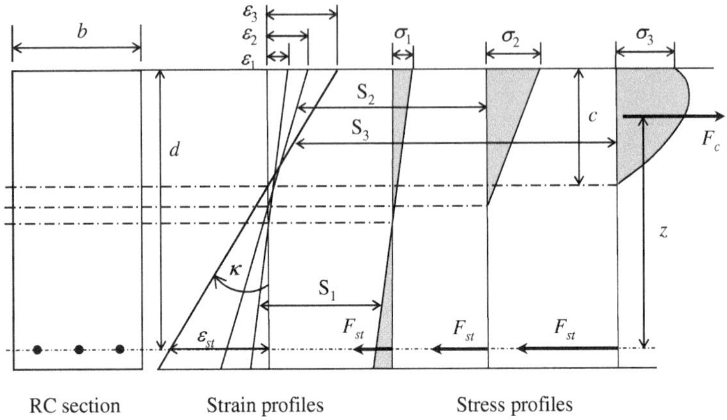

Fig. 2.3 Stress and strain profiles.

given in Section 2.8.1. When the beam deformation increases, the cross-sectional curvature κ increases gradually. Increasing curvature is reflected by the change of strain profiles from S_1 to S_3 in Fig. 2.3. When the loading is smaller than that corresponding to point A in Fig. 2.1, the concrete in both the tension zone and compression zone is linear elastic (approximately) and its strain and stress profiles are shown by S_1 in Fig. 2.3. The neutral axis (the dash dotted line), where the stress and strain are equal to zero, divides the cross-section into the tension zone and compression zone. When the load further increases, concrete will crack in the tension zone. As the tensile resistance of concrete is small, it can generally be neglected; and hence the tensile stress in the tension zone is considered to be zero after cracking. The stress and strain profiles in this stage are shown by S_2. Further increase in the curvature leads to nonlinearity of the concrete and its stress profile in this stage is represented by S_3.

It can be seen from Fig. 2.3 that increasing curvature causes the increase of compressive and tensile strains at the top and bottom of the beam, respectively. In the meantime, the neutral axis moves up in the deformation process, leading to an increasing tension zone and a decreasing compression zone. The tensile strain ε_{st} in the tension bars also increases gradually in the process and the tension bars will yield when it reaches its yield strain ε_y (point B in Fig. 2.1). When the compression zone reduces to a certain size, the total compressive resistance of concrete is not sufficient to balance the tensile force F_{st} caused by the tension bars. Concrete will crush in this case, which corresponds to point D in Fig. 2.1. As the tension force from the tension bars does not change after its yielding, the decrease of moment resistance in the descending branch is caused by the softening (stress reduction of concrete after the peak point of its stress-strain curve) of concrete material, which causes the reduction of the lever arm z. Therefore, any measure that can delay or inhibit concrete softening can maintain the postpeak moment resistance of an RC member, and hence increase the ductility of the member. It will be seen in Chapter 8 that confining concrete is a very effective way of increasing concrete strength and slowing down the softening of concrete.

The typical moment versus curvature relationship is shown in Fig. 2.4, which is qualitatively identical to the load-displacement curve in Fig. 2.1.

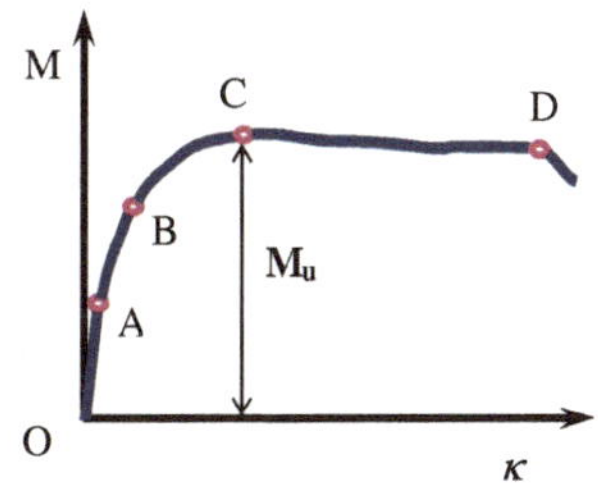

Fig. 2.4 Moment-curvature relationship.

The moment-curvature curve for an RC section with a constant axial force N can be produced through the following procedure:

(1) For a given value of curvature κ, calculate the depth of concrete compression zone c based on the balance of the forces perpendicular to the cross-section ($\sum F = 0$ in Fig. 2.3). There is only one unknown c in the equation, and hence c is solvable. When κ and c are given, the strain profile is fixed. For a given RC section, the stress-strain relationships of the materials are known and hence the stress profile is also fixed. The use of $\sum F = 0$ in this step is to find the position of the neutral axis so that the compression forces and tension forces are balanced.
(2) Take the moment of all the cross-sectional forces using the equation $\sum M = 0$ to calculate the moment resistance of the section and produce one point on the moment M vs. curvature κ curve. Moment taking is discussed in Section 2.6.1.
(3) A complete $M-\kappa$ curve can be produced by increasing the value of κ gradually and repeating the preceding process.

The flexural design of RC members in ultimate limit states mainly concerns the determination of:

(1) The moment resistance at point C, namely the flexural strength, or moment capacity, or ultimate strength or ultimate moment, and;
(2) The displacement limit at point D, namely the maximum deformation capacity, or ultimate displacement.

2.4 Typical response curves

The moment-curvature curves of a few typical RC beam and column sections are shown in Figs. 2.5 and 2.6, respectively. These $M-\kappa$ curves are produced using the general approach described in Section 2.3, in which the compression force of concrete is obtained by numerical integration of the nonlinear stress block using a conventional layered method (Wu et al. 2004). The layered method discretizes an RC section into a large number of constant strain layers. The M and N are obtained from the numerical summation of the forces/moments contributed by all layers across the cross-section. More details of the numerical modeling can be found in Wu et al. (2004).

The RC sections in Figs. 2.5 and 2.6 have a cross-section of 300 mm (breadth) × 600 mm (depth), with a concrete compressive strength $f_{co} = 40$ MPa, and steel yield strength $f_y = 500$ MPa. The Popovics model (Section 3.2.1) is used as the stress-strain relationship of concrete. Five cases of the tension steel bars, 4T25, 4T32, 6T32, 8T32,

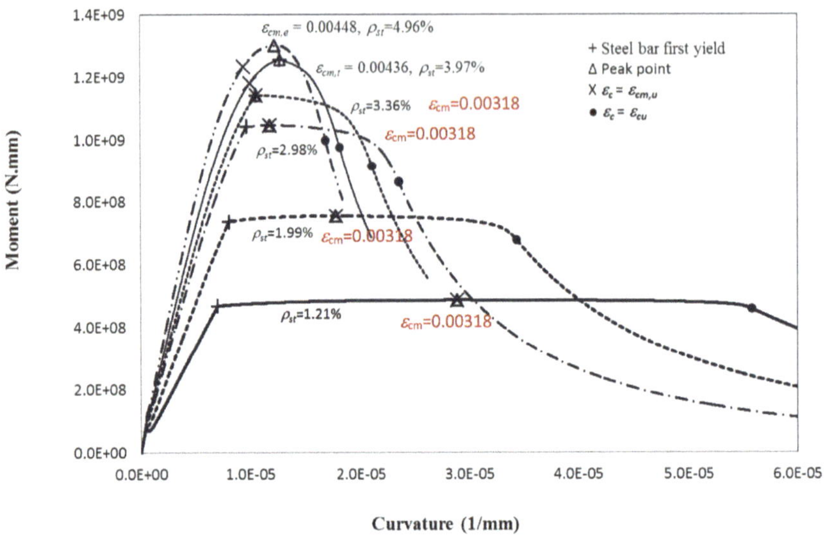

Fig. 2.5 Moment-curvature curves of RC beams.
From Wu, Y. F. 2016. "Theorems for flexural design of RC members." *J. Struct. Eng.* 142 (5), 04015172.

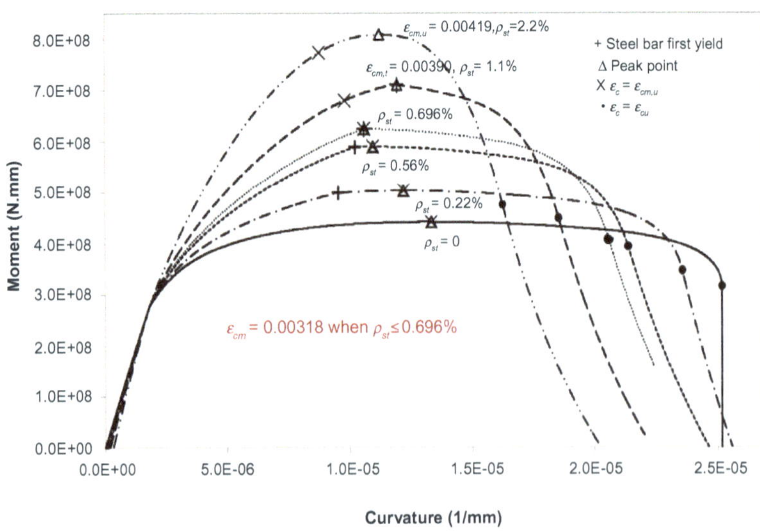

Fig. 2.6 Moment-curvature curves of RC columns with $N/(bdf_{co}) = 1/3$.
From Wu, Y. F. 2016. "Theorems for flexural design of RC members." *J. Struct. Eng.* 142 (5), 04015172.

and 10T32, corresponding to reinforcement ratios ($\rho_{st} = A_{st}/bd$) of 1.21%, 1.99%, 2.98%, 3.97%, and 4.96%, respectively, and a special case of $\rho_{st} = 3.36\%$, are numerically simulated in Fig. 2.5. To provide a uniform benchmark for comparison without the effect due to the variation in the effective depth d, all bars are assumed to be in one layer although two or more layers are required for the cases of high reinforcement ratios. The concrete cover thickness from the external face to the center of steel reinforcement is 60 mm. The cases of RC columns are assumed to have an axial force ratio of $N/(bdf_{co}) = 1/3$ and have the same RC section and material properties as the RC beams but different reinforcement ratios.

It can be seen from Figs. 2.5 and 2.6 that the reinforcement ratio and axial force significantly affect the flexural response of the RC beams and columns, including the flexural strength and ductility. Based on the discussions in Chapter 1, the reinforcement ratio and/or axial force must be limited in RC designs to ensure a sufficient ductility and safe failure mode of RC structures.

2.5 Failure modes

When the deformation of the flexural member increases, its load resistance increases first and then decreases after the peak point at which the flexural strength occurs (Fig. 2.7A). The concrete strain of the extreme compression face continues to increase throughout the deformation increase (Fig. 2.3). In the current reinforced concrete theory, it is assumed that the flexural strength or peak moment occurs at the exact moment when the concrete strain of the extreme compression face reaches a specific value ε_u, as shown in Fig. 2.7B. This assumption was made based on extensive research findings. Engineering experiences have found that the assumption gives sufficiently accurate results in designs, except in some unusual cases (see Section 2.7.3 and Chapter 3).

Depending on the strain condition of the tension steel at the onset of flexural strength (point C in Fig. 2.4), the flexural failure mode is categorized into the following three types (Fig. 2.7B):

(1) Underreinforced failure—the tension steel bars have yielded before the onset of flexural strength;
(2) Balanced failure—the tension steel bars just yield at the onset of flexural strength; and
(3) Overreinforced failure—the tension steel bars have not yielded at the onset of flexural strength.

RC members with the preceding three failure modes are referred to as underreinforced members, balanced members, and overreinforced members, respectively.

From the strain profile of an arbitrary RC section in Fig. 2.7B, one has

$$\frac{\varepsilon_u}{c} = \frac{\varepsilon_{st}}{d - c} \tag{2.1}$$

Defining the neutral axis parameter k_u as

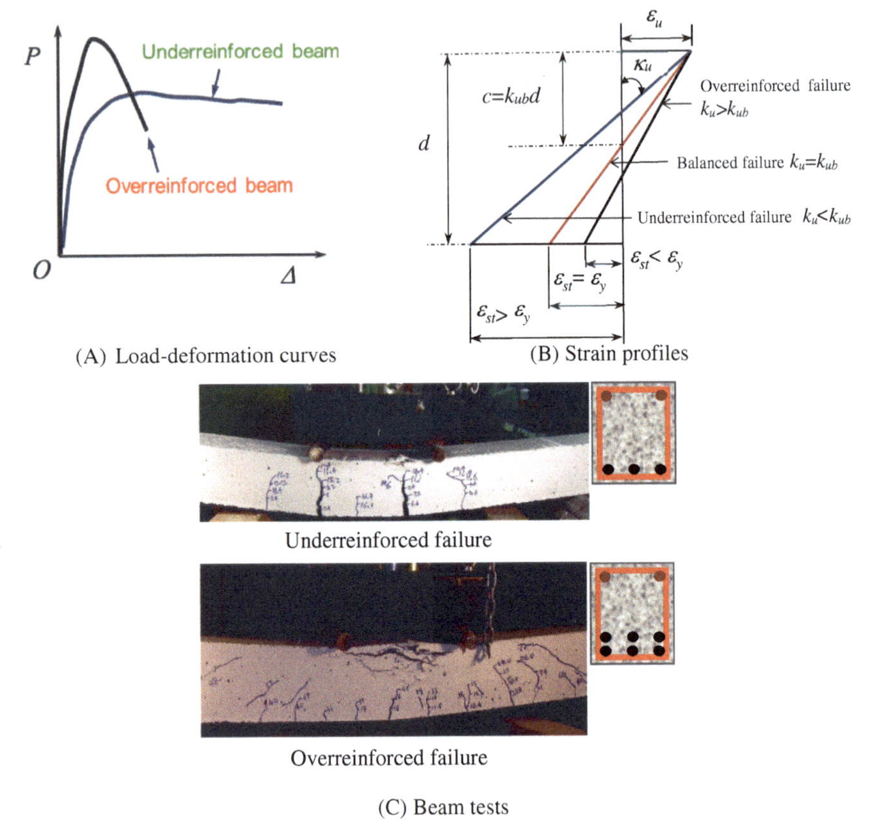

Fig. 2.7 Flexural failure modes.

$$k_u = \frac{c}{d} = \frac{\varepsilon_u}{\varepsilon_u + \varepsilon_{st}} \tag{2.2}$$

The neutral axis parameter of the balanced failure mode is referred to as k_{ub}, given by

$$k_{ub} = \frac{\varepsilon_u}{\varepsilon_u + \varepsilon_y} \tag{2.3}$$

As ε_u and ε_y are constants, k_{ub} is a constant used in RC designs to divide RC members into underreinforced members and overreinforced members (Fig. 2.7B). To ensure sufficient ductility of RC members, RC design codes limit k_u to a maximum allowable value k_{u0} that is significantly smaller than k_{ub}, which physically requires significant yielding of the tension reinforcement before the onset of flexural strength. Limiting a small value of k_u is also equivalent to ensuring a sufficiently large curvature κ_u at the onset of flexural strength since $\kappa_u = \varepsilon_u/(k_u d)$. Therefore, overreinforced members and some underreinforced members with insufficient yielding of tension bars at the onset of flexural strength are not allowed in practice, due to insufficient warning that cannot be observed clearly before failure.

From the balance of forces in Fig. 2.3, one has

$$F_c = N + F_{st} \tag{2.4}$$

where F_c is the resultant of the compressive stress of concrete at the peak moment, which is a function of concrete strength and the compression zone depth c, or

$$F_c = f(d \cdot k_u, f_{co}) \tag{2.5}$$

The preceding two equations show that, for a given RC section with a constant axial force N and a certain value of k_u, the force of the tensile rebars at the onset of the peak moment, F_{st}, is a fixed value. As the tensile strain is fixed under a certain k_u (Fig. 2.7B), the tension steel area is fixed. In other words, one k_u value corresponds to one A_{st}, and hence, limiting the value of k_u is equivalent to limiting the tension reinforcement ratio. Therefore, an overreinforced member with $k_u > k_{ub}$ means that too much tension steel is provided so that the tension steel does not yield at the onset of flexural strength.

Based on the preceding discussion, it is concluded that, when the reinforcement ratio increases, the curvature κ_u decreases, and hence the ductility of RC members decreases, which can be observed in Figs. 2.5 and 2.6. The preceding study also reveals that the flexural failure modes are directly related to (1) the strain of the tension steel, and hence the crack width, (2) the depth of compression zone of concrete or the extent of concrete crushing zone, and (3) the failure curvature κ_u, i.e., the deformation of the RC member at failure. Therefore, the flexural failure mode can be judged by observing its crack width, crushing zone depth, and deflection when a member fails; see Fig. 2.7C.

2.6 Calculation of design moment

2.6.1 Moment taking

Fig. 2.8 shows the typical case of internal forces in the cross-section of a flexural member, where T represents the resultant of all forces on the tension side of a flexural member and C the compression resultant of all forces in the compression zone. R is the resultant of C and T. Point O is an arbitrary point that locates at a distance of x from C. The moment of C and T about O is given by

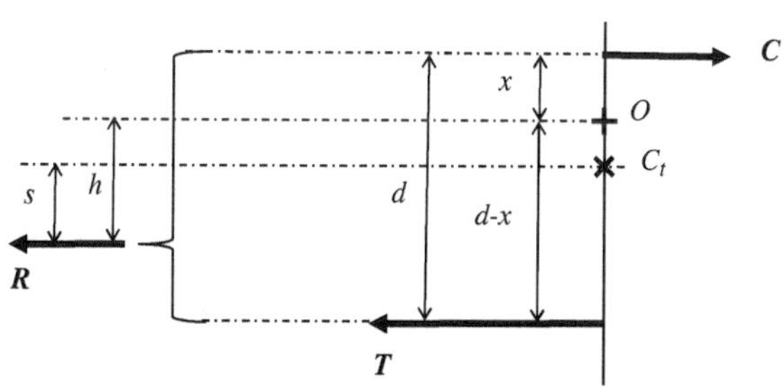

Fig. 2.8 Moment taking.

$$M = Cx + T(d - x) = Cx - Tx + Td \tag{2.6}$$

For beams, $N = 0$ or $T = C$. Eq. (2.6) gives

$$M = Td \tag{2.7}$$

As x is an arbitrary value, Eqs. (2.6), (2.7) show:

(1) For beams where the tension resultant is always equal and opposite to the compression resultant, moment can be taken about any point and the result will be the same.
(2) For columns where $N \neq 0$ or $C \neq T$, the result of the moment is directly related to the point about which the moment is taken.

For columns, the only correct place to take the moment is the centroid of a cross-section. This is because moment is directly related to curvature in an engineering sense and a zero curvature corresponds to a zero moment, as shown in Fig. 2.4. When the resultant R of C and T is exactly located at the centroid C_t of the cross-section, or $s = 0$ (Fig. 2.8), it will not cause bending of the member, i.e., the curvature $\kappa = 0$. The consistency of $\kappa = 0$ when $M = 0$ can be achieved only for C and T to take the moment about the centroid.

Another more general approach to take the moment for columns is to use the resultant force. The resultant R is fully equivalent to the combination of T and C statically. Therefore, the moment of R taken about any point must be equal to the moment of T and C taken about the same point, i.e.,

$$Rh = Cx + T(d - x) \tag{2.8}$$

For a given T and C and a selected point O, the right-hand side of Eq. (2.8) is known. As $R = T - C$, the location of the resultant, h, can be calculated from Eq. (2.8). Then the moment can be calculated by taking the moment of the resultant R about the cross-section centroid:

$$M = R \times s \tag{2.9}$$

In summary, for beams, moment taking can be done conveniently using Eq. (2.6) about any point. For columns, the moment must be taken about the cross-sectional centroid if Eq. (2.6) is used. Another way to take the moment for columns is to use Eqs. (2.8) and (2.9), which can avoid mistakes due to misconceptions.

2.6.2 Moment redistribution

Bending moments in different parts of an RC structure can be redistributed to suit design needs. One of the reasons to make use of moment redistribution is to facilitate more uniform distribution of reinforcement in a structure. For example, the moments at the supports of continuous beams are usually larger than those at spans. As a result, more rebars are required at supports (top bars) than at spans (bottom bars). To provide more uniform distribution of rebars and more economical design, design codes allow engineers to reduce support moments while increasing span moments appropriately. The rules for moment redistribution are closely related to the moment-curvature relationship as well as the moment envelope.

2.6.2.1 The mechanism of moment redistribution

The mechanism of moment redistribution is illustrated by the beam with fix-end supports in Fig. 2.9. Both the support moments and the span moment increase when the load w increases in its elastic deformation stage. As the elastic support moments are larger than that of the span moment, the supports reach the moment capacity M_u (Fig. 2.4) first before the span. When the deformation of the member continues to increase, the plastic rotation of the member occurs at the supports, while maintaining the support moment at M_u (slightly decreasing in reality; see Fig. 2.4). Further deformation causes the span moment to continue to increase as it is still elastic when the supports reach the moment capacity, until it also reaches its moment capacity. After all support and span moments reach their moment capacities, further increase in loading will cause the collapse of the member. The yielding of supports causes the elastic moment diagram as shown by the dashed line in Fig. 2.9 redistributed to the solid line as shown.

Fig. 2.10 shows the change in the moment diagram due to the change of support moments. In case C, the moment diagram can be moved up and down or rotated when the support moments M_A and M_B change. For a certain load w, $M_0 = wL^2/8$ remains unchanged, but the span moment $M_{mid} = M_0 - (M_A + M_B)/2$ can be changed when the support moments M_A and M_B vary. The balance of the beam is maintained, and

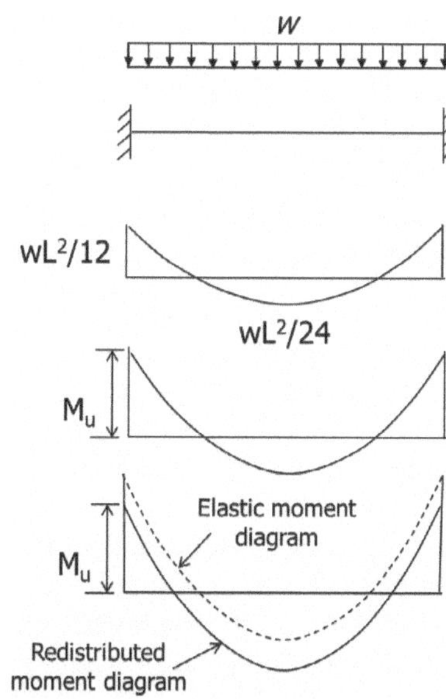

Fig. 2.9 Elastic and redistributed moment diagrams.

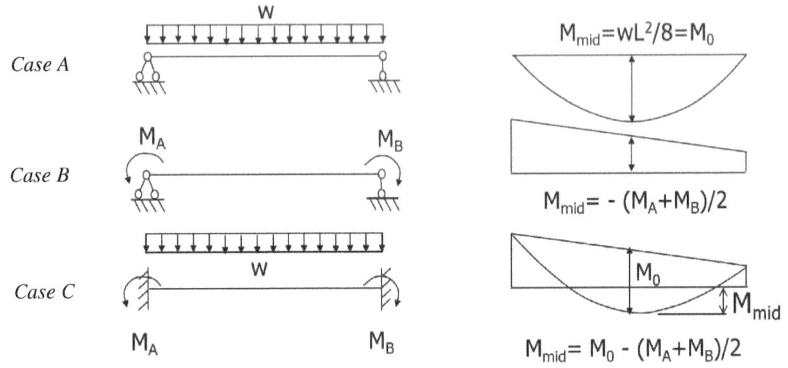

Fig. 2.10 Variation of moment diagram.

the member does not fail as long as the equation, $M_0 = M_{mid}+(M_A+M_B)/2$, is satisfied. This balance can be considered analogous to three persons, $I = M_{mid}$, $II = M_A/2$, and $III = M_B/2$, carrying a total load of M_0. The individual loads shared by, or distributed to, each person can be changed as long as $I + II + III = M_0$. Therefore, the moment resistances at different places can be redistributed by engineers according to design needs, provided the summation of moment resistance is larger than the required value.

Engineers adjust M_{mid}, M_A, and M_B by adjusting rebars at different locations. For example, an engineer can put less rebar at supports than that calculated from elastic moments. In this case, the supports yield earlier and a subsequent increase in loading will only cause an increase in span moment, thereby requiring more rebars at the span. The engineer can also do the opposite by allocating more rebars at the supports than are required by elastic moments, and meanwhile reducing rebar at the span. Therefore, redistribution of moments physically means relocation of rebars from one place to another.

Although engineers have the liberty to redistribute moment, there are rules that have to be obeyed. These rules are:

(1) the equation $M_{mid}+(M_A+M_B)/2 \geq M_0$ is satisfied, and
(2) the maximum amount of redistributed moment is limited to the code allowable values.

The amount of moment that can be redistributed is limited by the physical plastic hinge rotation capacity of an RC section, as shown in Fig. 2.4. The moment redistribution at a certain place takes place only when a plastic hinge forms at this place so that its moment resistance does not increase with further increase of loading. However, a plastic hinge has rotational capacity, as shown in Fig. 2.4. The plastic hinge will collapse when its deformation reaches point D in Fig. 2.4. The limit in plastic hinge rotation capacity determines the amount of moment that can be redistributed to other places. For normal RC continuous beams, this limit provided in a design code is usually around 30% of the elastic moment. For multistory sway frames, the limit is smaller, at about 10% of elastic moment to ensure stability of the frame.

The concept of moment redistribution can be generalized to load redistribution in structures or materials. In a frame, the internal forces can be redistributed between different members. For example, the use of fixed-end supports at column bases can reduce the moment in columns above the base, thereby reducing the costs of superstructure, but increasing the foundation moment and its costs (see Fig. 11.5A). The use of pin-end supports at column bases will reduce the foundation moment and its costs but increase the superstructural costs (Fig. 11.5B). In a building, the loads carried by different parts of the structure can be redistributed by adjusting their RC designs. Even in the compression test of a concrete cube or cylinder, a local internal cracking causes the release of its stress and redistribution of the stress to adjacent vicinities. In fact, the cube/cylinder test is not a "material test" but a "structural test," and hence structural redistribution applies. Consequently, the size and shape of test specimen, boundary condition, loading rate, etc. affect the test results (see Sections 8.2 and 11.10).

2.6.2.2 Effect of redistribution on moment envelope

Different arrangements of loads, or load patterns, produce different bending moment diagrams, as illustrated in Fig. 2.11. Members should be designed to worst-case scenarios of moment that are enclosed by the moment envelope, i.e., the outline of the bending moment diagrams of all possible load patterns. Fig. 2.11 shows the typical load patterns, bending moment diagrams, and the moment envelope of a three-span continuous beam. It is noted that the dead load and live load should be factored, and the load factors are different in different design codes.

Moment redistribution is applicable to the bending moment envelope. The reduction of support moments may not necessarily cause the increase in span moments, as illustrated in Fig. 2.12. The black solid lines in Fig. 2.12 show the moment envelope of the case in Fig. 2.11. It is assumed that the original design moment at support B is redistributed from its original value of M_{B2} to M_{B3}. Let's look at the moment redistribution in span 1. Since the support moment corresponding to load pattern 3 does not change, its bending moment diagram at span 1 remains unchanged. The support moment of load pattern 2 drops from M_{B2} to M_{B3}, causing the relevant bending moment diagram in span 1 to drop to the position shown by the dash-dotted line that

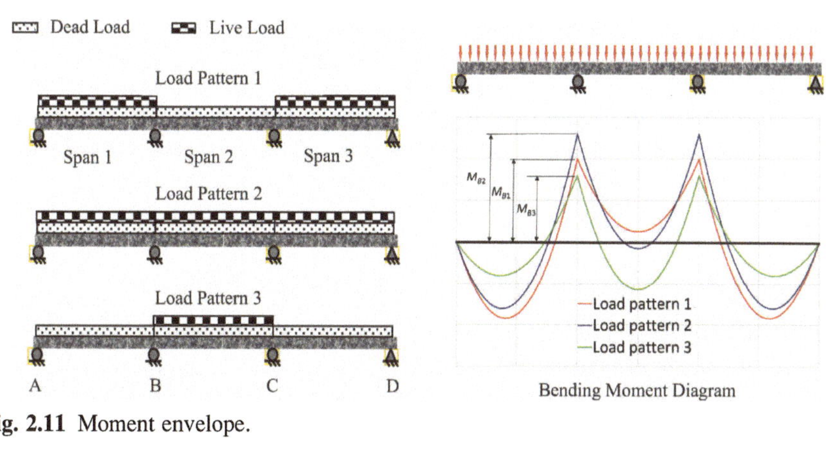

Fig. 2.11 Moment envelope.

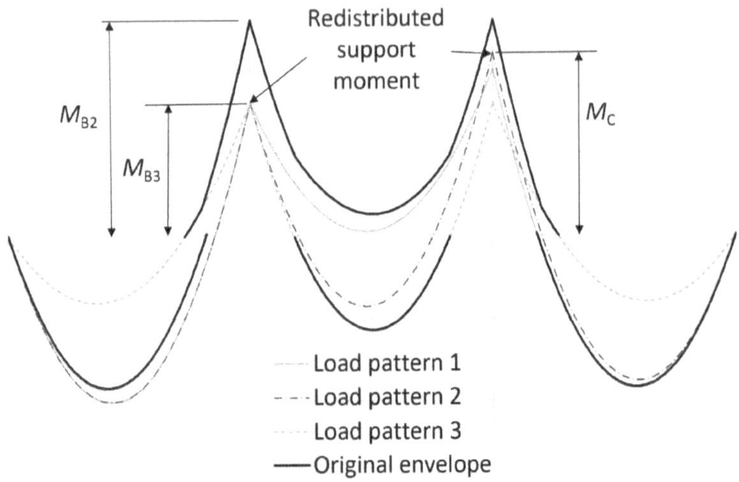

Fig. 2.12 Moment redistribution.

is below the original envelope. The support moment corresponding to load pattern 1 also drops from M_{B1} to M_{B3}, and its bending moment diagram drops to the same position as that shown by the dash-dotted line. Therefore, the span 1 moment is increased due to the moment redistribution at support B from M_{B2} to M_{B3}. However, the redistribution of the support B moment does not change the span 2 moment because the dropped bending moment diagrams of load patterns 1 and 2 remain above the original envelope, as shown in Fig. 2.12. Therefore, it does not change the design moment at span 2.

Now assume the moment at support C drops from M_{B2} to M_C, where $M_C > M_{B1}$ (Fig. 2.12). The bending moment diagram of load pattern 2 in span 3 drops to the position shown by the dash-dotted line above the original envelope. Therefore, it does not increase the design moment at span 3. For the same reason, the drop of support C moment does not change the design moment at span 2.

It can be seen from the preceding examples that the redistribution of moment is applicable to individual load pattern(s), and the reduction of the moment envelope at supports may not affect all span moment envelopes. In general, the moment redistribution is shifting (up, down, or rotating) of the individual moment diagram of a particular load pattern in a span. As design moment is determined from the moment envelope, shifting of the envelope moment in some places (such as supports) may or may not increase the original envelope moments in other places (such as spans), because the moment envelope at different places is determined by different load patterns.

2.6.2.3 Determination of moment envelope

Moment envelope is determined by overlapping moment diagrams of all possible load patterns together. When the number of spans is large, the number of possible load patterns is excessive. To avoid excessive workload, design codes and guidelines provide typical load patterns for producing the moment envelope. However, these typical load

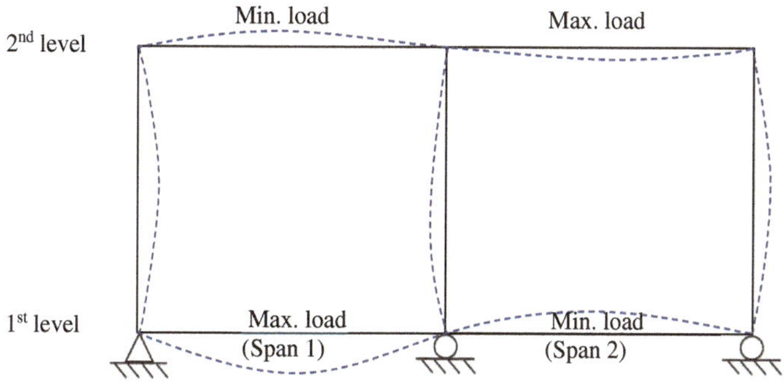

Fig. 2.13 Load pattern causing maximum span 1 moment at 1st level.

patterns are only applicable to simple and typical structures, such as normal continuous beams. Engineers have difficulty in determining load patterns for general structures such as that shown in Fig. 2.13.

A simple way to determine the specific load pattern for the design moment at a specific place is introduced here. The continuous beam shown in Fig. 2.14 is used to illustrate the approach. The load pattern causing the maximum span 3 moment of the continuous beam is shown in the figure. To cause a maximum moment at span 3, maximum load must be applied at span 3 as shown. The blue solid line shows the deformed shape of the beam when the load is applied at span 3 and no other load is applied elsewhere. The continuity of the beam requires the deformation of span 2 and span 4 moving upward and that of span 1 downward. If a load is applied at span 2 as shown, the deformation at span 2 will be pushed downward and the deformation of the whole beam is shown by the dashed line. Therefore, loading at span 2 will cause a reduction of deformation, and hence bending, at span 3. The load at span 2 causes a beneficial effect to the moment at span 3, and hence is called beneficial load. For the same reason, the load at span 4 is also beneficial to span 3 and a load at span 1 is adverse. Therefore, to produce the maximum possible bending moment at span 3, maximum loads should be arranged at all adverse spans and minimum loads should be placed at all beneficial spans, as shown.

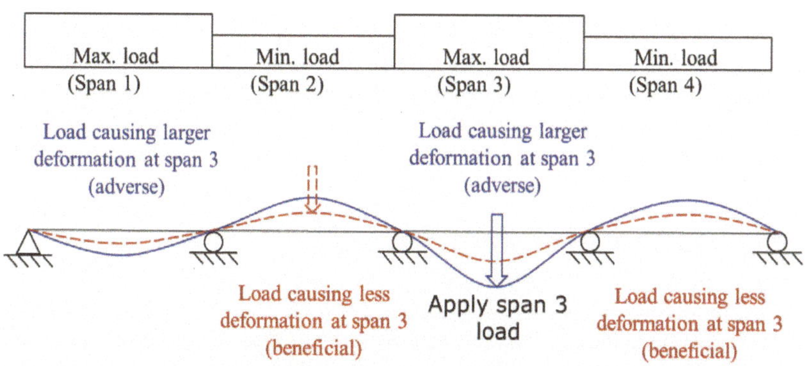

Fig. 2.14 Load pattern causing maximum span 3 moment.

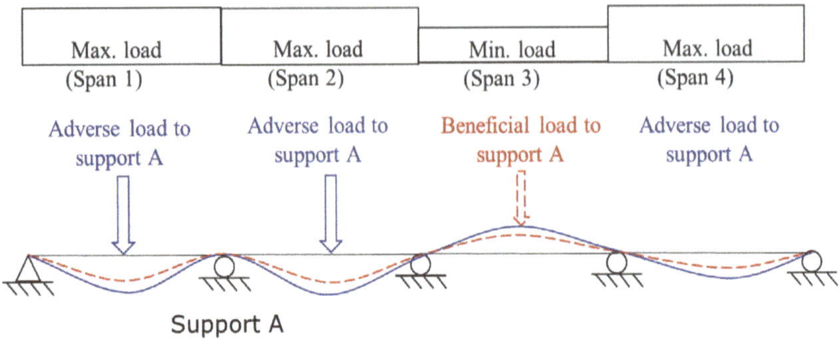

Fig. 2.15 Load pattern causing maximum support A moment.

Fig. 2.15 gives an example of load patterning to produce maximum moment at a support. To produce the maximum moment at support A, the adjacent spans of support A must be loaded with maximum loads so that support A is bent to its maximum degree. From the deformed shape of the beam, it can be clearly seen that the load at span 3 is beneficial and that at span 4 is adverse, and therefore the load pattern causing a maximum moment at support A is as shown.

It can be seen from Figs. 2.14 and 2.15 that, after drawing the deformed shape of the initially applied loading, the critical load pattern (worst-case scenario) can be easily determined: the adverse spans are those with a downward deformation, and the spans with upward deformation are beneficial spans. This analysis shows that the load pattern 2 in Fig. 2.11 is actually not the critical load pattern for the support moments at B and C. The exact critical load pattern for a maximum support B moment is maximum loads at spans 1 and 2 and minimum load at span 3, while for support C, it is maximum loads at spans 2 and 3 and minimum load at span 1. Nevertheless, the difference in support moments between these exact patterns and the load pattern 2 in Fig. 2.11 is insignificant, and hence, design codes allow people to use the load pattern 2 (fully load all spans) for calculation of maximum moments at all supports.

This approach can be easily applied for load patterning of more complicated structures, such as that shown in Fig. 2.13. The load patterning is not obvious even for such a simple frame structure. By applying the approach, the load pattern for the design moment at any place can be easily determined. For example, to produce a maximum moment at 1st level span 1, a maximum load is first applied at 1st level span 1. Based on the deformed shape as shown by the dashed line, maximum loads are applied at all places with a downward deformation and minimum loads are applied at all spans with an upward displacement.

2.7 Conventional flexural theory

The determination of the flexural strength M_u of an RC section is not straightforward, as no clear feature is identifiable at the peak moment resistance (point C in Fig. 2.4) apart from knowing that it is a peak point of the moment-curvature curve.

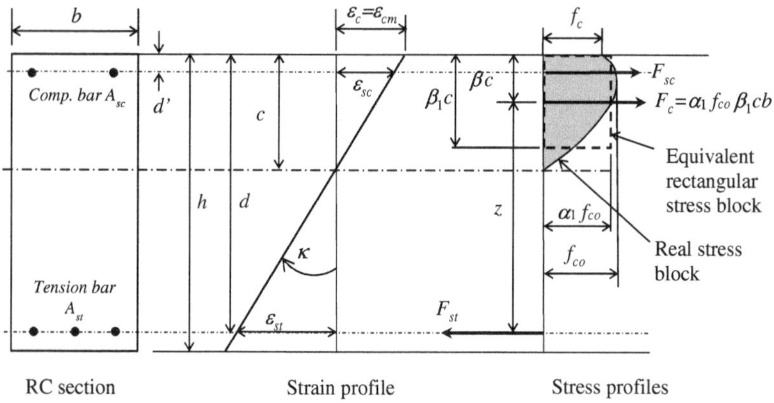

Fig. 2.16 Determination of flexural strength.

The development of the flexural design theory has experienced three major stages in its history:

(1) The ultimate strength design approach based on empirical design formulas derived from experiments;
(2) working stress design approach based on elasticity and allowable stresses; and
(3) a return to ultimate strength design approach based on plasticity using rectangular stress block (see Fig. 2.16).

The first rational flexural theory for reinforced concrete was developed by Koenen in 1886 and the first significant engineering tests of RC beams were reported by Wayss and Koenen in 1887 (Siess et al. 1952), after which numerous similar works were carried out. A major advancement in the flexural theory was made by von Emperger (1931) and Stussi (1932) who established the basis of the modern flexural design methodology. The comprehensive experimental studies from the 1930s to 1960s were the basis for the modern flexural theory (Whitney 1937, 1942; Jensen 1943; Hognestad 1951, 1952, 1956; Baker 1953; Hognestad et al. 1955; Janney et al. 1956; Mattock and Kriz 1961) that led to the current design practice and codes. The principle of the ultimate strength design theory based on plasticity using rectangular stress block is introduced in this section.

For a typical reinforced concrete section shown in Fig. 2.16 and under the assumption that plane section remains plane, the strain profile at the onset of flexural strength is shown by a straight line. The concrete strain ε_c at the extreme compression fiber at the onset of maximum moment is noted as ε_{cm} in this section. The stress distribution corresponding to the strain profile is nonlinear as shown by the shaded area. For convenience of calculation, the actual stress block of the shaded area is represented by an equivalent rectangular stress block that has the same area and centroid as those of the real stress block. The area of the stress block gives the magnitude of the resultant compression force of concrete. The centroid of the area gives the location of the resultant force. As a result, the real stress block and the equivalent rectangular stress block have exactly the same resultant force, which means that they are statically equivalent for calculation of bending moment.

At the onset of flexural strength, the stress of the equivalent rectangular stress block can be conveniently written as a fraction of concrete strength f_{co}, or $\alpha_1 f_{co}$. The location of the stress block centroid or the resultant force is also noted as a fraction of the compression zone depth c, or $\beta^* c$ so that the depth of the rectangular stress block is $\beta_1 c = 2\beta c$. From the stress profile in Fig. 2.16, one has

$$F_c = \alpha_1 f_{co} \beta_1 cb \tag{2.10}$$

$$F_{sc} = \begin{cases} A_{sc} f_y & (if\ \varepsilon_{sc} \geq \varepsilon_y) \\ \varepsilon_{sc} E_s A_{sc} & (if\ \varepsilon_{sc} < \varepsilon_y) \end{cases} \tag{2.11}$$

$$F_{st} = \begin{cases} A_{st} f_y & (if\ \varepsilon_{st} \geq \varepsilon_y) \\ \varepsilon_{st} E_s A_{st} & (if\ \varepsilon_{st} < \varepsilon_y) \end{cases} \tag{2.12}$$

The following relationships can be obtained from the strain profile:

$$\frac{\varepsilon_{sc}}{c - d'} = \frac{\varepsilon_{cm}}{c} \tag{2.13}$$

$$\frac{\varepsilon_{st}}{d - c} = \frac{\varepsilon_{cm}}{c} \tag{2.14}$$

The summation of the internal forces is equal to the total external axial force N, i.e.,

$$F_c + F_{sc} - F_{st} = N \tag{2.15}$$

Taking the moment of the internal forces about the cross-sectional centroid gives

$$M_u = F_c\left(\frac{h}{2} - \beta c\right) + F_{sc}\left(\frac{h}{2} - d'\right) + F_{st}\left(d - \frac{h}{2}\right) \tag{2.16}$$

2.7.1 Solution of the equations

In structural engineering, there are two typical design problems:

(1) Evaluation of load-carrying capacity of existing structures. For a given RC flexural member, it is the calculation of M_u under a particular N, or the calculation of N_u under a certain M.

(2) Design of members. For new RC flexural members, it is the calculations of A_{sc} and A_{st} for given M and N.

For the first type of problem, the geometrical dimensions b, d', d, reinforcement areas A_{sc} and A_{st}, concrete strength f_{co} and rebar material properties E_s, f_y, and ε_y are known. We have 10 unknowns in the preceding equations: the compression zone depth c, strain of compression reinforcement ε_{sc}, strain of tension reinforcement ε_{st}, stress block parameters α_1 and β_1, the strain of concrete at the extreme compression fiber at onset of flexural strength ε_{cm}, and the cross-sectional forces F_c, F_{sc}, F_{st} and M_u

(or N_u). However, we have only seven equations (Eqs. 2.10–2.16). Another three equations are needed to solve the problem. The additional three equations will be discussed in the next section. It is noted that M_u and N_u are not independent, and hence cannot be solved independently. In fact, M_u and N_u are related to each other in RC design through the $M-N$ interaction diagram. In other words, for a given value of N, a unique value of M_u can be obtained from the solutions, and for a given value of M, a unique value of N_u can be found.

For the second type of problem, the geometrical dimensions b, d', d, concrete strength f_{co} and rebar material properties E_s, f_y, and ε_y, and loading effects M and N, are known. We have 11 unknowns in the preceding equations: compression zone depth c, strain of compression reinforcement ε_{sc}, strain of tension reinforcement ε_{st}, stress block parameters α_1 and β_1, the strain of concrete at onset of flexural strength ε_{cm}, and the cross-sectional forces F_c, F_{sc}, and F_{st}, and reinforcement areas A_{sc} and A_{st}. Apart from the preceding seven equations plus the additional three equations to be discussed in the next section, another equation is required to get the solution of the unknowns. This extra equation is normally provided in RC design by selecting a compression steel area (or omitting the compression steel area in design, i.e., $A_{sc}=0$). In other words, by selecting a value of A_{sc}, the value of A_{st} can be obtained from the solution. Therefore, the RC design is not unique.

It is noted that Eq. (2.11) (Eq. 2.12 is similar) is not two equations but one. It comes from the steel stress σ_s vs. steel strain ε_s relationship, $\sigma_s = f(\varepsilon_s)$, which is usually expressed by the following elastoplastic model:

$$\sigma_s = \begin{cases} f_y & (if\ \varepsilon_s \geq \varepsilon_y) \\ \varepsilon_s E_s & (if\ \varepsilon_s < \varepsilon_y) \end{cases} \tag{2.17}$$

To solve equations involving such different relationships for different parts of a curve, an easy and common way is by trial and error. By assuming one case, such as $\sigma_{sc} = f_y$ (or $\sigma_{sc} = \varepsilon_{sc}E_s$) and $\sigma_{st} = f_y$ (or $\sigma_{st} = \varepsilon_{st}E_s$), the equation set (Eqs. 2.10–2.16) can be solved from which the solutions of ε_{sc} and ε_{st} are then checked with their assumptions of $\varepsilon_{sc} \geq \varepsilon_y$ and $\varepsilon_{st} \geq \varepsilon_y$. If the assumptions are true, the solutions are correct; otherwise, another case is assumed, and the equation sets are solved again until the assumptions meet the solutions.

2.7.2 Stress block parameters

As mentioned in the previous section, another three equations are required to solve the RC design equations. The current approach to obtain the additional equations is by providing empirical models for ε_{cm}, α_1 and β_1. Extensive experimental and theoretical studies have been carried out from the 1930s to the 1960s in studying the stress block parameters to develop the RC flexural design theory.

The stress block parameters ε_{cm}, α_1 and β_1 can be obtained from flexural tests of RC members. Structural response curves, such as that in Fig. 2.1, can be recorded in a flexural test. The deformations and loads at the peak point C can be measured in the tests, including ε_{sc}, ε_{st}, ε_{cm}, M_u, and N. F_{sc}, F_{st} and c can be calculated by Eqs. (2.11), (2.12), and (2.13), respectively. In this case, we have three unknowns α_1 and β_1, and F_c, in the remaining three equations, Eqs. (2.10), (2.15), (2.16), from which the unknowns can be resolved. One flexural test can provide one set of values of ε_{cm}, α_1, and β_1. Empirical models for ε_{cm}, α_1 and β_1 can be regressed using a sufficiently large test database. This is how the design models for these stress block parameters have been developed in design codes.

Different stress block parameters have been adopted in the literature. An example of the stress block parameter system is shown in Fig. 2.17 (Hognestad et al. 1955). The development of empirical models for the stress block parameters in Fig. 2.17 is shown in Fig. 2.18. The stress block parameters used by Hognestad et al. (1955) are ε_u, k_1, k_2, and k_3, as shown in Fig. 2.17. The parameter ε_u is the same as ε_{cm}, and $k_2 = \beta_1/2$. The other parameters k_1 and k_3 are defined differently from α_1 in Fig. 2.16. The parameter k_3 is introduced to relate the cylinder compressive strength of concrete f_c' to the flexural compressive strength of concrete f_{co} (the maximum stress of concrete under flexural bending in Fig. 2.16), i.e.,

$$f_{co} = k_3 f_c' \tag{2.18}$$

The average stress value of the concrete stress block at the onset of maximum moment is given by $\alpha_1 f_{co}$ (Fig. 2.16) that can be written as $\alpha_1 f_{co} = \alpha_1 k_3 f_c' = \alpha_1' f_c'$. This means that $\alpha_1 k_3$ can be replaced by another factor α_1'. In other words, if cylinder concrete strength f_c' is directly used in Fig. 2.16, $\alpha_1 f_{co}$ is simply replaced by $\alpha_1' f_c'$.

The test results collected by Hognestad et al. (1955) are shown in Fig. 2.18, which shows that the stress block parameters are functions of concrete strength only. Therefore, empirical equations of the stress block parameters as a function of concrete strength were regressed from the test results, as given by:

$$\varepsilon_u = 0.004 - \frac{f_c'}{6.5 \times 10^6} \quad (\text{in psi}) \tag{2.19}$$

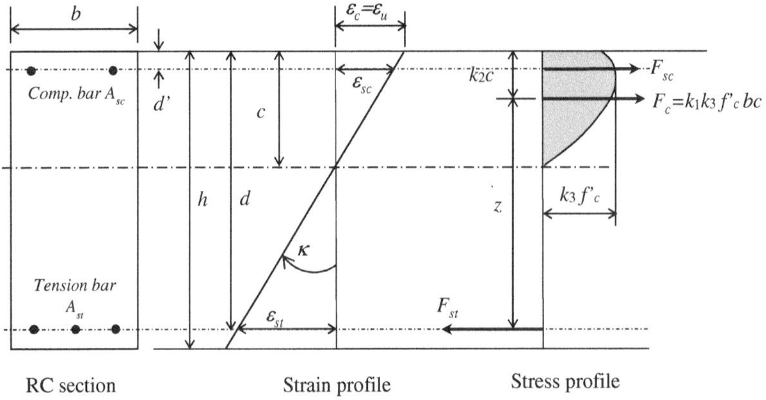

Fig. 2.17 Stress block parameters of Hognestad et al. (1955).

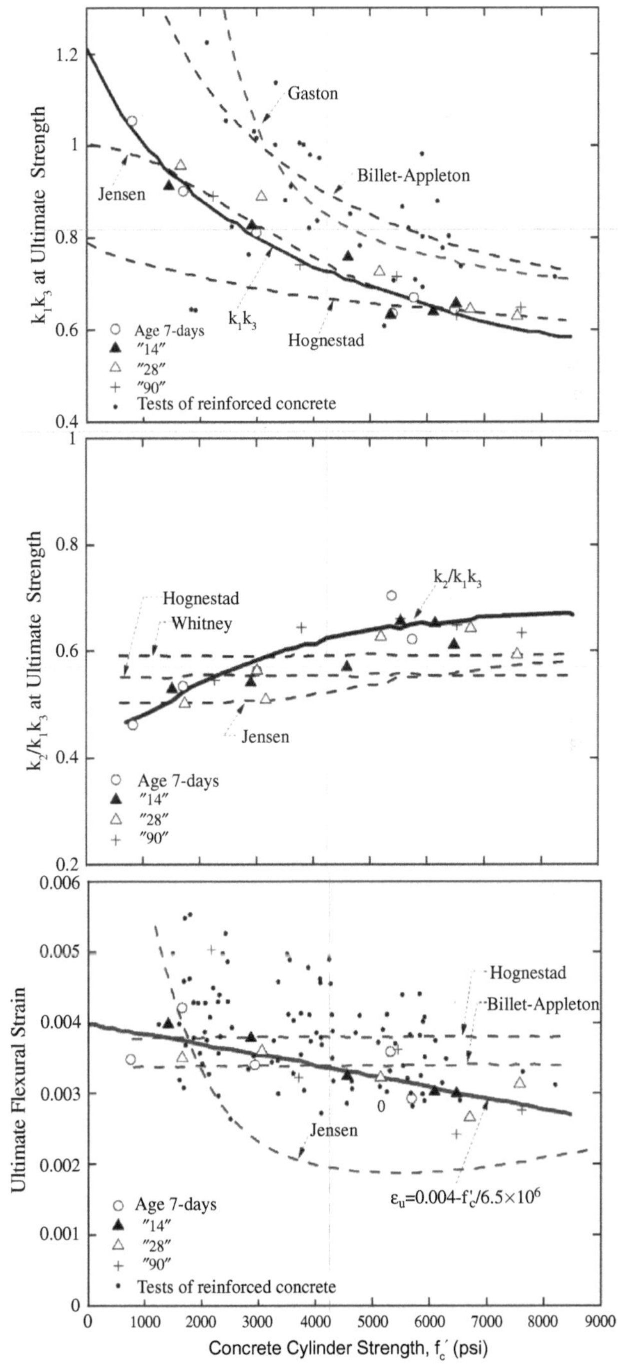

Fig. 2.18 Hognestad's stress block parameters.
Based on Hognestad, E., N. W. Hanson, and D. McHenry. 1955. "Concrete stress distribution in ultimate strength design." *ACI J. Proc.* 52 (12), 455–479.

$$k_1 = 0.94 - \frac{f_c'}{26,000} \text{ (in psi)} \tag{2.20}$$

$$k_2 = 0.5 - \frac{f_c'}{80,000} \text{ (in psi)} \tag{2.21}$$

$$k_3 = \frac{3900 + 0.35 f_c'}{3000 + 0.82 f_c' - f_c'/26,000} \text{ (in psi)} \tag{2.22}$$

It was also concluded that for design purposes a constant value $\varepsilon_u = 0.003$ is sufficient because it was found that the computed ultimate strength of reinforced concrete members is usually insensitive to the numerical value of ε_u.

A few other empirical models of the stress block parameters are also shown in Fig. 2.18. Clearly, significant differences exist among the models. This is because each model was developed using its own database. Test results could be different due to bias and differences in research focus, materials used, test conditions and facility, measurement and interpretation method, etc. Therefore, it is critical to use a database with unbiased, accurate, and comprehensive test results for regression of empirical models. The stress block parameter system in Fig. 2.16 is widely adopted in current design codes. However, the models of the stress block parameters do have differences in different countries. More discussions on the stress block parameters are given in the next section.

2.7.3 Discussions on stress block parameters

The stress block parameter approach has been applied in the design of RC structures ever since the ultimate strength design method was adopted in the 1930s. However, the equations or models of the stress block parameters vary from time to time and place to place, depending on the particular experimental database used for regression of the models. With the change of construction materials, e.g., use of higher strength concrete, as well as availability of more test data, these empirical models are updated continuously. Engineering practice and continuous research work have demonstrated that this approach is generally satisfactory and adequate for engineering purposes.

Although the existing design codes on the ultimate strength of reinforced concrete members were empirical, largely based on test results, the experimental works were considered comprehensive even in the 1950s, including thousands of tests covering practically all known variables (Siess et al. 1952). Hence, the developed design codes were believed well validated. The subsequent theoretical studies based on mathematical modeling (Kriz 1961) and numerical simulations (Kriz and Lee 1960), although few in the early days, further substantiated the appropriateness and accuracy of this experimentally based flexural theory.

An important implication (hypothesis) of this flexural theory is that the flexural strength is reached when the concrete strain at the extreme compression fiber reaches a given value, namely the "ultimate strain" ε_u. In other words, it is deemed that point C in Fig. 2.4 is reached when the concrete strain at the extreme compression fiber ε_c reaches ε_u (Fig. 2.17). The value of ε_u is generally considered a constant in the designs of ordinary RC members, but different values are used in different design codes: e.g., it is 0.003 in ACI 318 (ACI Committee 318 2005), 0.0035 in Eurocode 2 (CEN 2004), and 0.0033 in the Chinese concrete design code GB50010 (MOHURD 2010). This value may change for high-strength concrete in some codes: for example, in Eurocode 2 and GB50010, the value of ε_u is related to concrete strength when it is higher than 50 MPa. The variation of ε_u with concrete strength has been observed in early studies as shown in Fig. 2.18 and Eq. (2.19). However, the effect of concrete strength has become more significant in recent years due to the use of high-strength concrete.

The current RC theory assumes that the value of ε_u is a constant or a function of concrete strength for high-strength concrete, but not a function of reinforcement ratio. Although there is no strict theoretical proof of this conclusion in the conventional RC flexural theory, extensive experimental and theoretical investigations have so far shown that this flexural design approach can provide a reasonably accurate evaluation of actual flexural strength for normal RC members. However, it has been realized that there are cases where the moment resistance calculated based on a fixed ultimate strain is significantly underestimated compared with the actual one. Rusch (1960) pointed out in his early theoretical and experimental studies that reinforcement ratio and axial load affect the value of ε_u. Numerical simulations in the modern literature have also found that the value of ε_u can be different for RC members with different reinforcement ratios and axial force ratios. Fig. 2.19 shows the variations of the extreme compression strain during the bending process of the same RC column sections as that for Fig. 2.6 (with different reinforcement ratios and axial load ratios). It can be seen in Fig. 2.19 that the values of ε_u (value of ε_c at the peak point of the curve) are different for RC sections with different reinforcement ratios

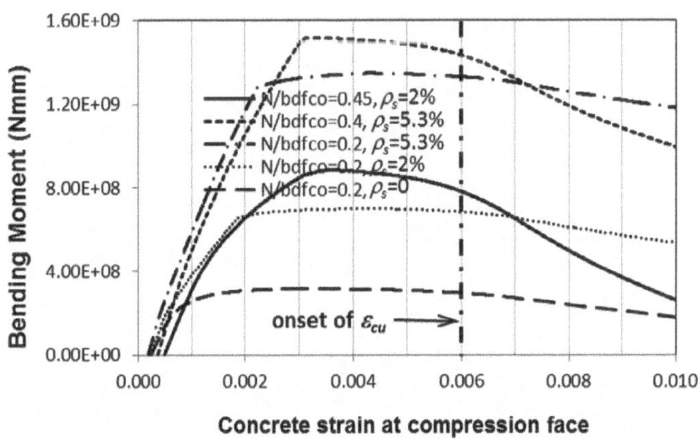

Fig. 2.19 *M* vs. ε_c relationship.
From Wu, Y. F. 2016. "Theorems for flexural design of RC members." *J. Struct. Eng.* 142 (5), 04015172.

and under different axial force ratios. However, this is not considered a problem so far in both the literature and practice, mainly due to the following reasons:

(1) Even when the approach is inaccurate for some cases, it gives a conservative result. This is because the accurate value of ε_u gives the exact result at point C and an inaccurate ε_u gives a result around point C that has a smaller value of moment resistance (Fig. 2.4).
(2) The approach is simple to use and easy to understand, which suits engineering applications.
(3) The accurate calculation of point C is difficult.

Therefore, the conventional approach developed in the early 1900s is still widely used in engineering all over the world.

The conventional flexural design approach is developed from experimental observations. The design equations/parameters are regressed from test results. As the test data cannot cover all possible cases, this approach is inductive in nature and may produce significant error in some special cases. A deductive flexural theory purely based on mathematics has been developed recently, which gives an exact solution of the flexural strength for all flexural members and is applicable to all cases, as long as the basic assumptions are met. The deductive approach is introduced in Chapter 3.

2.8 Miscellaneous relationships

2.8.1 Curvature of section

When a flexural member of uniform cross-section with a length of L is bent under a constant bending moment M, its curvature is also constant along the length. The length of the neutral axis, where there is no strain, remains its original length L. Based on definition, the curvature of a member is equal to the inverse of the radius of the neutral axis, i.e.,

$$\kappa = \frac{1}{r_m} \tag{2.23}$$

For a constant curvature, the radius is also a constant along the length. This means that the deformed shape of the member forms part of a circle with a constant radius, as shown in Fig. 2.20, where r_e, r_m, and r_i are the radii of the external face, the neutral axis, and the internal face, respectively. Based on the plane section assumption, all cross-sections remain plane after bending.

From the geometrical relationship in Fig. 2.20, one has the following equations

$$L_e = \alpha \cdot r_e \tag{2.24}$$

$$L = \alpha \cdot r_m \tag{2.25}$$

where L_e is the length of the external face after deformation, which has an original length of L. The elongation of the fiber at the external face is $\Delta L = L_e - L$. Based on the definition of strain, the strain at the bottom is given by

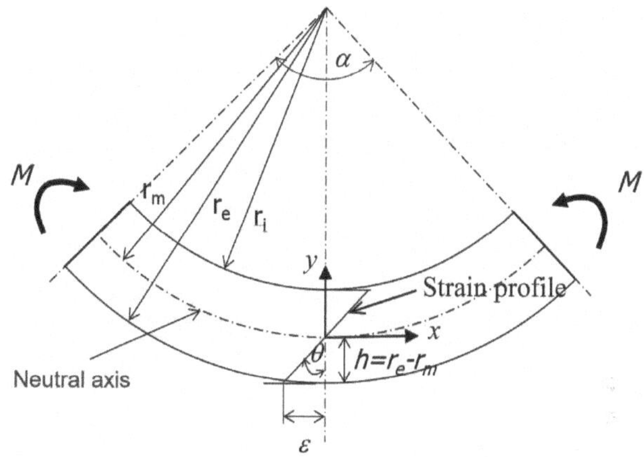

Fig. 2.20 Relationship between strain profile and curvature.

$$\varepsilon = \frac{L_e - L}{L} = \frac{\alpha \cdot r_e - \alpha \cdot r_m}{\alpha \cdot r_m} = \frac{r_e - r_m}{r_m} \tag{2.26}$$

The angle of the strain profile θ is given by

$$\theta = \frac{\varepsilon}{h} = \frac{\frac{r_e - r_m}{r_m}}{r_e - r_m} = \frac{1}{r_m} = \kappa \tag{2.27}$$

Eq. (2.27) shows that the angle or gradient of the strain profile is the curvature of the cross-section.

It is noted that the mathematical definition of curvature is given by Eq. (2.28), which is mathematically equivalent to Eq. (2.23).

$$\kappa = \frac{d^2 y}{dx^2} \tag{2.28}$$

For general flexural members, Eq. (2.28) gives a different value at different x coordinates. In other words, the curvature is for a particular cross-section, not for a member or a certain length of the member. For the case in Fig. 2.20, the curvature is equal at all cross-sections. For RC members, cracking causes significant fluctuation of curvature longitudinally (see Fig. 7.13). Therefore, the following two types of curvature are often used in reinforced concrete structures:

(1) Local curvature—the average curvature for the segment of member between two adjacent cracks.
(2) Average curvature—the average value of curvature for a certain length of a member which may include many cracks.

In these two cases, the curvature is not for a cross-section but the average curvature of a certain length of RC member.

2.8.2 The theorem of plane section

The assumption that plane sections remain plane (the strain distribution is linear about the neutral axis) and normal to the axis of the beam, often referred to as Bernoulli's hypothesis or the Euler-Bernoulli hypothesis, is widely accepted for analyses of slender flexural members in engineering. This assumption is based on comprehensive experimental observations from all sorts of different structures, including reinforced concrete structures (Park and Paulay 1975). The assumption was made initially for homogeneous beams under pure bending with small and linear elastic deformation. It was subsequently found from experimental evidence that it is also applicable to beams with nonlinear and nonelastic deformation (Park and Paulay 1975). Elastic theory has been used to evaluate the validity of this assumption for prismatic bars under a pure bending condition, which proved that it is strictly true under elastic conditions (Timoshenko and Goodier 1951). Qian (1950) has given a more general theoretical proof of the hypothesis using complementary energy theory, which concluded that it is true under pure bending regardless of the elastic or plastic stress condition. However, Qian's proof was based on the assumption that the deformation is small. A simpler and more general theoretical proof of the hypothesis is given in this section. The proof is based on the compatibility condition only and there is no limitation to a small deflection. It is also easier for engineers and researchers with an engineering background to understand.

Let's have the following thought experiment: consider a straight flexural member consisting of different materials but with a prismatic or uniform cross-section (i.e., constant cross-section), as shown in Fig. 2.21A and B, along the whole length. The member is gradually bent in one of its principal planes by a constant bending moment along the span (Fig. 2.21A) so that the curvature increases until the two ends meet, as shown in Fig. 2.21D. When the Saint-Venant's effect due to local stress distribution at the two ends is omitted, the only possibility of the deformed shape is a perfect circle, because both the member and loading are axially symmetrical about its center point C (Fig. 2.21D), including a perfect continuity at the joint of the two ends. In other words, the strain profiles of all cross-sections including that at the jointed ends must be the same due to the axial symmetry. Therefore, there is only one identical strain profile of the member along the length. This strain profile is assumed to be represented by a general function as shown in Fig. 2.21C.

Now the following equations can be obtained:

$$r_m = \frac{1}{\kappa} \tag{2.29}$$

$$L_m = L = 2\pi r_m = 2\pi \frac{1}{\kappa} \tag{2.30}$$

where L_m is the length of the neutral axis, which is equal to the original length of the member L; r_m is the radius of the neutral axis; and κ is the curvature of the typical cross-section. Based on the geometrical relationship, one has

$$L(y) = 2\pi(r_m + y) \tag{2.31}$$

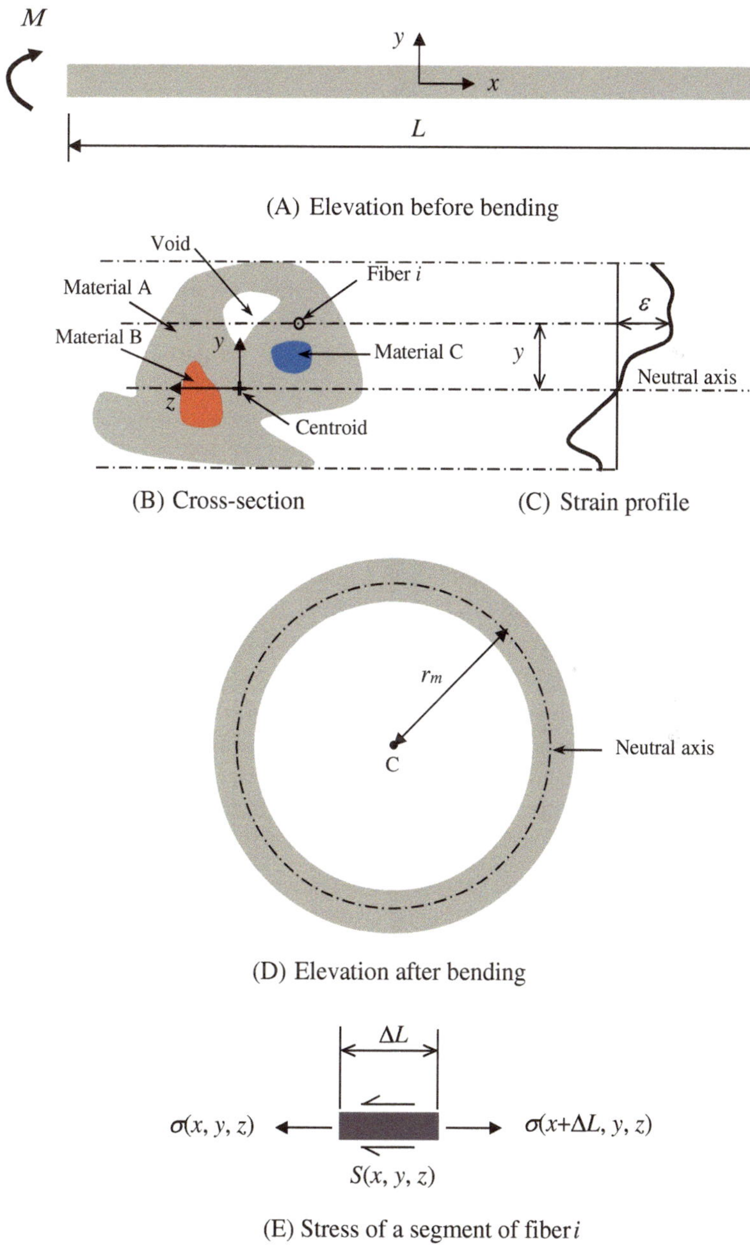

Fig. 2.21 Bending of member with a uniform cross-section under constant moment.

where $L(y)$ is the perimeter length of the fiber i that is located at y distance from the neutral axis (Fig. 2.21B). Based on the strain profile in Fig. 2.21C, one has

$$L(y) = [1 + \varepsilon(y)] \cdot L_{y0} = [1 + \varepsilon(y)] \cdot L \tag{2.32}$$

where L_{y0} is the original length of the fiber before bending, which is equal to the original length of the member L. Substituting Eq. (2.31) into Eq. (2.32) and considering Eq. (2.30) and Eq. (2.29), one gets

$$\varepsilon(y) = \frac{y}{r_m} = \kappa y \tag{2.33}$$

Eq. (2.33) represents a linear strain distribution, i.e., the plane section after bending remains plane. This conclusion is applicable regardless of the material properties, being homogeneous or heterogeneous, regardless of stress conditions, being elastic or plastic, linear or nonlinear. It is applicable to any member with a prismatic cross-section. For elastic materials, the stress changes in proportion to strain, and hence the neutral axis does not change during bending. For materials with nonlinear stress-strain relationships, the neutral axis may change in order to maintain the balance of the internal forces. It is noted that, when a constant axial force is applied to the centroid of the member, these derivations and conclusions are the same, except that the length L is the length of the member after application of the axial force and the strain profile in Fig. 2.21C is due to bending only.

For a flexural member with a certain length L and a given bending moment or curvature κ, it may not form a closed circle. In this case, one simply imagines that the member is extended to a length of $L = 2\pi\frac{1}{\kappa}$. It forms a full circle after bending in this case and the preceding proof is still applicable. Since $\kappa = \frac{2\pi}{L}$ and L can be any value, the conclusion is applicable to any curvature κ, regardless of small or large value.

No bond exists between fibers in a flexural member under pure bending. The reason is given here. For an arbitrary fiber i in the cross-section (Fig. 2.21B), the stress condition of a segment of the fiber is shown in Fig. 2.21E. Due to axial symmetry, the stress condition of all cross-sections is identical, which gives

$$\sigma(x, y, z) = \sigma(x + \Delta L, y, z) \tag{2.34}$$

where ΔL can be any value. From the balance of the forces on the segment in the x-direction, the total bond force on the surface of the fiber must be zero, or

$$S(x, y, z) = 0 \tag{2.35}$$

Eq. (2.35) is true for any fiber i with any cross-sectional shape and for any value of ΔL. Considering symmetry, the only possibility is

$$\tau_{xy} = \tau_{xz} = \tau_{zx} = \tau_{yx} = 0 \tag{2.36}$$

Based on Eq. (2.36), there is no bond stress between any adjacent fibers. In other words, the plane section behavior does not rely on, and cannot have, shear bond between fibers. Eq. (2.36) also means that the cross-sectional shear stresses are zero.

It should be noted that the local strain field near the two ends of the member may not be linear, depending on the way the bending moment is applied at the ends, that is, the local stress field at the ends. However, according to Saint-Venant's principle, the

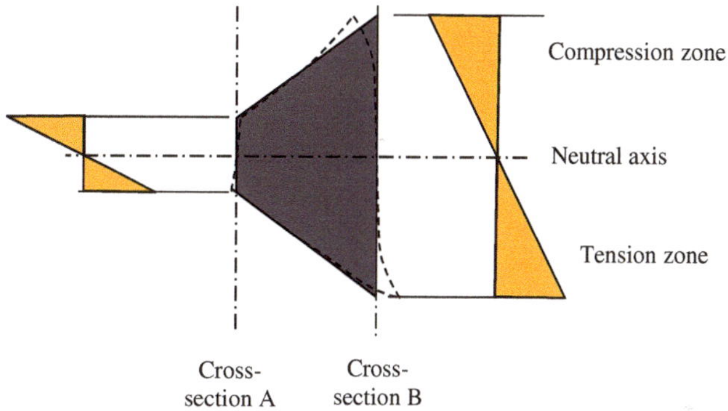

Fig. 2.22 Effect of nonuniformity of cross-section on plane strain condition.

local stress distribution at the ends only affects the small vicinity near the two ends. Wu (2022) gave a more stringent proof of the plane section hypothesis using a cylindrical coordinate system and considering the Saint-Venant's effect.

For a member with a shear stress in the cross-section, a plane section does not remain plane due to shear distortion. For a member with a nonuniform cross-section along the span, the plane section condition is also invalid, strictly speaking, for the reason shown in Fig. 2.22. If a plane strain condition is assumed, it can be seen in Fig. 2.22 that the stress on the cross-section B will cause the deformation as shown by the dashed line, which invalidates the plane strain condition even in the elastic state.

Based on the preceding thought experiment, the following theorem is derived:

Theorem of plane section. *For flexural members with a prismatic or constant cross-section consisting of any number of different materials, homogeneous or heterogeneous, and the local effect described by Saint-Venant's principle is omitted near the ends of the member, a plane section before loading remains plane after loading when it is under a constant bending moment along the span, regardless of the elastic or plastic stress condition, small or large deflection. The bond or shear stress between fibers is always zero throughout the bending process under the plane strain condition. The presence of shear stress or variation of cross-section longitudinally invalidates the plane strain condition.*

Although the plane section condition is strictly true only for members with a prismatic cross-section and under constant bending moment, this condition is applied as a hypothesis to structural analysis of various structures under different types of loads in engineering. Schlaich et al. (1987) introduced the concept of B and D regions of a structure, where B stands for beam or Bernoulli's hypothesis and D stands for disturbed. The strain distributions in the B regions are close to a linear or plane section while in the D regions the distributions of strain are significantly nonlinear. Typical B and D regions in a frame are illustrated in Fig. 2.23, where the D regions extend a distance equal to the member depth from a discontinuity (such as a change in cross-section or the presence of a point load). The portions of the member between D regions are B regions. A structural member can consist entirely of D regions, but in most cases, B and D regions coexist in one structure.

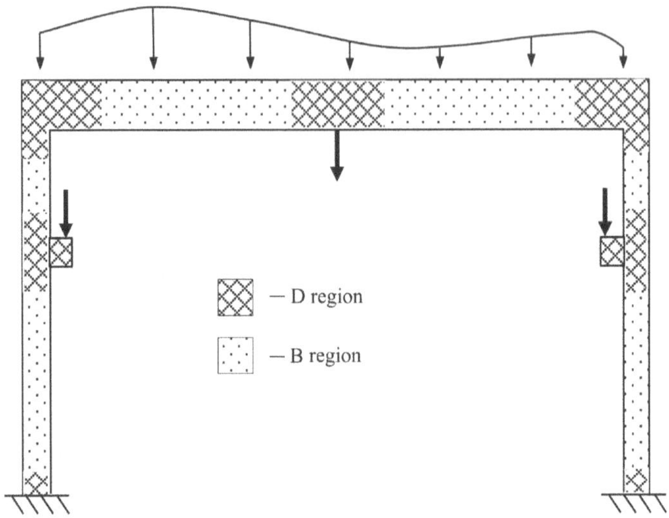

Fig. 2.23 B and D regions in a frame.
Based on Schlaich, J., I. Schäfer, and M. Jennewein. 1987. "Towards a consistent design of structural concrete." *J. Prestressed Concr. Inst.* 32 (3), 74–150.

Extensive experimental observations, practical experience, mathematical studies, and numerical simulations show that the Bernoulli's hypothesis is usually close enough to the actual strain distributions in most common structures. The hypothesis significantly simplifies the analyses, calculations, and designs of structures. Extensive practical experience has demonstrated that, apart from some special cases such as deep beams or members under high shear stress, the application of Bernoulli's hypothesis to general flexural members under various loading conditions results in acceptable errors in calculating structural responses from an engineering standpoint, in both B and D regions. As the plane strain condition is not strictly true for general flexural members and under general loadings, it is treated as a model. The scope or extent of validity of this model can only be determined from experimental evidence or practical observations. Reliable mathematical and numerical modeling can also be used to evaluate the validity of the plane section assumption; however, the mathematical or numerical model itself requires validation by experimental results or practical observations. It is widely accepted in structural engineering that the Bernoulli's hypothesis should only be applied to beams with a shear span-to-depth ratio (referred to as aspect ratio) of greater than 2.5. However, this is only a rough empirical guideline based on practical experience and it does not strictly prohibit the application of Bernoulli's hypothesis to beams with a smaller aspect ratio than 2.5. It simply means that, if Bernoulli's hypothesis is applied to deep beams with an aspect ratio less than 2.5, the accuracy of calculated results may be lower than that of slender beams. In fact, even for deep beams, Bernoulli's hypothesis is still widely used in engineering practice.

For reinforced concrete structures, cracking invalidates the condition of uniformity and axial symmetry of an RC member under pure bending. Experimental evidence has shown that the plane section assumption is still valid in the compression zone of RC members (Park and Paulay 1975). The strain profile at a cracked section is certainly different from that of an uncracked section. However, if a good bond exists and the concrete strain is measured over a gauge length that includes several cracks, it is found that Bernoulli's hypothesis applies to the average tensile strain (Park and Paulay 1975). This is because, for RC members with good bond, cracking can be smeared to many small cracks to reduce the nonuniformity of tensile strain along the span (see Fig. 1.6 and more discussions in Chapter 5). Therefore, it is generally accepted that, for reinforced concrete flexural members, the plane section assumption "is very nearly correct at all stages of loading up to flexural failure, provided good bond exists between the concrete and the steel" (Park and Paulay 1975). The bond in this case acts in opposite directions on the two sides of a crack (see Fig. 5.17C) and is local, which suppresses or restrains crack opening (see Fig. 5.16). Therefore, the average bond stress over a sufficient length along the span is zero. In this case, Eq. (2.35) is still true in an average sense, and consequently, the plane strain condition is still valid in an average sense.

When the plane section assumption is applicable, it can be used to calculate both the flexural strength and the curvature of an RC member. Attention should be paid not to apply it for calculation of the curvature of a section located between two cracks, because tensile stress exists in the concrete between two cracks that does not satisfy the plane section assumption. It will be seen later that, for a flexural member, e.g., the cantilever column shown in Fig. 7.13, the plane section assumption can be applied to calculate:

(1) The curvature of uncracked zones (the top region of the column in Fig. 7.13 where there is no cracking). The uncracked cross-section (Fig. 7.3) with the uncracked flexural rigidity E_cI_g is used for the calculation (see Chapter 7).
(2) The cross-sectional curvatures exactly at the crack positions, where the cracked cross-section (Fig. 7.4) and the cracked flexural rigidity E_cI_{cr} are used for calculation.
(3) The average curvature of the cracked part (the dashed line in Fig. 7.13C) where the "average" or "effective" flexural rigidity E_cI_e that has a value between E_cI_g and E_cI_{cr} is used for calculation.

It is noted that the plane section assumption cannot be used to calculate the actual curvature of the cross-sections located between two cracks that is shown by the solid line between cracks in Fig. 7.13C.

In summary, for slender RC members where flexure deformation dominates, the plane section assumption is

- applicable to members before cracking;
- applicable to a cracked RC section where the concrete stress in the tension zone is zero;
- applicable to the calculation of average curvature and average tensile strain of concrete for a segment of cracked RC member that includes several cracks; and
- inapplicable to a section between cracks.

References

ACI Committee 318. 2005. *Building code requirements for structural concrete and commentary (ACI 318-05/ACI 381R-05)*. Detroit, MI: American Concrete Institute.

Baker, A. L. L. 1953. "Further research in reinforced concrete, and its application to ultimate load design.". *Proc. Inst. Civil Eng.* 2 (3), 269–310.

CEN. 2004. *Eurocode 2: Design of concrete structures, part 1, general rules and rules for buildings (EN 1992-1-1:2004)*. Brussels: Comité Européen de Normalisation.

Hognestad, E. 1951. *A study of combined bending and axial load in reinforced concrete member. Bulletin No. 399*, University of Illinois Engineering Experiment Station.

Hognestad, E. 1952. "Fundamental concept in ultimate load design of reinforced concrete members.". *ACI J. Proc.* 48 (6), 809–830.

Hognestad, E. 1956. *Ultimate strength of reinforced concrete in American design practice. Bulletin D12*, Portland Cement Association Development Department.

Hognestad, E., N. W. Hanson, and D. McHenry. 1955. "Concrete stress distribution in ultimate strength design.". *ACI J. Proc.* 52 (12), 455–479.

Janney, J. R., E. Hognestad, and D. McHenry. 1956. "Ultimate flexural strength of prestressed and conventionally reinforced concrete beams.". *ACI J. Proc.* 52 (2), 601–620.

Jensen, V. P. 1943. "The plasticity ratio of concrete and its effect on the ultimate strength of beams.". *ACI J. Proc.* 39, 565–582.

Kriz, L. B. 1961. "Ultimate strength criteria for reinforced concrete.". *Trans. Am. Soc. Civil Eng.* 126 (1), 439–453.

Kriz, L. B., and S. L. Lee. 1960. "Ultimate strength of over-reinforced beams.". *J. Eng. Mech. Division, Proc. ASCE* 86 (EM3), 95–105.

Mattock, A. H., and L. B. Kriz. 1961. "Ultimate strength of nonrectangular structural concrete members.". *ACI J. Proc.* 57 (1), 737–766.

MOHURD (Ministry of Housing and Urban-Rural Development of the People's Republic of China). 2010. *GB50010-2010: Code for design of concrete structures.* Beijing, China: China Architecture & Building Press.

Park, R., and T. Paulay. 1975. *Reinforced concrete structures*. New York: John Wiley & Sons.

Qian, L. X. 1950. "The theory of complementary energy.". *Science China* 1 (2–4), 449–456.

Rusch, H. 1960. "Researches toward a general flexural theory for structural concrete.". *ACI J. Proc.* 57 (7), 1–28.

Schlaich, J., I. Schäfer, and M. Jennewein. 1987. "Towards a consistent design of structural concrete.". *J. Prestressed Concr. Inst.* 32 (3), 74–150.

Siess, C. P., G. C. Ernst, and A. L. Parme. 1952. "Review of research on ultimate strength of reinforced concrete members.". *ACI J. Proc.* 48 (6), 833–859.

Stussi, F. 1932. "Ueber die Sicherheit des einfach bewehrten Eisenbenton-Rechteckbalens.". *Publ. Int. Assoc. Bridge Struct. Eng. Zurich* 1, 487–495.

Timoshenko, S., and J. N. Goodier. 1951. *Theory of elasticity*. New York: McGraw-Hill.

von Emperger, F. 1931. "Der beiwert n = 15 und die zulassigen biegungs-spannungen.". *Beton und Eisen* 30 (19), 340–346.

Whitney, C. S. 1937. "Design of reinforced concrete members under flexure or combined flexure and direct compression.". *ACI J. Proc.* 33 (3), 483–498.

Whitney, C. S. 1942. "Plastic theory of reinforced concrete design.". *Trans. ASCE* 107, 251–282.

Wu, Y. F., M. C. Griffith, and D. J. Oehlers. 2004. "Numerical simulation of steel plated RC columns.". *Comput. Struct.* 82 (4-5), 359–371.

Deductive approach to flexural theory

3

A new flexural theory of RC members has recently been developed purely from mathematical derivation (Wu 2016). Three theorems are obtained from the new theory for the flexural design of RC members. The new theory provides a more rational and accurate approach to the flexural design of concrete structures. It also facilitates better understanding of the flexural behavior of reinforced concrete. The new theory further leads to mathematically based rational and precise quantification of flexural design indexes, including the definition of ultimate curvature of RC flexural members. An alternative flexural design approach is introduced based on the new flexural theory. The theorems can be applied to derive the stress block parameters of the conventional flexural design approach discussed in Chapter 2 without doing experimental tests of structural members. The new theory is introduced in this chapter. Examples of applications of the new theory are given in Chapter 4.

3.1 General assumptions

The flexural theorems are derived based on the following general assumptions:

1. Plane sections before bending remain plane after bending;
2. The tensile strength of concrete is neglected; and
3. The stresses of concrete and steel can be evaluated from given stress–strain relationships.

For simplicity of derivation, the RC sections are assumed to be rectangular and a perfectly elastoplastic stress–strain relationship is assumed for steel. For convenience of mathematical derivations, the stress block parameters defined in Fig. 3.1 are adopted in this chapter.

Detailed mathematical derivations are given in Section 3.5. The derived theorems are summarized and discussed in the following section. It is noted that the ultimate concrete strain ε_u at the onset of the flexural strength commonly used in conventional flexural theory in Chapter 2 is referred to as ε_{cm} in this chapter.

3.2 RC flexural design theorems

3.2.1 The first theorem

Theorem 3.1 *Underreinforced RC sections with both tension and compression reinforcements yielded before the onset of the flexural strength achieve the flexural strength at the moment when the actual concrete stress at the extreme compression fiber is equal to the stress of the equivalent rectangular stress block. The extreme*

Analytical Approaches for Reinforced Concrete. https://doi.org/10.1016/B978-0-12-821164-9.00003-6

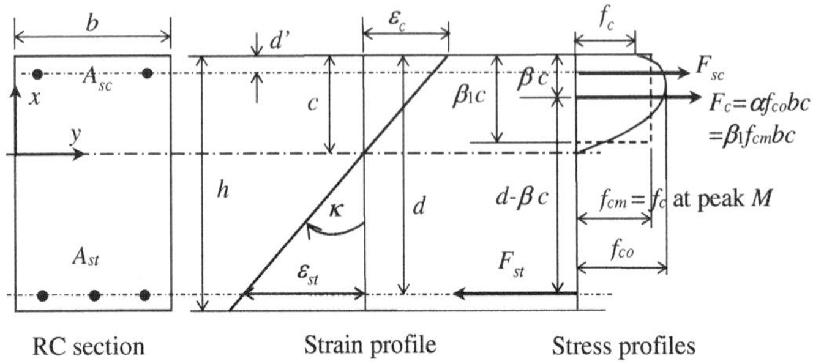

Fig. 3.1 Derivation of flexural theorems.
From Wu, Y. F. 2016. "Theorems for flexural design of RC members." *J. Struct. Eng.* 142 (5), 04015172.

concrete strain at the onset of the flexural strength, i.e., the flexural design strain ε_{cm}, is determined only by the shape of the stress-strain curve of concrete material.

This theorem states that at the onset of the flexural strength M_u (point C of the moment–curvature curve in Fig. 2.4), the concrete stress at the extreme compression fiber must satisfy the condition of $f_c = f_{cm}$ (Fig. 3.1). This condition leads to the following mathematical expression (for derivations see Section 3.5):

$$\frac{\int_0^{\varepsilon_{cm,u}} f(\xi)d\xi}{2f(\varepsilon_{cm,u})} + \frac{\int_0^{\varepsilon_{cm,u}} f(\xi)\xi d\xi}{\int_0^{\varepsilon_{cm,u}} f(\xi)d\xi} = \varepsilon_{cm,u} \tag{3.1}$$

where $\varepsilon_{cm,u}$ is the ε_{cm} for the case of underreinforced members, and $f(\xi)$ is the stress–strain relationship of concrete in which ξ is concrete strain. The solution of Eq. (3.1) gives the flexural design strain. It can be seen from Eq. (3.1) that the only factor that affects $\varepsilon_{cm,u}$ is the function $f(\xi)$, i.e., the shape of the stress–strain curve of concrete. Therefore, $\varepsilon_{cm,u}$ is only a function of $f(\xi)$, and hence Eq. (3.1) can be written as:

$$\varepsilon_{cm,u} = \mathcal{F}[f(\xi)] \tag{3.2}$$

Eq. (3.1) is general and applicable to any stress–strain relationship of concrete, $f(\xi)$. However, it is a transcendental equation and its unique solution $\varepsilon_{cm,u}$ can only be obtained numerically when a stress–strain relationship $f(\xi)$ of concrete is given.

For the two typical stress–strain models of concrete shown in Fig. 3.2A, the numerical solutions of Eq. (3.1) are obtained and shown in Fig. 3.2B. The two stress–strain models of concrete used in Fig. 3.2 are described in the following text. For convenience of discussions, the stress–strain relationship proposed by Popovics (1973) is adopted in all numerical examples in this chapter unless otherwise noted (UON).

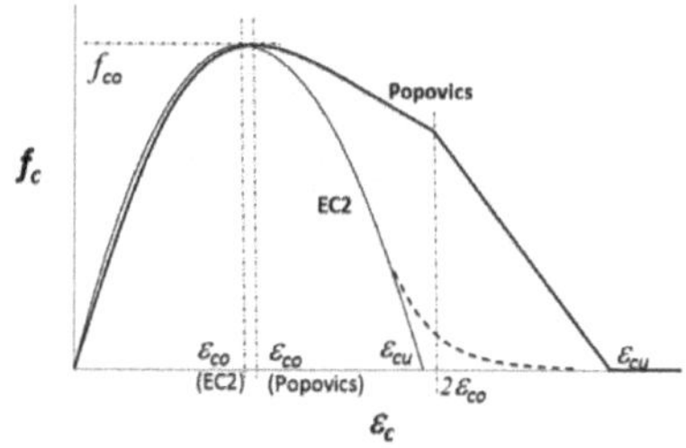

(A) Typical stress-strain relationships

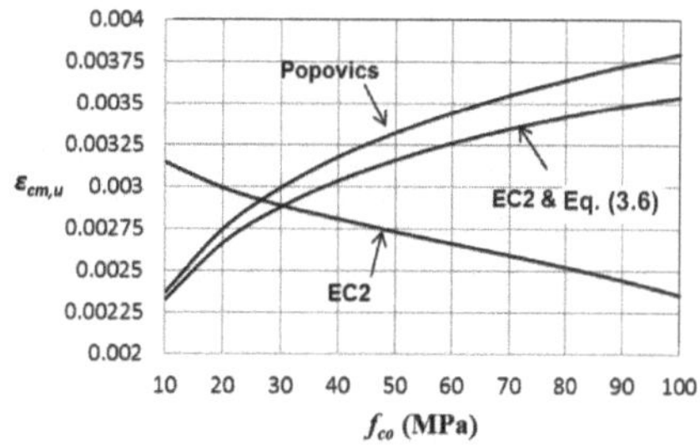

(B) $\varepsilon_{cm,u}$

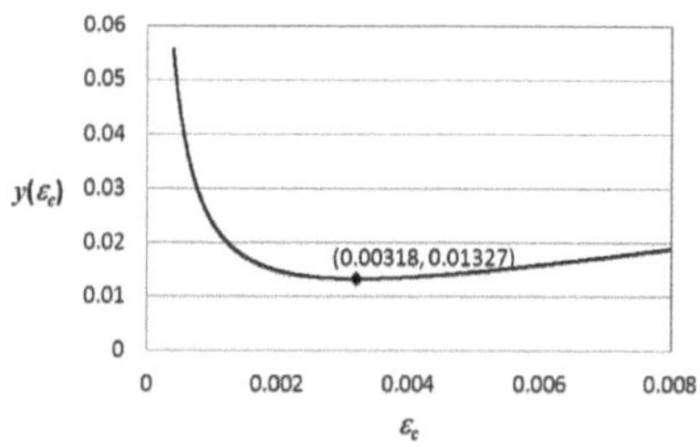

(C) $y(\varepsilon_c)$ for f_{co}=40 MPa

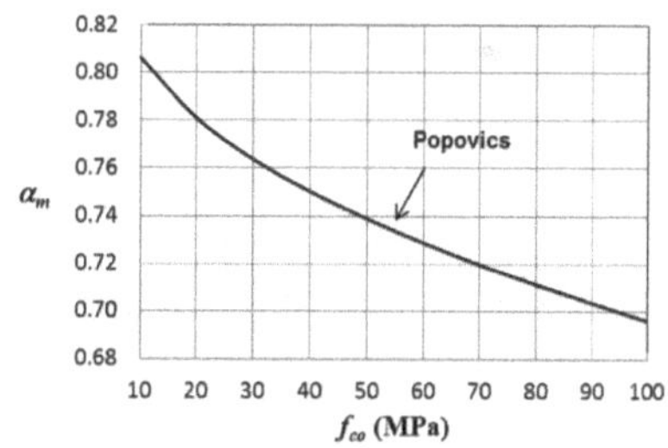

(D) α_m

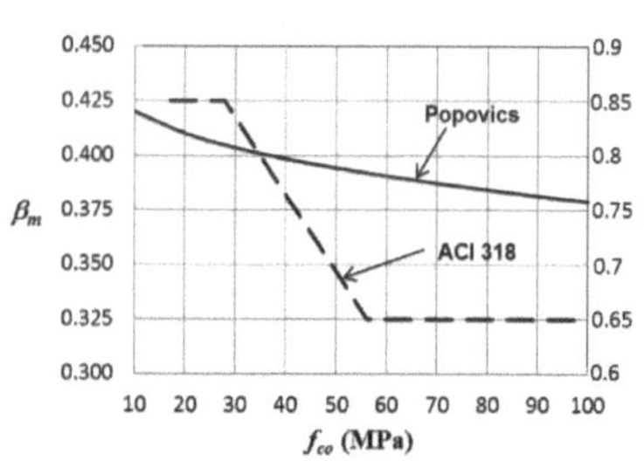

(E) β_m and β_1

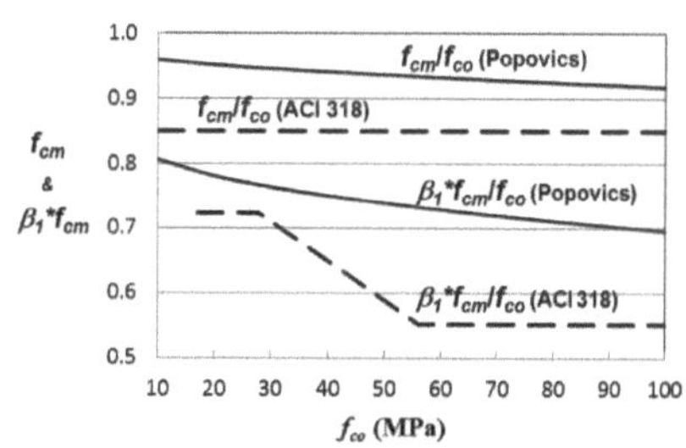

(F) f_{cm} and $\beta_1 * f_{cm}$

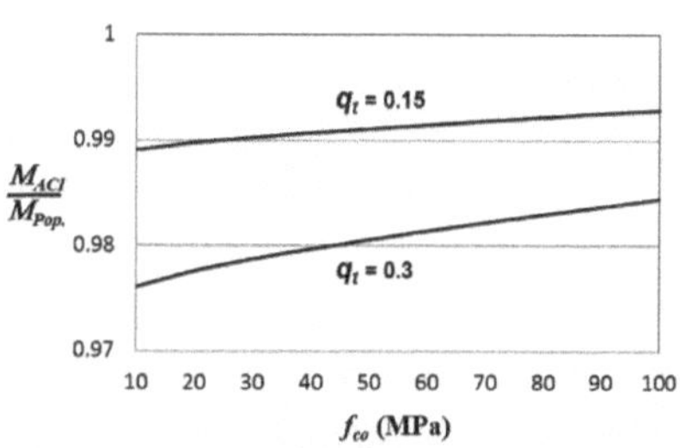

(G) $M_{ACI}/M_{Pop.}$

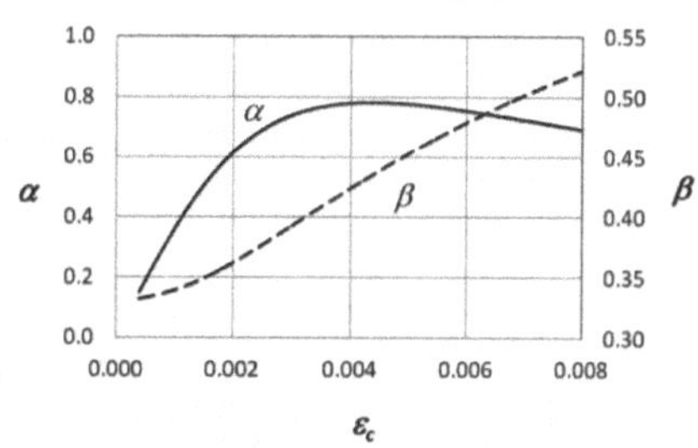

(H) α and β for f_{co} = 40 MPa

Fig. 3.2 Results of the new flexural theory.
From Wu, Y. F. 2016. "Theorems for flexural design of RC members." *J. Struct. Eng.* 142 (5), 04015172.

(1) Popovics model (Popovics 1973)

The stress–strain model proposed by Popovics (1973) is given by

$$f_c = \frac{f_{co} \cdot x \cdot r}{r - 1 + x^r} \tag{3.3}$$

in which

$$x = \frac{\varepsilon}{\varepsilon_{co}} \tag{3.4}$$

$$r = \frac{E_c}{E_c - f_{co}/\varepsilon_{co}} \tag{3.5}$$

$$\varepsilon_{co} = 9.37 \times 10^{-4} f_{co}^{0.25} \ \left(\text{in N/mm}^2\right) \tag{3.6}$$

$$E_c = 4730\sqrt{f_{co}} \ \left(\text{in N/mm}^2\right) \tag{3.7}$$

where ε is concrete strain and $f_c(\varepsilon) = f(\xi)$; ε_{co} is the strain at the peak stress f_{co}; and E_c is the elastic modulus of concrete. Eq. (3.7) is based on ACI 318–05 (ACI 2005).

(2) Eurocode 2 (EC2) model (CEN 2004)

The CEB-FIP/EC2 model (CEN 2004) is given by

$$f_c = \frac{f_{co} x(k - x)}{1 + (k - 2)x} \tag{3.8}$$

$$k = 1.1 E_c \frac{\varepsilon_{co}}{f_{co}} \tag{3.9}$$

$$\varepsilon_{co} = 0.0022 \tag{3.10}$$

$$E_c = 9500(f_{co} + 8)^{\frac{1}{3}} \ \left(\text{in N/mm}^2\right) \tag{3.11}$$

where x is given by Eq. (3.4). It is noted that another equation that is different from Eq. (3.8) is used in EC2 when the concrete is about to crush (the dashed line in Fig. 3.2A). However, that part of the curve is far beyond ε_{cm} and hence does not affect the calculation of the flexural design strain ε_{cm}.

It can be seen from Fig. 3.2B and Eq. (3.1) that:

(1) the flexural design strain $\varepsilon_{cm,u}$ is a function of concrete strength;
(2) apart from the shape of the concrete stress–strain curve, no other factor including the reinforcement ratio affects the value of $\varepsilon_{cm,u}$;
(3) for the same concrete strength, different stress–strain models give different shapes of the stress–strain curve $f_c(\varepsilon)$, and consequently affect the value of $\varepsilon_{cm,u}$;

(4) for normal concrete with a strength from 30 to 60 MPa, the flexural design strain is from 0.003 to 0.0034 if Popovics model is considered appropriate in reflecting the stress–strain relationship of concrete. The value of 0.003 recommended by American concrete design codes is a lower bound and that given in the Eurocodes of 0.0035 is the upper bound. The value of 0.0033 recommended by the Chinese design codes is closer to the average value in this range of concrete strength. The EC2 model of concrete gives smaller and less accurate values of the flexural design strain based on Fig. 3.2B; and

(5) for higher strength concrete, it is not suitable to use a constant value of $\varepsilon_{cm,u}$. Its value should be related to concrete strength.

Theorem 3.1 is obtained by converting the problem of finding the peak point of the M-κ curve into a study of the shape of the concrete stress–strain relationship (Section 3.5.1). A mathematical transformation converts the bending moment resistance of an RC section into the following equation:

$$M = C1 - C2 \times y(\varepsilon_c) \tag{3.12}$$

where $C1$ and $C2$ are constants for a particular RC section and $y(\varepsilon_c)$ is given by

$$y(\varepsilon_c) = \frac{\varepsilon_c \int_0^{\varepsilon_c} f(\xi)d\xi - \int_0^{\varepsilon_c} f(\xi)\xi d\xi}{\left(\int_0^{\varepsilon_c} f(\xi)d\xi\right)^2} \tag{3.13}$$

A typical curve of $y(\varepsilon_c)$ is shown in Fig. 3.2C. The peak moment occurs at the point where $y(\varepsilon_c)$ takes the minimum value. Theorem 3.1 is the result of minimization of $y(\varepsilon_c)$.

The peaking off process of the M-κ curve is actually the process of shape optimization of the concrete stress block. For members with both tension and compression reinforcements yielded before the onset of peak moment, the force of tension reinforcement F_{st} and the force of compression reinforcement F_{sc} remain constant after yielding. As a result, the resultant force of the concrete, $F_c = N + F_{st} - F_{sc}$ (Fig. 3.1), remains constant afterward for members with a constant axial force N. The typical moment–curvature curve for such an RC member is shown in Fig. 2.4 where the tension reinforcement yields at point B. The stress blocks for points A, B, C, and D in Fig. 2.4 are illustrated in Fig. 3.3. Since F_c is a constant, the area of concrete stress block stays constant after point B, but its shape and centroid position change when the curvature further increases, as shown in Fig. 3.3, with the lever arm changing from z_B to z_C and then z_D. At the point C, an optimal shape of the stress block forms with the maximum lever arm z_C (or minimum $y(\varepsilon_c)$), resulting in a maximum moment. The optimal shape of the stress block with the maximum lever arm forms at the exact moment that $\varepsilon_c = \varepsilon_{cm,u}$ at the extreme compression fiber, as shown in the strain profiles of Fig. 3.3. The stress profile corresponding to the point C is referred to as the optimal stress profile hereafter.

The condition of Theorem 3.1 that the compression bar has also yielded before the onset of flexural strength can be relieved with only insignificant error from an engineering design point of view. The reason will be given later.

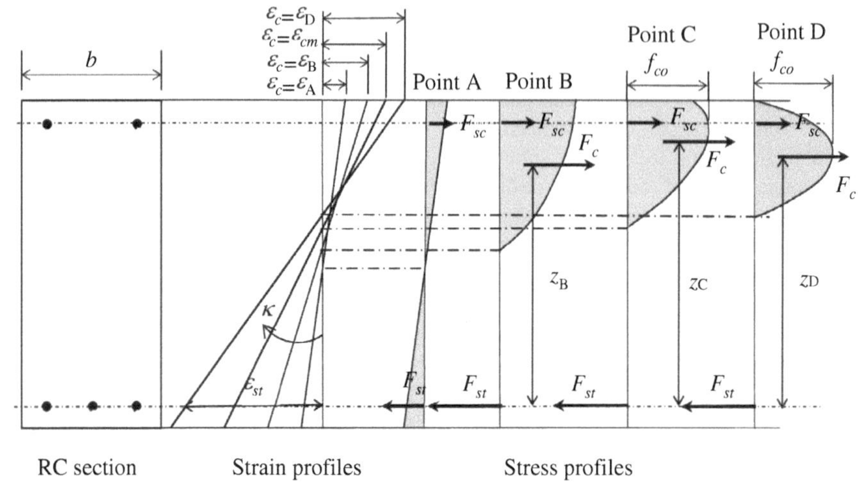

Fig. 3.3 Variation of stress blocks.

3.2.2 The second theorem

Theorem 3.2 *The flexural design strain is constant when the tensile reinforcement ratio ρ_{st} is smaller than a certain value, $\rho_{st} \leq \rho_m$. When $\rho_m < \rho_{st} \leq \rho_c$, the flexural strength occurs at the first yield of the tension reinforcement, and the flexural design strain increases when ρ_{st} increases. When $\rho_{st} > \rho_c$, the tension reinforcement is elastic at the onset of the flexural strength, and the flexural design strain reduces when ρ_{st} increases.*

A graphical interpretation of Theorem 3.2 is given in Fig. 3.4, which shows the overall picture of the flexural design strain ε_{cm} in relation to the tension reinforcement ratio ρ_{st}. Fig. 3.4 is produced using the Popovics model as the constitutive relationship of concrete for the case of $f_{co} = 40$ MPa. Both beams and columns are shown in the

Fig. 3.4 Graphical interpretation of Theorem 3.2. From Wu, Y. F. 2016. "Theorems for flexural design of RC members." *J. Struct. Eng.* 142 (5), 04015172.

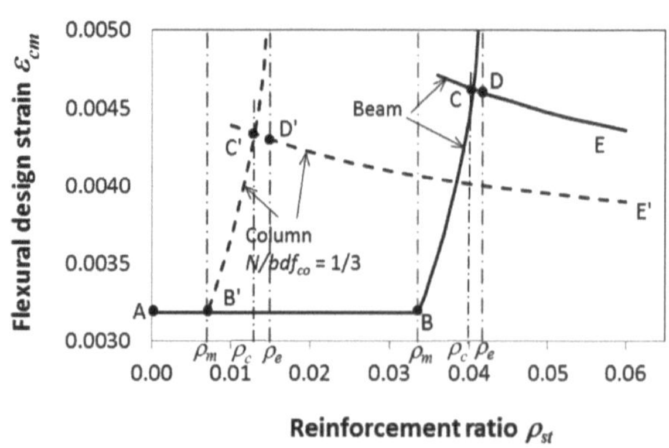

figure, where for columns a special case of axial load ratio $n' = N/(bdf_{co}) = 1/3$ is shown.

The whole extent of ε_{cm} is divided into the following three regions (cases):

(1) *Underreinforced sections*:

When the tension reinforcement ratio ρ_{st} is not greater than a certain value of ρ_m ($\rho_{st} \leq \rho_m$) given by Eq. (3.46), the flexural design strain is a constant that is irrelevant to the tension reinforcement ratio, as represented by the horizontal line AB for RC beams and AB′ for RC columns. In this case, the tension reinforcement yields before the onset of the flexural strength. This case of ε_{cm} is covered by Theorem 3.1 and the value of the flexural design strain is only a function of the concrete stress–strain relationship, as given by Eq. (3.1).

(2) *Overreinforced sections*:

The case when the tension reinforcement ratio ρ_{st} is greater than the value ρ_e given by Eq. (3.49), i.e., $\rho_{st} \geq \rho_e$, occurs in compression-controlled sections where the tension reinforcement does not yield throughout the bending process. The tension reinforcement ratio affects the value of the flexural design strain in this case. When the tension reinforcement ratio increases, the flexural design strain decreases, as shown by the line CE for beams and C′E′ for columns in Fig. 3.4. The flexural design strain is identified as $\varepsilon_{cm,e}$ in this case and is obtained from the solution of Eq. (3.63).

(3) *Balanced sections*:

There exists a transition zone between the previous two cases where the value of the flexural design strain increases when the tension reinforcement increases in the range of $\rho_m < \rho_{st} \leq \rho_c$ but reduces after $\rho_{st} > \rho_c$. In the range of $\rho_m < \rho_{st} \leq \rho_c$, the flexural strength is achieved at the exact moment when the tension reinforcement yields, which is called the balanced failure mode in Chapter 2. In conventional flexural theory, the flexural design strain is one specific value. There is only one specific reinforcement ratio for the case of the balanced failure. However, the flexural design strain is not a constant in this case: as shown in Fig. 3.4, each value of the flexural design strain corresponds to a certain reinforcement ratio. Therefore, there is a range of values of the reinforcement ratio for the balanced failure mode.

Lines BC for beams and B′C′ for columns in Fig. 3.4 represent the balanced failure mode. There is a small range of reinforcement ratio between the point C and the point D for beams or C′ and D′ for columns in Fig. 3.4 ($\rho_e > \rho_{st} > \rho_c$), where the tension reinforcement yields after the onset of the peak moment, i.e., the peak moment occurs when the tension reinforcement is elastic, and it yields afterward in the descending branch of the *M*-*κ* curve. As moment is equal to force multiplied by lever arm, RC beams with a reinforcement ratio in this range have a faster deterioration in lever arch than the increase in internal tension force after the curvature passes the peak moment. The flexural design strain in the range of $\rho_m < \rho_{st} \leq \rho_e$ is referred to as $\varepsilon_{cm,t}$ and is given by Eq. (3.53). The reinforcement ratio at the point C, ρ_c, can be obtained from Eq. (3.51) = Eq. (3.63).

The three lines in Fig. 3.4 reflect three types of internal stress conditions at the onset of the flexural strength:

1. Section type 1—For RC sections with a tension reinforcement ratio less than ρ_m, i.e., Line AB for beams or AB′ for columns, the tension reinforcement yields before the onset of the peak flexural resistance so that concrete resultant force stays constant afterward. With

further increase in curvature (or extreme concrete strain), the concrete stress block is able to adjust its shape to maximize its lever arm (optimal shape) at $\varepsilon_c = \varepsilon_{cm,u}$ so that the term $y(\varepsilon_c)$ in Eq. (3.13) is minimized to give the maximum moment.

2. Section type 2—from B to C (or B′ to C′), the tension reinforcement has not yielded when the concrete stress block reaches its optimal shape at $\varepsilon_c = \varepsilon_{cm,u}$. Although the shape of the concrete stress block is less efficient afterward (lever arm not maximized), the resultant force of concrete keeps increasing until the tension reinforcement yields. After that, the resultant force remains constant, but the stress block continues to deteriorate ($y(\varepsilon_c)$ increasing in Fig. 3.2C) so that the bending moment keeps decreasing after passing the yield point. Therefore, the yield point gives the peak moment.
3. Section type 3—after point C, the tension reinforcement is so heavy that the concrete stress block is not able to cause the yielding of the reinforcement at the peak moment.

Section type 1 makes the best use of both concrete resistance and steel material. Type 2 fully utilizes steel material, but the concrete resistance is used less efficiently (lever arm not maximized). Type 3 is unable to make full use of either concrete or steel materials.

The line B′C′ for columns in Fig. 3.4 is for illustration only and can be omitted if an equivalent reinforcement ratio $\rho' = \rho_{st} - \rho_{sc} + \frac{N}{f_y bd}$ is used to replace ρ_{st} as the horizontal axis in Fig. 3.4. In this case, the line BC is directly adopted for ρ' in the transition zone for both beams and columns. Therefore, for the general case of columns with both tension and compression reinforcements, line BC can be directly used, but for the equivalent reinforcement ratio ρ'. For section type 3 or overreinforced sections of columns, the flexural design strain of columns with different axial load ratios refers to Fig. 3.22.

3.2.3 The third theorem

Theorem 3.3 *For RC sections with a constant axial load, the ultimate curvature is attained when the concrete stress at the extreme compression fiber reaches zero. The RC section loses its integrity in axial load carrying capacity after the onset of this ultimate curvature.*

To explain this theorem, let's first look at an idealized case—a plain concrete section, as shown in Fig. 3.5. The example demonstrates that a plain concrete column will crush immediately after the cross-sectional curvature exceeds a certain value.

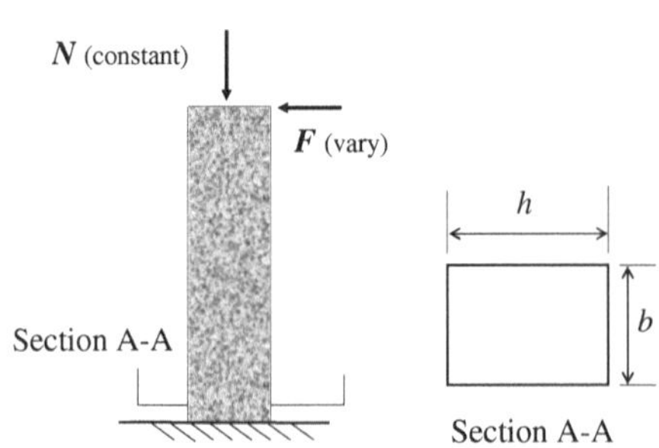

Fig. 3.5 Plain concrete column.

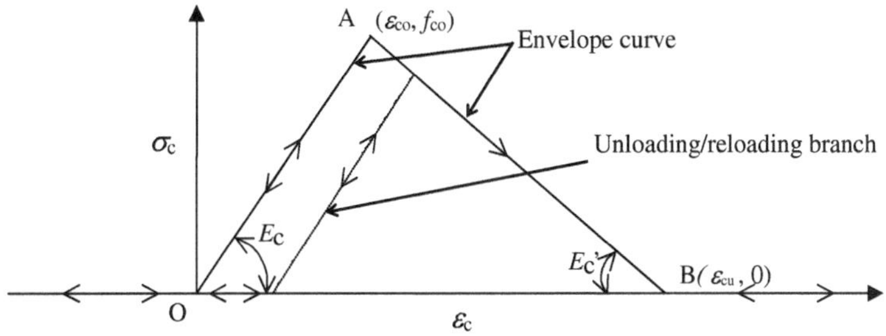

Fig. 3.6 Idealized stress–strain model of the concrete.

The column is loaded under displacement control by monotonically increasing the curvature at the bottom cross-section A-A. For simplicity of discussion, the stress–strain relationship of the concrete is idealized by the bilinear model O-A-B of Fig. 3.6. The variation of the stress profile of the typical section A-A in Fig. 3.5 with monotonic increase in the curvature of the section is shown in Fig. 3.7. As the column is initially loaded with a constant axial load N with a zero lateral force or moment, the initial stress profile is a uniform stress distribution shown by the stress distribution A in Fig. 3.7. When the curvature increases, the stress distribution changes from distribution A to distributions A′, B, C, D and E successively, as shown. In all these cases, the area of the stress blocks remains unchanged before it crushes to satisfy the condition of constant axial load.

For the distribution D in Fig. 3.7, the depth of compression zone is given by $x = \varepsilon_{cu}/\kappa_d$, where ε_{cu} is the ultimate concrete strain as shown in Fig. 3.6 and κ_d is the curvature when the extreme concrete fiber just reaches the strain ε_{cu}. The axial force is given by

$$N = b \cdot f_{co} \cdot x/2 = \frac{b \cdot \varepsilon_{cu} \cdot f_{co}}{2\kappa_d} \tag{3.14}$$

or

$$\kappa_d = \frac{f_{co} \cdot \varepsilon_{cu} \cdot b}{2N} = \frac{\varepsilon_{cu}}{2h \cdot (N/N_c)} = \kappa_{cu} \tag{3.15}$$

where N_c is the crushing axial load given by $N_c = f_{co} \cdot b \cdot h$; κ_{cu} is defined as the ultimate curvature.

The stress distribution E in Fig. 3.7 gives

$$x_e = \varepsilon_{cu}/\kappa_e \tag{3.16}$$

$$N_e = b \cdot f_{co} \cdot x_e/2 = \frac{b \cdot \varepsilon_{cu} \cdot f_{co}}{2\kappa_e} \tag{3.17}$$

in which κ_e is the curvature of the section for distribution E, and x_e is the distance from the neutral axis to the point where the compressive strain equals ε_{cu}, as shown in

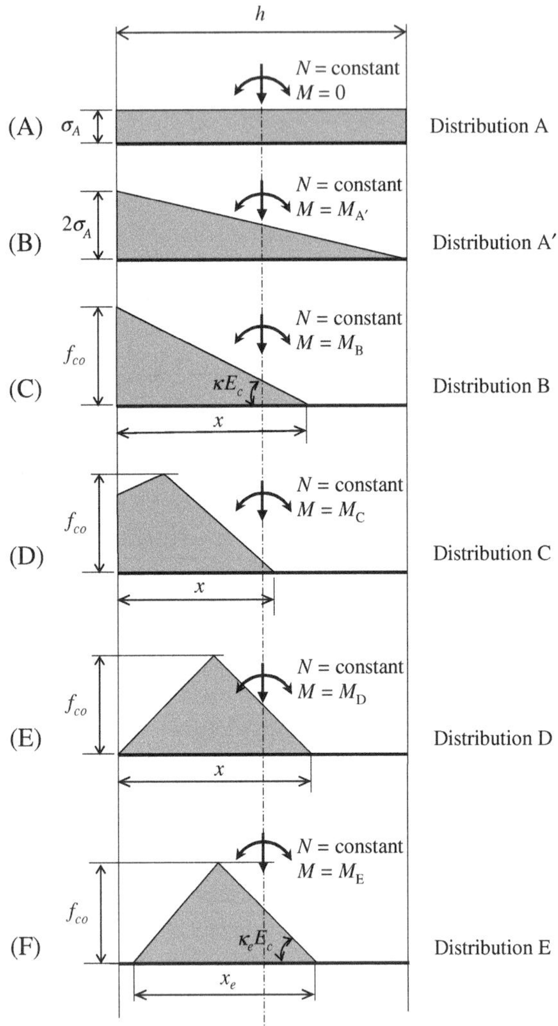

Fig. 3.7 Variation of stress blocks in section A-A: (A) stress distribution A; (B) stress distribution A′; (C) stress distribution B; (D) stress distribution C; (E) stress distribution D; and (F) stress distribution E.

Fig. 3.7, and N_e is the compression resistance of concrete for stress distribution E. Comparing Eq. (3.14) with Eq. (3.17),

$$\because \kappa_e > \kappa_d$$

$$\therefore N_e < N \tag{3.18}$$

Eq. (3.18) means that the axial force at distribution E must be smaller than N. In other words, the axial force N cannot be sustained at distribution E. As a result, the column will crush immediately after distribution D, or when the cross-sectional curvature κ exceeds the ultimate curvature κ_{cu} given by Eq. (3.15).

The physical explanation behind Eq. (3.18) is that the horizontal length of the triangular stress block reduces when the curvature further increases from distribution D. While the maximum stress of concrete remains constant at f_{co}, the area of the stress block must decrease. If the axial load remains constant at this stage, the whole concrete cross-section crushes suddenly. It can be seen from Eq. (3.15) that the ultimate curvature is in reverse proportion to the axial force N.

To further illustrate the results derived previously, the plain concrete column in Fig. 3.5 is numerically simulated and the results are shown in Fig. 3.8. The column has a cross-section of 400 mm × 400 mm, with the following material properties: $f_{co}=44$ N/mm^2; $\varepsilon_{co}=0.002$; $\varepsilon_{cu}=0.006$; and $N=0.3 \cdot f_{co} \cdot b \cdot h = 2112$ kN. The same stress–strain relationship as in Fig. 3.6 is used in the numerical simulation in order to provide the same benchmark for comparison with the analytical results. From Eq. (3.15), $\kappa_{cu} = \frac{f_{co} \cdot \varepsilon_{cu} \cdot b}{2N} = \frac{44 \times 0.006 \times 400}{2 \times 2112000} = 2.5 \times 10^{-5}$. The analytical response curve from distribution B to distribution D can also be derived and shown in Fig. 3.8. Details of the analytical solution refer to Wu et al. (2004b). Both the numerical and analytical results show that the column collapses suddenly at the curvature of $\kappa = 2.5 \times 10^{-5}$, as shown in Fig. 3.8. The conclusion is the same when a general nonlinear stress–strain relationship is used, as shown in Fig. 2.6.

From the previous discussions it can be seen that the concrete section completely crushes when the compressive strain of concrete at the extreme fiber exceeds ε_{cu}, i.e., the concrete at the extreme compression fiber completely crushes. Therefore, the ultimate deformation of a plain concrete column is reached when the curvature of a cross-section reaches a value given by Eq. (3.15). This curvature is defined as the ultimate curvature of an RC section in this book.

The results and conclusions derived from this idealized case of a plain concrete section with the bilinear stress–strain relationship of concrete are applicable to general

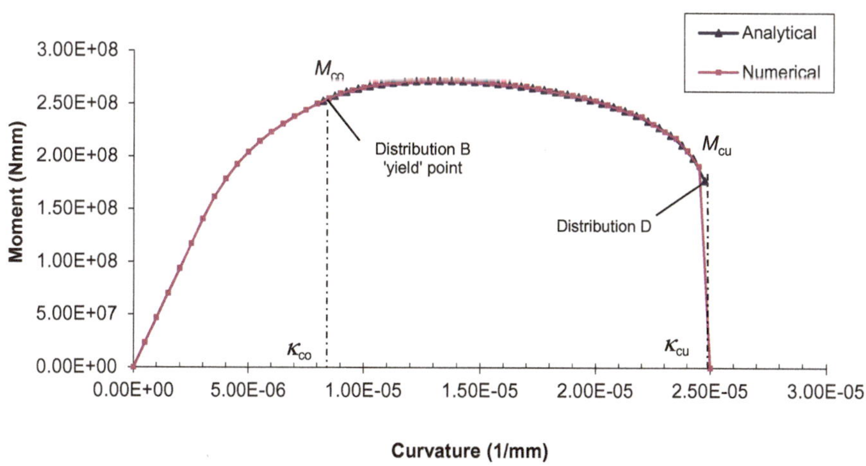

Fig. 3.8 Sudden collapse of plain concrete section.
From Wu, Y. F., D. J. Oehlers, and M. C. Griffith. 2004b. "Rational definition of the flexural deformation capacity of RC column sections." *Eng. Struct.* 26 (5), 641–650.

RC sections, which is described by Theorem 3.3. Detailed derivation of Theorem 3.3 for general RC sections with an arbitrary stress–strain relationship of concrete is given in Section 3.5.7. Theorem 3.3 expresses that the ultimate curvature of any RC section is achieved when the concrete fiber at the extreme compression face is softened to reach a zero stress, regardless of the shape of the stress–strain curve of concrete. For RC columns with reinforcement, sudden collapse of the whole section at onset of the ultimate curvature will not occur due to the existence of rebar, as shown in Fig. 2.6. However, it will be shown in Section 3.4 that an RC member loses its integrity in compressive resistance after the onset of this ultimate curvature; and hence the ultimate curvature should be treated as the maximum curvature that it is allowed to occur.

Exact zero strength $f_c = 0$ may correspond to a very large ε_{cu}, as shown by the dashed line in Fig. 3.2A. However, it does not mean that the collapse occurs at a large curvature. This is because the enclosed area under the dashed line in Fig. 3.2A is very small and a small further increase in curvature will exhaust this extra area and the column collapses shortly. Therefore, it is sensible to define a strain close to the complete crushing of concrete as the onset condition of the ultimate curvature. A value of ε_{cu} around 0.005–0.008, depending on concrete grade, can be selected as the ultimate strain. Detailed and more general mathematical derivation and rigorous interpretation are given in Section 3.5.7.

3.3 Numerical illustrations

Numerical case studies are used in this section for better understanding of the three theorems.

3.3.1 *Flexural design parameters*

For the typical flexural response curves of RC beams and columns in Figs. 2.5 and 2.6, the following four points are shown on the response curves:

(1) the first yield of the tension reinforcement, denoted as "+",
(2) the peak point of a M-κ curve "Δ",
(3) the onset of $\varepsilon_c = \varepsilon_{cm,u}$ at the extreme compression fiber "×", and
(4) the onset of $\varepsilon_c = \varepsilon_{cu}$ at the extreme compression fiber "·"

The following observations are made from the numerical case studies in Figs. 2.5 and 2.6:

(1) Flexural strength

For a curve with the point "+" occurring before the point "Δ" (underreinforced members based on definition), the peak point "Δ" overlaps with the point "×", which means that the peak moment occurs exactly at $\varepsilon_c = \varepsilon_{cm,u}$, as predicted by Theorem 3.1. For these members, Eq. (3.1) gives the result of $\varepsilon_{cm,u} = 0.00318$ (see Fig. 3.2B). The values of the strain ε_c at the extreme compression fiber at the onset of peak moment identified from the numerical curves are shown in Figs. 2.5 and

2.6. For all underreinforced members, the values are the same: $\varepsilon_c = 0.00318$, which matches $\varepsilon_{cm,u} = 0.00318$ from Theorem 3.1.

As underreinforced members have a larger yield plateau, the moment resistance is not sensitive to the value of the flexural design strain ε_{cm}. Therefore, the existing design codes can give very similar results of M_u that are actually very close to the exact peak moment, regardless of the difference in the recommended value of ε_u by different codes (i.e., 0.003 by ACI 318, 0.0035 by EC2, and 0.0033 by Chinese codes). For other members where the point "×" does not overlap with the point "Δ" (RC sections with overreinforced and balanced failure modes), the flexural strength calculated using $\varepsilon_c = \varepsilon_{cm,u}$ (point "×") can be significantly smaller than the peak moment if a constant value of $\varepsilon_{cm,u}$ is used, as recommended in the current design codes. In these cases, the flexural design strain ε_{cm} is not a constant but varies depending on the reinforcement ratio and the axial load ratio as described by Theorem 3.2 and depicted in Fig. 3.4.

(2) RC section types

Theorem 3.2 divides the full range of tension reinforcement ratio into three regions (Fig. 3.4):

- AB (or AB′) for section type 1,
- BC (or B′C′) for section type 2, and
- CE (or C′E′) for section type 3.

The maximum reinforcement ratio for section type 1 of RC beams is given by Eq. (3.46):

$$\rho_m = \frac{\alpha_m f_{co} \varepsilon_{cm,u}}{f_y\left(\varepsilon_y + \varepsilon_{cm,u}\right)} = 0.75*40*0.00318/(500*(0.0025+0.00318)) = 3.36\%$$

where α_m is given by Eq. (3.33) when $\varepsilon_c = \varepsilon_{cm,u}$, which is depicted in Fig. 3.2D. For RC columns with an axial load ratio of $N/(bdf_{co}) = 1/3$, $\rho_m = \frac{\alpha_m f_{co} \varepsilon_{cm,u}}{f_y\left(\varepsilon_y + \varepsilon_{cm,u}\right)} - \frac{1}{f_y}\frac{N}{bd} = 0.696\%$. These two cases are exactly at the border between type 1 and type 2 sections, which means the two cases of (1) the peak moment occurs at onset of $\varepsilon_{cm,u}$, and (2) yielding of the tension bars coincides with the peak moment, merging into one point. This condition is exactly the case in the numerical case studies in Figs. 2.5 and 2.6, where it can be seen that the peak point, yield point, and the onset of $\varepsilon_{cm,u}$ become one point (i.e., the three points "×", "+", and "Δ" overlap) when $\rho_{st} = \rho_m$.

The region BC (or B′C′) in Fig. 3.4 belongs to section type 2. For members with this type of section, the peak moment occurs exactly at the initial yielding of tension reinforcement but not at the onset of the strain $\varepsilon_{cm,u}$ at the extreme compression face, as shown by the case of $\rho_{st} = 3.97\% > \rho_m = 3.36\%$ for beams and $\rho_{st} = 1.1\% > \rho_m = 0.696\%$ for columns. It can be seen in Figs. 2.5 and 2.6 that, for these two cases, when the extreme compression face reaches the strain of $\varepsilon_c = \varepsilon_{cm,u}$ (point "×"), the moment resistance has not reached the peak point. Therefore, the flexural design strain for these cases is greater than $\varepsilon_{cm,u}$. The flexural design strain for type 2 sections is $\varepsilon_{cm,t}$ given by Eq. (3.53). The calculation of Eq. (3.53) for the beam with $\rho_{st} = 3.97\%$ gives a value of 0.00436, which is exactly the same as that shown in Fig. 2.5 for this curve that is obtained from numerical simulation by the layered method (Wu et al. 2004a). Clearly, $\varepsilon_{cm,t}$ is much greater than $\varepsilon_{cm,u}$, and the use of $\varepsilon_{cm,u}$ for calculation of flexural strength as recommended by current RC design codes may lead to significant error, as shown. It is the same case for the column with $\rho_{st} = 1.1\%$ in Fig. 2.6.

The line CE (or C′E′) is for type 3 sections where the tension reinforcement does not yield at the onset of the peak moment. When excessive amount of steel bar is provided, the steel reinforcement will not yield, which belongs to this case. The point D (or D′) locates at another boundary reinforcement ratio ρ_e above which tension rebar will not yield on the whole M-κ curve. For the RC section with exactly the reinforcement ratio of ρ_e, there is only one point on the M-κ curve where the tension rebar yields. The solution of ρ_e is given by Eq. (3.49). For $f_{co} = 40$ MPa and $\rho_{sc} = 0$, Eq. (3.49) gives $\rho_e = 4.16\%$ for beam, while for a column with $N/(bdf_{co}) = 1/3$, $\rho_e = 1.49\%$. The flexural responses of these two RC sections are numerically simulated by the layered method. The extreme concrete strain ε_c vs. the tension steel stress f_{st} curves are numerically generated and shown in Fig. 3.9. It can be seen that both cases reach the yield point at its peak, and it is the only point at which the tension bar yields. This demonstrates the physical meaning of ρ_e in Fig. 3.4. It is noted that for RC members with a reinforcement ratio of ρ_e, the peak moment occurs before yielding the tension reinforcement, i.e., the peak point in Fig. 3.9 does not match with the onset of flexural strength but occurs after that.

When $\rho_{st} > \rho_e$, the tension bars do not yield throughout the moment–curvature curves. The RC beam with a tension reinforcement ratio of 4.96% ($>\rho_e = 4.16\%$) in Fig. 2.5 and the column with a reinforcement ratio of 2.2% ($>\rho_e = 1.49\%$) belong to this case. The numerical moment–curvature curve of the beam in Fig. 2.5 shows that the peak moment onsets at the extreme concrete strain of 0.00448 that is identical

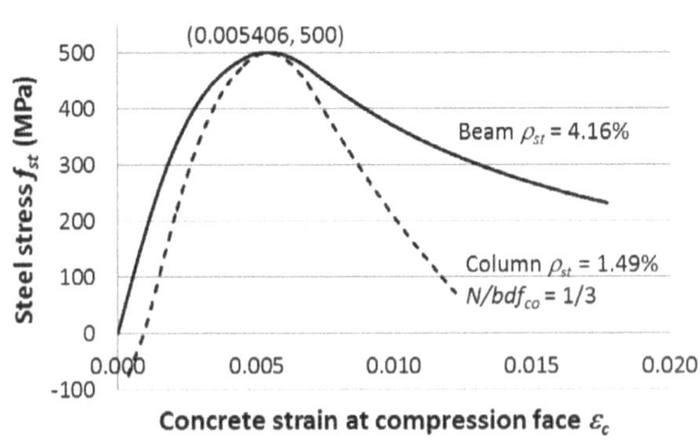

Fig. 3.9 Cases with a reinforcement ratio of ρ_e.
From Wu, Y. F. 2016. "Theorems for flexural design of RC members." *J. Struct. Eng.* 142 (5), 04015172.

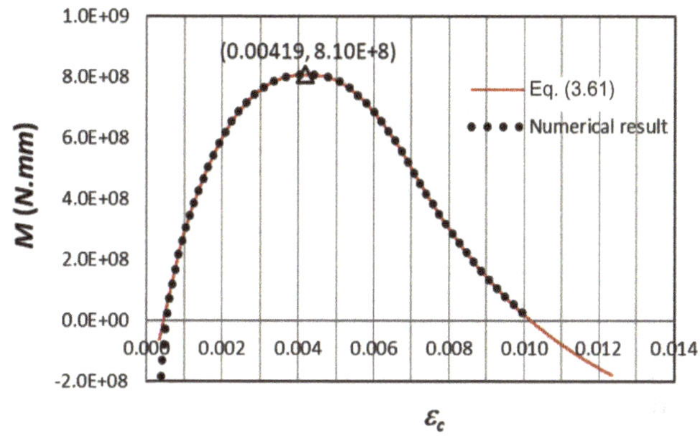

Fig. 3.10 Eq. (3.61) compared with numerical results.
From Wu, Y. F. 2016. "Theorems for flexural design of RC members." *J. Struct. Eng.* 142 (5), 04015172.

to the solution from Eq. (3.63) for $\varepsilon_{cm,e}$. Fig. 3.10 shows the moment vs. ε_c curve for the column in Fig. 2.6 with the tension reinforcement ratio of 2.2%. For columns, it is assumed that an axial load is applied before bending. Therefore, a compressive strain in concrete exists before bending (at $M = 0$), as shown in the figure. The tension reinforcement of this column stays elastic throughout the bending process. Its elastic solution of the M-ε_c relation is given by Eq. (3.61), which is also shown in Fig. 3.10. The analytical solution of the flexural design strain is given by Eq. (3.63), which gives $\varepsilon_{cm,e} = 0.00419$ for this column. It can be seen from Figs. 3.10 and 2.6 that the value of ε_c at the onset of the peak moment is 0.00419, which is identical to the result from Eq. (3.63). It can also be seen from Fig. 3.10 that the flexural resistance M calculated from a flexural design strain from 0.003 to 0.0035 recommended by current RC design codes does make a difference from that at the peak point.

A few M-ε_c curves for columns with symmetrical reinforcement ($\rho_{st} = \rho_{sc}$, $\rho_s = \rho_{st} + \rho_{sc}$) and different axial load ratios are shown in Fig. 3.11. It can be seen from the figure that the sensitivity of the flexural strength to the value of ε_c is generally dependent on the axial force ratio and reinforcement ratio.

3.3.2 *Elastic compression reinforcement*

It is assumed in the RC theorems that the compression reinforcement has yielded at the onset of the flexural strength. In the case of an RC section with elastic compression reinforcement, the tension reinforcement certainly yields before the onset of the flexural strength. Otherwise, both the compression bar and tension bar can continue to increase their forces at further increase of curvature, and this will cause further increase in moment resistance and subsequent yielding of the tension bars sooner or later. Due to yielding of the tension reinforcement, a moment–curvature curve with elastic compression bar at the peak moment has a quite flat yield plateau. In this case, the moment resistance does not change much over a large range of ε_c as shown by the

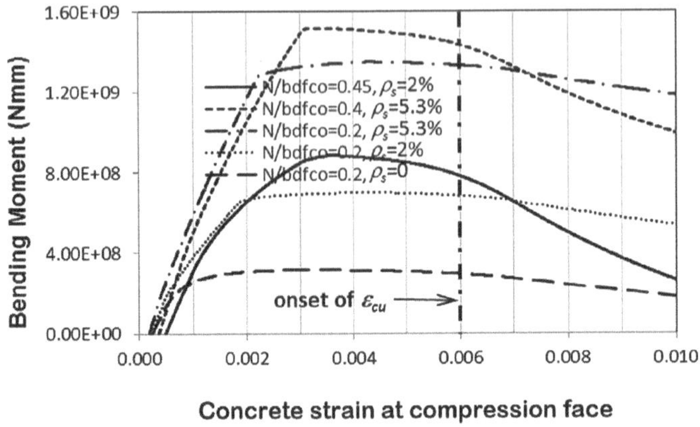

Fig. 3.11 Typical column response curves.
From Wu, Y. F. 2016. "Theorems for flexural design of RC members." *J. Struct. Eng.* 142 (5), 04015172.

case study in Fig. 3.12. As a result, the flexural design strain $\varepsilon_{cm,u}$ given by Theorem 3.1 (Eq. 3.1) can be used to calculate the flexural strength with only insignificant error. This means that the condition in Theorem 3.1 that the compression bar is yielded before the onset of the flexural strength can be relieved from an engineering design point of view.

The analytical solution of the M-ε_c relation for the preceding case with elastic compression bar (Eq. 3.66) is compared with the numerical result in Fig. 3.12. It can be seen from the numerical result that the compression reinforcement yields after the onset of the peak moment, so the solution with elastic compression bar,

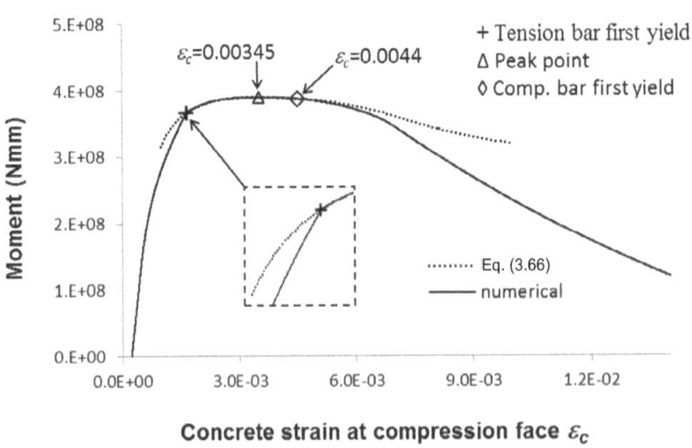

Fig. 3.12 RC section with compression bars, $f_{co} = 40$ MPa, $f_y = 500$ MPa, $\rho_{st} = \rho_{sc} = 0.2\%$, $N/(bd'f_{co}) = 1.8$.
From Wu, Y. F. 2016. "Theorems for flexural design of RC members." *J. Struct. Eng.* 142 (5), 04015172.

Eq. (3.66), is therefore applicable at the peak. It can be seen from the figure that Eq. (3.66) gives results identical to the numerical ones between the point "+" (tension yielding) and the point "◇" (compression yielding). The value of ε_c at the point "◇" obtained from Eq. (3.54) is $\varepsilon_c = 0.0044$. The peak point for RC sections with elastic compression bars is given by Eq. (3.70), which gives $\varepsilon_{cm,c} = 0.00345$. Both points match the numerical results exactly, as shown in Fig. 3.12.

3.3.3 Ultimate failure point

The points at the onset of ε_{cu} are indicated on the moment–curvature curves by the point "•" in Figs. 2.5 and 2.6. Unlike the pure concrete section in Fig. 3.8, an RC section with reinforcement does not collapse suddenly when ε_c exceeds ε_{cu} (curvature $\kappa > \kappa_{cu}$), due to the presence of reinforcing bars, provided they do not buckle. This is because crushing of concrete causes extra compression strain of the section that reduces the strain of the tensile steel ε_{st} caused by bending, as shown in Fig. 3.13. The axial force demand on the concrete $F_c = N - F_{sc} + F_{st}$ (Fig. 3.1) therefore reduces after further increase in curvature, so as to balance the drop of F_c caused by the reduced compression resistance of concrete.

Although a sudden collapse does not occur when $\kappa > \kappa_{cu}$ in RC sections with reinforcement, the internal compression resistance of an RC section qualitatively changes to a monotonically decreasing trend, and the reserve in F_c is exhausted, after this point (see distribution D in Fig. 3.7). The compression capacity of concrete physically drops due to the reduction in the area of the compression stress block after $\kappa > \kappa_{cu}$ (stress distribution E in Fig. 3.7). The column must therefore rely on crushing into deeper core concrete (neutral axis moving downward in Fig. 3.13) to provide the balance of internal forces. In the meantime, the onset of ultimate strain on the concrete compression face causes spalling of the concrete cover. As a result, the longitudinal reinforcement will lose its support and likely buckle, causing a sudden reduction in F_{sc}, consequent increase in the demand of F_c, and subsequent collapse of the RC section at rebar buckling. For these reasons, the absolute ultimate curvature of plain concrete sections, κ_{cu}, that is defined at the onset of ε_{cu} on the extreme compression face, should still be adopted as the ultimate curvature of an RC section.

In the literature and current practice, the ultimate deformation is usually defined at a point where the moment resistance of an RC section drops a certain percentage of the

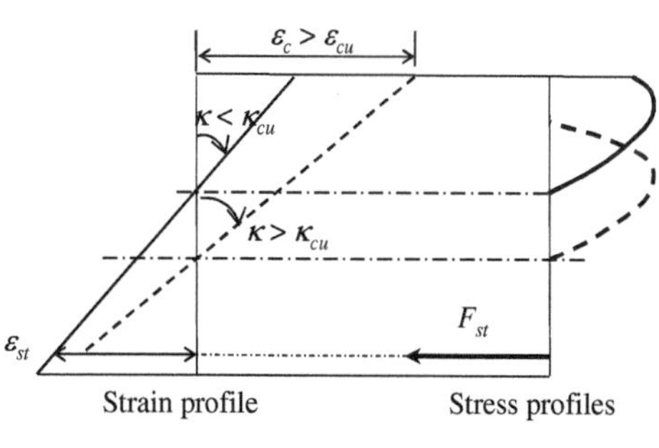

Fig. 3.13 Strain and stress blocks after $\kappa > \kappa_{cu}$.
From Wu, Y. F. 2016. "Theorems for flexural design of RC members." *J. Struct. Eng.* 142 (5), 04015172.

peak moment (Fig. 1.11). Different values of the drop, such as 10%, 15%, or 20%, have been used, based on engineering consensus and having little theoretical ground. Such definition of ultimate deformation is based on strength concerns and is not the true ultimate deformation. Furthermore, it is subjective, inaccurate, and irrational. However, the drop of moment at the ultimate deformation defined by Theorem 3.3 is not very different from that based on the conventional definition. The drop of moment resistance at the onset of ε_{cu} is about 10%–30% for the RC beams without compression reinforcement in Fig. 2.5. Compression reinforcement in beams provides additional compressive resistance, and hence a smaller drop of moment at crushing of the concrete if buckling of the bar does not occur. The moment drop of RC columns without compression reinforcement (unusual in practice) is larger at the onset of ε_{cu} (up to 40% in Fig. 2.6), due to higher demand in compressive resistance of concrete. However, the drop is small for normal RC columns that have symmetrical reinforcement on both tension and compression sides (less than 10% at onset of ε_{cu}), as shown by the typical cases in Fig. 3.11. Therefore, for practical RC members the moment drop at the onset of ε_{cu} is generally similar to that of the conventional definition of the ultimate deformation. Nevertheless, the ultimate curvature values based on the conventional definition and Theorem 3.3 can be very different, particularly when a postpeak response curve is relatively flat, as shown in Figs. 2.5, 2.6, and Fig. 3.11.

In fact a concrete spalling strain typically around 0.5% has been suggested in the literature to define the ultimate curvature of RC sections (Chai et al. 1991). This kind of definition is consistent with Theorem 3.3. However, their criterion came from experimental observations and experience. Theorem 3.3 provides the theoretical ground and rationale for the criterion.

3.4 Ultimate curvature and curvature ductility of RC sections

Theorem 3.3 defines the ultimate curvature of an RC section. This section examines the factors that affect ultimate curvature and curvature ductility of RC sections.

3.4.1 Curvature ductility of plain concrete sections

In Fig. 3.7, the moment corresponding to a given curvature κ from distribution B to distribution D is given by (Wu et al. 2004b):

$$\frac{M}{b}=f_{co}\cdot\left[\frac{\varepsilon_{co}^2\cdot\beta\cdot(1+\beta)}{6\kappa^2}+\frac{\varepsilon_{co}}{2\kappa}\cdot\left(\frac{h}{2}-x\right)\cdot\beta-\frac{1}{6}\cdot\left(1+\alpha-\frac{\alpha\cdot\kappa\cdot x}{\varepsilon_{co}}\right)\cdot\left(\frac{\beta\cdot\varepsilon_{co}}{\kappa}-x\right)\cdot\left(\frac{3h}{2}+\frac{\beta\cdot\varepsilon_{co}}{\kappa}-x\right)\right] \tag{3.19}$$

where M is the moment of the section at curvature κ, and x is the compression zone depth given by

$$x=\frac{\varepsilon_{co}}{\kappa}+\frac{\varepsilon_{co}}{\alpha\cdot\kappa}\left[1-\left(1+\alpha-\frac{2\kappa\cdot\alpha\cdot N}{f_{co}\cdot\varepsilon_{co}\cdot b}\right)^{1/2}\right] \tag{3.20}$$

and

$$\begin{cases} \alpha=\dfrac{E_c'}{E_c} \\ \beta=\dfrac{\varepsilon_{cu}}{\varepsilon_{co}}=1+\dfrac{1}{\alpha} \end{cases} \tag{3.21}$$

where E_c and $E_c{}'$ are the elastic modulus of the ascending branch and slope of the softening branch of the concrete model, respectively, as shown in Fig. 3.6.

The moment resistance at the ultimate curvature can be obtained by substituting Eq. (3.15), that is $\kappa_{cu}=f_{co}\cdot\varepsilon_{cu}\cdot b/2N$, into Eq. (3.19), giving

$$M_{cu}=N\cdot\left(\frac{h}{2}-\frac{4\beta-2}{3\beta}\cdot\frac{N}{f_{co}\cdot b}\right) \tag{3.22}$$

The moment resistance M_{co} corresponding to the distribution B is obtained by substituting $\kappa_{co}=f_{co}\cdot\varepsilon_{co}\cdot b/2N$ into Eq. (3.19), giving

$$M_{co}=N\cdot\left(\frac{h}{2}-\frac{2}{3}\cdot\frac{N}{f_{co}\cdot b}\right) \tag{3.23}$$

The points of M_{cu} and M_{co} are plotted in Fig. 3.8 where it can be seen that the point at distribution B can be treated as a "yielding" point. The drop of moment at the ultimate curvature relative to the "yield" point at distribution B is given by

$$\frac{M_{co}-M_{cu}}{M_{co}}=\frac{2(\beta-1)}{3\beta}\cdot\frac{\left(\dfrac{N}{N_c}\right)}{\left[\dfrac{1}{2}-\dfrac{2}{3}\cdot\left(\dfrac{N}{N_c}\right)\right]} \tag{3.24}$$

where N/N_c gives the axial load level. It can be seen from Eq. (3.24) that the moment drop at the ultimate curvature relative to the "yield" point is only determined by the material property β (Eq. 3.21) and the axial load level. Typical curves of Eq. (3.24) are shown in Fig. 3.14. However, it should be noted that the moment drop given by Eq. (3.24) is relative to the "yield" strength M_{co}, not the peak strength.

As shown in Fig. 3.14, the moment drop from the "yield" point to the ultimate point varies from 10% to 20% of the "yield" moment for axial load levels of 0.1 to 0.2 of the crushing load N_c. Therefore, the traditional ultimate displacement defined at 10% of

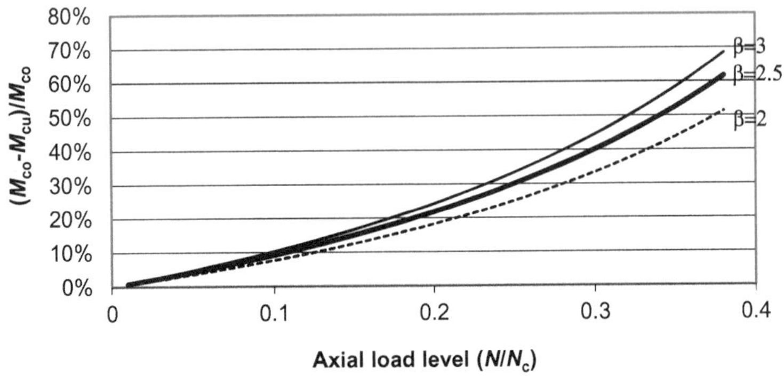

Fig. 3.14 Moment drop vs. axial load.
From Wu, Y. F., D. J. Oehlers, and M. C. Griffith. 2004b. "Rational definition of the flexural deformation capacity of RC column sections." *Eng. Struct.* 26 (5), 641–650.

moment drop is equivalent to the ultimate displacement of a column with an axial load level of 0.1. For other axial load levels, the traditional ultimate displacement cannot reflect the true deformation capacity of a column.

From Fig. 3.8, it can be seen that the curve from the "yield" point to the ultimate point can be defined as a yield plateau. With this definition, the maximum curvature ductility of the concrete section is given by

$$\Delta\kappa = \kappa_{cu} - \kappa_{co} = \frac{f_{co} \cdot b}{2N}(\varepsilon_{cu} - \varepsilon_{co}) = \frac{\varepsilon_{cu} - \varepsilon_{co}}{2(N/N_c) \cdot h} \tag{3.25}$$

This definition of curvature ductility is different from the traditional definition where a certain percentage drop from the peak moment is used to locate the point of the ultimate curvature, regardless of the axial load level. In the literature, it is in fact well established that the deformation behavior of a concrete column is closely related to its axial load level. From this point of view, the traditional definition of ultimate curvature and curvature ductility is inadequate as it does not take into consideration the axial load level and therefore cannot reflect the true extent of physical deterioration of a column at ultimate displacement. The definition given by Eq. (3.25) adequately reflects the degree of damage to the concrete of a column for specific axial load level. The yield plateau starts at the very first point at which the extreme compressive fiber just starts to yield (distribution B in Fig. 3.7) and ends at the true ultimate curvature where any further curvature increase will result in the complete crushing of the whole concrete section (distribution D). Therefore, distributions B and D have clear physical meanings that are the qualitative turning points of the concrete material on the surface of the column from complete elastic to yielding and from yielding to completely crushing, respectively.

Eq. (3.25) shows that the curvature ductility of a plain concrete section is:

(a) directly proportional to the plastic deformation capacity of concrete material, $\varepsilon_{cu} - \varepsilon_{co}$,
(b) inversely proportional to the axial load level N/N_c, and

(c) inversely proportional to the cross-sectional depth h. This is consistent with the common sense that a deeper column is stiffer and more brittle than a thinner one. This effect can also be seen from the numerical simulation results in Fig. 4.9B.

The preceding conclusions are based on the simplified stress–strain model of Fig. 3.6. Another assumption made in this study is that the neutral axis is within the depth of the cross-section, as shown for distributions B to D in Fig. 3.7. When the axial load level is very high ($N > 0.5f_{co}A_g$), the neutral axis will fall outside the right-hand side of the cross-section at stress distribution B. However, this is outside the practical range of axial load level allowed in RC designs.

3.4.2 Effect of longitudinal reinforcement on ductility of RC column sections

From the previous section, it can be seen that the deformation capacity of a plain concrete section is determined by three factors: the material property or the stress–strain relationship of concrete (ε_{co} and ε_{cu}), the depth of cross-section (h), and the axial load level (N/N_c). For RC sections with reinforcement, there exist two constituent materials of concrete and longitudinal reinforcing bars. If an RC section is treated as a pure concrete section and the longitudinal reinforcing bars as external members that apply axial force to the concrete, then the effect of the longitudinal reinforcing bars can be studied by considering it as the axial load effect. Since the presence of the longitudinal reinforcement will not change the material and geometric property of a concrete section, the only way it affects the deformation capacity of the concrete section is through changing the axial load level applied to the concrete section. This conclusion comes from a cross-sectional level. For deformation at a member level, there are other factors that affect longitudinal deformation distribution, such as cracking of concrete, debonding of reinforcement, and plastic hinge zone, which affect member deformation through tension stiffening and plastic hinge length.

The axial load produced by the longitudinal reinforcement varies with the change of cross-sectional curvature. However, the ultimate curvature of a section is only affected by the instantaneous axial load when the strain of the extreme compressive fiber just reaches ε_{cu}. Therefore, the effect of longitudinal reinforcement on the ultimate curvature of a concrete section can be studied with the ultimate strain profiles shown in Fig. 3.15. For convenience of study, the reinforcement is assumed to be symmetrical in Fig. 3.15, i.e., $A_{st} = A_{sc} = 0.5A_s$. For members with nonsymmetrical reinforcements including RC beams, the rationale discussed in the following sections is equally applicable while the conclusions may not be the same.

3.4.2.1 At critical axial load

For the strain profile Case A shown in Fig. 3.15 where the depth of the compression zone is exactly half of the cross-sectional depth, the strain of the compression bars ε_{sc} is the same as that of the tension bars ε_{st}. If the same stress–strain relationship is assumed for both the compression and the tension reinforcement and ignoring the strain history effect, the net additional axial load produced by the longitudinal

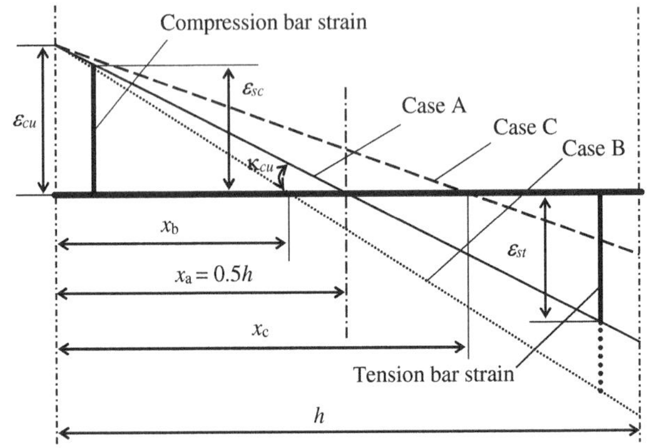

Fig. 3.15 Ultimate strain profiles of RC section. From Wu, Y. F., D. J. Oehlers, and M. C. Griffith. 2004b. "Rational definition of the flexural deformation capacity of RC column sections." *Eng. Struct.* 26 (5), 641–650.

reinforcement is zero. In other words, the existence of the longitudinal reinforcement will not affect the axial load level; hence it will not affect the ultimate deformation of the RC section. In this case, the axial load satisfies

$$N = 0.5 \cdot (0.5h) \cdot b \cdot f_{co} = 0.25N_c \tag{3.26}$$

which is 25% of the concrete crushing load N_c. Therefore, this is a critical axial load level at which, no matter how much longitudinal reinforcement is used, the deformation capacity of the RC sections remains the same as that of the plain concrete column. This phenomenon will be seen in the full nonlinear numerical simulations in the next section.

Eq. (3.26) is applicable only when the bilinear triangular concrete model is used. When a different stress–strain model of concrete is used, this critical load level will be different due to the difference in the area of the stress-blocks at the ultimate curvature with different models. For a general stress–strain relationship, the critical axial load is given by

$$N = \lambda \cdot f_{co} \cdot \left(\frac{h}{2}\right) \cdot b = 0.5\lambda \cdot N_c \tag{3.27}$$

where λ is the area coefficient of the stress block corresponding to the ultimate curvature when the ultimate concrete strain ε_{cu} is reached at the extreme compression face. When Popovics model (Eq. 3.3) is used, the area coefficient is only a function of concrete strength. Numerical integration of the area of the stress–strain curve gives Fig. 3.16 (Wu et al. 2004b). For $f_{co} = 30$ MPa, Fig. 3.16 gives $\lambda = 0.65$, and Eq. (3.27) gives $0.325N_c$.

3.4.2.2 For axial load levels lower than the critical load

When the axial load level is lower than the critical load, the compression zone depth will be smaller than half the sectional depth at the ultimate curvature κ_{cu}, as shown by the Case B in Fig. 3.15. In this case, the tensile strain in the tensile reinforcement will

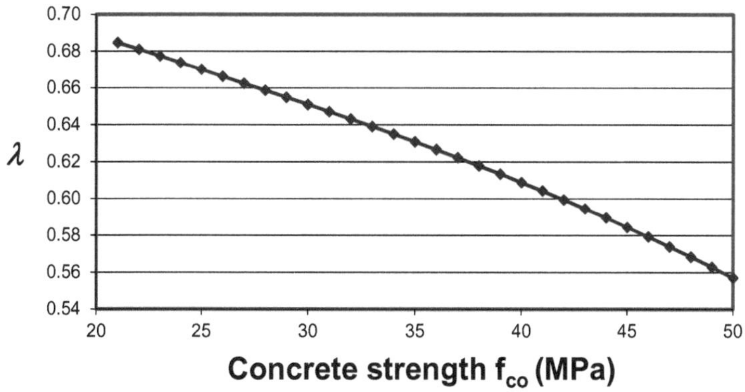

Fig. 3.16 Area coefficient λ based on Popovics model of concrete.
From Wu, Y. F., D. J. Oehlers, and M. C. Griffith. 2004b. "Rational definition of the flexural deformation capacity of RC column sections." *Eng. Struct.* 26 (5), 641–650.

be greater than that in the compression reinforcement, and therefore there would be an extra compression force applied to the concrete on top of the external axial load N. In other words, the presence of longitudinal reinforcement may cause an increase in the axial load level to the concrete section. Although the tensile reinforcement and the compression reinforcement may both yield at the ultimate deformation, the tensile force in the tensile reinforcement may still be significantly greater than the compressive force in the compression reinforcement if the strain of the tensile rebars is much greater than that of the compression bars due to the effect of strain hardening of rebar.

Based on Eqs. (3.15) and (3.25), the deformation capacity and ductility of the concrete section will be reduced by the presence of longitudinal reinforcement. The more longitudinal reinforcement, the greater the reduction of the deformation capacity of the concrete section. This phenomenon was observed from the experimental studies by Berwanger (1975): "the ductility of beam-column decreased with increase in the longitudinal steel percentage." The observation is also consistent with the well-known conclusion that the ductility of an RC beam (where the axial load is zero) reduces with the increase in the longitudinal reinforcement ratio (Fig. 2.5), simply because the longitudinal bars apply extra axial load to the concrete section. From this point of view, the principle used to study the deformation capacity and ductility for columns with symmetrical reinforcement in this section is also applicable to RC members with nonsymmetrical reinforcement.

3.4.2.3 For axial load levels higher than the critical axial load

The compression zone depth will be more than half h at the ultimate curvature when the axial load level is higher than the critical load as shown by Case C in Fig. 3.15. Therefore, the compression force in the reinforcement will be larger than that in the tension reinforcement. The resultant force of the longitudinal reinforcement will be such that it reduces the axial load level applied to the concrete section. Based on Eqs. (3.15) and (3.25), the ultimate curvature and curvature ductility of the concrete section will be increased by the presence of the longitudinal reinforcement. However, the increase in the ultimate curvature by the longitudinal bars is limited due to the high axial load level. This will be seen in the numerical study in the next section.

It can be concluded from the study in this section that:

- there is a critical axial load level of approximately $0.3N_c$ given by Eq. (3.27) that determines the effect of the longitudinal bars on the deformation capacity of an RC column section;
- for an axial force below the critical load level, the presence of the longitudinal reinforcement reduces column deformation capacity and ductility; and
- for an axial force above the critical axial load level, the longitudinal reinforcement helps to increase the deformation capacity and ductility.

3.4.3 Numerical studies

Numerical case studies are provided in this section to visualize and further understand the conclusions from the analytical studies in the previous sections. The layered method is used for the numerical simulations (Wu et al. 2004a). The analytical studies in the previous sections are based on simplified models and assumptions. For more general conclusions, the numerical simulations in this section are based on full nonlinear models, such as Popovics model for concrete (Eq. 3.3) and Menegotto-Pinto's model for steel bars (Gomes and Appleton 1997), and take into account separate stress–strain history of every individual fiber of the cross-section.

The RC column cross-section to be studied is 400 mm × 400 mm with 12 reinforcing bars (4 bars each face). The concrete cross-section is discretized into 25 layers. Each layer retains its own stress–strain history in the numerical simulations. No stirrups and confinement are considered in the study in order to investigate the net effect of the longitudinal bars only. The following material properties are used for the concrete and reinforcing bars: $f_{co}=44$ N/mm^2; $\varepsilon_{co}=0.002$; $\varepsilon_{cu}=0.006$; $f_{ct}=5$ N/mm^2; $f_y=446$ MPa; $E_s=200{,}000$ N/mm^2; $E_h=600$ N/mm^2; $R_0=20.0$; $a_1=19.0$; $a_2=0.3$. Definitions of the parameters are given in the notations.

To vary the reinforcement ratio, different bar diameters are used: 16 mm, 20 mm, 25 mm, and 32 mm corresponding to reinforcement ratios of 1.5%, 2.36%, 3.68%, and 6.03%, respectively. Concrete cover thickness (to the center of the reinforcing bars) is 29 mm for the case of 16 mm bars. To provide a uniform benchmark for comparison, this cover thickness is adopted for all the other reinforcement arrangements regardless of bar diameter, because changing cover thickness effectively changes the lever arm of longitudinal bars, which is not desirable for comparison. A plain concrete section without reinforcement is also included in the study.

The numerical results for the cases with the axial load levels of $40\%N_c$, $30\%N_c$, $20\%N_c$, and $5\%N_c$ are shown in Figs. 3.17–3.20, respectively. The points of ultimate displacement at the onset of the ultimate concrete strain ε_{cu} on the extreme compression face are indicated by "⋄" in the response curves, and the "yielding" point of concrete at the onset of ε_{co} on the extreme compression face is indicated by "×". The regions between the points "×" and "⋄", as calculated by Eq. (3.25) for plain concrete sections, are highlighted by the two broken lines. From these figures, it can be seen that the distance between the broken lines is fairly constant, as indicated by Eq. (3.25),

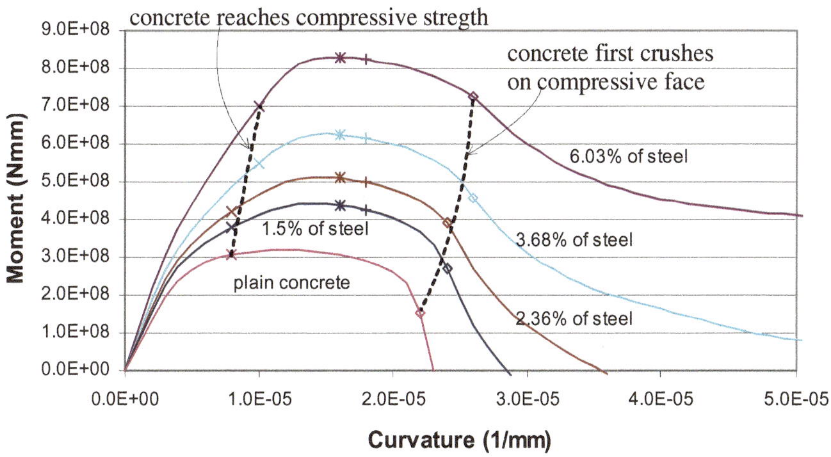

Legend: × - conc. reaches comp. strength ✱ - comp. reinf. yields

+ - tension reinf. yields ◊ - conc. first crushes on col. face

Fig. 3.17 Moment–curvature curves of RC sections with $N = 40\%N_c$.
From Wu, Y. F., D. J. Oehlers, and M. C. Griffith. 2004b. "Rational definition of the flexural deformation capacity of RC column sections." *Eng. Struct.* 26 (5), 641–650.

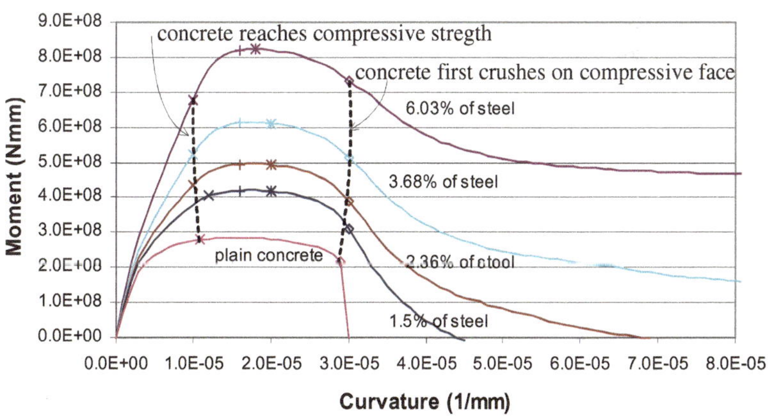

Fig. 3.18 Moment–curvature curves of RC sections with $N = 30\%N_c$.
From Wu, Y. F., D. J. Oehlers, and M. C. Griffith. 2004b. "Rational definition of the flexural deformation capacity of RC column sections." *Eng. Struct.* 26 (5), 641–650.

and gives a good indication of the yield plateau, except for the case with $5\%N_c$ in Fig. 3.20. From the appearance of all the response curves for all the axial load levels, it looks reasonable to define the point ◊ as the ultimate curvature of the RC sections. Although reasonable strength remains after the points ◊ for curves with $5\%N_c$, the remaining strength after points ◊ is only theoretical and not reliable, as discussed in the previous sections.

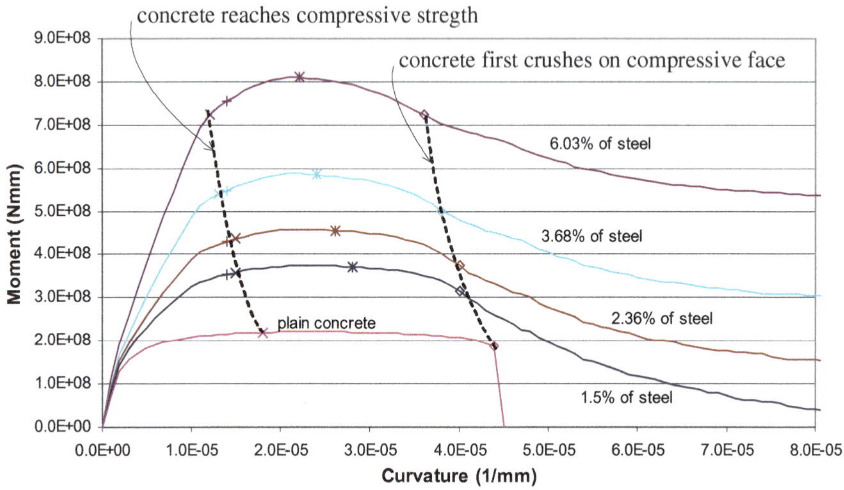

Fig. 3.19 Moment–curvature curves of RC sections with $N = 20\% N_c$.
From Wu, Y. F., D. J. Oehlers, and M. C. Griffith. 2004b. "Rational definition of the flexural deformation capacity of RC column sections." *Eng. Struct.* 26 (5), 641–650.

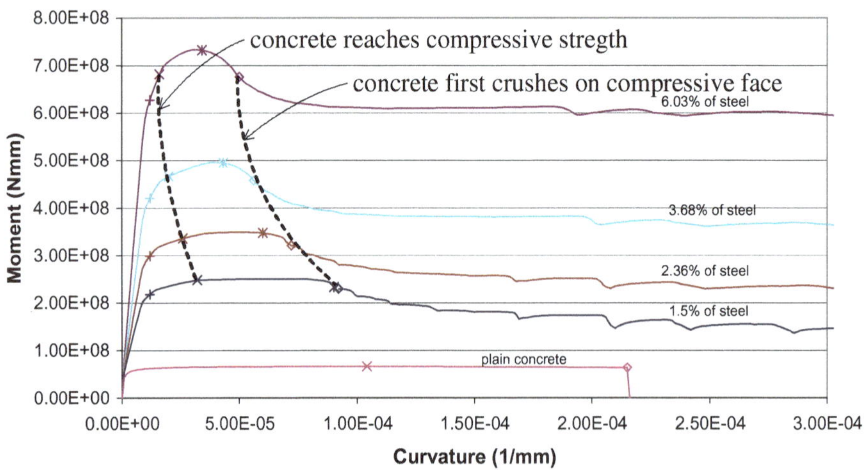

Fig. 3.20 Moment–curvature curves of RC sections with $N = 5\% N_c$.
From Wu, Y. F., D. J. Oehlers, and M. C. Griffith. 2004b. "Rational definition of the flexural deformation capacity of RC column sections." *Eng. Struct.* 26 (5), 641–650.

For convenience of study, all the points ◇ from the response curves are summarized in Fig. 3.21 in which more axial load levels are calculated and illustrated. It was previously concluded theoretically that the ultimate curvature κ_{cu} decreases when the steel content increases for axial loads less than the critical axial load and the result is the opposite when axial loads are higher than the critical load. The critical axial load is given by Eq. (3.27) as $N = 0.5\lambda \cdot N_c = 0.5 \times 0.59 N_c = 0.295 N_c$, where $\lambda = 0.59$

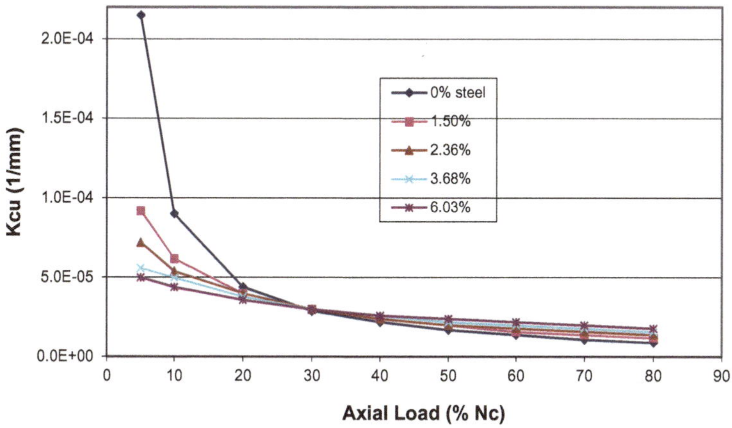

Fig. 3.21 Axial load-κ_{cu} relations.
From Wu, Y. F., D. J. Oehlers, and M. C. Griffith. 2004b. "Rational definition of the flexural deformation capacity of RC column sections." *Eng. Struct.* 26 (5), 641–650.

from Fig. 3.16 for a concrete strength of 44 MPa. The theoretical results are confirmed by the numerical results in Fig. 3.21 where it can be seen that the reinforcement content does not affect the ultimate curvature at the axial load level of about $30\%N_c$. In other words, the numerically calculated critical axial load is about $30\%N_c$, which is the same as the theoretical critical load given by Eq. (3.27).

It can be seen from Figs. 3.17–3.20 that the definition of yielding point exhibits some complexities. From Fig. 3.20 for the low axial load of $5\%N_c$ it is apparent that the point + (onset of yielding of tension bars) gives a better indication of the yielding for columns with a low axial load. However, it is more reasonable to define the yield strength at the point × (onset of peak concrete strain ε_{co} at the compression face) for columns with a higher axial load, as shown in Figs. 3.17–3.19 where the axial load varies from 20% to $40\%N_c$. In fact, it was found that the point + (yielding of tension bars) does not occur for any of the columns for the axial load levels of $50\%N_c$ and above. Clearly, the yielding of the response curve must be defined at the onset of concrete "yielding" for columns with a significant axial load in which concrete softening and crushing dominate the softening of the response curve. The two different cases of yielding of the response curve, i.e., due to the tensile yielding of reinforcement and due to the compressive yielding of concrete, reflect the two different softening mechanisms that are governed by the axial load level. Therefore, the yield plateau shall be defined from the point + to the point ◇ for RC columns with insignificant axial load or beams, and be defined from the point × to the point ◇ for columns with a significant axial load level.

3.5 Derivation of the theorems

In Fig. 3.1, the compression force of concrete F_c is given by

$$F_c = \alpha f_{co} bc = A_{st} f_{st} - A_{sc} f_{sc} + N = \beta_1 c f_{cm} b \tag{3.28}$$

where c is the depth of the compression zone; b is the breadth of the section; f_{co} is the strength of concrete; α is the ratio of average compressive stress of the whole compression zone to f_{co}; A_{st} and f_{st} are the area and stress of the tension reinforcement, respectively; A_{sc} and f_{sc} are the area and stress of the compression reinforcement, respectively; N is the axial force assumed to be constant during flexural bending; and f_{cm} is the stress of the equivalent rectangular stress block. f_{st} is positive when it is in tension and f_{sc} and N are positive in compression. Integration of the concrete stress block gives:

$$F_c = b\int_0^c f\left(\frac{x\varepsilon_c}{c}\right)dx \tag{3.29}$$

where $f(\varepsilon)$ gives the stress–strain relationship of concrete; x is the distance from a compression fiber to the neutral axis (Fig. 3.1); and ε_c is the concrete strain at the extreme compression fiber (referred to as extreme concrete strain hereafter). The moment of the RC section is given by:

$$\begin{aligned} M &= F_c\left(\frac{h}{2}-\beta c\right)+A_{st}f_{st}\left(d-\frac{h}{2}\right)+A_{sc}f_{sc}\left(\frac{h}{2}-d'\right) \\ &= N\frac{h}{2}+A_{st}f_{st}d-A_{sc}f_{sc}d'-\frac{\beta}{\alpha}\frac{(A_{st}f_{st}-A_{sc}f_{sc}+N)^2}{f_{co}b} \end{aligned} \tag{3.30}$$

where h is the depth of the section, d the effective depth, d' thickness of the concrete cover from the compression face to the centroid of the compression reinforcement; and β^*c gives the distance from the compression centroid to the extreme compression fiber (Fig. 3.1).

On introducing the following coordinate transformation:

$$\xi = \frac{x\varepsilon_c}{c} \tag{3.31}$$

Eq. (3.29) becomes

$$F_c = \frac{bc}{\varepsilon_c}\int_0^{\varepsilon_c} f(\xi)d\xi \tag{3.32}$$

It can be seen from Eq. (3.32) that the introduction of Eq. (3.31) changes the integration of the stress block in Fig. 3.1 (involving both ε_c and the compression zone depth c in the integration of Eq. (3.29)) into the integration of the stress–strain curve of concrete. It is seen later that this coordinate transformation converts the study of RC sections into the analysis of the stress–strain curve of concrete.

Substituting Eq. (3.28) into Eq. (3.32) gives:

$$\alpha = \frac{1}{f_{co}\varepsilon_c}\int_0^{\varepsilon_c} f(\xi)d\xi \tag{3.33}$$

The position of the resultant compression can be determined from:

$$F_c \cdot (c - \beta c) = b\int_0^c f\left(\frac{x\varepsilon_c}{c}\right)xdx = b\int_0^{\varepsilon_c} f(\xi)\frac{\xi c}{\varepsilon_c}d\frac{\xi c}{\varepsilon_c} = \frac{bc^2}{\varepsilon_c^2}\int_0^{\varepsilon_c} f(\xi)\xi d\xi \tag{3.34}$$

which leads to

$$\beta = 1 - \frac{bc}{F_c\varepsilon_c^2}\int_0^{\varepsilon_c} f(\xi)\xi d\xi = 1 - \frac{\int_0^{\varepsilon_c} f(\xi)\xi d\xi}{\varepsilon_c\int_0^{\varepsilon_c} f(\xi)d\xi} \tag{3.35}$$

Substituting Eqs. (3.33) and (3.35) into Eq. (3.30) gives

$$M = N\frac{h}{2} + A_{st}f_{st}d - A_{sc}f_{sc}d' - \frac{(A_{st}f_{st} - A_{sc}f_{sc} + N)^2}{b}y(\varepsilon_c) \tag{3.36}$$

where $y(\varepsilon_c)$ is a function of ε_c given by:

$$y(\varepsilon_c) = \frac{\varepsilon_c\int_0^{\varepsilon_c} f(\xi)d\xi - \int_0^{\varepsilon_c} f(\xi)\xi d\xi}{\left(\int_0^{\varepsilon_c} f(\xi)d\xi\right)^2} = \frac{\beta}{\alpha f_{co}} \tag{3.37}$$

In Eq. (3.36), f_{st} and f_{sc} increase when the curvature or ε_c increases. Therefore, Eq. (3.36) is a function of ε_c, or $M = M(\varepsilon_c)$. By taking the derivative of Eq. (3.36) with respect to ε_c, the extreme concrete strain at onset of the flexural strength, ε_{cm}, namely the flexural design strain, can be obtained by:

$$\varepsilon_{cm} = \varepsilon_c \left| \frac{dM}{d\varepsilon_c} = 0 \right. \tag{3.38}$$

3.5.1 Underreinforced sections

The compression reinforcement is assumed to have yielded before the onset of flexural strength in all cases, unless otherwise noted. Elastic compression reinforcement is examined later. It can be seen from Eq. (3.36) that for underreinforced sections with the reinforcement yielded before the onset of the flexural strength, i.e., $f_{st} = f_{sc} = f_y$, M reaches a maximum when $y(\varepsilon_c)$ reaches its minimum. In this case, the flexural design strain for underreinforced sections, $\varepsilon_{cm,u}$, is given by:

$$\varepsilon_{cm,u} = \varepsilon_c \left| \frac{dy(\varepsilon_c)}{d\varepsilon_c} = 0 \right. \tag{3.39}$$

Eqs. (3.36) and (3.39) show that $\varepsilon_{cm,u}$ is governed by the stress–strain relationship of concrete only.

Taking the derivative of Eq. (3.37) with respect to ε_c to get $\varepsilon_{cm,u}$ and substituting it back into Eq. (3.37) gives:

$$y(\varepsilon_{cm,u}) = \frac{1}{2f(\varepsilon_{cm,u})} \tag{3.40}$$

Substituting Eq. (3.40) into Eq. (3.36) and comparing Eq. (3.36) with Eq. (3.30) leads to:

$$\frac{\beta}{\alpha} = \frac{f_{co}}{2f(\varepsilon_{cm,u})} \tag{3.41}$$

Substituting Eq. (3.41) into Eq. (3.28) gives:

$$F_c = 2\beta c \cdot f(\varepsilon_{cm,u}) \cdot b \tag{3.42}$$

F_c in Eq. (3.42) can be interpreted as a force calculated from a rectangular stress block with stress f_{cm} equal to $f(\varepsilon_{cm,u})$ and a depth equal to $2\beta c$ (Fig. 3.1). As the rectangular stress block has a resultant force that locates at half of its depth or βc from the extreme compression fiber, the rectangular stress block is equivalent to the actual nonlinear stress block not only in its magnitude but also in the location of the resultant. Interpretation of Eq. (3.42) leads to Theorem 3.1 as given in Section 3.2.1.

Substituting Eqs. (3.32) and (3.35) into Eq. (3.42) gives:

$$\frac{\int_0^{\varepsilon_{cm,u}} f(\xi)d\xi}{2f(\varepsilon_{cm,u})} + \frac{\int_0^{\varepsilon_{cm,u}} f(\xi)\xi d\xi}{\int_0^{\varepsilon_{cm,u}} f(\xi)d\xi} = \varepsilon_{cm,u} \tag{3.43}$$

3.5.2 Maximum reinforcement ratio for $\varepsilon_{cm} = \varepsilon_{cm,u}$

The derivation of Eq. (3.43) is based on the assumption that all reinforcements yield before the onset of the flexural strength. When tension reinforcement is above a certain ratio, it may not yield before the peak moment. Therefore, the maximum reinforcement ratio allowed by Eq. (3.43) can be derived with the condition that at the onset of $\varepsilon_{cm,u}$ the tension bar just starts to yield. Based on Fig. 3.1, one has:

$$\frac{\varepsilon_{st}}{d-c} = \frac{\varepsilon_c}{c} \tag{3.44}$$

where ε_{st} is the strain of the tensile reinforcement. Substituting $\varepsilon_c = \varepsilon_{cm,u}$ and $\varepsilon_{st} = \varepsilon_y$ into Eq. (3.44) gives:

$$c=\frac{d\cdot\varepsilon_{cm,u}}{\varepsilon_y+\varepsilon_{cm,u}} \tag{3.45}$$

Substituting Eq. (3.45) into Eq. (3.28) leads to:

$$\rho_m=\frac{A_{st}}{bd}=\frac{\alpha_m f_{co}\varepsilon_{cm,u}}{f_y\left(\varepsilon_y+\varepsilon_{cm,u}\right)}-\frac{1}{f_y}\frac{N}{bd}+\rho_{sc} \tag{3.46}$$

where ρ_m is the maximum tension reinforcement ratio allowed by Eq. (3.43); $\rho_{sc}=A_{sc}/(bd)$, the compression reinforcement ratio; and α_m is the value of α at $\varepsilon_c=\varepsilon_{cm,u}$ given by Eq. (3.33) (Fig. 3.2D). It is noted that the compression reinforcement is assumed to have yielded in Eq. (3.46).

3.5.3 Elastic solutions

When the tension reinforcement is greater than a certain value, it does not yield throughout the bending process. The minimum reinforcement ratio for this case can be calculated by assuming that the tension reinforcement is elastic, calculating the maximum stress of the reinforcement during bending, and then letting the maximum stress equal the yield stress. However, this derivation process is tedious and cannot lead to a simple analytical solution. The following process is used to derive the minimum elastic reinforcement ratio.

Substituting Eq. (3.44) into Eq. (3.32) and letting $\varepsilon_{st}=\varepsilon_y$ gives $F_c=\frac{bd}{\left(\varepsilon_y+\varepsilon_c\right)}\int_0^{\varepsilon_c}f(\xi)d\xi$. Letting $\frac{dF_c}{d\varepsilon_c}=0$ gives:

$$f\left(\varepsilon_{cy}\right)-\frac{A\left(\varepsilon_{cy}\right)}{\left(\varepsilon_y+\varepsilon_{cy}\right)}=0 \tag{3.47}$$

where

$$A\left(\varepsilon_{cy}\right)=\int_0^{\varepsilon_{cy}}f(\xi)d\xi \tag{3.48}$$

and ε_{cy} is the concrete strain at $\frac{dF_c}{d\varepsilon_c}=0$. Eq. (3.47) gives the value of ε_c at which F_c is maximized for all strain distributions that pass the yield point of the tension reinforcement. This can be interpreted with Fig. 3.1: when the reinforcement strain ε_{st} is fixed at ε_y, the variation of ε_c at the extreme compression fiber changes the concrete compression force F_c, and the solution of Eq. (3.47) corresponds to the one that gives the maximum compressive resistance of concrete, which corresponds to the maximum steel area $A_{st,e}$ that can be loaded to yield. If the reinforcement area is greater than $A_{st,e}$, the tension reinforcement will not be able to yield any time during bending. Replacing $\varepsilon_{cm,u}$ with ε_{cy} in Eq. (3.46), the minimum elastic reinforcement ratio is given by:

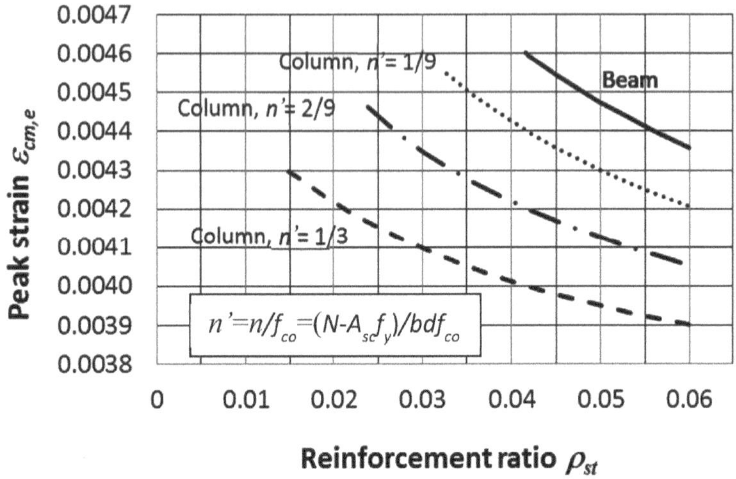

Fig. 3.22 Flexural design strain for sections with elastic tension reinforcement. From Wu, Y. F. 2016. "Theorems for flexural design of RC members." *J. Struct. Eng.* 142 (5), 04015172.

$$\rho_e = \frac{A(\varepsilon_{cy})}{f_y(\varepsilon_y + \varepsilon_{cy})} - \frac{1}{f_y}\frac{N}{bd} + \rho_{sc} \tag{3.49}$$

The solution of ε_{cy} in Eq. (3.47) is only a function of the concrete stress–strain relationship and the steel yield strain.

When the tension reinforcement is elastic, the flexural strength can be derived by taking the derivative of Eq. (3.36) with respect to ε_c and letting it equal zero. Detailed derivations are given in Section 3.5.8 and the solution for the extreme concrete strain at the peak moment, $\varepsilon_{cm,e}$, is given by Eq. (3.63), which is a function of the net axial stress $n = (N - A_{sc}f_y)/bd$ and the tension bar rigidity ratio $R_{st} = A_{st}E_s/bd$. The numerical solutions of Eq. (3.63) for RC beams and a few cases of columns are as shown in Fig. 3.22.

3.5.4 Transition stage

When $\rho_{st} = \rho_m$, the tension reinforcement yields at $\varepsilon_c = \varepsilon_{cm,u}$. When $\rho_{st} = \rho_e$, the tension reinforcement yields at $\varepsilon_c = \varepsilon_{cy}$. Therefore, when $\rho_m < \rho_{st} < \rho_e$ the tension reinforcement will yield at a value in between, or $\varepsilon_{cm,u} < \varepsilon_c < \varepsilon_{cy}$. Assume the tension reinforcement yields at $\varepsilon_c = \varepsilon_{cp}$. It can be concluded from Eq. (3.37) that when $\varepsilon_c > \varepsilon_{cp}$, the moment resistance drops when ε_c further increases because $y(\varepsilon_c)$ increases after it exceeds the minimum value at $\varepsilon_{cm,u}$ (Fig. 3.2C). Therefore, the maximum moment can only occur at or before the tension reinforcement yields.

At first yield of the tension reinforcement, the compression zone depth can be obtained from Eq. (3.44):

$$c = \frac{d\varepsilon_{cp}}{(\varepsilon_y + \varepsilon_{cp})} \tag{3.50}$$

Substituting Eq. (3.50) and Eq. (3.33) into Eq. (3.28) gives

$$\left(\varepsilon_y+\varepsilon_{cp}\right)\sigma_{sn}-\int_0^{\varepsilon_{cp}} f(\xi)d\xi=0 \tag{3.51}$$

where

$$\sigma_{sn}=\frac{A_{st}f_y-A_{sc}f_y+N}{bd} \tag{3.52}$$

The solution of ε_{cp} in Eq. (3.51) gives the flexural design strain in the transition stage when the maximum moment occurs at the yielding of the tension bars. The bending moment can be determined by letting $\varepsilon_c=\varepsilon_{cp}$ in Eq. (3.61).

Since it is possible that the peak moment occurs before the tension reinforcement yields, it is necessary to use Eq. (3.63) to find the concrete strain at the onset of the peak moment for the elastic case. If the result from Eq. (3.63) satisfies $\varepsilon_{cm,e}<\varepsilon_{cp}$, the maximum moment occurs in the elastic stage and hence is given by substituting $\varepsilon_{cm,e}$ into Eq. (3.61). Therefore, the flexural design strain in the transition stage $\varepsilon_{cm,t}$ is given by:

$$\varepsilon_{cm,t}=\min\left(\varepsilon_{cp},\varepsilon_{cm,e}\right) \tag{3.53}$$

The reinforcement ratio at point C and C′ in Fig. 3.4, ρ_c, can be determined by letting Eq. (3.51) = Eq. (3.63).

3.5.5 RC sections with elastic compression bars

In previous derivations, the compression reinforcement is assumed to yield before the onset of the flexural strength. If the compression reinforcement is still elastic at the onset of the peak moment, the tension reinforcement must have already yielded. Otherwise, the increase in curvature will cause continuous increase of the forces in tension and compression reinforcements and hence the bending moment. Derivation of the flexural design for this case is given in Section 3.5.8. The flexural design strain for this case is identified as $\varepsilon_{cm,c}$ and is given by Eq. (3.70).

Eq. (3.70) is derived based on the assumption that the compression reinforcement is elastic. At the onset of yielding of the compression reinforcement, one has $c=d'\,\varepsilon_c/(\varepsilon_c-\varepsilon_y)$. Substituting it into Eq. (3.28) and considering Eq. (3.33) leads to:

$$\frac{(A_{st}-A_{sc})f_y+N}{bd'}-\frac{\int_0^{\varepsilon_c} f(\xi)d\xi}{\varepsilon_c-\varepsilon_y}=0 \tag{3.54}$$

Eq. (3.54) gives the extreme concrete strain at the onset of yielding of the compression reinforcement. If a solution of ε_c cannot be found from Eq. (3.54), it means that the compression reinforcement does not yield, and Eq. (3.70) is applicable. If the solution of ε_c from Eq. (3.54) is larger than that from Eq. (3.70), the peak moment occurs before the compression reinforcement yields and Eq. (3.70) is still applicable. Otherwise, the compression bars would have yielded before the onset of the peak moment and the theories developed in the previous sections are applicable.

3.5.6 Equivalent reinforcement ratio

The solution for columns at the transitional stage ($\rho_m < \rho_{st} < \rho_e$) in Fig. 3.4 can be obtained simply by shifting the solution for beams (line BC) horizontally using Eq. (3.52). Because the solution to Eq. (3.51) only depends on the value of σ_{sn} as given by Eq. (3.52), the same solution applies to beams as well as columns when $\sigma_{sn} = (\rho_{st} - \rho_{sc})f_y + N/(bd) = \text{constant} = \rho' f_y$. Therefore, line BC is in fact the solution for ρ'. Therefore, for the general case of columns with both tension and compression reinforcements, line BC can be directly used, but for an equivalent reinforcement ratio ρ' calculated by:

$$\rho' = \rho_{st} - \rho_{sc} + \frac{N}{f_y bd} \tag{3.55}$$

3.5.7 Ultimate curvature

Substituting Eq. (3.44) into Eq. (3.32), one gets:

$$F_c = \frac{bd}{(\varepsilon_{st} + \varepsilon_c)}\int_0^{\varepsilon_c} f(\xi)d\xi = \frac{b}{\kappa}\int_0^{\varepsilon_c} f(\xi)d\xi = \frac{b}{\kappa}A(\varepsilon_c) \tag{3.56}$$

where κ is the curvature of the section. Substituting Eq. (3.56) into Eq. (3.28) gives:

$$A_{st}f_{st} - A_{sc}f_{sc} + N = \frac{b}{\kappa}A(\varepsilon_c) \tag{3.57}$$

where $A(\varepsilon_c) = \int_0^{\varepsilon_c} f(\xi)d\xi$, the area enclosed by the stress–strain curve up to the strain value of ε_c. Assume concrete strength drops to zero when ε_c reaches ε_{cu}, that is $f(\xi) = 0$ when $\xi \geq \varepsilon_{cu}$ (Fig. 3.2A). In this case, the enclosure area of the stress–strain curve $A(\varepsilon_c)$ reaches its maximum at $\varepsilon_c = \varepsilon_{cu}$. When ε_c further increases beyond ε_{cu}, $A(\varepsilon_c)$ remains constant.

By defining the curvature of the section as κ_{cu} at $\varepsilon_c = \varepsilon_{cu}$, one has $A(\varepsilon_c) = \text{constant}$ when $\kappa \geq \kappa_{cu}$. For the special case of plain concrete columns with $A_{st} = A_{sc} = 0$ and $N = \text{constant} \neq 0$, Eq. (3.57) becomes $N = b \cdot A(\varepsilon_c)/\kappa$. After κ increases to κ_{cu} and beyond, the equilibrium of Eq. (3.57) cannot be sustained. This is because N on the left-hand side remains constant while the right-hand side reduces when the curvature increases beyond $\kappa = \kappa_{cu}$. A sudden collapse of the concrete section therefore occurs at $\kappa = \kappa_{cu}$ (Fig. 3.8).

Due to the slower degradation of concrete near $f_c = 0$ (see the dashed line in Fig. 3.2A), the concrete strain at $f_c = 0$ can be very large. However, this does not mean that the collapse of the section occurs at a large curvature. When the remaining concrete stress f_c is small at the extreme fiber, further increase in $A(\varepsilon_c)$ is very limited (the small area under the dashed line in Fig. 3.2A) at further increase of ε_c. To satisfy the equation $N = b \cdot A(\varepsilon_c)/\kappa$, possible increase of κ is therefore very small, because a small increase in κ will exhaust the remaining area of $A(\varepsilon_c)$. In other words, the moment resistance of the section drops steeply, although not vertically, when the concrete stress at the extreme fiber reaches a stress close to zero.

For RC sections with both tension and compression bars, the left-hand side of Eq. (3.57), $A_{st}f_{st} - A_{sc}f_{sc} + N$, must reduce together with the right-hand side when the curvature increases beyond $\kappa = \kappa_{cu}$. With a constant axial load, $A_{st}f_{st} - A_{sc}f_{sc}$ must reduce after further increase of the curvature. Since $A_{st}f_{st} - A_{sc}f_{sc}$ is the net tension force, the only possible scenario for the net tension force to decrease when curvature increases is shown in Fig. 3.13 by the dashed line where the neutral axis moves toward the tension side when κ increases further. In this case, the increase in κ causes the strain in the tension reinforcement to reduce, and in the meantime, it leads to the crushing of concrete at the top of the compression zone. It can be clearly seen in Fig. 3.13 that this results in the reduction of lever arm of the concrete resultant force.

Overall, the force in tension bars decreases while that in the compression bars remains constant (yielded). The concrete resultant force not only has a decrease in magnitude due to crushing and consequent decreased effective compression area, but also reduces its lever arm. These three effects cause a fast degradation of moment resistance of the section.

Although an RC section does not collapse suddenly at $\kappa = \kappa_{cu}$, it can be seen in Fig. 3.13 that the concrete compression reserve is exhausted in the compression zone, and a qualitative change of the compression resistance from "surplus" to "deficit" occurs after this point. Furthermore, the compression bars can buckle any time after the crushing of concrete at the compression zone. Therefore, the onset of $\varepsilon_c = \varepsilon_{cu}$ can be considered as the last moment of a stable process.

3.5.8 Derivation of cases with elastic bars

3.5.8.1 Flexural design with elastic tension reinforcement

Substituting Eq. (3.44) into Eq. (3.28) gives

$$\alpha f_{co} bc = A_{st}\frac{\varepsilon_c(d-c)}{c}E_s + N - A_{sc}f_y \tag{3.58}$$

where the tension bars are elastic, and the compression reinforcement is assumed to have yielded before the onset of the flexural strength. The solution of Eq. (3.58) is given by

$$c = \frac{N - A_{sc}f_y - A_{st}E_s\varepsilon_c + \sqrt{(A_{st}E_s\varepsilon_c - N + A_{sc}f_y)^2 + 4bA_{st}E_s d\int_0^{\varepsilon_c} f(\xi)d\xi}}{2\dfrac{b}{\varepsilon_c}\displaystyle\int_0^{\varepsilon_c} f(\xi)d\xi} \tag{3.59}$$

Substituting Eq. (3.44) into Eq. (3.36) gives

$$M = N\frac{h}{2} - A_{sc}f_y d' + A_{st}E_s d\left[\varepsilon_c\frac{d}{c} - \varepsilon_c - \frac{A_{st}E_s}{bd}\left(\varepsilon_c\frac{d}{c} - \varepsilon_c + \frac{N - A_{sc}f_y}{A_{st}E_s}\right)^2 y(\varepsilon_c)\right] \tag{3.60}$$

Substituting Eq. (3.59) into Eq. (3.60) leads to

$$M = N\frac{h}{2} - A_{sc}f_y d' + A_{st}E_s d \times \eta\left(\varepsilon_c, \frac{N - A_{sc}f_y}{bd}, \frac{A_{st}E_s}{bd}\right) \tag{3.61}$$

where η is a function of ε_c, the net axial stress $(N - A_{sc}f_y)/bd$, and the tension bar rigidity ratio $R_{st} = A_{st}E_s/(bd)$, given by

$$\eta = \frac{2\int_0^{\varepsilon_c} f(\xi)\mathrm{d}\xi}{\frac{N - A_{sc}f_y}{bd} - \frac{A_{st}E_s}{bd}\varepsilon_c + \sqrt{\left(\frac{A_{st}E_s}{bd}\varepsilon_c - \frac{N - A_{sc}f_y}{bd}\right)^2 + 4\frac{A_{st}E_s}{bd}\int_0^{\varepsilon_c} f(\xi)d\xi}} - \varepsilon_c - \frac{A_{st}E_s}{bd}\cdot y(\varepsilon_c) \times \left(\frac{2\int_0^{\varepsilon_c} f(\xi)d\xi}{\frac{N - A_{sc}f_y}{bd} - \frac{A_{st}E_s}{bd}\varepsilon_c + \sqrt{\left(\frac{A_{st}E_s}{bd}\varepsilon_c - \frac{N - A_{sc}f_y}{bd}\right)^2 + 4\frac{A_{st}E_s}{bd}\int_0^{\varepsilon_c} f(\xi)d\xi}} - \varepsilon_c + \frac{N - A_{sc}f_y}{A_{st}E_s}\right)^2 \tag{3.62}$$

The maximum bending moment is achieved when $\frac{d\eta}{d\varepsilon_c} = 0$, or

$$\frac{d\eta}{d\varepsilon_c}\left(\varepsilon_c, \frac{N - A_{sc}f_y}{bd}, \frac{A_{st}E_s}{bd}\right) = 0 \tag{3.63}$$

The solution of Eq. (3.63) gives the extreme concrete strain $\varepsilon_{cm,e}$ at the onset of the peak moment. The elastic design moment can be obtained by substituting $\varepsilon_{cm,e}$ into Eq. (3.61).

3.5.8.2 *Flexural design with elastic compression reinforcement*

After yielding the tension reinforcement, Eq. (3.36) becomes

$$M = N\frac{h}{2} + A_{st}f_y d - A_{sc}E_s\varepsilon_{sc}d' - \frac{(A_{st}f_y - A_{sc}E_s\varepsilon_{sc} + N)^2}{b}y(\varepsilon_c) \tag{3.64}$$

From the strain profile in Fig. 3.1, one gets $c = d'\,\varepsilon_c/(\varepsilon_c - \varepsilon_{sc})$. Substituting it into Eq. (3.28) and considering Eq. (3.33), the following equation is obtained

$$bd'\int_0^{\varepsilon_c} f(\xi)d\xi = \left(A_{st}f_y + N\right)\varepsilon_c - A_{sc}E_s\varepsilon_{sc}\varepsilon_c - \left(A_{st}f_y + N\right)\varepsilon_{sc} + A_{sc}E_s{\varepsilon_{sc}}^2 \tag{3.65}$$

Solving for ε_{sc} from Eq. (3.65) and substituting it into Eq. (3.64) gives

$$M = N\frac{h}{2} + A_{st}f_y d - b(d')^2 \times \mu(\varepsilon_c, n_1, E_1) \tag{3.66}$$

where μ is a function ε_c with two parameters, n_1 and E_1, given by;

$$\mu = \frac{n_1 + E_1\varepsilon_c}{2} - \frac{1}{2}\sqrt{(n_1 + E_1\varepsilon_c)^2 - 4E_1\left(n_1\varepsilon_c - \int_0^{\varepsilon_c} f(\xi)d\xi\right)} + y(\varepsilon_c)\left(\frac{n_1 - E_1\varepsilon_c}{2} + \frac{1}{2}\sqrt{(n_1 + E_1\varepsilon_c)^2 - 4E_1\left(n_1\varepsilon_c - \int_0^{\varepsilon_c} f(\xi)d\xi\right)}\right)^2 \tag{3.67}$$

$$n_1 = \frac{A_{st}f_y + N}{bd'} \tag{3.68}$$

$$E_1 = \frac{A_{sc}E_s}{bd'} \tag{3.69}$$

The flexural design strain $\varepsilon_{cm,c}$ at the onset of peak moment can be obtained from

$$\frac{d\mu}{d\varepsilon_c}(\varepsilon_c, n_1, E_1) = 0 \tag{3.70}$$

3.6 Alternative flexural design procedure

When the value of $\varepsilon_{cm,u}$ is determined from Eq. (3.43), the corresponding β value at $\varepsilon_c = \varepsilon_{cm,u}$, namely, β_m, can be calculated from Eq. (3.35). Design charts of $\varepsilon_{cm,u}$ (Fig. 3.2B) and β_m (Fig. 3.2E), as functions of concrete strength, can be generated for design purposes. Therefore, RC members with underreinforced sections can be designed, using the equivalent rectangular stress block that has the stress $f_{cm} = f(\varepsilon_{cm,u})$ and the depth $2\beta_m c$. Compared with the current design procedure that needs to specify flexural design strain ε_{cm} and strength f_{cm} as well as the depth of the equivalent stress block $\beta_1 c$ (Fig. 3.1), the only required information in this procedure is the stress–strain relationship of concrete, $f_c = f(\varepsilon_c)$, which is available in the extant literature.

The stress–strain relationships proposed by Popovics (1973) and CEB-FIP/Eurocode 2 (CEN 2004), both widely adopted in design practices and research work, are used here to produce the design charts and demonstrate the alternative design procedure in this section. Details of the two stress–strain models are given in Section 3.2.1. Using the two stress–strain models (Eqs. 3.3 and 3.8), the flexural design strain $\varepsilon_{cm,u}$ is numerically calculated and depicted in Fig. 3.2B. It can be seen from the figure that $\varepsilon_{cm,u}$ is related to the stress–strain model and is particularly sensitive to the concrete peak strain ε_{co}. For normal concrete, the peak strain in the Popovics model (Eq. 3.6) is a function of concrete strength that is more accurate than the constant ε_{co} in the EC2 model (Eq. 3.10). If Eq. (3.6) is used to replace Eq. (3.10) of EC2, the results from the two models are close, as shown in Fig. 3.2B. Saliger (1950) indicated that $\varepsilon_{cm,u}$ is proportional to concrete strength, while Jensen (1943) and Hognestad et al. (1955) suggested an opposite trend. It can be clearly seen in Fig. 3.2B that it depends on the concrete stress–strain relationship. Although the value of $\varepsilon_{cm,u}$ is significantly different in the literature and design codes, it is also well understood that the flexural strength of underreinforced RC members is insensitive to the value of $\varepsilon_{cm,u}$.

Using the Popovics model, the values of β_m and f_{cm} are calculated and shown in Fig. 3.2E and F, respectively. These charts in Fig. 3.2 can be conveniently used for routine design to avoid direct calculations of the stress block parameters using the complex equations.

3.6.1 Comparison with ACI 318

For normal strength concrete (30–70 MPa), the value of $\varepsilon_{cm,u}$ in Fig. 3.2B is in the range of 0.003 to 0.0035, which is consistent with the values in existing design codes (ACI 318 = 0.003, EC2 = 0.0035 for concrete class $\leq$ C56/60). In ACI 318, the equivalent rectangular stress block has a stress of $f_{cm} = 0.85 f_{co}$ and a depth of $\beta_1 \times c = 2\beta_m \times c$. The stress block parameters β_1 and f_{cm}/f_{co} in ACI 318 are compared with those from the analytical solutions in Fig. 3.2E and F, respectively. Using the equivalent rectangular stress block in Fig. 3.1, the following relationships can be derived for singly reinforced RC beams:

$$\frac{F_c}{bcf_{co}} = \beta_1 \frac{f_{cm}}{f_{co}} \tag{3.71}$$

$$\frac{M}{A_{st} f_y d} = 1 - \frac{f_{co}}{2 f_{cm}} q_t \tag{3.72}$$

where $q_t = \frac{A_{st} f_y}{bdf_{co}}$. For stress blocks with a certain c value, Eq. (3.71) gives the effect of stress block parameters on the resultant concrete force. Eq. (3.72) gives the effect of stress block parameters on the flexural strength of RC sections with a certain reinforcement ratio. Eq. (3.72) shows that for beams with a certain reinforcement ratio, the flexural strength is determined by the stress of the equivalent rectangular stress block f_{cm}/f_{co} only. The depth of the stress block, controlled by β_1, does not affect the bending moment. This is because the balance of forces in the cross-section requires $\beta_1 c = A_{st}f_y/(bf_{cm})$. Therefore, the depth of the rectangular stress block is fixed for a certain f_{cm} value. Variation of β_1 changes only the value of c, not the depth of the stress block $\beta_1 c$.

The results of Eq. (3.71) in Fig. 3.2E and F show that for stress blocks with a certain c value, the difference in the stress blocks between ACI 318 and the Popovics model is significant. In other words, the shapes of the two equivalent stress blocks are significantly different. For the same c value, the ACI 318 stress block generally has a smaller concrete stress (Fig. 3.2F) but a larger lever arm that is in reverse proportion to β_1 (Fig. 3.2E). However, the moments calculated from the two methods are very similar, as shown in Fig. 3.2G. M_{ACI} and $M_{Pop.}$ in Fig. 3.2G, the flexural strength from the ACI 318 and the Popovics model, respectively, are calculated using Eq. (3.72) with their own stress block parameters. The difference between the two for $q_t = 0.15$ and 0.3 (corresponding to reinforcement ratios of about 1% and 2%, respectively, for $f_{co} = 30$ MPa and $f_y = 500$ MPa) is about 1% and 2%, respectively, which is negligible.

The stress block parameters in ACI 318 were back calculated from a comprehensive test database of experimentally measured bending moments (Hognestad et al. 1955). Therefore, ACI 318 can reflect flexural strength accurately. However, the stress-block parameters are insensitive to the flexural strength, which means the ACI stress block parameters indirectly calculated from bending moment resistance could be less accurate than those obtained from the direct mathematical integration of the stress block, provided the stress–strain relationship of concrete is reasonably accurate.

The consequence of underestimating f_{cm} and β_1 is an overestimation in the depth of the concrete compression zone, as $c = A_{st}f_y/(bf_{cm}\beta_1)$. For a certain value of $\varepsilon_{cm,u}$, the tensile strain of reinforcement gets underestimated in this case; however, it does not affect the tensile force as the reinforcement yields in underreinforced members.

3.6.2 Design procedure

To apply Theorem 3.2 for RC flexural design, the following procedures can be adopted. It is noted that the safety factor or strength reduction factor are omitted in the following procedure, which should be allowed in practical designs.

1. To calculate the flexural strength of a given RC section
   (a) Determine ε_{cm} from the chart in Fig. 3.4 based on the tension reinforcement ratio ρ_{st}.
   (b) If $\rho_{st} \le \rho_m$, determine α_m and β_m from the design charts in Fig. 3.2D and E, calculate the compression zone depth c from the balance of forces in Fig. 3.1 or using Eq. (3.28) with $f_{st} = f_{sc} = f_y$. With the calculated c and the strain profile (Fig. 3.1) check whether $\varepsilon_{sc} > \varepsilon_y$; if not, let $f_{sc} = E_s\varepsilon_{sc}$ and recalculate c. Calculate the design moment from the stress block in Fig. 3.1 or using Eq. (3.30).
   (c) If $\rho_{st} > \rho_m$, determine α and β from the charts in Fig. 3.2H based on $\varepsilon_c = \varepsilon_{cm}$, calculate c from cross-sectional force balance or using Eq. (3.28) with $f_{sc} = f_y$, and $f_{st} = f_y$ if $\rho_{st} \le \rho_c$ or $f_{st} = E_s\varepsilon_{st}$ if $\rho_{st} > \rho_c$. Check whether the compression bars are yielded and recalculate c using an elastic relationship of the compression bar if not. Calculate the design moment from the stress block in Fig. 3.1 or using Eq. (3.30).

2. To design an RC section based on a given bending moment
   (a) Assume an underreinforced section and $\varepsilon_{cm} = \varepsilon_{cm,u}$. Determine α_m and β_m from the design charts in Fig. 3.2D and E and calculate $f_{cm} = \alpha_m f_{co}/2\beta_m$, or directly from the stress–strain relationship $f_{cm} = f(\varepsilon_{cm,u})$. Let $f_{sc} = f_{st} = f_y$.
   (b) Use $\sum F = 0$ and $\sum M = 0$ (Eqs. 3.28 and 3.30) to solve the unknown c and A_{st}.
   (c) If $\rho_{st} \le \rho_m$, check $\varepsilon_{sc} > \varepsilon_y$; if yes, design is correct and finished; if not, let $f_{sc} = E_s\varepsilon_{sc}$ and go to Step b.
   (d) If $\rho_{st} > \rho_m$, select ε_{cm} from the design chart in Fig. 3.4 based on the calculated ρ_{st} from Step b. Determine α and β from the design chart in Fig. 3.2H based on ε_{cm}. Let $f_{sc} = f_{st} = f_y$ if $\rho_{st} \le \rho_c$ or $f_{st} = E_s\varepsilon_{st}$ and $f_{sc} = f_y$ if $\rho_{st} > \rho_c$. Go to Step b and recalculate A_{st}. Repeat Step d until A_{st} converges. Two iterations should be sufficient for engineering design purposes.

Using the preceding procedure, no *M-N* interaction chart is needed for RC column design.

Example A
Calculate the flexural strength of the RC beams in Fig. 2.5 with the reinforcement ratios of (1) $\rho_{st} = 1.99\%$, and (2) $\rho_{st} = 4.96\%$.
Case 1 solution:

$\rho_{st} < \rho_m$, the tension reinforcement would have yielded at the peak moment. From Fig. 3.4, $\varepsilon_{cm} = 0.00318$.

Using Fig. 3.2D and E, one gets $\alpha_m = 0.75$ and $\beta_m = 0.398$, respectively.

Solving Eq. (3.28) gives $c = A_{st}f_{st}/\alpha f_{co}b = \rho_{st}df_{st}/\alpha f_{co} =$ 0.0199 × 540 × 500/(0.75 × 40) = 178.9 mm.

From Eq. (3.30), $M = A_{st}f_{st}d - \frac{\beta}{\alpha}\frac{(A_{st}f_{st})^2}{f_{co}b} = \rho_{st}bd^2f_{st}\left(1 - \frac{\beta}{\alpha f_{co}}\rho_{st}f_{st}\right) = 0.0199 \times 300 \times 540^2 \times 500 \times \left(1 - \frac{0.398}{0.75 \times 40} \times 0.0199 \times 500\right) = 754.8\,\text{kNm}$, which is identical to the numerical solution of the peak moment in Fig. 2.5.

It is noted that the design charts are provided in spreadsheets and, therefore, precise values of the parameters are used in these calculations.

Case 2 solution:

$\rho_{st} > \rho_m$, Fig. 3.4 gives $\varepsilon_{cm} = 0.00448$.

From Fig. 3.2H, $\alpha = 0.781$ and $\beta = 0.438$.

Eq. (3.28) gives $\alpha f_{co}bc = A_{st}E_s\varepsilon_{st} = A_{st}E_s\frac{d-c}{c}\varepsilon_{cm}$, from which $c = 365.9$ mm.

Using Eq. (3.30), $M = 1301.3$ kNm, which is identical to the numerical result of the peak moment in Fig. 2.5.

Example B
Design a symmetrically reinforced RC column ($A_{st} = A_{sc}$) with $N/bdf_{co} = 0.45$ and $M = 885$ kNm. The geometry and material properties of the column are the same as those in Example A.
Solution:

From Eq. (3.55) $\rho' = \rho_{st} - \rho_{sc} + \frac{N}{f_y bd} = 0.036 > \rho_m$.

In Fig. 3.4, treat the horizontal axis as ρ' and read flexural design strain from the line ABC, which gives $\varepsilon_{cm} = 0.00355$.

From Fig. 3.2H, when $\varepsilon_c = 0.00355$, $\alpha = 0.7674$ and $\beta = 0.4098$.

Using Eq. (3.28) or $\alpha f_{co}bc = N$, $c = 316.7$ mm.

Check and find that both the tension and compression bars have yielded by calculating the strain of the steel bars from the strain profile.

From Eq. (3.30) or $M = N\frac{h}{2} + A_{st}f_y(d-d') - \frac{\beta N^2}{\alpha f_{co} b}$,

$A_{st} = 1615\ \text{mm}^2$,
$A_s = 2A_{st} = 3230\ \text{mm}^2$, and
$\rho_s = \rho_{st} + \rho_{sc} = 2\%$.

The results of ρ_s and ε_{cm} exactly match the numerical ones in Fig. 3.11.

If ρ' is beyond point C in Fig. 3.4, the tension reinforcement would not yield at the peak moment and Fig. 3.22 has to be used for evaluation of ε_{cm}. For RC beam design, A_{st} can be calculated first from Eq. (3.30) based on $\varepsilon_{cm} = \varepsilon_{cm,u}$. If the obtained reinforcement ratio ρ_{st} is greater than ρ_m, it is an overreinforced section (which is rare and not allowed in practice) and an iteration procedure can be used to revise ε_{cm} and recalculate A_{st}. It is noted that iterations are required only when the tension reinforcement does not yield, which is uncommon in practice. This approach suits spreadsheet-based design practice.

References

ACI (American Concrete Institute). 2005. *Building code requirements for structural concrete and commentary*, ACI 318-05/ACI 381R-05. Farmington Hills, MI.

Berwanger, C. 1975. *Effect of axial load on the moment-curvature relationship of reinforced concrete members*, ACI SP 50-11. Detroit: American Concrete Institute, 263–288.

CEN (European Committee for Standardization). 2004. *Eurocode 2: Design of concrete structures. Part 1: general rules and rules for buildings*, EN 1992-1-1:2004. Brussels, Belgium: Comité Européen de Normalisation.

Chai, Y. H., M. J. N. Priestley, and F. Seible. 1991. "Seismic retrofit of circular bridge columns for enhanced flexural performance." *ACI Struct. J.* 88 (5), 572–584.

Gomes, A., and J. Appleton. 1997. "Nonlinear cyclic stress-strain relationship of reinforcing bars including buckling." *Eng. Struct.* 9 (10), 822–826.

Hognestad, E., N. W. Hanson, and D. McHenry. 1955. "Concrete stress distribution in ultimate strength design." *ACI J. Proc.* 52 (12), 455–479.

Jensen, V. P. 1943. "The plasticity ratio of concrete and its effect on the ultimate strength of beams." *ACI J. Proc.* 39, 565–582.

Popovics, S. 1973. "A numerical approach to the complete stress-strain curve of concrete." *Cem. Concr. Res.* 3 (5), 583–599.

Saliger, R. 1950. *Die neue Theorie des Stahlbetons auf Grund der Bildsamkeit vor dem Bruch*, 3rd Ed. Vienna, Austria: Franz Deuticke.

Wu, Y. F. 2016. "Theorems for flexural design of RC members." *J. Struct. Eng.* 142 (5), 04015172.

Wu, Y. F., M. C. Griffith, and D. J. Oehlers. 2004a. "Numerical simulation of steel plated RC columns." *Comput. Struct.* 82 (4–5), 359–371.

Wu, Y. F., D. J. Oehlers, and M. C. Griffith. 2004b. "Rational definition of the flexural deformation capacity of RC column sections." *Eng. Struct.* 26 (5), 641–650.

Applications of the flexural theorems

4

This chapter gives examples of application of the RC flexural theorems for derivation of stress block parameters used in the conventional flexural design approach. The conventional stress block parameters defined in Fig. 2.16 are adopted in this chapter.

4.1 Application to design of RC members under elevated temperature

It is difficult to carry out flexural tests of RC members in fire or under elevated temperature. Therefore, the stress block parameters for the design of RC members under elevated temperature are difficult to derive through experimental tests. Numerous stress-strain models of concrete under elevated temperature have been reported. Such a model can be used for derivation of the stress block parameters through the application of the flexural theorems (Yuan and Wu 2019a).

4.1.1 Stress-strain model of concrete under elevated temperature

Elevated temperature during fire has a great effect on the stress-strain relationship of concrete. Extensive research has shown that concrete strength decreases and the corresponding strain at peak stress increases as temperature increases (Purkiss and Li 2013). It has also been reported that preloading reduces the effect of elevated temperature on concrete strength and the corresponding strain (Lie 1992) while it results in additional strain, termed transient creep strain (Khoury et al. 1985). The magnitude of the instantaneous stress-related strain and transient creep strain is closely related to the heating rate, concrete mix, and the preloading stress. The confinement by transverse steel bars on core concrete is affected by the deterioration of the bond between the reinforcing bars and the surrounding concrete and the reduction in the yield stress of transverse reinforcement during fire (Khennane and Baker 1993).

Numerous models have been proposed in the past few decades for the tension and compression stress-strain relationship of unconfined and confined concrete at elevated temperatures (Anderberg and Thelandersson 1976; Li and Purkiss 2005; Youssef and Moftah 2007). In Section 4.1, the stress-strain model proposed by Popovics (Fig. 3.2A, Eq. (3.3)) is adopted for concrete under ambient temperature, and the model proposed by Schneider (1986) is adopted for concrete under elevated temperature, expressed by

Analytical Approaches for Reinforced Concrete. https://doi.org/10.1016/B978-0-12-821164-9.00004-8

$$f_{cT} = \frac{f'_{cT}\varepsilon_{cT}r}{\varepsilon_{oT}\left(r - 1 + \left(\frac{\varepsilon_{cT}}{\varepsilon_{oT}}\right)^r\right)} \tag{4.1}$$

in which

$$r = \frac{E_{cT}}{E_{cT} - \frac{f'_{cT}}{\varepsilon_{oT}}} \tag{4.2}$$

where f_{cT} and ε_{cT} are the concrete compressive stress and strain at elevated temperature, respectively; f'_{cT} and ε_{oT} are the concrete strength at elevated temperature and the strain at peak stress, respectively; E_{cT} is the elastic modulus of concrete at elevated temperature, given by:

$$E_{cT} = 4730\sqrt{f'_{cT}} \tag{4.3}$$

Schneider (1986) recommended an r value of 2.5 and 3 for lightweight and normal-weight concrete, respectively, whereas Terro (1998) and Youssef and Moftah (2007) recommended an r value of 2. Schneider's model (Schneider 1986) in Eq. (4.1) returns to Popovics' model for concrete at ambient temperature.

The model proposed by Hertz (2005) is adopted for f'_{cT}

$$f'_{cT} = f_{co}\left(\frac{1}{1 + \frac{T}{T_1} + \left(\frac{T}{T_2}\right)^2 + \left(\frac{T}{T_8}\right)^8 + \left(\frac{T}{T_{64}}\right)^{64}}\right) \tag{4.4}$$

where f_{co} is the concrete compressive strength at ambient temperature; T is the elevated temperature in degree Celsius ($\geq 20°C$); T_1, T_2, T_8 and T_{64} are constants determining the reduction in the concrete compressive strength for different aggregate types, and $T_1 = 15{,}000$, $T_2 = 800$, $T_8 = 570$, and $T_{64} = 100{,}000$ for siliceous aggregate. ε_{oT} is given by (Terro 1998):

$$\varepsilon_{oT} = 2.05 \times 10^{-3} + 3.08 \times 10^{-6}T + 6.17 \times 10^{-9}T^2 + 6.58 \times 10^{-12}T^3 \tag{4.5}$$

4.1.2 Stress-strain relationship of steel bar under elevated temperature

Elevated temperature reduces the yield strength as well as the elastic modulus of steel reinforcement. A perfect elastoplastic model is adopted for deriving the flexural theorems and hence is also assumed for steel reinforcement in this chapter. The AS 4100 model (Standard Australia 1998) is adopted for the elastic modulus of a steel bar at elevated temperature E_{sT}:

$$\frac{E_{sT}}{E_s}=\begin{cases} 1+\dfrac{T}{\left(2000\ \ln\left(\dfrac{T}{1100}\right)\right)} & 0<T\leq 600°\text{C} \\ \dfrac{(690-0.69T)}{(T-53.5)} & 600<T\leq 1000°\text{C} \end{cases} \tag{4.6}$$

where E_s is the elastic modulus of steel at ambient temperature. The yield strength of steel bar at elevated temperature f_{yT} is given by (Lie 1992):

$$f_{yT}=\begin{cases} \left(1+\dfrac{T}{900\ \ln(T/1750)}\right)f_y & 0<T\leq 600°\text{C} \\ \left(\dfrac{340-0.34T}{T-240}\right)f_y & 600<T\leq 1000°\text{C} \end{cases} \tag{4.7}$$

These models have been widely used in the literature for the stress-strain behavior of steel under elevated temperature.

4.1.3 Stress block parameters

Using the flexural theorems, the flexural design strain ε_{cm} and the equivalent rectangular stress block parameters α_1 and β_1 at the onset of the flexural strength of RC members can be calculated. The conventional stress block parameters in Fig. 2.16 and those used for the derivation of the flexural theorems in Fig. 3.1 are related through the following relationships:

$$F_c=\alpha_1 f_{co}\beta_1 cb=\alpha f_{co}bc$$

which gives

$$\alpha_1=\alpha/\beta_1=\frac{\alpha}{2\beta} \tag{4.8}$$

Substituting Eqs. (3.33) and (3.35) into Eq. (4.8) and letting $\varepsilon_c=\varepsilon_{cm}$ gives

$$\alpha_1=\frac{\left[\int_0^{\varepsilon_{cm}} f(\varepsilon)d\varepsilon\right]^2}{2f_{co}\left[\varepsilon_{cm}\int_0^{\varepsilon_{cm}} f(\varepsilon)d\varepsilon-\int_0^{\varepsilon_{cm}} f(\varepsilon)\varepsilon d\varepsilon\right]} \tag{4.9}$$

Based on Fig. 3.1 and Eq. (3.35), one has $\beta_1=2\beta$, or

$$\beta_1 = 2 - \frac{2\int_0^{\varepsilon_{cm}} f(\varepsilon)\varepsilon d\varepsilon}{\varepsilon_{cm}\int_0^{\varepsilon_{cm}} f(\varepsilon) d\varepsilon} \tag{4.10}$$

Using the stress-strain model of concrete under elevated temperature, these stress-block parameters can be calculated. Substituting Eq. (4.1) into Eqs. (3.1), (4.9), and (4.10), the values of ε_{cm}, α_1 and β_1 for different concrete strengths under different temperatures are calculated to form a database. Through regression analysis using the database (Yuan and Wu 2019a), the following relationships are obtained for ε_{cm}, α_1, β_1 and α_m:

$$\varepsilon_{cm} = 0.566 f_c^{0.152}\left(4*10^{-9}T^2 + 0.00001T + 0.0027\right) \tag{4.11}$$

$$\alpha_1 = 1.066 f_c^{-0.014}\left(-6*10^{-8}T^2 + 0.0001T + 0.935\right) \tag{4.12}$$

$$\beta_1 = 1.09 f_c^{-0.025}\left(-2*10^{-7}T^2 + 0.0003T + 0.794\right) \tag{4.13}$$

$$\alpha_m = 1.164 f_c^{-0.039}\left(-2*10^{-7}T^2 + 0.00035T + 0.740\right) \tag{4.14}$$

It is noted that the database used for the regression of these equations includes concrete strength f_{co} in the range of 20 MPa to 60 MPa and temperature T from 20°C to 800°C. Therefore, Eqs. (4.11)–(4.14) are only applicable within these ranges of concrete strength and temperature. It is further noted that Popovics' model and its extended form Eq. (4.1) are only applicable to normal-strength concrete. For high-strength concrete, these stress-strain models are not applicable and other suitable models must be used to re-derive the equivalent stress block parameters.

4.1.4 Design example

The application of the equivalent stress block method for the design of RC members under elevated temperature is illustrated through an example. The design approach and procedure are identical to those of current design practice, except that the stress block parameters are calculated from Eqs. (4.11)–(4.14).

Question

Calculate the flexural strength of an RC beam at the temperature of 400°C. The RC beam has a cross section of b (breadth) $\times$ h (depth) $=$ 300 mm $\times$ 600 mm. The material properties are as follows: concrete compressive strength $f_{co} = 40$ MPa, steel yield strength $f_y = 500$ MPa, steel elastic modulus $E_s = 200$ GPa, steel reinforcement ratio $\rho_{st} = 1.99\%$, concrete cover $=$ 60 mm from the tension face to the center of the tensile reinforcement.

Solution:
The typical stress and strain blocks at the onset of flexural strength of an RC section are shown in Fig. 2.16. The concrete strength and steel material properties at 400°C are given by Eqs. (4.4), (4.6), and (4.7), respectively, which gives

$f'_{cT} = 29.95$ MPa,
$E_{sT} = 160.46$ GPa, and
$f_{yT} = 349.43$ MPa.

The flexural design strain ε_{cm} is calculated using Eq. (4.11):

$\varepsilon_{cm} = 0.007278.$

The stress block parameters are calculated from Eqs. (4.12) and (4.13):

$\alpha_1 = 0.9773$, and
$\beta_1 = 0.8713.$

The balance of the internal force in the cross section gives

$$\alpha_1 f'_{cT} \beta_1 cb = A_{st} f_{yT} \tag{4.15}$$

Substituting all known parameters and material properties into the preceding equation gives:

$c = 147.2$ mm.

From the strain profile in Fig. 2.16, one gets

$$\frac{\varepsilon_{cm}}{c} = \frac{\varepsilon_{st}}{d-c} \tag{4.16}$$

where $d = 540$ mm.
Eq. (4.16) gives:

$\varepsilon_{st} = 0.01942$

The yield strain of steel at 400°C is

$\varepsilon_{yT} = f_{yT}/E_{sT} = 0.002178 < \varepsilon_{st};$

hence the steel has yielded, and it is an underreinforced section.
The flexural strength of the RC section is given by:

$$M_{max} = A_{st} f_{yT} \left(d - \frac{\beta_1 c}{2} \right) \tag{4.17}$$

which gives $M_{max} = 536.1$ kN m.
The check of steel yielding by Eq. (4.16) can be omitted if the critical reinforcement ratio ρ_m at 400°C is calculated using Eq. (3.46), which gives
$\rho_m = 5.8\% > \rho_{st} = 1.99\%$. Hence it is an underreinforced section.

4.1.5 Validation of results

The calculated results with the approach described in the previous section are further compared with experimental results in this section. Flexural tests under elevated temperatures are rare in the literature. Test results of two RC beams and 12 RC columns were collected from Sun et al. (2001, 2002) and Yang et al. (2001). These RC members had the same cross-sectional dimensions of $b \times h = 100\ \text{mm} \times 200\ \text{mm}$ and two longitudinal steel bars of 12-mm diameter at both the tension and compression sides. R4@80 mm c/c steel stirrups were used as shear reinforcement. For half of the specimens (one beam and six columns), the compression face was covered with heat insulation material. As a result, only the tension face and two sides were exposed to fire. For the other half of the members, the tension face was isolated from fire and only the compression face and the two sides were exposed to fire. The midspan region of the specimens was heated with a high-temperature furnace. The specimens were loaded as the furnace was heated up to 800°C. All beams were tested under four-point bending with a pure bending region of 1400 mm. The columns were tested under eccentric load with the column length of 1440 mm.

The temperatures measured at the locations of the tension and compression reinforcement are used as the temperatures of the respective steel bars. The concrete temperature is taken as the average measured temperature of the compression zone. The concrete compressive strength and the steel yield strength at room temperature are 33.5 MPa and 233.9 MPa, respectively.

Fig. 4.1 shows the comparison between the flexural strength from the tests and the calculated ones. The average error, standard deviation, and maximum error are 6.06%, 2.77%, and 11.54%, respectively. Overall, a reasonable agreement is observed between the test results and the calculated results.

4.1.6 Flexural analysis of RC members under elevated temperature

The flexural behavior in terms of flexural strength, flexural design strain, and ultimate curvature of RC beams and columns at elevated temperatures is analyzed in this section using the flexural theorems.

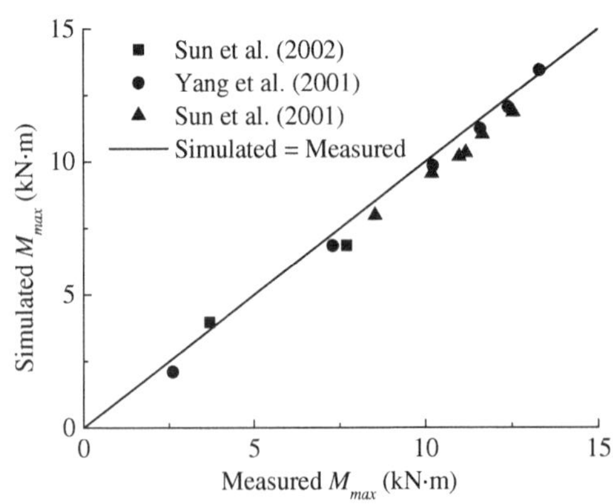

Fig. 4.1 Comparisons between calculated flexural strengths and measured results.
From Yuan, F., and Y. F. Wu. 2019a. "Application of RC flexural theorems for member design under elevated temperature." *Eng. Struct.* 201, 109762.

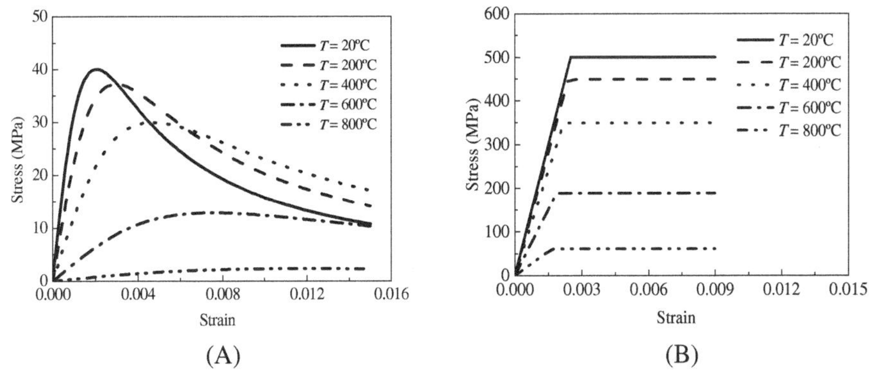

Fig. 4.2 Stress-strain relationships at elevated temperature: (A) concrete, (B) steel bars. From Yuan, F., and Y. F. Wu. 2019a. "Application of RC flexural theorems for member design under elevated temperature." *Eng. Struct.* 201, 109762.

A typical RC section is used in this study, which has a cross section of $b \times h = 300$ mm $\times$ 600 mm, with a concrete compressive strength of $f_{co} = 40$ MPa, steel yield strength of $f_y = 500$ MPa, and a concrete cover thickness of 60 mm from the tensile face to the center of the tension steel reinforcement. Five cases of tensile reinforcement ratios of 1.21%, 1.99%, 2.98%, 3.97%, and 4.96% are studied. The compressive reinforcement is ignored for beams. The temperatures studied in this section vary from ambient temperature (20°C) to 900°C. Six cases of RC columns with the reinforcement ratios $\rho_{st} = A_{st}/(bd) = \rho_{sc}$ of 0.0, 0.22%, 0.56%, 0.7%, 1.1% and 2.2% and a fixed net axial force ratio $n' = (N - A_{sc}f_y)/bdf_{co} = 1/3$ are also simulated, where A_{sc} and A_{st} are the area of the compression and tension reinforcement, respectively, for columns. Based on the flexural theorems, the axial load effect is often determined by the net axial stress $n = (N - A_{sc}f_y)/bd$ rather than N as illustrated by Eqs. (3.46), (3.49), (3.55), and (3.63). The columns have the same cross section and material properties as the RC beams.

The stress-strain relationships of the concrete and steel bars at different temperatures are shown in Fig. 4.2.

4.1.6.1 Flexural strength

Fig. 4.3 shows the effect of temperature on the flexural strength M_{max} of RC members at different reinforcement ratios. It can be seen from the figure that the moment capacity of both RC beams and columns shows a significant reduction as temperature increases. Based on the results in Fig. 4.3, the loss of load-carrying capacity of a structural member will be sufficiently large to cause a collapse of the structure when the temperature is sufficiently high. For an RC column, a critical temperature exists at which the column totally loses its moment resistance, or crushes, as shown in Fig. 4.3B. The critical temperatures for the RC columns studied in this section vary from 630°C to 663°C, depending upon the reinforcement ratio.

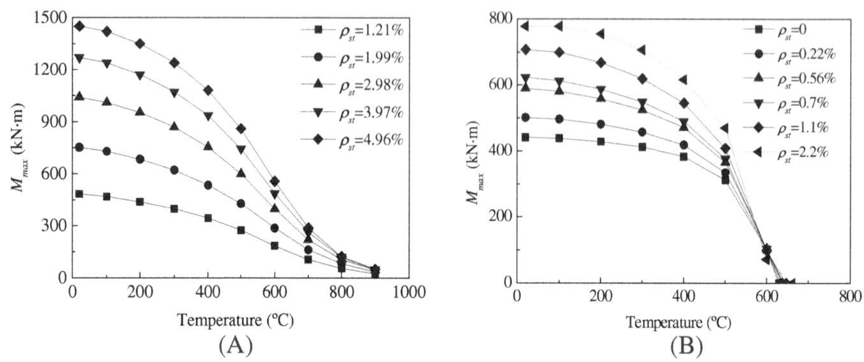

Fig. 4.3 Effect of temperature on peak moment: (A) RC beams, (B) RC columns.
From Yuan, F., and Y. F. Wu. 2019a. "Application of RC flexural theorems for member design under elevated temperature." *Eng. Struct.* 201, 109762.

4.1.6.2 Flexural design strain

Fig. 4.4 shows the effect of temperature on flexural design strains ε_{cm} and the maximum reinforcement ratio for underreinforced section ρ_m (Eq. (3.46)) of the RC sections. It is clearly seen from Fig. 4.4 that ε_{cm} shows an ascending trend as the temperature increases for both the beams and columns. The results of ρ_m in Fig. 4.4 show that it first increases and then decreases with increasing temperature. For RC columns, ρ_m drops below zero at a critical temperature between 500°C and 600°C. This means that at a temperature above the critical value, the constant axial force causes such a large axial strain of the column that the tension reinforcement does not yield at the onset of the flexural strength.

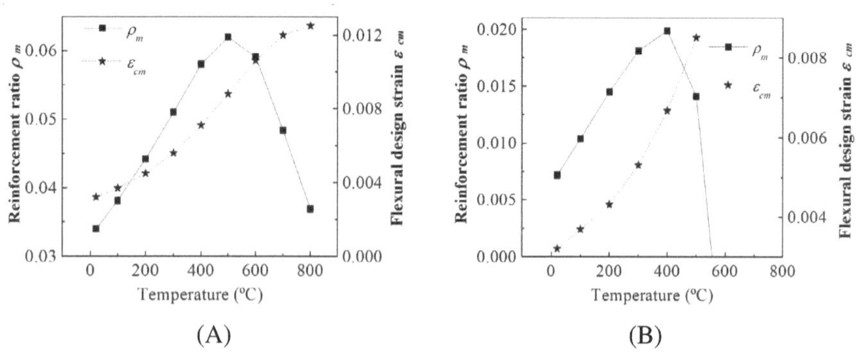

Fig. 4.4 Effect of temperature on ε_{cm} and ρ_m: (A) RC beams, (B) RC columns.
From Yuan, F., and Y. F. Wu. 2019a. "Application of RC flexural theorems for member design under elevated temperature." *Eng. Struct.* 201, 109762.

4.1.6.3 Ultimate curvature

According to Theorem 3.3, the ultimate curvature is achieved when the concrete strain at the extreme compression fiber reaches the ultimate strain of concrete (ε_{cu}) at which concrete is completely crushed. This strain is about three times the peak strain (ε_{co}) where concrete reaches its peak stress at ambient temperature. In the following study, ε_{cu} is assumed to be three times the strain at peak stress for concrete at both ambient and elevated temperatures, or $\varepsilon_{cu} = 3\varepsilon_{coT}$. The influence of elevated temperature on the ultimate curvature (κ_{cu}) of RC sections is shown in Fig. 4.5. Fig. 4.5A shows that the ultimate curvature of RC beams increases significantly with the increase of temperature when the temperature is lower than 600°C and then remains almost constant when the temperature goes higher.

For RC columns, κ_{cu} increases rapidly with increase in temperature at temperatures lower than 500°C. However, it starts decreasing when the temperature increases further. The resultant force provided by the steel bars changes from tension to compression when the temperature changes from 500°C to 600°C, leading to an increase in the depth of the compression zone and thus a smaller κ_{cu}.

The reinforcement ratio has an adverse effect on κ_{cu} for both RC beams and RC columns. The mechanism is: the tension force provided by the steel reinforcement increases as the reinforcement ratio increases. As a result, the depth of compression zone becomes larger in order to balance the larger tension force, leading to a larger c and thus a smaller κ_{cu}.

4.1.6.4 Parametric studies

The effects of member parameters, including compressive strength of concrete f_{co}, yield strength of steel reinforcement f_y, net axial force ratio $n' = (N - A_{sc}f_y)/bdf_c$ and member size $b \times h$, on M_{max}, κ_{cu}, and ρ_m are studied in this section. The flexural design strain ε_{cm} is a function of concrete stress-strain relationship only (Eq. (3.1)),

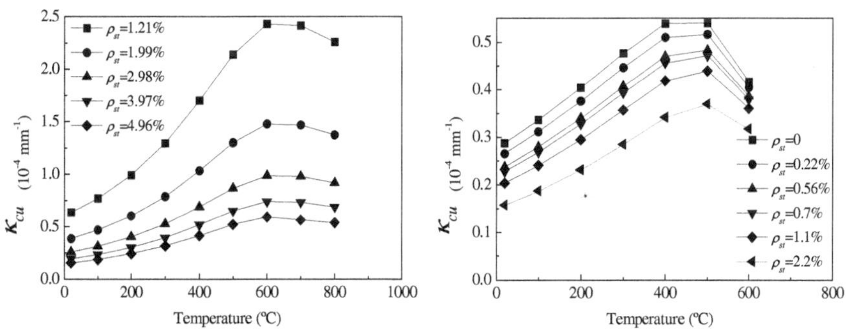

(A) (B)

Fig. 4.5 Effect of temperature on ultimate curvature: (A) RC beams, (B) RC columns. From Yuan, F., and Y. F. Wu. 2019a. "Application of RC flexural theorems for member design under elevated temperature." *Eng. Struct.* 201, 109762.

and hence is not affected by these parameters. Another important parameter, reinforcement ratio, is studied in the previous section (Figs. 4.3 and 4.5).

Variation of temperature is from 20°C to 800°C in the parametric studies. Variations of other parameters are relative to the control specimen that has a cross section of 400 mm × 400 mm, concrete compressive strength of $f_{co} = 40$ MPa, steel yield strength of $f_y = 460$ MPa, steel reinforcement ratio of 1%, and net axial force ratio of 0.3. When one factor is studied, other factors are kept at the constant values of the control specimen.

Fig. 4.6 shows the variations of peak moment M_{max}, ultimate curvature κ_{cu} and the maximum reinforcement ratio of underreinforced sections ρ_m as the concrete grade f_{co} changes from 20 MPa to 60 MPa. An increase in M_{max} and κ_{cu} is observed as f_{co} increases, and the increasing rate decreases as temperature increases. The reinforcement ratio ρ_m increases quickly with the increase in f_{co}. It can be seen from the figure that, to achieve a good flexural performance of RC sections exposed to fire, a higher concrete strength f_{co} is preferable.

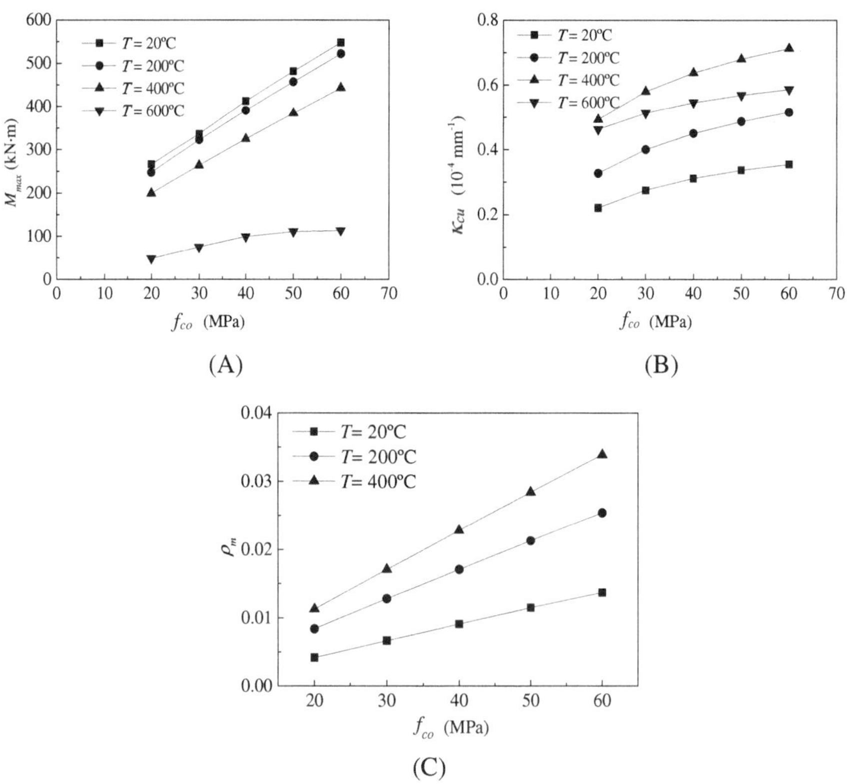

Fig. 4.6 Effect of concrete compressive strength f_{co} on: (A) peak moment, (B) ultimate curvature, (C) critical reinforcement ratio.
From Yuan, F., and Y. F. Wu. 2019a. "Application of RC flexural theorems for member design under elevated temperature." *Eng. Struct.* 201, 109762.

Fig. 4.7 shows the variations of M_{max}, κ_{cu}, and ρ_m as f_y varies from 250 MPa to 550 MPa. It can be seen from Fig. 4.7A that the increase of f_y leads to an increase of M_{max} when the temperature is not higher than 400°C. When the temperature reaches 600°C, the increase of f_y has no effect on M_{max}. A descending trend of κ_{cu} is observed as f_y increases for all temperatures (Fig. 4.7B). This is attributed to the fact that the increase in f_y leads to a larger compression resultant force, thus a larger depth of compression zone and consequent reduction of κ_{cu}. Fig. 4.7C shows that ρ_m decreases as f_y increases at all temperatures. In summary, a larger f_y results in a higher flexural strength but lower ductility. Therefore, the steel yield strength should be carefully selected in a fire design based on design need while considering a good balance between strength and ductility.

The effects of n' on the three parameters are shown in Fig. 4.8. The value of n' varies from 0 to 0.5 in the parametric study and is determined from the concrete strength at ambient temperature. The axial force remains constant during the loading process. It can be found from Fig. 4.8A that M_{max} first increases and then decreases with the increase of axial force ratio, which is well understood for RC columns under ambient

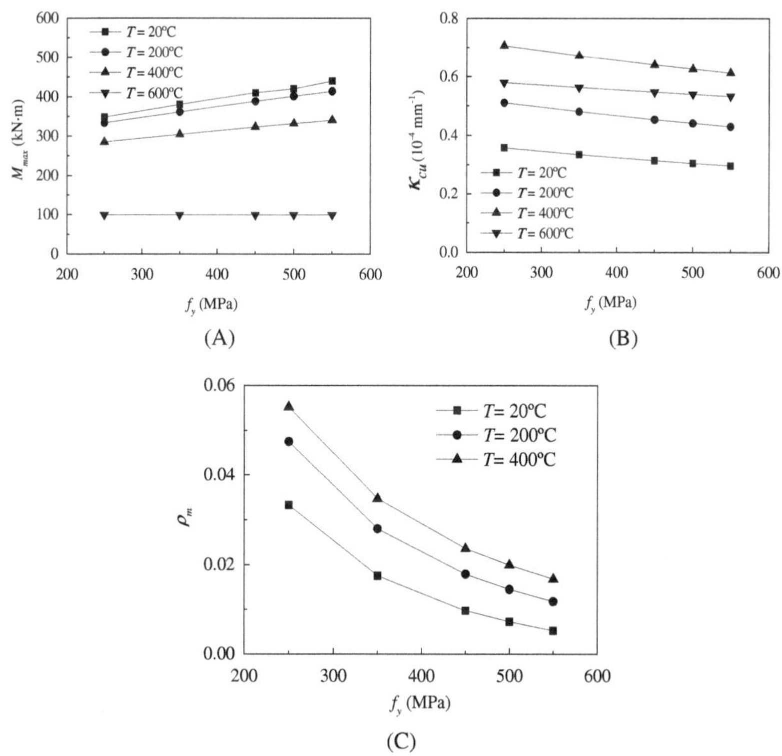

Fig. 4.7 Effect of steel yield strength f_y on: (A) peak moment, (B) ultimate curvature, (C) critical reinforcement ratio.
From Yuan, F., and Y. F. Wu. 2019a. "Application of RC flexural theorems for member design under elevated temperature." *Eng. Struct.* 201, 109762.

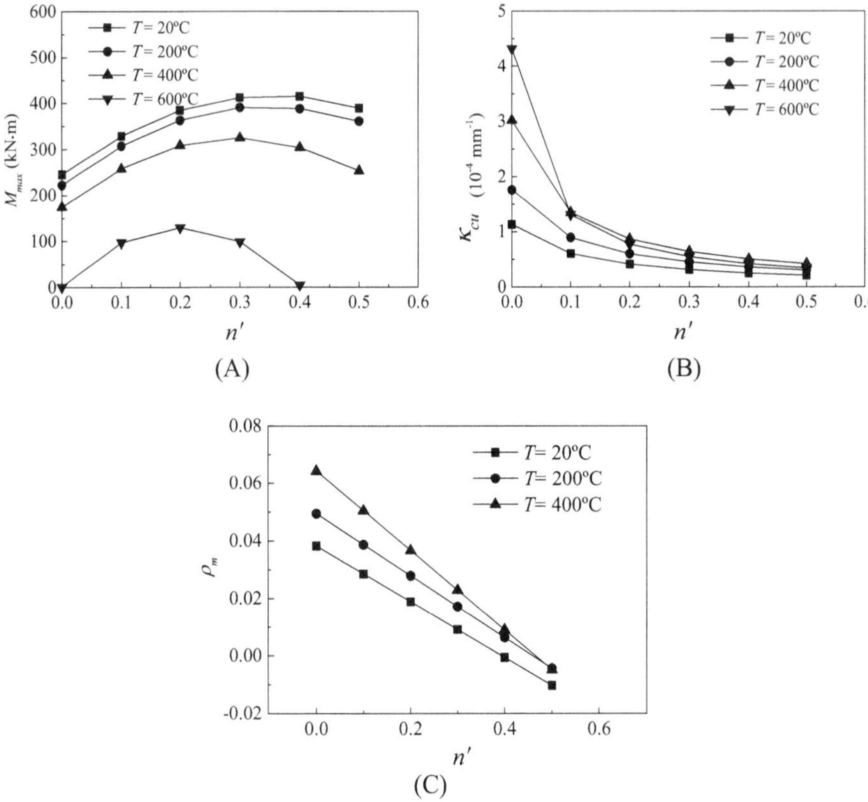

Fig. 4.8 Effect of n' on: (A) peak moment, (B) ultimate curvature, (C) critical reinforcement ratio.
From Yuan, F., and Y. F. Wu. 2019a. "Application of RC flexural theorems for member design under elevated temperature." *Eng. Struct.* 201, 109762.

temperature. At the net axial force ratio above 0.3, the reduction of flexural strength with increasing axial force ratio is more pronounced for members subjected to higher temperatures. A descending trend is observed for κ_{cu} as the axial force ratio increases (Fig. 4.8B). Fig. 4.8C shows that ρ_m decreases rapidly with increasing value of n' at both ambient and higher temperatures.

In summary, a high net axial force ratio above 0.3 results in a smaller flexural strength and deformation capacity, especially at higher temperatures. Therefore, to achieve a good flexural performance of RC members subjected to fire, a high level of axial force should be avoided in the design.

Member size has a significant effect on the behavior of the RC member under fire. The temperature increases much slower in a larger member than a smaller one, which significantly improves the fire resistance of RC members. Fig. 4.9 shows the effects of member dimensions on M_{max}, κ_{cu}, and ρ_m when the temperature rises to a certain

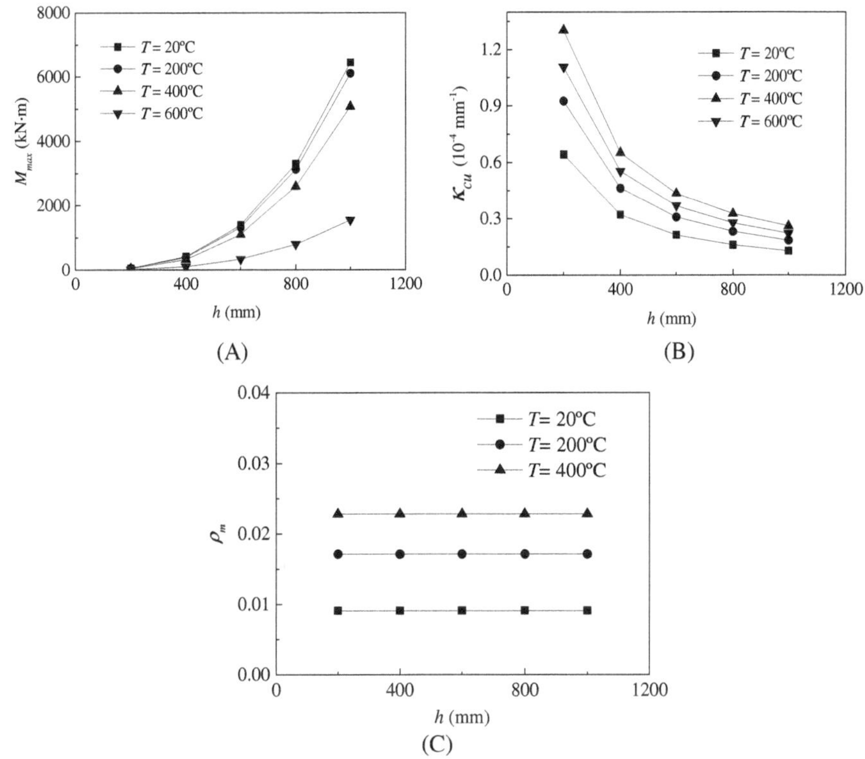

Fig. 4.9 Effect of member size on: (A) peak moment, (B) ultimate curvature, (C) critical reinforcement ratio.
From Yuan, F., and Y. F. Wu. 2019a. "Application of RC flexural theorems for member design under elevated temperature." *Eng. Struct.* 201, 109762.

value. The member sections are assumed to have five different dimensions of $b \times h = 200 \times 200$, 400×400, 600×600, 800×800, and 1000×1000 mm. The results in Fig. 4.9A show that the member size has a significant effect on the moment capacity M_{max}; however, this size effect on M_{max} decreases significantly as temperature increases. This result means that, although a larger member takes a longer time to be heated to a certain temperature, its member capacity also drops a lot more compared with smaller members in a fire. A descending trend of κ_{cu} is observed from Fig. 4.9B as the member size increases. This result is consistent with Eq. (3.15). Fig. 4.9C shows that the member size has no effect on ρ_m at both ambient and high temperatures. This is expected as ρ_m is given by Eq. (3.46), which is not related to member size at a given net axial force ratio n'. For safe design of RC sections, the geometric dimensions should be large enough to resist loads. But on the other hand, attention should also be paid to the adverse effect of the geometric dimension on the deformation capacity and ductility of the RC section in the design.

4.2 Stress block parameters for RC members reinforced with FRP bar

This section provides the derivation of the stress block parameters for flexural designs of RC members reinforced with FRP bars based on the work by Yuan and Wu (2019b). As FRP is always elastic, the elastic solutions of the flexural theorems apply. The solution for $\varepsilon_{cm,e}$ given by Eq. (3.63) is applicable in this case as the flexural design strain.

4.2.1 *Flexural failure of FRP-reinforced concrete beams*

Different from steel-reinforced concrete sections that are commonly designed to ensure tension-controlled failure featured by yielding of steel bars before crushing of concrete, the elastic behavior of FRP reinforcement causes a different failure mode that necessitates redevelopment of the flexural design approach for FRP-reinforced concrete sections. A compression failure mode manifested by concrete crushing is more desirable for flexural members reinforced with FRP bars than the brittle tension failure mode manifested by sudden rupture of FRP bars. Through concrete softening/crushing prior to the tensile rupture of FRP reinforcement, a flexural member does exhibit more inelastic deformation before failure than the failure mode of FRP rupture without concrete softening. To resolve the ductility problem of FRP-reinforced concrete beams, only compression failure mode is recommended in ACI 440.1R-15 (ACI 2015). This failure mode corresponds to concrete crushing without reinforcement yielding, i.e., the failure mode of overreinforced RC members.

Different FRP materials may have significantly different elastic moduli and ultimate strains. For instance, the elastic modulus of glass fiber-reinforced polymer (GFRP) is about 1/4–1/3 times that of carbon fiber-reinforced polymer (CFRP) and the ultimate strain of GFRP reinforcement is about 1.5 times that of CFRP reinforcement. Therefore, the reinforcement ratios of different FRP materials are not comparable, which causes complications in applying the theorems (e.g., unable to use the reinforcement ratios in Fig. 3.4). This problem can be overcome by using the effective reinforcement ratio (Qu et al. 2009), defined by:

$$\rho = \frac{E_f A_f}{E_s bd} = \frac{E_f}{E_s}\rho_f \tag{4.18}$$

where E_f and A_f are the elastic modulus and cross-sectional area of FRP reinforcement, respectively; E_s is the elastic modulus of steel taken as 200 GPa; $\rho_f = A_f/(bd)$, the FRP reinforcement ratio. All reinforcement ratios denoted as ρ in this section refer to the effective reinforcement ratio.

Fig. 4.10 illustrates the three typical cases of flexural failure of FRP-reinforced concrete beams:

(1) When the tensile resistance of FRP reinforcement is greater than a certain amount, the moment–curvature curve reaches a peak point before FRP rupture (the thick solid line in Fig. 4.10).

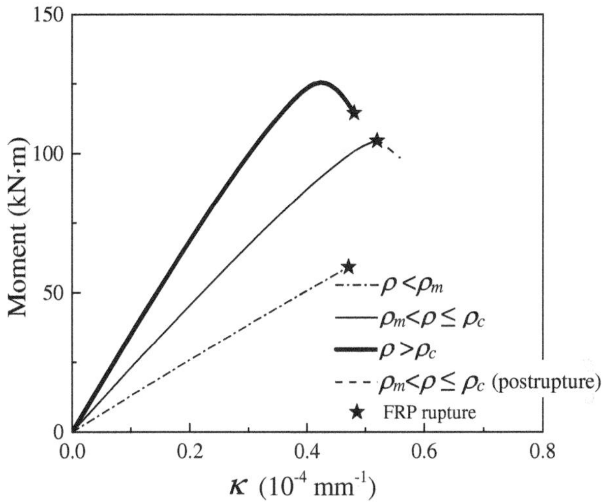

Fig. 4.10 Moment vs. curvature curves for members with different reinforcement ratios. From Yuan, F., and Y. F. Wu. 2019b. "Analytical method for derivation of stress block parameters for flexural design of FRP reinforced concrete members." *Compos. Struct.* 229, 111459.

(2) When the reinforcement ratio is at a certain range, the moment–curvature curve peaks off (with a zero slope) and ruptures exactly at the same point. This case is illustrated in Fig. 4.10 by the thin solid line.

(3) When the reinforcement ratio is less than a certain amount, the moment–curvature curve stops at FRP rupture before peaking off, as shown by the dash-dotted line in Fig. 4.10.

The following material properties and models are used to produce the curves in Fig. 4.10: $f_{co} = 30$ MPa, $E_f = 160$ GPa, the FRP strength $f_{fu} = 2400$ MPa, CEB-FIP model for concrete, $b \times d = 200 \times 360$ mm.

The preceding cases (1) and (2) are covered by the flexural theorems. Case (1) is applicable when $\rho > \rho_c$ as shown in Fig. 3.4. The moment–curvature curve reaches the peak point when the extreme compressive strain ε_{cm} reaches the value $\varepsilon_{cm,e}$ given by Eq. (3.63). Case (2) is applicable when $\rho_m < \rho \leq \rho_c$ as shown in Fig. 3.4. Case (3) is applicable when $\rho \leq \rho_m$. Typical values of ρ_m and ρ_c for typical FRP bars are shown in Fig. 4.11. It can be seen from Fig. 4.11 that, for the case of $\rho \leq \rho_m$, the reinforcement ratio is generally below the practical reinforcement ratio adopted in practice. Case (2) ($\rho_m < \rho \leq \rho_c$) is in a narrow band (Fig. 4.11). As discussed earlier, the brittle failure mode of FRP rupture before the onset of flexural strength is not allowed in practice (ACI 2015). Therefore, only the case of $\rho > \rho_c$ is studied in the following.

4.2.2 *Stress–strain relationships*

As FRP bar is elastic, the elastic solution of the flexural design strain is given by Eq. (3.63). The other two stress block parameters α_1 and β_1 are given by Eqs. (4.9) and (4.10), respectively. Since the flexural design strain ε_{cm} is related to the

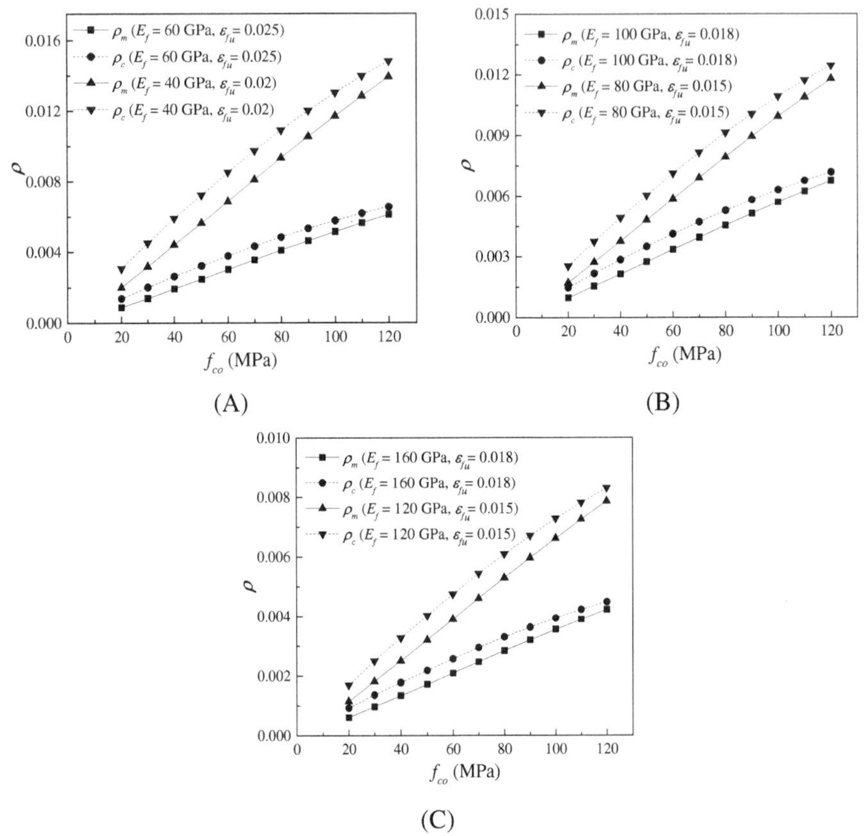

Fig. 4.11 Values of ρ_m and ρ_c for: (A) typical GFRP bars; (B) typical AFRP bars; and (C) typical CFRP bars.
From Yuan, F., and Y. F. Wu. 2019b. "Analytical method for derivation of stress block parameters for flexural design of FRP reinforced concrete members." *Compos. Struct.* 229, 111459.

reinforcement ratio for elastic reinforcement (Fig. 3.4), it can be seen from Eqs. (4.9) and (4.10) that both α_1 and β_1 are related to the reinforcement ratio. Therefore, for FRP-reinforced concrete members, the three stress block parameters ε_{cm}, α_1 and β_1 are functions of the reinforcement ratio as well as the concrete stress-strain relationship. Using a stress-strain model of concrete, these stress block parameters can be calculated.

In this part of the study, the typical compressive stress-strain relationship of concrete in CEB-FIP/Eurocode 2 is used as the stress-strain model of concrete. This model can cover a larger range of concrete strength compared with Popovics' model and has been widely used in the literature. Details of the stress-strain model are given as follows:

$$f_c = \begin{cases} \dfrac{f_{co}x(k-x)}{1+(k-2)x} & \varepsilon \le \varepsilon_{max} \\ \dfrac{f_{co}}{\left[\dfrac{\varepsilon_{co}}{\varepsilon_{max}}\xi - 2\left(\dfrac{\varepsilon_{co}}{\varepsilon_{max}}\right)^2\right]x^2 + \left[4\dfrac{\varepsilon_{co}}{\varepsilon_{max}} - \xi\right]x} & \varepsilon > \varepsilon_{max} \end{cases} \tag{4.19}$$

in which

$$x = \frac{\varepsilon}{\varepsilon_{co}} \tag{4.20}$$

$$\xi = \frac{4\left[\left(\dfrac{\varepsilon_{max}}{\varepsilon_{co}}\right)^2(k-2) + 2\left(\dfrac{\varepsilon_{max}}{\varepsilon_{co}}\right) - k\right]}{\left[\left(\dfrac{\varepsilon_{max}}{\varepsilon_{co}}\right)(k-2)+1\right]^2} \tag{4.21}$$

$$\varepsilon_{max} = \varepsilon_{co}\left[0.5(0.5k+1) + \sqrt{0.25(0.5k+1)^2 - 0.5}\right] \tag{4.22}$$

$$k = 1.05E_c\frac{\varepsilon_{co}}{f_{co}} \tag{4.23}$$

$$\varepsilon_{co} = 0.0007(f_{co})^{0.31} \tag{4.24}$$

$$E_c = 22000\left(\frac{f_{co}}{10}\right)^{0.3} \tag{4.25}$$

where ε_{max} = concrete strain when concrete stress is equal to $0.5f_{co}$ on the descending part of the stress-strain curve.

4.2.3 Parametric studies

Parametric studies are conducted to evaluate the stress block parameters of FRP-reinforced concrete sections. As mentioned previously, the only factors that affect the stress block parameters are the stress-strain relationship of concrete and reinforcement ratio. It can be seen from Eqs. (4.19)–(4.25) that the only factor that affects the stress-strain relationship of concrete is the compressive strength of concrete. The effects of these two factors (f_{co} and ρ) on the stress block parameters are systematically studied through the following parametric studies.

4.2.3.1 Effects of compressive strength of concrete

In this parametric study, the concrete strength f_{co} varies from 20 to 120 MPa. Numerical results of ε_{cm}, α_1 and β_1 for different concrete strengths under different reinforcement ratios are calculated by Eqs. (3.63), (4.9), and (4.10), respectively, and shown in

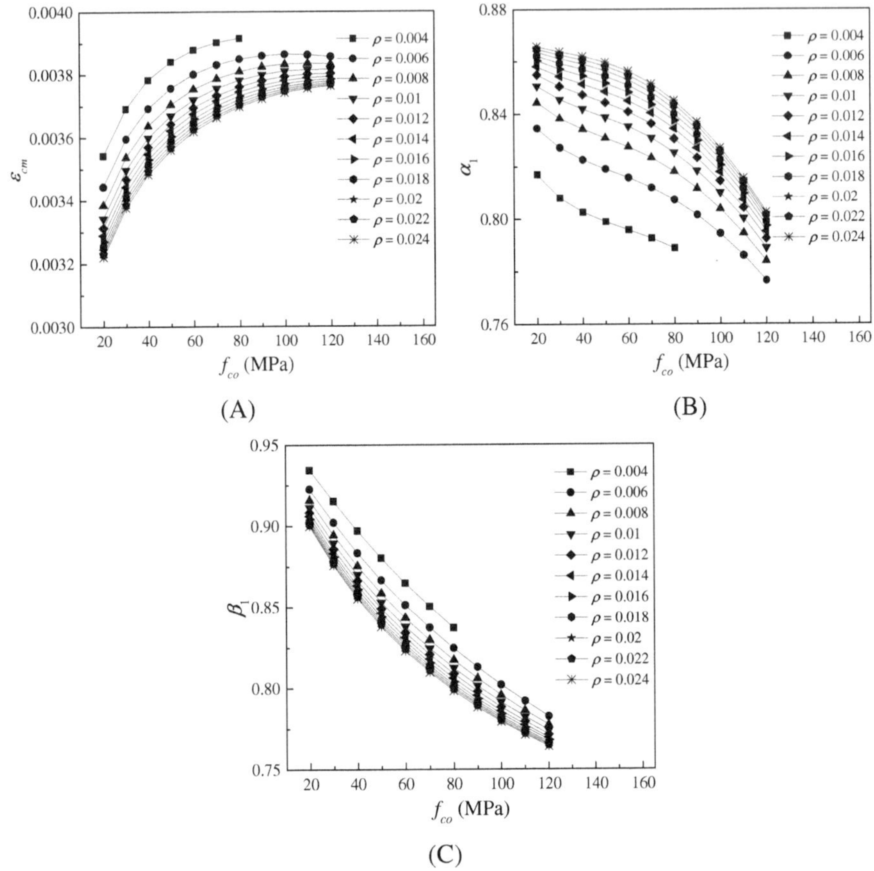

Fig. 4.12 Effect of concrete compressive strength on the stress block parameters of: (A) ε_{cm}, (B) α_1, and (C) β_1.
From Yuan, F., and Y. F. Wu. 2019b. "Analytical method for derivation of stress block parameters for flexural design of FRP reinforced concrete members." *Compos. Struct.* 229, 111459.

Fig. 4.12. It can be seen from Fig. 4.12 that the flexural design strain ε_{cm} shows an increasing trend as f_{co} increases, while both α_1 and β_1 exhibit decreasing trends with increasing f_{co}.

4.2.3.2 Effects of effective reinforcement ratio

It can be clearly seen from the flexural theorems that the longitudinal reinforcement affects the value of the stress block parameters when the longitudinal reinforcement does not yield (Fig. 3.4). Previous test results also showed that the longitudinal reinforcement ratio ρ has evident influence on the stress block parameters of FRP-reinforced concrete beams (Theriault and Benmokrane 1998; GangaRao et al. 2006). However, no existing design code has taken this factor into consideration. Fig. 4.13 shows the effect of effective reinforcement ratio on the stress block parameters of FRP-reinforced concrete sections, where ρ varies from the value of ρ_c to 3%. It

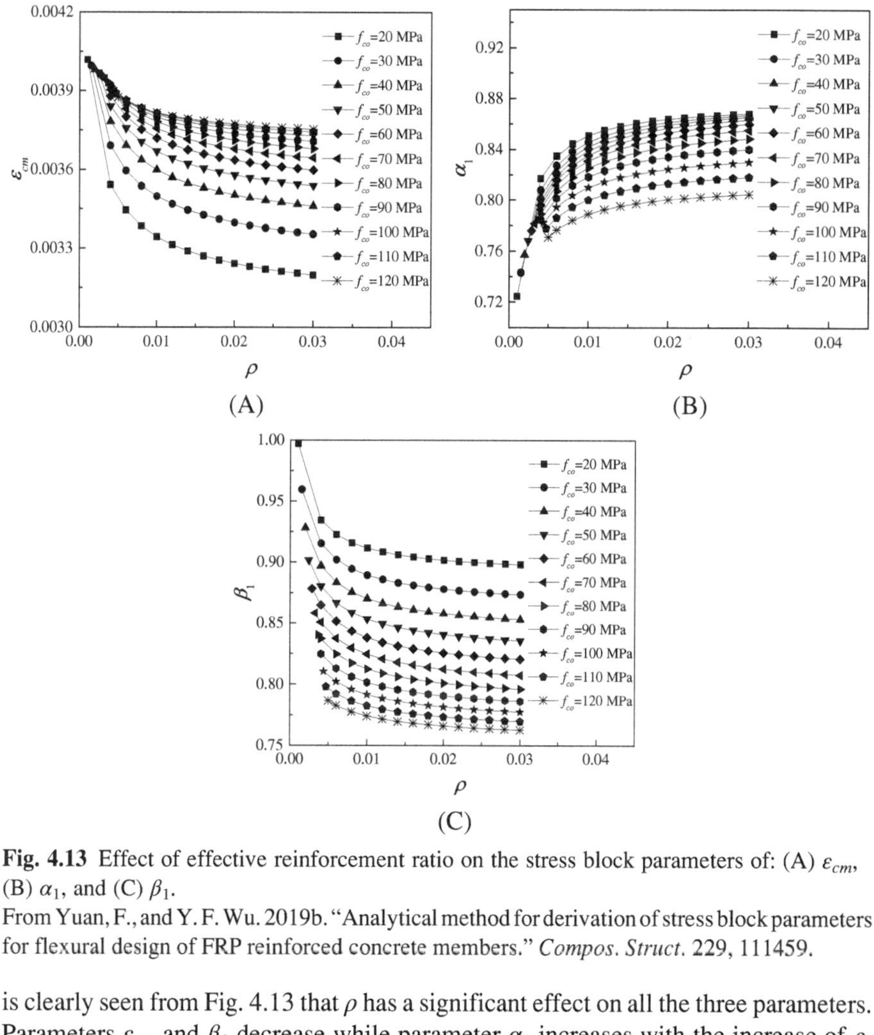

Fig. 4.13 Effect of effective reinforcement ratio on the stress block parameters of: (A) ε_{cm}, (B) α_1, and (C) β_1.
From Yuan, F., and Y. F. Wu. 2019b. "Analytical method for derivation of stress block parameters for flexural design of FRP reinforced concrete members." *Compos. Struct.* 229, 111459.

is clearly seen from Fig. 4.13 that ρ has a significant effect on all the three parameters. Parameters ε_{cm} and β_1 decrease while parameter α_1 increases with the increase of ρ.

4.2.4 Stress block parameters

By analyzing the trends of variation of the three stress block parameters obtained in the previous section, a power relationship is observed between both ε_{cm} and β_1 and ρ, while a logarithmic relationship is found between α_1 and ρ. All the three parameters exhibit a polynomial relationship with f_{co}. Through regression analyses, the following models of ε_{cm}, α_1 and β_1 are obtained:

$$\varepsilon_{cm} = 0.917\rho^{-0.038}\left(-7*10^{-8}f_{co}^2 + 10^{-5}f_{co} + 0.0031\right) \tag{4.26}$$

$$\alpha_1 = 1.177\left(-6*10^{-6}{f_{co}}^2 + 0.0002f_{co} + 0.855\right)\left(0.032\ln(\rho) + 0.986\right) \tag{4.27}$$

$$\beta_1 = 0.892\rho^{-0.026}\left(8*10^{-6}f_{co}^2 - 0.0024f_{co} + 0.950\right) \tag{4.28}$$

Eqs. (4.26)–(4.28) can be conveniently used by engineers for routine design work to avoid the complicated numerical integrations in Eqs. (3.63), (4.9), and (4.10). It is noted that Eqs. (4.26)–(4.28) are applicable to effective reinforcement ratio in the range of $\rho_c \leq \rho \leq 3\%$ and concrete strength in the range of 20 MPa $\leq f_{co} \leq$ 120 MPa. For effective reinforcement ratio higher than 3% or concrete strength greater than 120 MPa, these models of stress block parameters are not applicable and further parametric studies are required to reevaluate the equivalent stress block parameters.

4.2.5 Comparisons with existing design codes

Test results of 43 FRP-reinforced concrete beams were collected to verify the accuracy of the models. The test results are summarized in Table 4.1. All selected beams failed in concrete crushing. For comparison, the widely used design codes ACI 440.1R-15 (ACI 2015) and CSA-S806-12 (CSA 2012) are also used to calculate the flexural strength of these members. It should be noted that the material partial safety factors and all reduction factors are set to unity in the calculations using the design codes so that all comparisons are made under a uniform benchmark.

4.2.5.1 ACI 440.1R-15

According to ACI 440.1R-15, if the reinforcement ratio is larger than that for balanced failure, the flexural strength, M_{uc}, of an FRP-reinforced concrete section can be predicted as follows:

$$M_{uc} = A_f f_f\left(d - \frac{\beta_1 c}{2}\right) \tag{4.29}$$

in which

$$c = \frac{A_f f_f}{\alpha_1 \beta_1 f_{co} b} \tag{4.30}$$

$$f_f = \sqrt{\frac{\left(E_f\varepsilon_{cm}\right)^2}{4} + \frac{\alpha_1\beta_1 f_{co}}{\rho_f}E_f\varepsilon_{cm}} - 0.5E_f\varepsilon_{cm} \leq f_{fu} \tag{4.31}$$

In this model, the extreme compressive strain ε_{cm} for the calculation of M_{uc} is assumed to be 0.003. The rectangular stress block parameter α_1 is assumed to be 0.85 and parameter β_1 is given by

$$\beta_1 = \begin{cases} 0.85 & f_{co} \leq 28 \\ 0.85 - 0.05\left(\dfrac{f_{co} - 28}{7}\right) \geq 0.65 & f_{co} > 28 \end{cases} \tag{4.32}$$

Table 4.1 Evaluation of models.

Ref.	Speci. ID	E_f (GPa)	f_{fu} (MPa)	ρ_f (%)	f_{co} (MPa)	Exp. M_{ue} (kNm)	ACI			CSA			Eqs. (4.26)–(4.28)		
							M_{uc} (kNm)	Error (%)	M_{uc}/M_{ue}	M_{uc} (kNm)	Error (%)	M_{uc}/M_{ue}	M_{uc} (kNm)	Error (%)	M_{uc}/M_{ue}
Adam et al. 2015	A25–3	30	640	0.886	24.5	41.36	25.20	39.07	0.609	27.01	34.70	0.653	28.08	32.11	0.679
	A45–2	30	640	0.886	48	45.05	33.89	24.77	0.752	37.78	16.12	0.839	41.27	8.37	0.916
	A45–3	30	640	1.449	48	60.39	41.73	30.90	0.691	46.33	23.28	0.767	50.38	16.58	0.834
	A70–2	30	640	1.449	74.4	72.99	51.47	29.48	0.705	55.78	23.57	0.764	63.47	13.04	0.870
	A70–3	30	640	2.174	74.4	79.81	61.09	23.45	0.765	65.93	17.38	0.826	74.79	6.28	0.937
Kassem et al. 2011	C1-4b	114	1506	0.567	40.4	74.58	70.95	4.86	0.951	76.90	3.11	1.031	82.14	10.14	1.101
	C1–6	114	1506	0.905	39.3	83.13	73.87	11.14	0.889	79.35	4.54	0.955	84.38	1.50	1.015
	C1–8	114	1506	1.206	39.3	90.39	81.77	9.54	0.905	87.45	3.25	0.967	92.93	2.80	1.028
	C2–4	122	1988	0.542	39.9	78.75	63.23	19.73	0.803	68.42	13.11	0.869	73.00	7.29	0.927
	C2–4b	122	1988	0.510	39.9	78.18	69.82	10.69	0.893	75.62	3.28	0.967	80.72	3.24	1.032
	C2–6	122	1988	0.814	40.8	80.89	74.39	8.04	0.920	80.25	0.79	0.992	85.60	5.82	1.058
	C2–8	122	1988	1.085	40.8	89.39	82.45	7.76	0.922	88.57	0.92	0.991	94.40	5.60	1.056
	G1–6	40	617	1.666	39.05	77.47	60.87	21.43	0.786	65.73	15.16	0.848	70.01	9.63	0.904
	G1–8	40	617	2.221	39.05	89.76	67.87	24.39	0.756	73.02	18.65	0.814	77.65	13.49	0.865
	G2–6	36	747	1.455	39.05	71	55.71	21.53	0.785	60.33	15.02	0.850	64.37	9.34	0.907
	G2–8	36	747	1.940	39.05	84.54	62.35	26.25	0.738	67.29	20.40	0.796	71.65	15.25	0.848
	AR-6	71	1800	0.905	39.05	70.85	61.66	12.96	0.870	66.62	5.97	0.940	70.97	0.17	1.002
	AR-8	71	1800	1.206	39.05	71.75	68.80	4.11	0.959	74.05	3.21	1.032	78.76	9.77	1.098
Theriault and Benmokrane 1998	BC2NB	38	773	1.136	53.1	20	16.62	16.90	0.831	18.72	6.41	0.936	20.54	2.71	1.027
	BC2HA	38	773	1.136	57.2	19.7	16.96	13.91	0.861	19.35	1.80	0.982	21.39	8.57	1.086
	BC2HB	38	773	1.136	57.2	20.6	16.96	17.67	0.823	19.35	6.09	0.939	21.39	3.82	1.038
	BC4NB	38	773	2.717	46.2	20.6	16.78	18.55	0.815	18.35	10.94	0.891	19.77	4.02	0.960
	BC4HA	38	773	2.717	53.9	21	17.73	15.58	0.844	19.82	5.60	0.944	21.67	3.18	1.032
	BC4HB	38	773	2.717	54.9	21.4	17.83	16.68	0.833	20.00	6.53	0.935	21.90	2.34	1.023
	BC4VA	38	773	2.717	93.5	28.4	24.47	13.83	0.862	24.90	12.32	0.877	28.99	2.07	1.021
	BC4VB	38	773	2.717	93.5	29.5	24.47	17.04	0.829	24.90	15.59	0.844	28.99	1.73	0.983

Continued

Table 4.1 Continued

Ref.	Speci. ID	E_f (GPa)	f_{fu} (MPa)	ρ_f (%)	f_{co} (MPa)	Exp. M_{ue} (kNm)	ACI			CSA			Eqs. (4.26)–(4.28)		
							M_{uc} (kNm)	Error (%)	M_{uc}/M_{ue}	M_{uc} (kNm)	Error (%)	M_{uc}/M_{ue}	M_{uc} (kNm)	Error (%)	M_{uc}/M_{ue}
Rashid et al. 2005	AF2T1	53	1760	0.395	42.8	44.17	38.14	13.65	0.863	42.03	4.85	0.952	45.58	3.19	1.032
Grace et al. 1998	cb-st	147	2250	0.262	48.26	51.91	47.31	8.86	0.911	52.63	1.38	1.014	57.32	10.4	1.104
Wang and Belarbi 2005	P4G	41	690	2.202	48	47	41.70	11.27	0.887	45.91	2.31	0.977	49.67	5.69	1.057
	P8G	41	552	3.390	48	51	45.76	10.27	0.897	50.07	1.82	0.982	54.09	6.07	1.061
	P4C	124	2069	0.667	48	51	40.33	20.93	0.791	44.45	12.85	0.872	48.11	5.66	0.943
	F4G	41	690	2.202	30	40	33.74	15.65	0.844	35.55	11.13	0.889	37.10	7.25	0.928
	F8G	41	552	3.390	30	42	36.56	12.96	0.870	38.26	8.92	0.911	39.90	4.99	0.950
	F4C	124	2069	0.667	30	46	32.70	28.90	0.711	34.50	25.01	0.750	36.01	21.71	0.783
Toutanji and Saafi 2000	GB1–1	40	695	0.525	35	60	43.38	27.70	0.723	46.93	21.78	0.782	50.09	16.52	0.835
	GB1–2	40	695	0.525	35	59	43.38	26.47	0.735	46.93	20.45	0.795	50.09	15.10	0.849
	GB2–1	40	695	0.787	35	65	51.50	20.77	0.792	55.53	14.57	0.854	59.01	9.21	0.908
	GB2–2	40	695	0.787	35	64.3	51.50	19.91	0.801	55.53	13.64	0.864	59.01	8.22	0.918
	GB3–1	40	695	1.251	35	71	43.79	38.32	0.617	47.00	33.81	0.662	49.74	29.94	0.701
	GB3–2	40	695	1.251	35	70.5	43.79	37.89	0.621	47.00	33.34	0.667	49.74	29.44	0.706
Yang et al. 2012	GG	48.1	941	1.801	75.9	82.8	66.66	19.49	0.805	71.41	13.76	0.862	81.06	2.11	0.979
	GG-SN	48.1	941	1.801	89.3	92	73.40	20.22	0.798	75.77	17.65	0.824	87.75	4.62	0.954
	GG-ST	48.1	941	1.801	104.4	103.6	80.43	22.37	0.776	79.47	23.29	0.767	93.95	9.31	0.907
Average								**18.97**	**0.810**		**12.75**	**0.876**		**9.03**	**0.950**
SD (%)								**8.49**			**9.21**			**7.54**	
Max. error (%)								**39.07**			**34.70**			**32.11**	

4.2.5.2 CSA-S806-12

CSA-S806-12 also determines the flexural strength of FRP-reinforced concrete members by Eqs. (4.29)–(4.31). Different from ACI 440.1R-15, ε_{cm} is assumed to be 0.0035 and the rectangular stress block parameters α_1 and β_1 are given by the following equations:

$$\alpha_1 = 0.85 - 0.0015 f_{co} \geq 0.67 \tag{4.33}$$

$$\beta_1 = 0.97 - 0.0025 f_{co} \geq 0.67 \tag{4.34}$$

4.2.6 Design examples

Application of the models of the equivalent stress block parameters for design of FRP-reinforced concrete members is illustrated by way of examples in this section, in comparison with the designs using the preceding two design codes. The design approach and procedure are identical to those of current design practice, except that the stress block parameters are calculated from Eqs. (4.26)–(4.28).

Question 1

Calculate the flexural strength of an FRP-reinforced concrete beam with a cross section of b (breadth) × h (depth) = 300 mm × 600 mm. The material properties are: concrete compressive strength f_{co} = 60 MPa, FRP rupture strength f_{fu} = 2400 MPa, FRP elastic modulus E_f = 160 GPa, FRP reinforcement ratio ρ_f = 1.25%, concrete cover = 60 mm from the tension face to the center of the tensile reinforcement.

Solution:

The stress and strain blocks at the onset of flexural strength of the FRP-reinforced concrete section are shown in Fig. 4.14. The flexural design strain is calculated using Eq. (4.26):

$\varepsilon_{cm} = 0.003766$.

The stress block parameters are calculated from Eqs. (4.27) and (4.28) to give

$\alpha_1 = 0.8345$, and

$\beta_1 = 0.8393$.

The balance of the internal forces in the cross section gives

$$\alpha_1 f_{co} \beta_1 c b = A_f E_f \varepsilon_f \tag{4.35}$$

where ε_f is the strain of FRP reinforcement. From the strain profile in Fig. 4.14, one gets

$$\frac{\varepsilon_{cm}}{c} = \frac{\varepsilon_f}{d - c} \tag{4.36}$$

Substituting all known parameters and material properties into Eqs. (4.35) and (4.36) gives

$c = 185.3$ mm, and

$\varepsilon_f = 0.00721$.

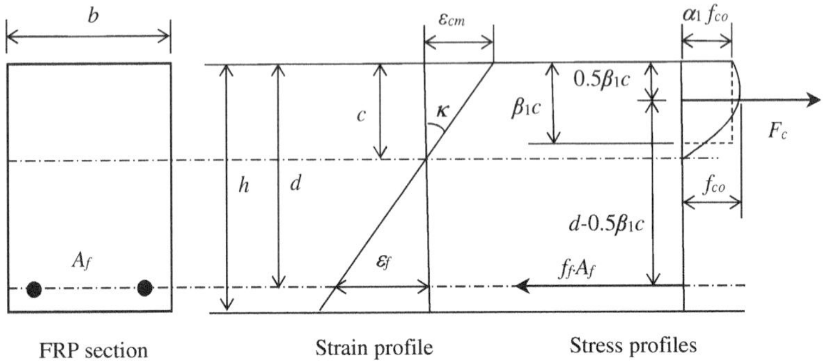

Fig. 4.14 Strain and stress blocks at onset of flexural strength of FRP-reinforced section. From Yuan, F., and Y. F. Wu. 2019b. "Analytical method for derivation of stress block parameters for flexural design of FRP reinforced concrete members." *Compos. Struct.* 229, 111459.

The rupture strain of FRP is

$$\varepsilon_{fu} = f_{fu}/E_f = 0.015 > \varepsilon_f,$$

hence the FRP bar has not ruptured at the onset of the flexural strength and it belongs to compression failure. The flexural strength of the FRP-reinforced concrete section is given by Eq. (4.29):

$$M_{max} = A_f f_f \left(d - \frac{\beta_1 c}{2} \right) = 1079.8 \text{ kN m}$$

Numerical simulation is used to validate these design results. The numerical results of ε_{cm}, α_1, and β_1 are calculated to be 0.00372, 0.8353, and 0.838, respectively, which are very close to the corresponding values calculated from Eqs. (4.26)–(4.28). The numerical result of the ultimate bending moment is $M_{max} = 1075.3$ kN·m, which is also very close to that calculated in the design example.

The ultimate bending moment calculated from ACI 440.1R-15 is

$M_{uc} = 887.1$ kN m,

and that from CSA-S806-12 is

$M_{uc} = 978.4$ kN m.

Both are smaller than the more accurate result calculated previously. This is because their stress block parameters are not related to reinforcement ratio and are less accurate than those given by Eqs. (4.26)–(4.28).

Question 2

Calculate the required area of FRP bars for an ultimate design moment of $M_{max} = 800$ kN · m. The beam has the same cross-sectional and material properties as those in Question 1.

Solution:
Substituting Eqs. (4.35) and (4.36) into Eq. (4.29) gives

$$M_{uc} = A_f f_f \left(d - \frac{A_f f_f}{2\alpha_1 f_{co} b} \right) = \frac{E_s}{E_f} bd^2 \rho f_f \left(1 - \frac{E_s}{E_f} \frac{\rho f_f}{2\alpha_1 f_{co}} \right) \tag{4.37}$$

where f_f is given by Eq. (4.31). The stress block parameters of ε_{cm}, α_1, and β_1 are related to the effective reinforcement ratio ρ by Eqs. (4.26)–(4.28). Therefore, a trial value of ρ is first assumed to evaluate the stress block parameter. The obtained stress block parameters are then used in Eq. (4.37) to calculate ρ. The obtained ρ is further used to reevaluate ε_{cm}, α_1, and β_1, and then recalculate ρ until it converges. Two to three iterations should give a sufficiently accurate result. It is noted that, for calculation of the required FRP area for a given design moment, iteration process is also required in the designs using ACI and CSA codes.

Through iteration the following result is obtained:

$$A_f = 867.9\,\text{mm}^2$$

The iteration process can be conveniently implemented using a computer program or spreadsheet. The numerical simulation result of FRP reinforcement area is $A_f = 876.0$ mm^2 which is close to the calculated one.

The required FRP areas calculated from ACI 440.1R-15 and CSA-S806-12 are 1514.2 mm^2 and 1135.5 mm^2, respectively, which show significant overdesign in the reinforcement area of 72.3% and 29.6%, respectively.

4.2.7 Comparison of models

Fig. 4.15 shows the comparison of ε_{cm}, α_1 and β_1 from Eqs. (4.26)–(4.28) with those from the two existing design codes. The average value of ε_{cm} calculated by Eq. (4.26) is closer to that recommended in CSA-S806-12 than that in ACI 440.1R-15, as shown in Fig. 4.15A. The parameter α_1 is constant in ACI 440.1R-15 and shows a decreasing trend as f_{co} increases in CSA-S806-12. The trend of α_1 given by Eq. (4.27) is consistent with that in CSA-S806-12, as shown in Fig. 4.15B. Fig. 4.15C shows that the parameter β_1 exhibits a decreasing trend with the increase in f_{co} in all the three models. Compared with ACI 440.1R-15, CSA-S806-12 gives closer results of β_1 to those calculated by Eq. (4.28).

The flexural strengths of the test beams in Table 4.1 calculated by ACI 440.1R-15 and CSA-S806-12 as well as those from Eqs. (4.26)–(4.28) are given in Table 4.1. The error of prediction in the table is calculated by:

$$Error(\%) = \frac{|M_{ue} - M_{uc}|}{M_{ue}} \times 100 \tag{4.38}$$

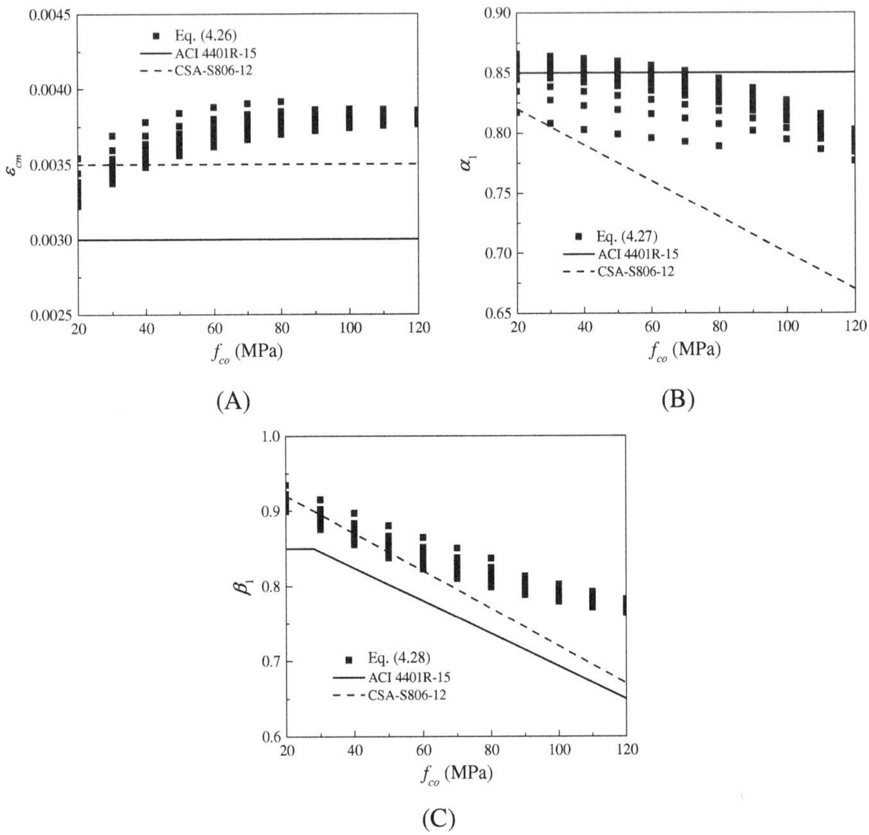

Fig. 4.15 Comparison of models: (A) ε_{cm}, (B) α_1, and (C) β_1.
From Yuan, F., and Y. F. Wu. 2019b. "Analytical method for derivation of stress block parameters for flexural design of FRP reinforced concrete members." *Compos. Struct.* 229, 111459.

where M_{ue} is the flexural strength from the experimental tests. Fig. 4.16 shows the experimental results versus the predicted ones from the three models. The errors of results calculated from Eqs. (4.26)–(4.28) are relatively smaller compared with the other two existing models. The average error, standard deviation, and maximum error of the flexural strengths predicted by Eqs. (4.26)–(4.28) are 9.03%, 7.54%, and 32.11%, respectively, compared with 18.97%, 8.49%, and 39.07% by ACI 440.1R-15 and 12.75%, 9.21%, and 34.70% by CSA-S806-12. Based on the test results, the stress block parameters produced from the flexural theorems are more accurate than those recommended in existing design codes. It is noted from the results in Table 4.1 that the accuracy of the stress block parameters is higher in predicting the flexural strength of FRP-reinforced concrete beams with high-strength concrete ($f_c > 60$ MPa), as shown in Fig. 4.17. The average error of the flexural strengths predicted by Eqs. (4.26)–(4.28) is 4.35%, compared with 19.40% by ACI 440.1R-15

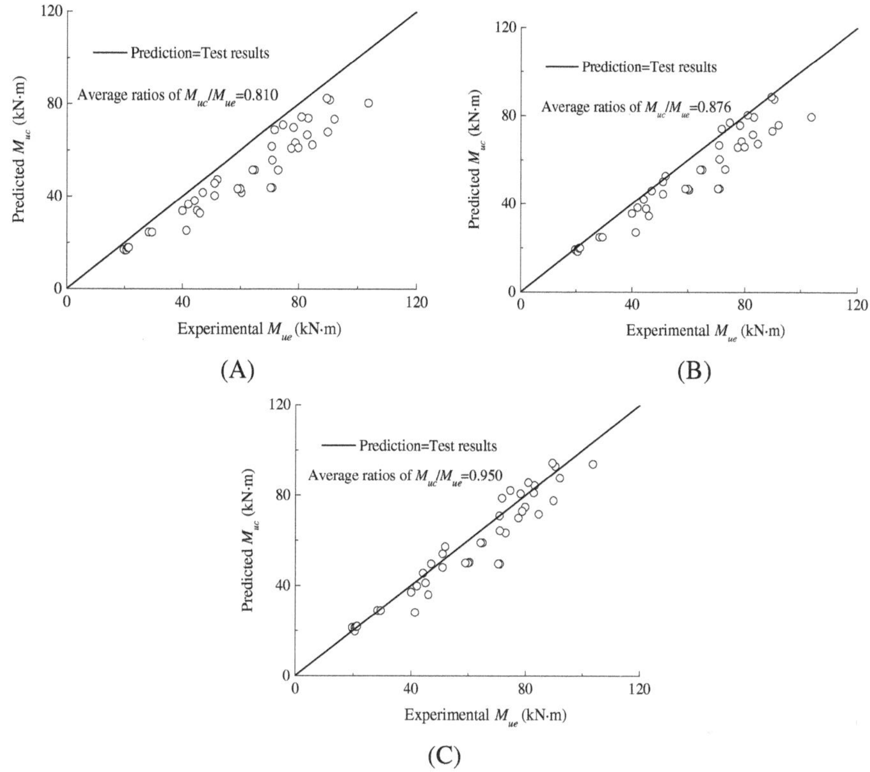

Fig. 4.16 Performance of models: (A) ACI 440.1R-15; (B) CSA-S806-12; (C) Eqs. (4.26)–(4.28).
From Yuan, F., and Y. F. Wu. 2019b. "Analytical method for derivation of stress block parameters for flexural design of FRP reinforced concrete members." *Compos. Struct.* 229, 111459.

and 16.67% by CSA-S806-12. This is because Eqs. (4.26)–(4.28) are derived using the CEB-FIP/EC 2 model of concrete and this model of concrete is relatively more accurate for higher-strength concrete. For normal-strength concrete with $f_{co} < 60$ MPa, Popovics' model of concrete is more accurate.

It can be found from Table 4.1 and Fig. 4.16 that the average ratios of M_{uc}/M_{ue} calculated by ACI 440.1R-15, CSA-S806-12, and Eqs. (4.26)–(4.28) are 0.810, 0.876, and 0.95, respectively. The flexural design strain ε_{cm} is fixed to be 0.003 in ACI 440.1R-15 and 0.0035 in CSA-S806-12. However, it is clearly shown by the flexural theorems that ε_{cm} is not only related to concrete strength but also reinforcement ratio for RC members with elastic reinforcement (Fig. 3.4). It can be seen from Fig. 4.12A that all values of ε_{cm} are greater than 0.003, especially for members with high concrete strength. When the concrete strength is higher than 40 MPa, the value of ε_{cm} is essentially greater than 0.0035 for all reinforcement ratios. The smaller value of the flexural design strain

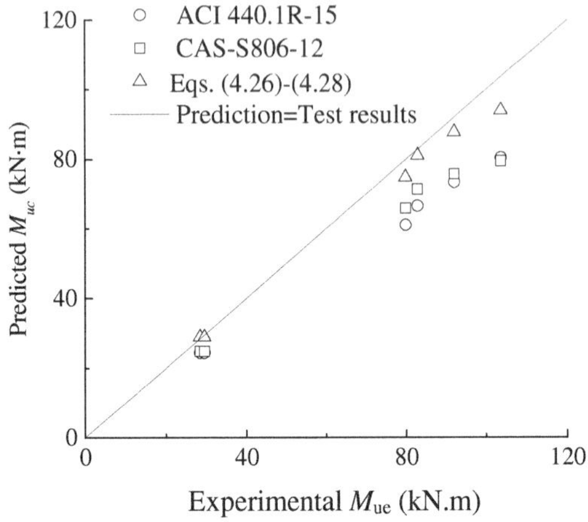

Fig. 4.17 Comparison of models for FRP-reinforced high-strength concrete beams. From Yuan, F., and Y. F. Wu. 2019b. "Analytical method for derivation of stress block parameters for flexural design of FRP reinforced concrete members." *Compos. Struct.* 229, 111459.

ε_{cm} results in substantial underestimation of the flexural strength, because it causes underestimation of not only the internal forces but also its lever arm. When ε_{cm} changes from 0.003 in ACI 440.1R-15 to 0.0035 in CSA-S806-12, the average ratio of M_{uc}/M_{ue} increases from 0.815 to 0.879. Actually, ε_{cm} varies, and a more reasonable way is to use a formula of ε_{cm} related to its key factors, as in Eq. (4.26). Apart from ε_{cm}, the parameters α_1 and β_1 are also closely related to not only concrete strength but also reinforcement ratio, as shown in Figs. 4.12 and 4.13. However, they are defined to be either fixed values or a function of concrete strength only in the existing design codes.

Eqs. (4.26)–(4.28) are derived directly from the flexural theorems. It considers all factors without omissions and approximation, and hence is mathematically precise. It is noted that the theorems are derived with only three assumptions: (1) plane section remains plane; (2) tensile strength of concrete is neglected; and (3) the stress-strain relationships of concrete and reinforcement are fixed and given. Therefore, the accuracy of the derived model, Eqs. (4.26)–(4.28), depends mainly on the accuracy of the stress-strain model of concrete because there is little error in the stress-strain relationship of FRP. For different problems or materials where the stress-strain relationships of concrete are different, these stress block parameters must be re-derived using the theorems and relevant stress-strain relationship.

For design problems where the area of FRP reinforcement is to be calculated for a given design moment, the error produced by existing design codes is even larger, as shown by the design Question 2 in Section 4.2.6. Fig. 4.18 shows the strain distributions for this design question at the onset of the flexural strength calculated with different models. It can be seen that the flexural design strain and the FRP bar strain calculated from Eqs. (4.26)–(4.28) are much larger than those calculated from the

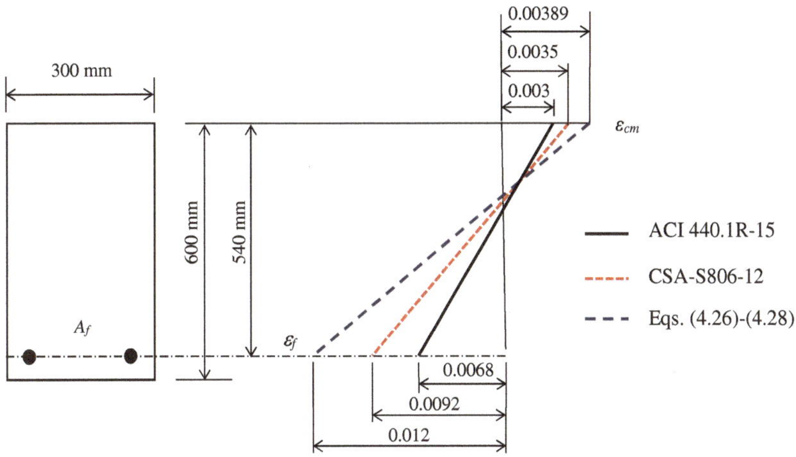

Fig. 4.18 Strain distributions at onset of flexural strength.
From Yuan, F., and Y. F. Wu. 2019b. "Analytical method for derivation of stress block parameters for flexural design of FRP reinforced concrete members." *Compos. Struct.* 229, 111459.

two design codes ACI 440.1R-15 and CSA-S806-12, resulting in more efficient use of FRP material (higher FRP stress at ultimate failure), and hence a smaller required area of FRP bars for a given flexural strength. The use of the more accurate model leads to savings in reinforcement area of 72.3% and 29.6%, respectively, compared with those designed with the ACI and CSA codes.

References

ACI (American Concrete Institute). 2015. *Guide for the design and construction of structural concrete reinforced with FRP bars, ACI 440.1R-15.* Farmington Hills, MI.

Adam, M. A., M. Said, A. A. Mahmoud, and A. S. Shanour. 2015. "Analytical and experimental flexural behavior of concrete beams reinforced with glass fiber reinforced polymers bars." *Construct. Build Mater.* 84, 354–366.

Anderberg, Y., and S. Thelandersson. 1976. *Stress and deformation characteristics of concrete at high temperatures*, Lund, Sweden: Lund Institute of Technology.

Canadian Standards Association (CSA). 2012. *Design and construction of building components with fibre-reinforced polymers*, S806-12. Toronto, Canada.

GangaRao, H. V., N. Taly, and P. V. Vijay. 2006. *Reinforced concrete design with FRP composites*, CRC Press, Taylor & Francis Group.

Grace, N. F., A. K. Soliman, G. Abdel-Sayed, and K. R. Saleh. 1998. "Behavior and ductility of simple and continuous FRP reinforced beams." *J. Compos. Constr.* 2 (4), 186–194.

Hertz, K. D. 2005. "Concrete strength for fire safety design." *Mag. Concr. Res.* 57 (8), 445–453.

Kassem, C., A. S. Farghaly, and B. Benmokrane. 2011. "Evaluation of flexural behavior and serviceability performance of concrete beams reinforced with FRP bars." *J. Compos. Constr.* 15 (5), 682–695.

Khennane, A., and G. Baker. 1993. "Uniaxial model for concrete under variable temperature and stress." *J. Eng. Mech.* 119 (8), 1507–1525.

Khoury, G. A., B. N. Grainger, and P. J. Sullivan. 1985. "Transient thermal strain of concrete: literature review, conditions within specimen and behavior of individual constituents." *Mag. Concr. Res.* 37 (132), 131–144.

Li, L. Y., and J. Purkiss. 2005. "Stress–strain constitutive equations of concrete material at elevated temperatures." *Fire Saf. J.* 40 (7), 669–686.

Lie, T. T. 1992. "Structural fire protection." In *Manuals and Reports on Engineering Practice No. 78*, New York, N.Y.: ASCE.

Purkiss, J. A., and L. Y. Li. 2013. *Fire safety engineering design of structures*, CRC Press.

Qu, W., X. Zhang, and H. Huang. 2009. "Flexural behavior of concrete beams reinforced with hybrid (GFRP and steel) bars." *J. Compos. Constr.* 13 (5), 350–359.

Rashid, M. A., M. A. Mansur, and P. Paramasivam. 2005. "Behavior of aramid fiber-reinforced polymer reinforced high strength concrete beams under bending." *J. Compos. Constr.* 9 (2), 117–127.

Schneider, U. 1986. "Modelling of concrete behavior at high temperatures." *Des. Struct. Against Fire*, 53–69.

Standards Australia (AS). 1998. *Steel structures*, AS 4100-1998. Sydney, Australia, reconfirmed 2016.

Sun, J. F., Y. Li, X. D. Shi, and Z. H. Guo. 2001. "Experimental investigation on behavior of eccentrically compressed reinforced concrete members heated on three faces under different temperature conditions." *J. Build. Struct.* 22 (4), 84–89 (in Chinese).

Sun, J. F., X. D. Shi, and Z. H. Guo. 2002. "Experimental study on mechanical behaviors of three surfaces heated RC beams at elevating temperature and after cooling." *Build Struct.* 32 (1), 34–37 (in Chinese).

Terro, M. J. 1998. "Numerical modeling of the behavior of concrete structures in fire." *ACI Struct. J.* 95 (2), 183–193.

Theriault, M., and B. Benmokrane. 1998. "Effects of FRP reinforcement ratio and concrete strength on flexural behavior of concrete beams." *J. Compos. Constr.* 2 (1), 7–16.

Toutanji, H. A., and M. Saafi. 2000. "Flexural behavior of concrete beams reinforced with glass fiber-reinforced polymer (GFRP) bars." *ACI Struct. J.* 97 (5), 712–719.

Wang, H., and A. Belarbi. 2005. "Flexural behavior of fiber-reinforced-concrete beams reinforced with FRP rebars." *ACI Struct. J.* 51 (230), 895–914.

Yang, J. P., X. D. Shi, and Z. H. Guo. 2001. "Experimental study on behavior of eccentrically compressed reinforced concrete members under different heat-loading paths." *Eng. Mech.* 18 (3), 81–90 (in Chinese).

Yang, J. M., K. H. Min, H. O. Shin, and Y. S. Yoon. 2012. "Effect of steel and synthetic fibers on flexural behavior of high-strength concrete beams reinforced with FRP bars." *Compos. Part B Eng.* 43 (3), 1077–1086.

Youssef, M. A., and M. Moftah. 2007. "General stress–strain relationship for concrete at elevated temperatures." *Eng. Struct.* 29 (10), 2618–2634.

Yuan, F., and Y. F. Wu. 2019a. "Application of RC flexural theorems for member design under elevated temperature." *Eng. Struct.* 201, 109762.

Yuan, F., and Y. F. Wu. 2019b. "Analytical method for derivation of stress block parameters for flexural design of FRP reinforced concrete members." *Compos. Struct.* 229, 111459.

Bond between reinforcement and concrete

5

The reinforcing effect in concrete relies on the bond between the reinforcement and the concrete. The interaction between the reinforcement and concrete substrate depends on the interfacial bond characteristics that control the strength, rigidity, ductility, durability, etc., of reinforced concrete members. Therefore, the bond is a key factor in reinforced concrete. Its fundamental mechanisms and the effects on concrete structures are discussed in this chapter.

5.1 Composite action

Fig. 5.1 shows two steel beams and their response curves under a point load. Beam A is exactly the same as Beam B except that it is cut into two separate parts at its midheight. The contact face between the two parts of Beam A is called the interface. The two parts of Beam A are free to slide or slip at the interface. The strain profile of Beam A is discontinuous at its midheight due to the interfacial slip. The difference of strain between the two parts at the interface is called slip strain ε_{slp}. When the two parts of Beam A are fully bonded together, there will be no slip at the interface, or $\varepsilon_{slp} = 0$, and the strain profile of Beam A becomes one straight line that is identical to that of Beam B. In this case, Beam A becomes Beam B. Therefore, the magnitude of the slip strain at the interface reflects the degree of bond at the interface. It can be seen from Fig. 5.1C that the interfacial bond plays an important role in the load-carrying capacity, rigidity, and ductility of the member.

Fig. 5.2 shows an RC column with one steel plate attached to its side. The bond resistance, or interfacial shear force, at the contact face between the RC column and the steel plate depends on the rigidity K_b of the bolts, which is called a shear connector in composite structures. The response curves of the column with different values of K_b are shown in Fig. 5.3. It can be seen from the figure that the interfacial bond significantly affects the behavior of the composite members. A stronger interfacial bond, or smaller interfacial slip, increases the composite action, which increases the stiffness and strength of the composite members.

The increase in the load-carrying capacity of composite members due to the interfacial bond is called the composite effect, as illustrated by the cantilever beam in Fig. 5.4. The internal bond or shear stress is exposed in Fig. 5.4B. The integration of the shear stress at one side is the total shear force T. Due to action and reaction, the total shear force at the two sides is equal, which does not produce an axial force

Analytical Approaches for Reinforced Concrete. https://doi.org/10.1016/B978-0-12-821164-9.00005-X

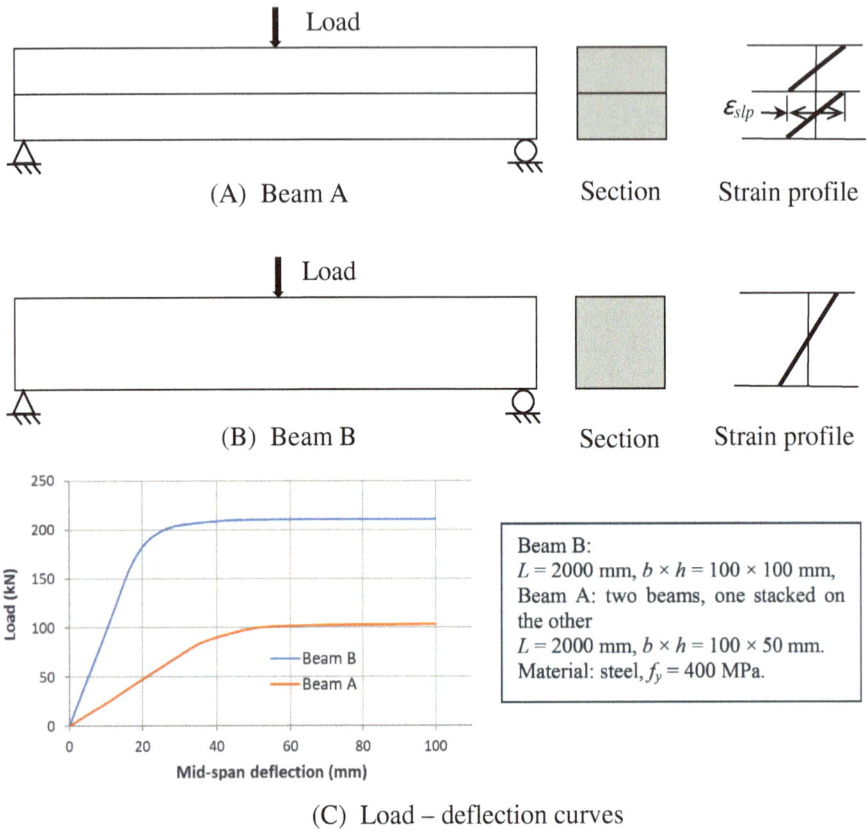

Fig. 5.1 Effect of composite action.

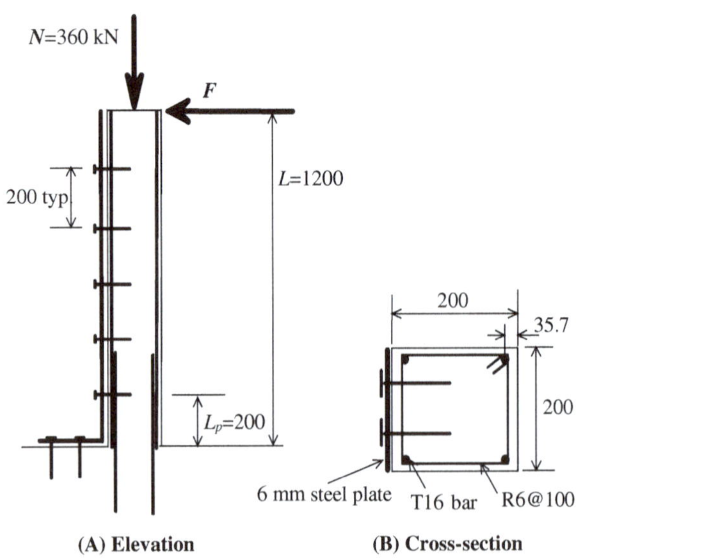

Fig. 5.2 Column details.
From Wu, Y. F. 2002. Seismic retrofitting of rectangular reinforced concrete columns with partial interaction plating. PhD Thesis. Australia: The University of Adelaide.

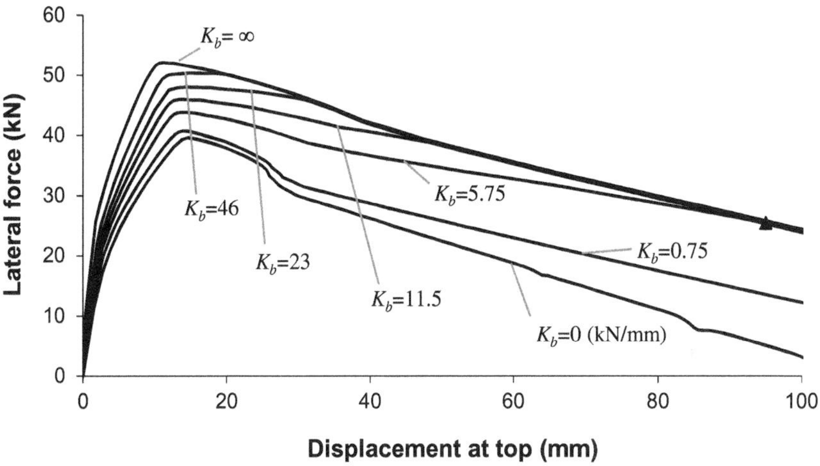

Fig. 5.3 Responses of columns with different bolt rigidities.

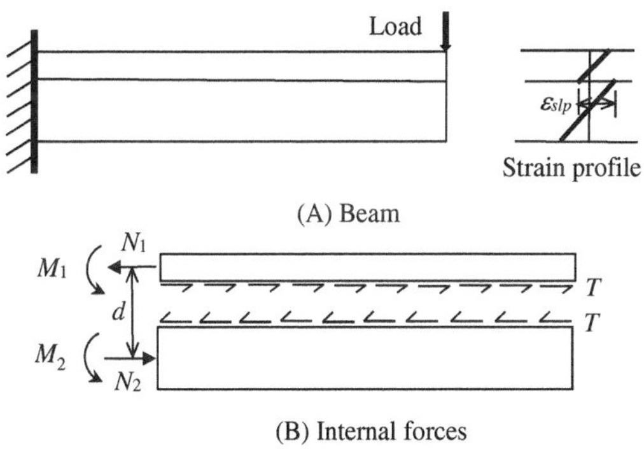

Fig. 5.4 Composite effect.

on the whole beam but induces axial forces in the two parts of the member as shown. From the equilibrium of a separate part in the horizontal direction, one has

$$T = N_1 = N_2 \tag{5.1}$$

The total bending moment of the member is given by

$$M = M_1 + M_2 + T \times d \tag{5.2}$$

It can be seen from Eq. (5.2) that the interfacial bond force T produces an additional moment resistance of $T \times d$, which is why a stronger shear connection (larger T) leads to a greater load-carrying capacity of a composite member, as shown in Figs. 5.1 and 5.3. In other words, the composite action comes from the term $T \times d$.

When the slip strain ε_{slp} is equal to zero in Fig. 5.1A, the two parts of Beam A are fully bonded together, and this case is called full interaction. When the bond stress τ_b

at the interface is zero, the two parts of Beam A can slip freely along the interface, and hence have no interaction. When the bond stress at the interface is between these two cases, it is called partial interaction. Partial interaction can be achieved by a bond at the interface through different mechanisms, including adhesion, friction, and mechanical interlocking. The partial interaction at an interface is also called composite action, and members with partial interaction are called composite members.

The interfacial slip brings another degree of freedom to the structure. Therefore, the structural behavior and analyses of a composite member are more complicated than for noncomposite structures. Full interaction members with no slip or zero slip strain at the interface are not composite members because this extra degree of freedom does not exist, and its strain profile is simplified to one continuous line (plane section). For example, if a structural member is composed of several parts of different materials that are fully bonded together without any interfacial slip, the member is not considered as a composite member from a structural analysis point of view, because there is no extra complexity in the structural analyses caused by the interfacial slip. Therefore, the sufficient and necessary condition to be a composite member is:

$$\begin{cases} \varepsilon_{slp} \neq 0 \\ \tau_b \neq 0 \end{cases} \tag{5.3}$$

With the preceding background, it can be seen that the magnitude of interfacial shear at the interface determines the magnitude of increase in the load-carrying capacity of a composite member. RC members are composed of two materials, concrete and steel, which are bonded together. Whether an RC member is designed as a composite member or not depends on whether it satisfies the sufficient and necessary condition of composite action, i.e., $\varepsilon_{slp} \neq 0$ and $\tau_b \neq 0$. It is later shown in this chapter that RC members are sometimes considered as composite members (e.g., in Sections 5.9 and 5.10) but are designed to be noncomposite members at other times (e.g., in Section 5.11), depending on the effect of the interfacial slip on the relevant problem.

5.2 Bond of reinforcement

In a composite member, separate elements are connected together at interfaces through the bond. Mechanically speaking, the degree of composite action of a structural member depends on the interfacial bond characteristics. Bond resistance at an interface generally involves two directions: (1) the resistance normal to the interface, namely normal bond, and (2) the resistance parallel to the interface, namely shear bond. For flexural members, although the normal bond is necessary to form a composite member, it does not affect structural behavior, and hence is usually not a concern in composite structures. It is the shear bond that makes a difference in the composite action.

Concrete is weak in tensile resistance. Engineers resolve the problem by providing reinforcement in concrete structures. The two materials of concrete and reinforcement are bonded together in a concrete structure to form composite action. In a new

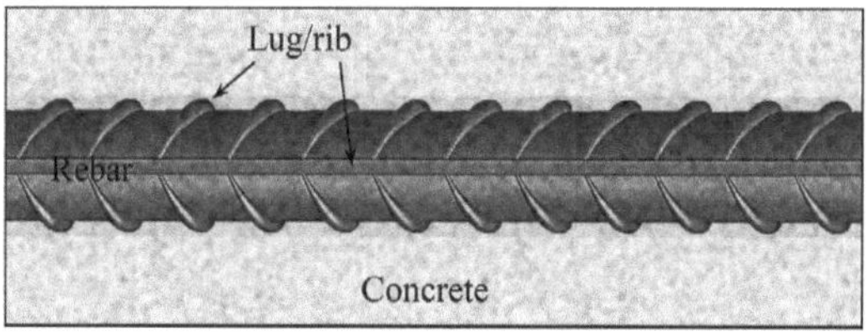

Fig. 5.5 Deformed bar for bond enhancement.

construction, reinforcement is usually provided in the form of reinforcing bars (rebar) cast into concrete (Fig. 5.5). For retrofitting/strengthening/repairing of existing structures, reinforcement is often attached to external faces of a concrete structure (Fig. 5.2). For both cases, the effectiveness of the reinforcement depends on the bond stiffness and strength, as illustrated in Fig. 5.3. In fact, the bond between reinforcement and concrete is one link of a structure, as shown in Fig. 1.9. In general, the design principle is to avoid failure at the link of the bond so that failure occurs at other links involving failure of reinforcing material and concrete. In this way, materials can be fully utilized to achieve more economical designs. However, it will be shown later that bond failure or debonding is sometimes unavoidable, as it is difficult to strengthen the link of the bond. Furthermore, people sometimes deliberately make use of debonding to achieve a certain design purpose such as increasing ductility in special cases (Section 7.3.2(5)).

When a rebar and concrete are cast together, a bond forms naturally. Shrinkage of concrete makes a firmer contact between the concrete and the rebar and hence is beneficial to the bond. To enhance the bond of rebar in concrete, deformation such as lugs/ribs are commonly made on the surface of the rebar to provide interlocking between the two materials, as shown in Fig. 5.5. Bond can be achieved through different mechanisms. Typical bond mechanisms are discussed in the following section.

5.3 Bond mechanisms and bond–slip relationship

There are three, and only three, shear bond mechanisms at a bond interface: adhesion, mechanical interlock, and friction. Any kind of interfacial bond can be considered as, and represented by, a combination of these three bond mechanisms. The stress and strain fields at a bond interface, which may involve bonding agent (adhesive) and a thin layer of adherent materials near the contact face, are much more complex than those in other parts of a structure. To avoid such complexity and facilitate the application of conventional structural theories to composite structures, a convenient and sensible approach is to lump all additional deformations that cannot be included in conventional structural theory (such as beam theory) into a concentrated displacement at the interface, namely, interfacial slip. The interaction between the connected adherents is subsequently simplified into a resistance to the interfacial slip by a bond-slip relationship, which can be identified from bond tests or pull-off tests for a particular type of interfacial joint. Such an ingenious treatment of bond interface converts a

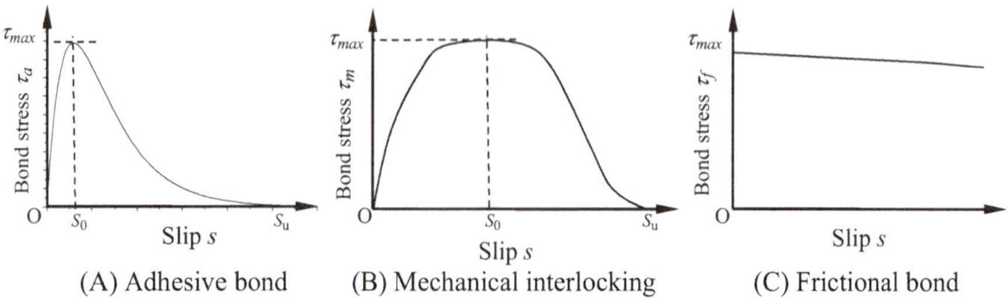

Fig. 5.6 Typical bond–slip relationships.

highly complex problem into one within the regime of conventional structural mechanics and forms the cornerstone of modern composite mechanics.

Therefore, the composite effect and the behavior of composite members are governed by the bond–slip relation. The typical interfacial bond–slip relationships for the three mechanisms are shown in Fig. 5.6 and discussed as follows.

5.3.1 Adhesion

Adhesion is a chemical bond often incurred by a bonding agent such as adhesive applied between two adherents. The possible debonding failure planes in an interfacial joint where a reinforcement as adherent is adhesively bonded to the concrete substrate are shown in Fig. 5.7, which include: (1) adhesive-to-adherent interfaces, (2) cohesive failure in adhesive, (3) cohesive failure in concrete, and (4) adherent failure. A properly designed reinforcing system should have a reinforcement that is sufficiently strong so that failure does not occur in the reinforcing material. A proper bond design requires the adhesive to be sufficiently strong and the bond interfaces be adequately treated so that debonding does not occur in the adhesive and at the adhesive-to-adherent interfaces. Therefore, debonding of an adequately designed reinforcing system of concrete structures usually occurs inside the concrete, i.e., by cohesive failure in concrete. The cohesive failure of concrete is determined by a tensile property of concrete material that is weak and brittle. As a result, the debonding resistance of reinforcement adhesively bonded to the concrete substrate is weak and unreliable.

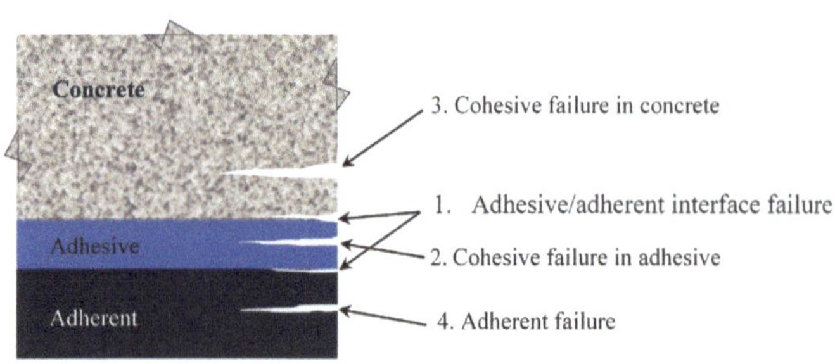

Fig. 5.7 Debonding failure planes.

The typical bond–slip relationship of adhesive bond is shown in Fig. 5.6A. With the increase of the interfacial slip, the adhesive bond increases first until its peak, then decreases gradually to zero. Fig. 5.8 shows the development of the bond stress in a typical pull-off test of an adhesively bonded joint. The qualitative behavior of the adhesive bond is generally similar for both internally bonded rebar and externally bonded reinforcement.

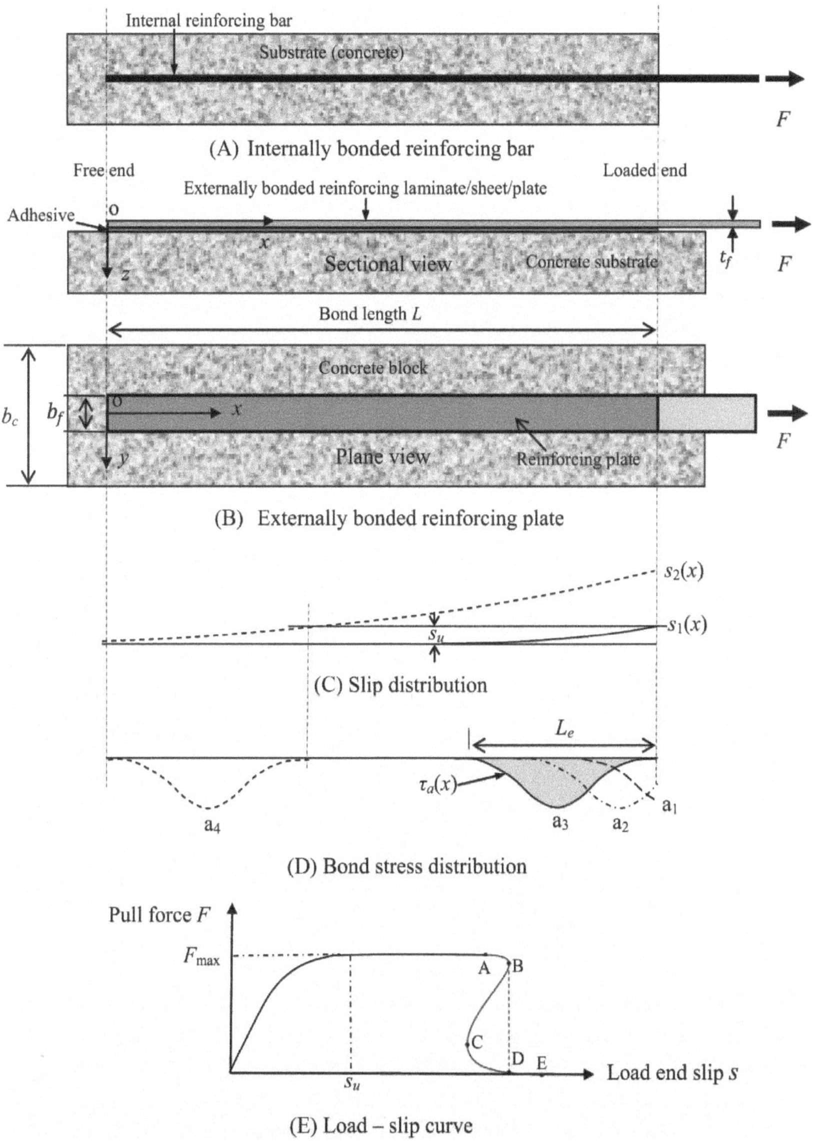

Fig. 5.8 Pull-off test of adhesively bonded joint.

Under a pull load F as shown in Fig. 5.8A and B, the reinforcement (adherent) tends to move, or slip, relative to the substrate concrete. The slip mobilizes the interfacial resistance that has a typical bond stress τ vs. slip s relationship as shown in Fig. 5.6A. When the slip increases from zero to s_u, Fig. 5.6A shows that the bond stress first increases from zero to its peak resistance τ_{max} at the slip of s_0, and then decreases to zero at the slip of s_u. When the slip at the loaded end in Fig. 5.8A and B increases to s_u, the slip distribution along the whole bond interface is shown by $s_1(x)$ in Fig. 5.8C. In this process, the adhesive bond stress distribution gradually develops, as illustrated by a1 → a2 → a3 in Fig. 5.8D, where a3 is a full bond stress block with $\tau = 0$ at the loaded end where the slip just reached s_u. A further pull after a3 causes a further increase in the interfacial slip, e.g., from distribution $s_1(x)$ to distribution $s_2(x)$. Based on the bond–slip relationship in Fig. 5.6A, this simply moves the bond stress block backward from a3 to the distribution shown by a4. The bonded area in front of a4 has reached zero bond stress or has been fully debonded. Therefore, the total adhesive bond force F, which is the area enclosed by the bond stress block, reaches its peak value of F_{max} at a3 and remains constant thereafter, as shown by Fig. 5.8E.

The length corresponding to the bond stress block a3, L_e, as shown in Fig. 5.8D, is called the effective bond length, which means that a bond length greater than L_e does not further increase the total bond force F further. However, an adhesive joint with bond length greater than L_e can increase the ductility of the bond (the plateau in Fig. 5.8E). If a bond length of $L = L_e$ is provided, the pull force will drop to zero immediately after reaching F_{max}.

When the bond stress block moves further beyond the position of a4, its area will reduce, causing a reduction of the bond force F. If the loading process is under load control where F increases monotonically, position a4 is the last moment of loading, corresponding to point A in Fig. 5.8E, which will be the last point of the curve. If it is under displacement control where the slip at the loaded end increases monotonically, point B in Fig. 5.8E will be the last point of the curve. If the pull-off test is controlled by the free-end slip where the slip at the free end of the plate is monotonically increased, the whole unloading process involving a "snap back" curve BCDE will be observed, which will be further discussed in Sections 6.4.3 and 6.4.4.

The effective bond length is a function of the properties of adhesive and adherents. For normal steel or fiber-reinforced polymer (FRP) reinforcing sheet/plate externally bonded to concrete with a thin coat of normal epoxy resin, L_e *is* usually in the range of 50–200 mm (Chen and Teng 2001), depending on properties of the joint. This length is small and, consequently, the adhesive bond strength is limited. Extensive studies on adhesively bonded systems have been reported in the literature. More details of the adhesive bond and its modeling are given in Sections 6.3 and 6.4. It can be seen in these sections that the bond capacity is affected by concrete strength rather than the adhesive strength. The reason is that debonding is governed by cohesive failure in concrete (Fig. 5.7), as discussed earlier. As a result, a stronger adhesive does not provide additional bond strength, and hence, adhesive property usually does not affect the interfacial properties unless a soft and thick layer of adhesive is used (Dai et al. 2005). On the other hand, adhesives and the surface material of adherents degrade significantly due to aging, weathering, differential thermal movement, etc., particularly on the exposed external surfaces. Therefore, the long-term performance of the adhesive bond is usually unreliable and should not be relied upon in engineering.

For internally bonded rebar, no adhesive is provided, and the adhesion depends on the chemical bond between cement mortar and the bar surface, which is weaker than the adhesive bond of epoxy. Therefore, the adhesion is weak and poor in internally bonded rebar. The bond between a plain bar with a smooth surface and concrete actually relies upon the frictional bond largely, which is discussed in the following section.

5.3.2 Mechanical interlock

Mechanical interlocking is normally more reliable and robust than an adhesive bond, and hence is widely used in engineering. Mechanical interlocking between adherents can be achieved through different means, such as anchoring/bolting (Fig. 5.2) and embedment of a protruded part from one adherent into another adherent (substrate), e.g., the bond of deformed bars in concrete in Fig. 5.5. Mechanical interlock is also called dowel action when anchors/bars are used to connect the interface. The dowel action provided by mechanical fasteners can provide strong interfacial shear resistance. The typical bond–slip relationship due to mechanical interlock is shown in Fig. 5.6B, where the bond stress is in an average sense, i.e., the bond stress $\tau =$ dowel force of one anchor divided by its tributary area.

Fig. 5.9 shows the typical dowel action provided by a bolt. The dowel force of an anchor is usually transmitted between adherents through two paths (Fig. 5.9A): (1) by direct bearing of the anchor shaft to the adherent as shown by F_b in Fig. 5.9A; and (2) the adhesion and friction, T_1, at the interface 1 that is transmitted through the bolt shaft into the substrate. The interfacial shear force T_2 at interface 2 is not transmitted through the anchor and is not considered part of the dowel action. Conventional steel plating (Fig. 5.2) and a mechanically fastened FRP plate (MF-FRP) as shown in Fig. 5.10A works largely through Path 1, whereas fiber anchoring as shown in Fig. 5.10B, where the fan of a fiber anchor is adhesively bonded to the top face of

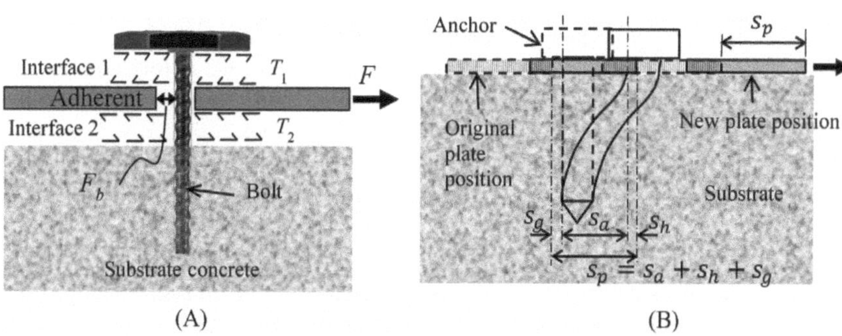

Fig. 5.9 Dowel action: (A) Shear resistances, and (B) Slip.
From Wu, Y. F., L. He, and L. C. Bank. 2016. "Bond test protocol for plate-to-concrete interface involving all mechanisms." *J. Compos. Constr.* 20 (1), 04015022.

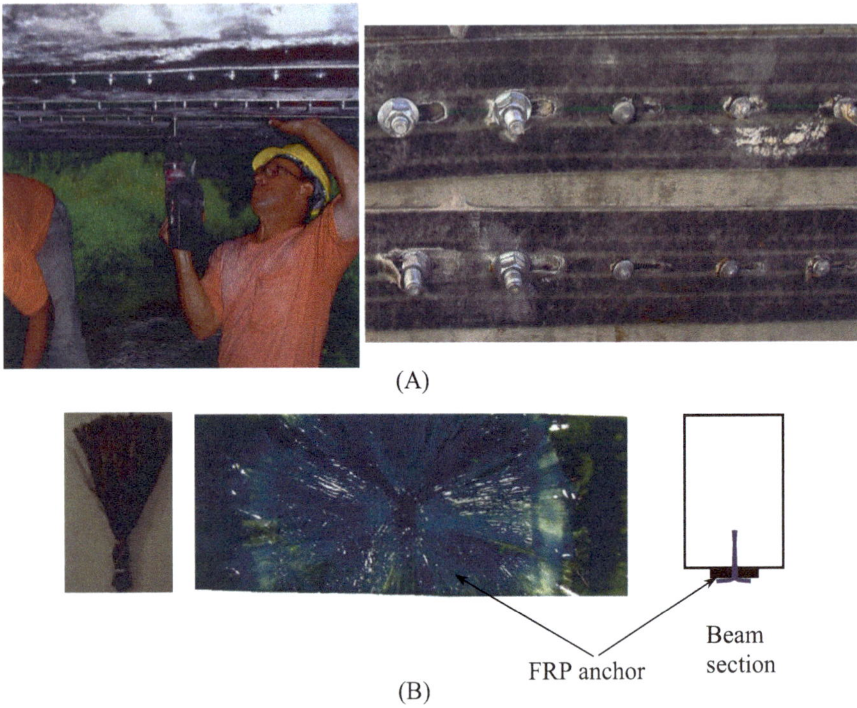

Fig. 5.10 Mechanical interlocking systems: (A) MF-FRP system, and (B) Fiber anchoring system.
(A) Courtesy: L. C. Bank, A. Lamanna, and D. Arora.

a reinforcing plate (adherent), works through Path 2. More details of these mechanical fastening systems are given in Sections 6.2 and 6.5.

The interfacial slip comes from the deformations of mechanical fasteners and the substrate materials (adherents), as well as the closing of the gap between the anchor shaft and the side of the hole in the adherent, as shown in Fig. 5.9B. The total interfacial slip s_p is the summation of the anchor deformation s_a, the deformation of the hole including the cut by the anchor into the plate s_h, and the initial gap between the hole and the anchor shaft s_g, i.e.,

$$s_p = s_a + s_h + s_g \tag{5.4}$$

As a result, the bond through mechanical interlocking is usually a bit looser in the beginning (at small slip) compared with the tight connection of adhesion and friction. However, it can be much stronger and more reliable than the adhesive bond when its resistance is fully mobilized. It is a common and popular choice to connect/bond elements in steel structures and conventional composite structures by mechanical interlocking. Modeling of the mechanical bonding system is given in Section 6.2.

5.3.3 Friction

A frictional bond is induced by interfacial slide that causes frictional resistance against the relative movement. The frictional bond can be very significant and tight, far more effective than one might imagine. A good example is the high strength friction grip (HSFG) bolts used for steel joint connections. HSFG bolts provide a much tighter and stronger connection than the ordinary bolts that rely on dowel action to transfer interfacial shear.

The typical bond–slip relationship of a frictional bond is shown in Fig. 5.6C. Frictional resistance τ_f is given by

$$\tau_f = p \times \mu \tag{5.5}$$

where p and μ are the normal pressure and frictional coefficient, respectively. Friction is usually considered as a constant during slipping because both normal pressure and frictional coefficient are considered to be constants. In reality, however, it may decrease gradually with the increase of interfacial slip for two reasons: (1) the relative sliding along the interface smoothes the originally rougher interface, which reduces the frictional coefficient, and (2) if the normal pressure is caused passively by the reinforcement passing through the interface (see Fig. 5.24), the wearing of the sliding interface also reduces the normal pressure.

A sliding interface can be quite rough, which may also induce mechanical interlocking apart from friction. However, it is not difficult to separate friction from mechanical interlocking by following the definition of friction. Friction is proportional to the normal pressure applied. The interfacial shear resistance without applying a normal pressure is the mechanical interlocking. The additional interfacial shear resistance after a normal pressure is applied is the friction. Therefore, for a particular interface or joint, two tests can be utilized to determine the bond mechanisms: the first without a normal pressure and the second with a normal pressure. If the total resistance from the first and second test is F_1 and F_2, respectively, the friction F_f and the mechanical interlock F_m are given by

$$F_m = F_1 \tag{5.6}$$

$$F_f = F_2 - F_1 \tag{5.7}$$

The frictional coefficient of the interface is given by

$$\mu = \frac{F_f}{N} \tag{5.8}$$

where N is the normal force applied to the interface. Two typical frictional bonding systems are introduced in Sections 5.12 and 6.5.

5.4 Bond design of rebar

All the preceding three mechanisms play an important role in the bond of rebar in concrete. For plain bars with a smooth surface, the chemical adhesion and friction dominate the bond. As the chemical bond is brittle, it breaks down when the reinforcing bars start to slip in concrete. Further slip of the bar is only resisted by the friction between the substrate concrete and the bar. For deformed bars with lugs/ribs on the surface, mechanical interlocking plays a significant role in resisting slip after the chemical bond breaks down. The bearing of lugs/ribs against concrete not only causes the mechanical interlocking action but also pushes surrounding concrete outward, which causes a normal internal pressure on the substrate concrete. This normal pressure may cause longitudinal splitting cracks of the substrate concrete (Fig. 5.11A). Thicker concrete cover and more transverse reinforcement (stirrups) provide higher confinement against longitudinal cracking. The confinement pressure induces frictional resistance to interfacial slip and its magnitude depends on the concrete cover thickness surrounding the bar as well as transverse reinforcement. The mechanical interlocking diminishes when the concrete between lugs of bar is sheared off by rebar slip (Fig. 5.5), after which only frictional resistance remains. More details on the bond mechanism of rebar can be found in the CEB-FIP report *Bond of Reinforcement in Concrete* (fib 2000).

Different bond failure modes occur when a rebar is under different confinement conditions. The bond–slip relationships of three typical bond failure modes are illustrated in Fig. 5.11B (Harajli et al. 2004). When a rebar is embedded inside concrete with sufficiently thick concrete cover surrounding the bar or heavy transverse reinforcement so that the concrete cover does not split (crack through) when the rebar debonds, the concrete between lugs of bar is pulled out when this failure mode occurs; hence it is referred to as pull-out failure. In this case, the surrounding concrete is still

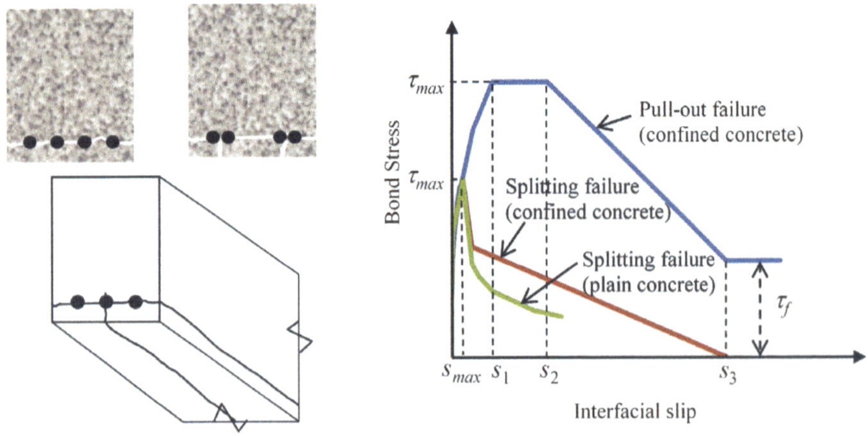

(A) Longitudinal splitting cracks (B) Bond-slip curves

Fig. 5.11 Steel bar-concrete interfacial bond–slip relationship.

intact when the bar is pulled out of the concrete. The residual bond strength (τ_f) is due to remaining frictional resistance when the bar slides in concrete. This is the preferred bond failure mode as the surrounding concrete provides the maximum possible bond resistance to the bar. The third failure mode shown is the splitting failure without significant confinement from the surrounding concrete, typically for a rebar embedded in plain concrete without a thick concrete cover. The bond failure occurs when the surrounding concrete splits into parts due to longitudinal cracks (Fig. 5.11A). Rebar loses its restraint and support from the surrounding concrete at complete splitting of the concrete. The second failure mode, splitting failure in confined concrete, occurs in the case of a confinement condition between the preceding two cases, typically when a certain restraint exists to prevent complete splitting of the cover concrete. It is applicable to cases with moderate transverse reinforcement that prevents complete loss of confinement after significant longitudinal splitting of the concrete. The remaining confinement provides certain residual friction and dowel action to the bar.

There are many factors that affect the bond–slip relationship, such as concrete strength, concrete cover thickness, rebar diameter, stirrups, rebar spacing and splicing, etc. (Wu and Zhao 2013). Due to the complexity of the bond problem, different bond test conditions in different research works, and the inherently unstable nature of debonding, bond test results are significantly scattered. Consequently, significant differences exist in test results reported in the literature and in bond models that are developed from test results. As a result of the large scattering, bond design must be more conservative.

The development of bond resistance is explained in Fig. 5.12. When the bar is pulled gradually at the loaded end, its bond stress block develops from a1 → a2 → a3,

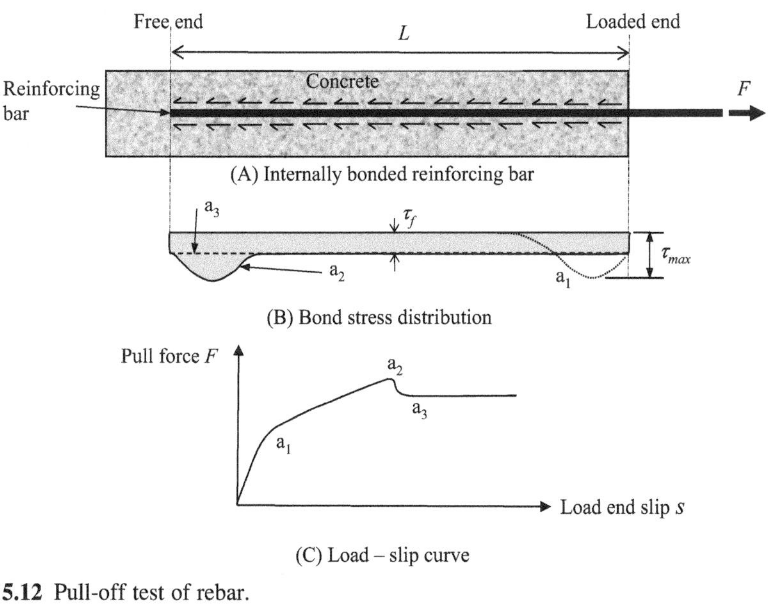

Fig. 5.12 Pull-off test of rebar.

as shown in Fig. 5.12B, where a2 refers to the maximum enclosed bond stress block (shown by the solid line) and a3 corresponds to the uniform bond stress distribution of τ_f (dashed line) throughout the bond length after the onset of the slip S_3 (Fig. 5.11B) at the free end of the bar. The bond stress along the whole bond length remains a constant value of τ_f with further increase in the interfacial slip. Therefore, the total bond resistance (the enclosed area of the bond stress block) or the pull force F stays constant after a3 as shown in Fig. 5.12C. Due to the residual bond τ_f, the total bond resistance increases when the bond length increases. In such a bond system, the total bond resistance can continuously increase with the increase of bond length. In other words, a sufficient bond strength can always be achieved if a sufficient bond length is provided.

The general requirement for a sufficient anchorage length is that the bar should not be pulled out at its design tensile force F, or

$$F = A_{st} f_{st} = \pi \frac{d_b^{\,2}}{4} f_{st} \leq \pi d_b L f_{bu} \tag{5.9}$$

where A_{st}, d_b, L, and f_{st} are the cross-sectional area, diameter, bond length and the stress of the rebar, respectively; f_{bu} is the average bond strength at the onset of maximum bond resistance (corresponding to a2 in Fig. 5.12B). It can be seen from Fig. 5.12B that the average bond stress for a2 varies when the bond length changes. Therefore, a conservative value of f_{bu} should be taken in a design code, considering other factors such as large scattering of bond test results, and also to limit local cracking surrounding rebar caused by debonding. Rearrangement of Eq. (5.9) gives

$$L_b \geq \frac{f_{st}}{4 f_{bu}} d_b = K d_b \tag{5.10}$$

where L_b is the required bond length; and $K = \frac{f_{st}}{4f_{bu}}$ which is related to the stress of the bar. Therefore, in reinforced concrete, sufficient anchorage/bond of a rebar in concrete is usually achieved by providing a bond length that satisfies Eq. (5.10), namely the minimum anchorage length. The maximum allowable tensile stress f_{st} of a bar is its design yield strength f_y. When $f_{st} = f_y$, $K = K_a = \frac{f_y}{4f_{bu}}$, which is a constant. The anchorage length corresponding to this value is called the development length L_d. Therefore, the bond design requirement of a rebar is given as

$$L_b \geq L_d = K_a d_b \tag{5.11}$$

In RC design, rebars are generally designed to yield at ultimate loads, and therefore the required bond length is given by multiples of bar diameter, $K_a d_b$. The values of K_a or its formula are provided in design codes for various bond scenarios. Depending on bar type, concrete strength, and other bond conditions, the value of K_a varies from around 10 to more than 70. Therefore, the required anchorage length can be large. When there is insufficient room to allow for a straight anchorage length, rebar can be bent/hooked or anchored, as shown in Fig. 5.13. A bend or hook will produce high

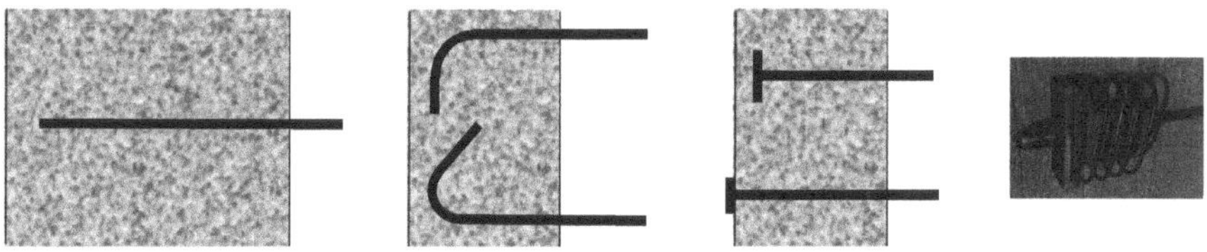

(A) Straight (B) Bend/hook of bar (C) End anchorage (D) End anchor

Fig. 5.13 Anchorage of rebar.
(D) Courtesy: F. Zhang.

stress in concrete in the vicinity of the bend that may damage the concrete material. A straight portion after a bend/hook has been found ineffective in increasing bond (Park and Paulay 1975). Design guidelines are given in design codes on bending angle, radius, and length of the straight portion after a bend, which are based on experimental tests. End anchorage will also produce a high stress concentration of concrete at the end. For a prestressed concrete tendon, the stress in the concrete could be so high that it causes a burst of concrete at the anchor position. In this case, concrete material surrounding an anchor can be confined, such as by using spiral transverse reinforcement (Fig. 5.13D), to increase concrete strength. Confinement can increase concrete strength several times, as discussed in Chapter 8.

5.5 Anchorage design of rebar for flexural members

The bond stress distribution of rebar in a flexural member is significantly different from that of the bar in a pull-off test shown in Fig. 5.12. This is because the rebar in Fig. 5.12 is only pulled at one end while a rebar inside a flexural member is subjected to tension all over the place. The tension force depends on the bending moment at a particular cross section. Fig. 5.14 shows a typical segment of a beam under single

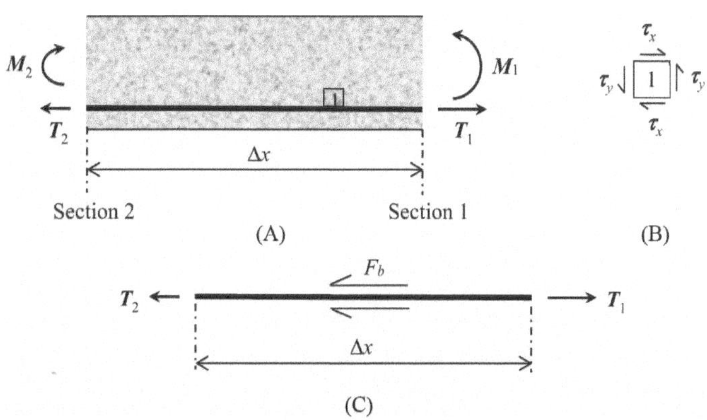

Fig. 5.14 Bond in a typical segment of flexural member.

curvature bending (M does not change sign inside Δx). Assume the beam has a uniform cross section along the span, and the bending moment at section 1 is greater than or equal to that of section 2, i.e., $M_1 \geq M_2$, hence the tensile force of the bar at the two sections satisfies $T_1 \geq T_2$. The bond stress of rebar inside the segment can be discussed for the following three cases:

(1) Case (a) $M_2 = 0$ and $T_2 = 0$:

This case is identical to that in Fig. 5.12, and hence the design of bond length can be based on Eq. (5.10).

(2) Case (b) $M_1 = M_2$ and $T_1 = T_2$:

This case happens in a segment of a member within the pure bending zone of flexural members, such as the region between the two point-loads in Fig. 1.3 where the shear force is zero. The shear stresses of a small concrete element 1 in direct contact with the rebar are enlarged and shown in Fig. 5.14B. τ_y is the shear stress on a cross section inside the segment that is caused by the shear force of the cross section. As the shear force is zero for all cross sections, τ_y is equal to zero anywhere inside the segment. From statics, we know $\tau_x = \tau_y$, hence τ_x must be zero too. On the other hand, the balance of the forces on the rebar in the x-direction in Fig. 5.14C requires that the total bond force F_b on the rebar be zero because $T_1 = T_2$. Because Δx in Fig. 5.14C can be any value, the only possible solution of this condition is that the bond stress τ is zero anywhere along the bar. As the bottom of element 1 is in contact with the rebar, τ_x of element 1 must be equal to the bond stress, and hence is zero too. This is consistent with the results from the preceding cross-sectional analysis.

The previous analyses show that no bond exists between concrete and rebar in a pure bending zone of a flexural member. In other words, the composite action between longitudinal rebars and concrete in the region does not require a bond between the two elements, which seems against the principle discussed in Section 5.1. However, careful analysis of the bond stress in a whole member reveals that the composite action is developed through the bond outside the pure bending zone, as discussed shortly.

(3) Case (c) $M_2 > 0$ and $T_2 > 0$:

From Fig. 5.14C, one has $F_b = T_1 - T_2$. Theoretically, the ultimate bond resistance F_{bu} for any segment in a beam has to be checked for $F_{bu} \geq F_b = T_1 - T_2$ in which $F_{bu} = \pi d_b \Delta x f_{bu}$. However, there are a large number of possible segments to be checked, which is impractical to be done in engineering designs. Therefore, case (c) is difficult to be designed according to Fig. 5.14.

As $T_2 > 0$, $F_b = T_1 - T_2$ leads to $F_b < T_1$. For a given value of T_1, the maximum bond length is required when $T_2 = 0$. This means that if T_2 is ignored and the bond is designed for T_1 only, the result will be conservative. In design practice of reinforced concrete structures, simple empirical and conservative rules are followed to ensure an adequate bonding condition. For example, each curtailed bar should extend beyond the point at which it is no longer needed for a distance that it is well anchored into the concrete. In other words, even when a rebar is not needed (i.e., $T_1 = 0$), it is still fully anchored with a length sufficient to develop its full strength (anchorage length requirements are different at different zones). In this way, for any cross section where $T_1 \neq 0$, the bond length from the cross section to the end of the bar always has a bond length that satisfies case (a). By adopting such simple approaches, the bond requirement of $F_b = T_1 - T_2$ is always and conservatively satisfied. In this way, debonding will not happen, no matter whether it is case (a), case (b), or case (c); meanwhile it provides a simple, practical, and conservative design approach for a very complicated bond design problem. This is the rationale of bond design practice for reinforced concrete.

5.6 Beam action and arch action

From structural mechanics one has:

$$V = \frac{dM}{dx} = \frac{dT \cdot z}{dx} = T \cdot \frac{dz}{dx} + z\frac{dT}{dx} = T \cdot \frac{dz}{dx} + z \cdot q \tag{5.12}$$

where T and z are the resultant tension force of the reinforcement and its lever arm from the tension force to the centroid of the compression force, respectively; $q = \frac{dT}{dx}$, the rate of change of force in the tension bar, which is equal to bond force per unit length. The first term in Eq. (5.12), $T \cdot \frac{dz}{dx}$, is called the arch action and the second term, $z \cdot q$, is referred to as the beam action (Park and Paulay 1975). Details of these actions are discussed in the following paragraphs.

A typical strut/arch action is shown in Fig. 5.15A, where the steel bar and the concrete are debonded in the span and only connected to the concrete strut at the supports. Under a point load, the internal forces of concrete form a triangular strut as shown. The intersecting point of a cross section with the concrete strut gives the location of the compression centroid of the cross section (Fig. 5.15A). The shear force V is obtained from Fig. 5.15A as

$$V = T \cdot \tan(\theta) = T \cdot \frac{dz}{dx} \tag{5.13}$$

where $\frac{dz}{dx}$ is the rate of change of the lever arm, i.e., the change of the vertical location of the compression centroid of concrete per unit length, which is the slope of the compression strut that is equal to $tan(\theta)$. When there is no bond between concrete and rebar, or $q = 0$, Eq. (5.12) degenerates into Eq. (5.13). In other words, the shear force of the member only transmits through the strut/arch action when rebars are not bonded to concrete.

For an ordinary RC beam under an arbitrary distributed load as shown in Fig. 5.15B where rebars and concrete are bonded together, there is another shear transfer mechanism, namely beam action. The thick arrows in Fig. 5.15B indicate the path of the compression resultant force, i.e., the location of the resultant compression force (F_c in Fig. 2.3) at different cross sections. Clearly, the path of the arrow forms a compression strut or arch. The internal forces are studied by isolating an arbitrary small element from Fig. 5.15B, as shown in Fig. 5.15C.

As the internal strut (dashed thick arrow in Fig. 5.15C) shows the position of the resultant compression force, the resultant force can be dissolved into a horizontal component and a vertical one. Due to the balance of horizontal forces at section A, the horizontal component of the strut force must be equal to the tension force of the rebar, T. The vertical component forms a shear force V_a as shown, which is given by

$$V_a = T \cdot \tan(\theta) = T \cdot \frac{dz}{dx} \tag{5.14}$$

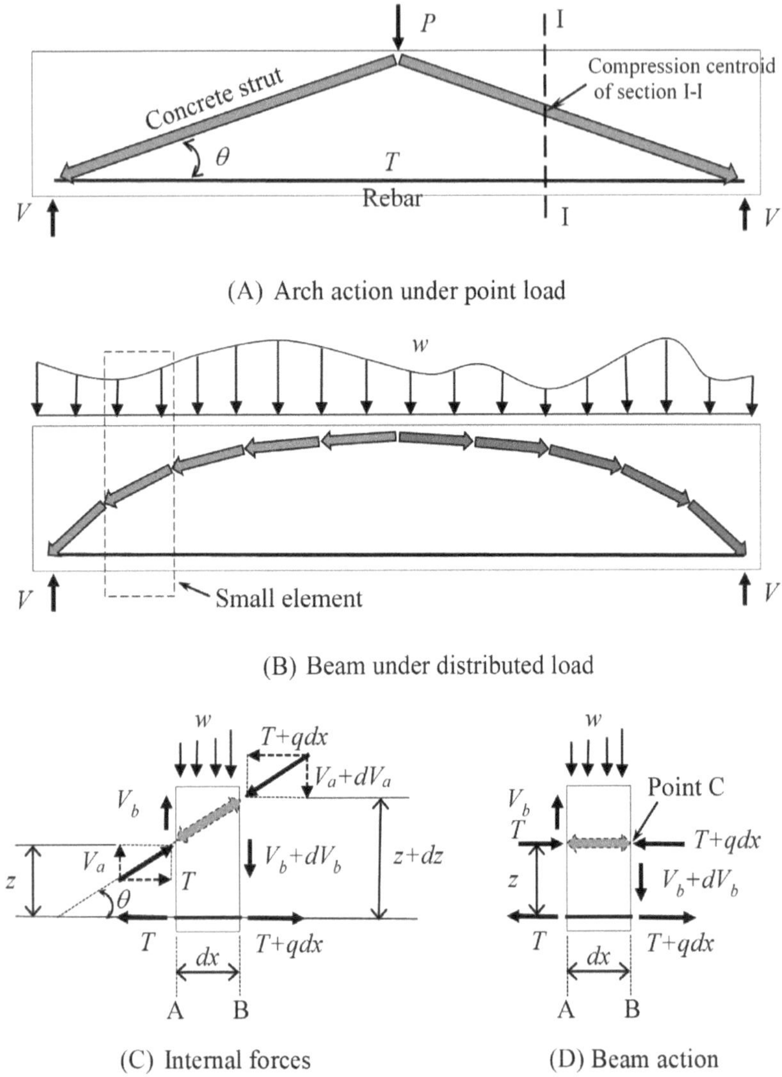

Fig. 5.15 Arch and beam actions.

In the meantime, the bending moment at section A is given by

$$M_A = T \cdot z \tag{5.15}$$

At section B

$$M_B = (T + qdx)(z + dz) \tag{5.16}$$

Substituting Eq. (5.15) into Eq. (5.16) and omitting higher order terms gives

$$M_B - M_A = dM = Tdz + qzdx \tag{5.17}$$

Dividing both sides of Eq. (5.17) by dx gives Eq. (5.12).

For the case of $dz = 0$ or $\theta = 0$, the internal forces are shown in Fig. 5.15D. Taking the moment about Point C in Fig. 5.15D and omitting higher order terms gives

$$zqdx - V_b dx = 0 \tag{5.18}$$

which gives

$$V_b = qz \tag{5.19}$$

Eq. (5.19) means that the unit bond force q will produce a rotating effect that must be counter-balanced by an additional shear force V_b. In other words, there is another shear force transfer mechanism, namely the beam action that acts through the bond between rebar and concrete.

Substituting Eqs. (5.14) and (5.19) into Eq. (5.12) gives

$$V = V_a + V_b \tag{5.20}$$

Therefore, the total shear force at a section is the summation of these two mechanisms, i.e., the arch/strut action and the beam action. The preceding analytical studies explain that transverse loading applied on a beam is transmitted to supports through two mechanisms: (1) the internal arch/strut action that is formed by the change of location of the resultant compression force, and (2) the beam action that is produced by the bond.

For very slender members (with long shear span), the arch/strut action is small due to the smaller value of $tan(\theta)$ or $\frac{dz}{dx}$. For beams with unbounded rebar or tendon, there is no beam action as q is zero. For normal RC beams, both arch and beam actions exist, and they act together to transmit applied lateral loads. It is noted that in a pure bending region of a beam where d$M = 0$, the only solution of Eq. (5.17) in this case is when $dz = 0$ and $q = 0$, i.e., both arch action and beam action are zero.

5.7 Effect of cracks on bond

Cracks affect bond distribution. As shown in Fig. 5.16, concrete on the two sides of a crack tends to separate while the rebar tries to pull them back through the bond. As the local pull forces on the two sides of a crack are in the opposite directions, the local bond stress on the two sides of a crack has an opposite sign. In other words, the distribution of bond stress in a cracked RC member is complicated. Fig. 5.17 illustrates the distribution of bond stress along the longitudinal bar of a simply supported RC slab, which is calculated from a finite element analysis of the member. Clearly, the

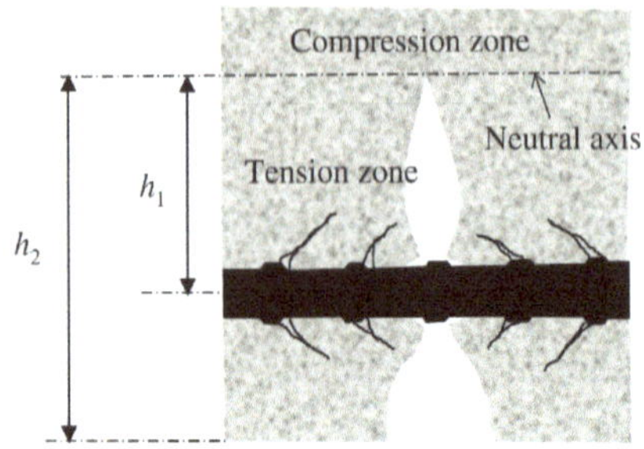

Fig. 5.16 Effect of cracking on bond.

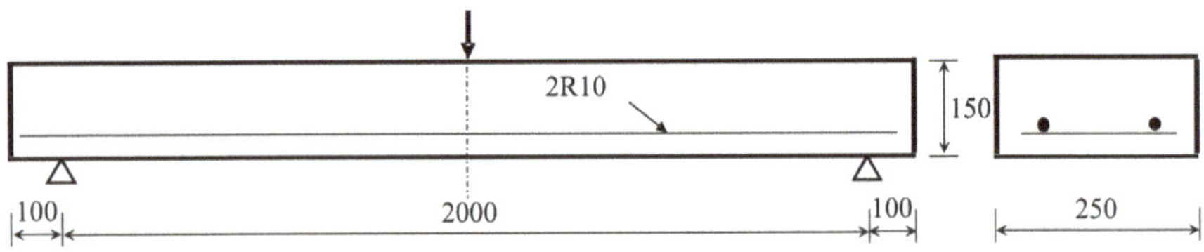

(A) Details of RC slab

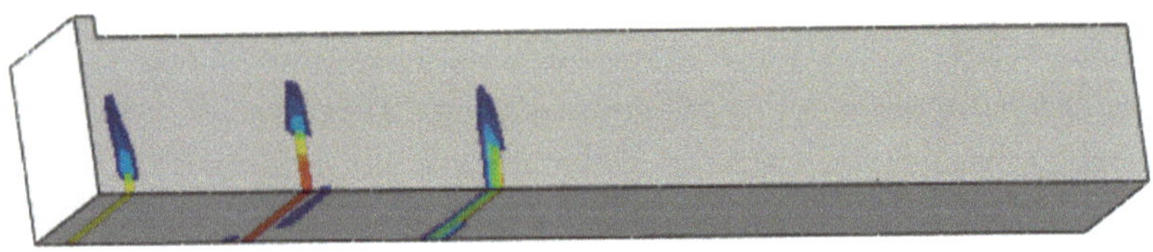

(B) Crack pattern

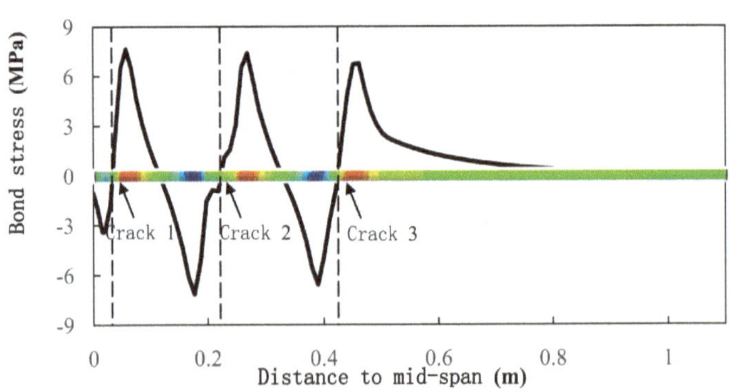

(C) Bond stress distribution

Fig. 5.17 Effect of crack on bond stress distribution.

bond stress distribution is very different from that in a pull-off test of bar (Fig. 5.12B). Cracking causes complications for bond stress calculation and design. Design codes take cracks into consideration in anchorage design by adopting a simple strategy: by recommending a longer anchorage length when the rebar is curtailed/cut in a tension zone of concrete. In other words, the rebar anchorage length is different for rebars that terminate in a tension zone from that of rebars terminated in a compression zone. The value of anchorage length recommended in design codes is generally empirical based on experimental results and engineering consensus.

5.8 Effect of bond on cracking

It can be seen from Fig. 1.6 that the crack width and crack distribution are very different among the six types of test beams. The number of cracks generally increases, and the crack width decreases from the Type 1 beam to the Type 4 beam. Type 2 and Type 3 beams are both reinforced with plain round bars, but the diameter of the Type 3 beam is larger. Therefore, the bond in the Type 3 beam is larger than that in the Type 2 beam. Type 4 to Type 6 beams are reinforced with deformed bars with ribs on the rebar surface, which provides a stronger bond than the Type 3 beam. Therefore, a stronger bond leads to smaller but more cracks while a weaker bond causes wider but fewer cracks. The mechanism of the bond effect on cracking is studied in this section.

5.8.1 Crack formation in RC ties

The relation between bond and cracking can be explained using an RC member under axial tension. The RC tie with a uniform cross section in Fig. 5.18A has a slightly different tensile strength at different cross sections due to nonhomogeneity of concrete material and construction error in the cross-sectional area. Before the occurrence of a crack, the tension force F applied to the member will produce two components, a force on the concrete section F_c and another on the steel bars F_s. F_c and F_s are uniformly distributed along the length after a small transition distance away from the two ends based on Saint-Venant's principle. The nominal stresses in concrete and steel are, respectively, $\sigma_c = F_c/A_c$ and $\sigma_s = F_s/A_s$, where A_c and A_s are the nominal area of concrete section and steel bars, respectively. Assume the cross-sectional tensile strength of concrete $A_c f_t$ varies along its length where f_t is the tensile strength of concrete material, as shown in Fig. 5.18B, and the weakest cross sections are A, B, C, D, with $(A_c f_t)_A < (A_c f_t)_B < (A_c f_t)_C < (A_c f_t)_D$. When the tension force F increases gradually, the first crack will occur at the weakest section A when F increases to a certain value, so that the concrete stress σ_t reaches the concrete tensile strength f_t. Further increase in F will move up the dashed line in Fig. 5.18B to touch the solid line at points B, C, and D successively and produce subsequent cracks at these sections.

Cracking at a cross-section will release the tensile stress in concrete that causes a drop of it on both sides of the crack, as shown in Fig. 5.18C after the occurrence of the crack at section A. The concrete stress on the surface of the member will return to its

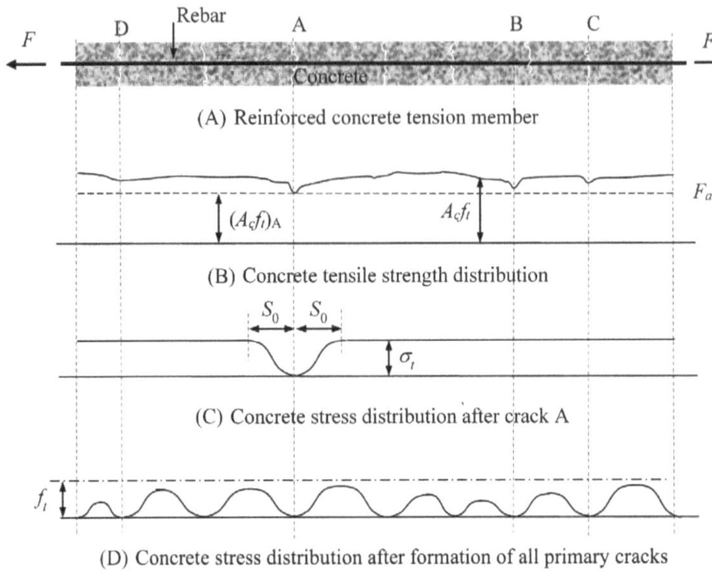

Fig. 5.18 Development of primary cracks.

nominal value σ_t at a certain distance S_0 away from the crack (Fig. 5.18C), referred to as slip length or transfer length, that is closely related to bond (Piyasena 2002). At the end of the slip length, the slip between the steel bar and surrounding concrete is zero and the concrete stress beyond the zero-slip point is not influenced by the formation of the crack. As the concrete surface stress is smaller within the slip length than that beyond the slip length, another crack always forms outside the slip length.

Another crack cannot be developed within a distance of S_0 from a crack because of the lower tensile stress within the slip length. Therefore, the closest crack that can be formed is at a distance of S_0 from crack A, giving a minimum crack spacing of $a_{min} = S_0$. If two cracks at a distance of less than $2S_0$ have formed, σ_t between the two cracks must be below f_t and no new crack can be formed between them. As a result, the crack spacing a must be in the range of $2S_0 > a \geqq S_0$. Stress release in concrete similar to that of crack A will occur at sections B, C, D, etc. successively, producing a final σ_t distribution, as shown in Fig. 5.18D, that is smaller than f_t at all sections. These cracks are referred to as primary cracks. If the variation of the solid line in Fig. 5.18B (actual strength $A_c f_t$) is not significant, all primary cracks are formed within a small load increment from the first crack. However, there could be significant variation of the actual strength in different cross-sections due to the variation of concrete strength. Therefore, the increment of load in the primary cracking stage could be significant even for an RC member with a uniform cross section (Wu and Gilbert 2008). Furthermore, as the strength distribution line in Fig. 5.18B is random, the crack distribution will also be random while satisfying the condition of

$$2S_0 > a \geq S_0 \tag{5.21}$$

Based on Piyasena (2002), the slip length can be evaluated with the following empirical formulas:

$$S_0 = 90 + \left(1.3 - \frac{f_{co}}{80}\right)\left(0.28 + \frac{d_b}{28}\right)\left(\frac{0.4 + 0.012 f_{co}}{\rho}\right) \quad \text{(in N, mm)} \tag{5.22}$$

For normal RC beams with a moderate reinforcement ratio ρ, the slip length S_0 varies from about 120 mm to 150 mm.

The crack formation process is further studied in Fig. 5.19 where a portion of the RC tie in Fig. 5.18 between two adjacent primary cracks is used for further analysis. As there is no crack inside the RC member between sections I and J, the loading condition and internal stress distribution of the member is similar to that of the rebar pull-off test shown in Fig. 5.12. Based on Fig. 5.12B, the bond stress in the tension test is shown in Fig. 5.19B. For the case $L > 2S_0$, there is a region between sections I_1 and J_1 where the concrete stress is not affected by the cracks at I and J. The tensile force/stress in concrete is induced by the bond. Integration of the bond stress from the crack position (end of the portion) to a cross-section gives the tensile force on concrete at that cross section. Therefore, the tensile stress distribution in concrete is as shown in Fig. 5.19C. As the slip or bond stress is zero at the end of the slip length at sections I_1 and J_1 as shown, the tensile stress of concrete σ_t reaches a maximum at these two sections and remains constant in between. Since the total tension force at any cross-

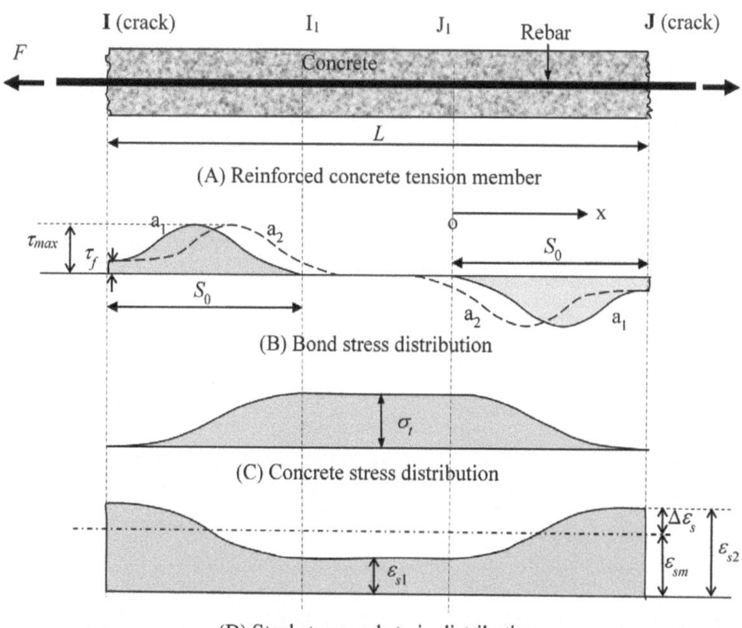

Fig. 5.19 Formation of primary crack.

section is F, the tension force in the rebar is equal to F minus the tension force of concrete. As a result, the tensile stress distribution of the rebar is as shown in Fig. 5.19D.

When the tension force F further increases slightly, the bond stress block moves from the loaded end toward the middle of the RC member, e.g., from a_1 (solid line) to a_2 (dashed line). The area of the bond stress block increases during the shift of the bond stress block, and hence the concrete stress in the middle plateau between the two bond stress blocks increases. A new crack forms at the weakest section in the middle plateau region when σ_t reaches f_t. This is how another primary crack forms between two primary cracks that have a spacing larger than $2S_0$.

For the other case of $L < 2S_0$, the internal stress distributions are shown in Fig. 5.20. After the formation of the primary cracks, the slip lengths from both sides overlap, causing a drop of the concrete stress σ_t below the tensile strength f_t everywhere, as shown by the solid line in Fig. 5.20C. Further pull will move the bond stress block toward the center as shown by the dashed line in Fig. 5.20B, keeping a zero bond stress at the center line due to a zero slip at this point. Depending on the total length L and bond characteristics, the area of the bond stress block may increase so that the concrete stress σ_t at the center line is increased to f_t as shown by the dashed line in Fig. 5.20C. A new crack is formed at the center line in this case, which is called a secondary crack. The process of subdivision by a new secondary crack can continue as long as the bond from the remaining space between two cracks is sufficient to develop a concrete stress that can reach the concrete tensile strength f_t. In the

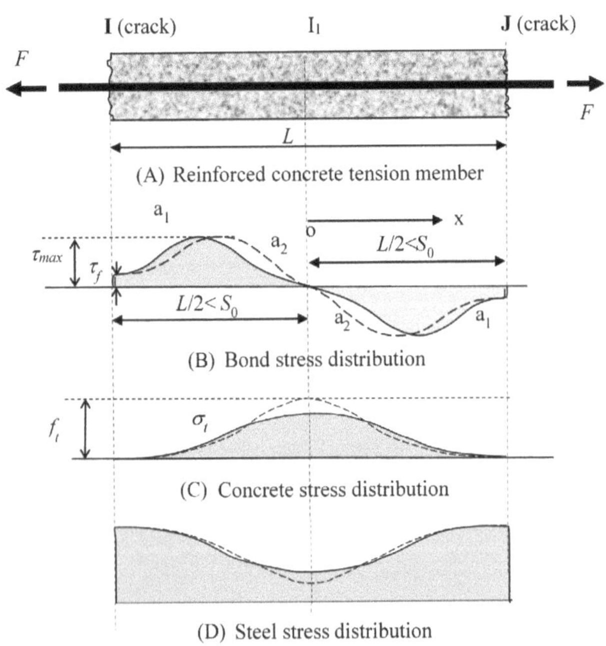

Fig. 5.20 Formation of secondary crack.

meantime, a secondary crack may not be able to form between two primary cracks if σ_t at the center line cannot reach f_t even if the steel bar is significantly stretched. Therefore, formation of a secondary crack is conditional: first, a secondary crack may not form if the spacing between two cracks, L, is too small; and second, the bond characteristics (bond–slip relationship) are not able to develop a sufficiently large bond force during further pull, e.g., a constant bond–slip relationship will not increase the area of the bond stress block at further increase of slip. Formation of a secondary crack may need significant further pull to develop sufficient bond force, which will significantly increase the crack width at the original primary cracks (I and J sections). Therefore, secondary cracks may form significantly later than the formation of primary cracks.

For a stronger interfacial bond with larger τ_{max} and τ_f values, sufficient σ_t can be developed to crack the concrete with a smaller bond length. Therefore, crack spacing is smaller for RC members with a stronger bond. When all remaining crack spacings in an RC member are not sufficient to form a new crack, further pull only widens the widths of existing cracks. Average crack width is smaller when crack spacing is smaller because there is a larger number of cracks to share the elongation of the member.

5.8.2 Crack formation in RC beams

The crack formation mechanism discussed in Section 5.8.1 is also applicable to flexural cracking of RC members in principle. For the RC beam shown in Fig. 5.21, the shaded part of the RC beam between the two point-loads is similar to an RC tension member in Fig. 5.18. The concrete area A_c of the shaded RC tension member should be determined in such a way that its centroid coincides with that of the rebars, so that the resultant forces of rebar and concrete are in line, and consequently the shaded part is under concentric tension (Park and Paulay 1975).

While the formation of cracks in the constant bending moment region is similar to an RC tension member, resulting in a crack spacing in the range of $2S_0 > a \geqq S_0$, the formation of cracks in a nonconstant moment region is somehow different. For the RC beam in Fig. 5.21, assume the RC section along the whole span is uniform as shown in

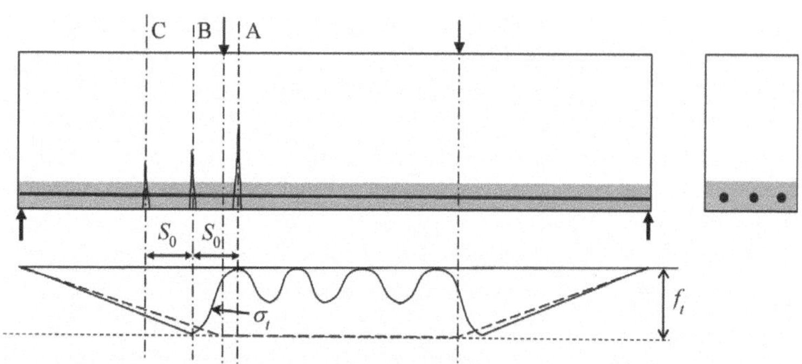

Fig. 5.21 Formation of cracks in a flexural member.

the figure. The tensile stress distribution of the uncracked concrete at the bottom of the beam is the same as that of the bending moment diagram. The tensile stress of concrete just before the formation of a primary crack is shown by the dashed line in Fig. 5.21. Assume the crack nearest the left shear span in the constant moment region is at line A. Formation of the crack A releases concrete stress in its surrounding concrete up to the distance of a slip length S_0 as shown. The concrete stress within the slip length is smaller and therefore cracking does not occur. When the applied load further increases, the concrete stress σ_t at the bottom of the member will increase, and when the stress distribution is increased to that shown by the solid line in Fig. 5.21, point B that has a distance of S_0 from point A will reach f_t first, and hence cracks. Since point B has the highest concrete stress, it is the only place that will crack when the σ_t line moves up and touches the f_t line (thin dashed line). A small variation of material strength and cross section such as that shown by the solid line in Fig. 5.18B will only slightly change the location of point B. For the same reason, further increase of loading will further move up the σ_t line and produce another crack at point C that has a distance of S_0 from point B, and so on. Therefore, the crack spacing at the shear span of the beam in Fig. 5.21 is fixed at $a = S_0$.

The preceding discussion is also applicable to the region of a beam that has a varying or nonconstant bending moment distribution, such as beams under distributed loadings. Therefore, constant crack spacing will occur in an RC flexural member or region of a member that has a varying bending moment diagram. This phenomenon has been observed by researchers from experimental tests. It is noteworthy that, although stirrups weaken concrete at a cross section, laboratory tests and numerical modeling have shown that cracks may not occur at stirrup locations. In other words, they have an insignificant effect on crack spacing and distribution. More detailed study on crack spacing in RC members can be found in Piyasena (2002).

5.9 Evaluation of crack width and spacing

5.9.1 Design for cracking

Cracking affects appearance, water tightness of RC members, and corrosion of reinforcement inside. Therefore, crack width needs to be controlled. From appearance point of view, the crack width that will not alert public is considered to be around 0.3 mm. Penetration of corrosion agents such as chlorides and carbon dioxide into steel reinforcement neutralizes the alkaline environment of concrete that protects steel bars from corrosion. Therefore, corrosion of steel reinforcement is closely related to the permeability of the concrete cover, including the permeability of concrete material (uncracked), thickness of concrete cover, geometric shape of cracks (surface crack width, crack length, and crack width at rebar surface; see Fig. 5.16), exposure environment, loading nature (related to opening and closing of cracks), etc. Due to the complexity of the problem, the current design for durability of RC structures related to corrosion only considers concrete material (concrete grade) and surface crack width for a given type of exposure conditions. While research on the effect of crack width on corrosion of reinforcement is inconclusive and even conflicting (Beeby 1978), it is prudent to limit crack width of an RC structure to hinder access of corrosion agents. Depending on exposure condition, the allowable surface crack width (referred to crack width hereafter unless otherwise noted) for durability design of RC structures is generally in the range of 0.1–0.4 mm at service load (Park and Paulay 1975), commonly 0.3 mm for normal exposure condition.

Cracking is affected by many factors, such as restrained shrinkage of concrete and thermal movement, tensile strain of structure due to loading and displacement, and distribution of reinforcement. Due to limited tensile strain capacity of concrete material, separation of concrete from reinforcement (i.e., cracking) is unavoidable when significant tensile strain occurs in RC structures. Cracking limiting strategy is not to avoid cracking but to distribute cracking to a larger number of cracks so that crack width is reduced. Cracking due to restrained shrinkage and thermal movement is usually controlled by empirical rules in RC designs. Cracking due to loading and displacement is generally controlled by two approaches, as recommended in design codes:

(1) deemed-to-satisfy empirical rules—by limiting key factors that affect crack width such as stress of rebar in cracking zone, concrete cover thickness, and bar spacing and diameter; and
(2) through calculation of crack width using models recommended in design codes.

It is noted that cracking of RC structures cannot be predicted with accuracy due to the randomness and the complex nature of cracking. Therefore, larger than normal error and scatter are expected in the calculation of cracks with a model. Different approaches have been developed in the literature for evaluation of crack width and spacing. Typical models are introduced in the following sections for better understanding of the cracking mechanism.

5.9.2 Slip theory

The slip theory is a classical theory of crack formation that was first introduced by Saliger (1936) and further developed by Watstein and Parsons (1943) for reinforced concrete ties. Based on the analysis in Section 5.8, the formation of new cracks must satisfy certain conditions for the RC tie in Fig. 5.20. When a new crack forms, it should have a minimum distance a_{min} away from the existing crack so that the bond force developed from the old crack to the new crack is sufficient to cause the cracking of the concrete section, i.e., $A_c f_t = q a_{min} = n\pi d_b a_{min} f_{bu}$, where n is the number of rebars. Therefore, the minimum crack spacing is given by

$$a_{min} = \frac{A_c f_t}{n\pi d_b f_{bu}} = \frac{d_b f_t}{4 f_{bu} \rho} \tag{5.23}$$

where $\rho = A_s/A_c = \frac{1}{4} n\pi d_b^2 / A_c$, the reinforcement ratio.

If the distance from crack I to crack J in Fig. 5.20 is greater than $2a_{min}$, a new crack can be formed in between. Otherwise, the new crack cannot be formed because the tension force to crack the concrete between I and J must be applied to both sides of the concrete member. In other words, a new crack can be formed only if $L \geqq 2a_{min}$.

Since no new crack can be formed when $L < 2a_{min}$ and the minimum crack spacing is a_{min}, any crack spacing between a_{min} and $2a_{min}$ is possible with an average value of $1.5a_{min}$. The maximum crack spacing is given by

$$a_{max} = 2a_{min} = \frac{d_b f_t}{2f_{bu}\rho} \tag{5.24}$$

The conclusion that $2a_{min} > a \geqq a_{min}$ from the preceding classical crack theory looks similar to Eq. (5.21). However, S_0 given by Eq. (5.22) is different from a_{min} given by Eq. (5.23). Primary crack spacing is determined by the drop of concrete surface stress caused by a crack, and interfacial bond will not affect this drop of stress as long as the bond stress is larger than a certain amount, while the secondary crack spacing is closely related to the bond, as given by Eq. (5.23).

The crack width w is the difference of elongation between steel bar and concrete between two adjacent cracks, i.e.,

$$w = \int_0^a [\varepsilon_s(x) - \varepsilon_c(x)]dx = a(\varepsilon_{sm} - \varepsilon_{cm}) \tag{5.25}$$

where a is the crack spacing; $\varepsilon_s(x)$ and $\varepsilon_c(x)$ are the steel and concrete strain at location x, respectively; ε_{sm} and ε_{cm} are the mean strain of steel and concrete, respectively, between the two cracks. A simple estimation of the maximum crack width w_{max} can be made by ignoring the deformation of concrete, which is assumed to be small compared with steel, as follows:

$$w_{max} = a_{max}\varepsilon_{sm} = a_{max}\frac{f_s}{E_s} \tag{5.26}$$

where f_s and E_s are the stress and Young's modulus of the rebar, respectively. The right-hand side gives the elongation of rebar between two cracks at the maximum crack spacing under the steel stress of f_s. Substituting Eq. (5.24) into Eq. (5.26) gives

$$w_{max} = K\frac{d_b f_s}{\rho} \tag{5.27}$$

where K is a constant related to material properties, given by

$$K = \frac{f_t}{2f_{bu}E_s} \tag{5.28}$$

Equation (5.27) shows that the maximum crack width is in direct proportion to the rebar stress and diameter, and in reverse proportion to the reinforcement ratio.

The preceding classic theory involves numerous assumptions, including that concrete stress and strain in a cross section is uniformly distributed, that bond strength is constant, that concrete strain is negligible, that the effect of concrete to rebar deformation is ignored (tension stiffening effect), that crack width is uniform at a cross section, etc. Although simple and approximate, the theory captures the main factors that affect the crack spacing and width.

5.9.3 No-slip theory

In contrary to the slip theory where slip at the bar-concrete interface is equal to the crack width, i.e., crack width is uniform in an elevation or the plane section remains plane, the no-slip theory proposed by Base et al. (1966) assumes no slip at the bar-concrete interface. They made this assumption based on the experimental observation that, for the range of crack widths normally permitted in reinforced concrete designs, there is essentially no slip of the rebar relative to concrete at the contact interface. The crack width is thus zero at the steel-concrete interface and increases toward the tension face of a beam, as shown in Fig. 5.22A. In the no-slip theory, the crack width is due to elastic shear deformation of the cover concrete or the differential deformation of concrete surrounding a longitudinal reinforcement, which implies that the plane section does not remain plane (Fig. 5.22A).

The stress of concrete at a crack is zero. The stress of concrete at the surface level increases with the distance from the crack and it reaches maximum at a certain distance from the crack, as shown by section A in Fig. 5.22B (Beeby 1979). Base et al. (1966) tried to estimate the location of section A using elasticity theory. Beeby (1979) used the simple engineering rule that a concentrated load as shown in Fig. 5.22B is dispersed in angle of $\theta = 45°$ to estimate the location of section A. The surface concrete will not be fully stressed before the force is dispersed to section A. Therefore, section A is the closest section where the surface concrete stress reaches a maximum. Based on this argument, Base et al. (1966) assumed that the nearest place a crack can be formed is at a distance close to c from the crack, where c is the clear concrete cover

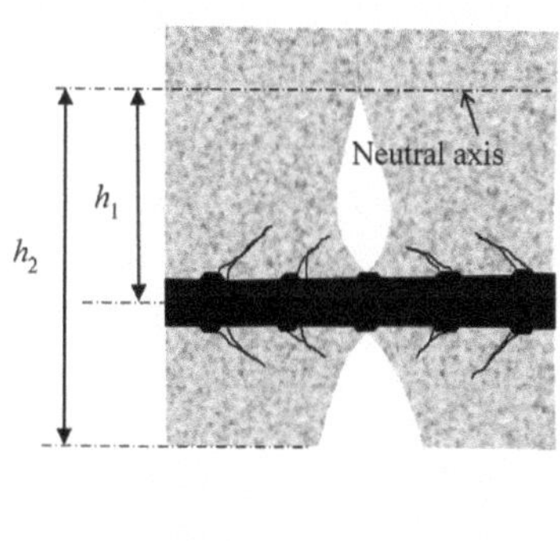

(A) Crack width distribution

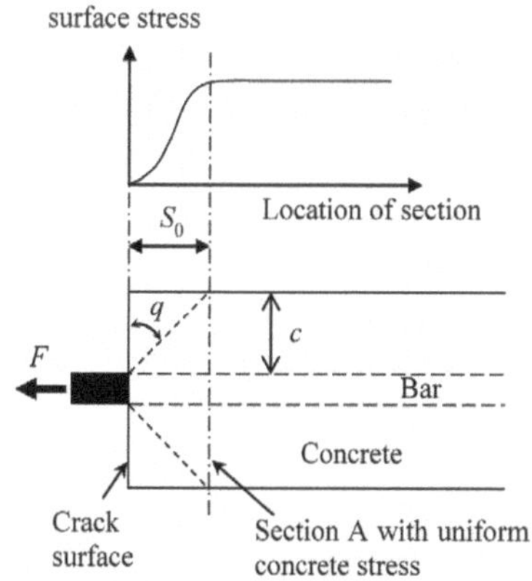

(B) Crack spacing

Fig. 5.22 No-slip model.

as shown in Fig. 5.22B. This argument was supported by their experimental test results as given by

$$S_0 = kc \tag{5.29}$$

where k is a coefficient to be determined from regression of experimental test results.

The crack spacing is given by Eq. (5.21). The maximum crack width is obtained by substituting the maximum crack spacing of $a_{max} = 2S_0$ into Eq. (5.26) to give $w_{max} = 2kc\frac{f_s}{E_s}$. Considering the fact that crack width is zero at the neutral axis and increases with the distance from the neutral axis (Fig. 5.22A), a factor of h_2/h_1 that is related to the level of maximum crack position is further allowed in their model, where h_1 and h_2 are shown in Fig. 5.22A. Therefore, the following formula for the prediction of the maximum crack width on the surface of concrete beams reinforced by deformed bars was proposed by Base et al. (1966):

$$w_{max} = 2kc\frac{f_s}{E_s}\frac{h_2}{h_1} = 3.3c\frac{f_s}{E_s}\frac{h_2}{h_1} \tag{5.30}$$

where $2k$ was determined to be 3.3 by the regression analysis of their extensive test results conducted at the Cement and Concrete Association.

Based on the classical slip theory and the no-slip theory, numerous models have been developed by considering more factors in the determination of crack spacing a and the mean strains $(\varepsilon_{sm} - \varepsilon_{cm})$ in Eq. (5.25). For example, Borges (1966) believed that the drop of surface concrete stress is related to both slip and concrete cover shear deformation, and hence the transfer length S_0 should be a combination of Eq. (5.23) and Eq. (5.29), which forms the basis of European codes/models. A more accurate calculation of $(\varepsilon_{sm} - \varepsilon_{cm})$ can be made by proper consideration of the tension stiffening, a concept that will be discussed in Section 5.10. More studies and relevant references on crack width can be found in Piyasena (2002), Borosnyói and Balász (2005), and Tan (2019).

5.9.4 Mathematical modeling

Solutions of crack spacing and crack width can be obtained mathematically from the differential equation based on the bond–slip relationship. For the RC tie shown in Figs. 5.19 and 5.20, one has the following equations:

$$\varepsilon_c = \frac{du_c}{dx} \tag{5.31}$$

$$\varepsilon_s = \frac{du_s}{dx} \tag{5.32}$$

$$\frac{ds}{dx} = \frac{d(u_s - u_c)}{dx} = \varepsilon_s - \varepsilon_c \tag{5.33}$$

$$\frac{d^2s}{dx^2} = \frac{d\varepsilon_s}{dx} - \frac{d\varepsilon_c}{dx} \tag{5.34}$$

where u_c and u_s are the displacements of concrete section and reinforcement, respectively, and ε_c and ε_s are their respective strain; and s is the interfacial slip. Assume the interfacial bond stress τ vs. slip s relationship is given by

$$\tau = g(s) \tag{5.35}$$

Since the variation of the tension force F_c in concrete comes from the bond, from the balance of an infinite small concrete segment with a length of dx, one has

$$dF_c = d(A_c\varepsilon_c E_c) = -n\pi d_b \tau dx \tag{5.36}$$

which gives

$$\frac{d\varepsilon_c}{dx} = -\frac{n\pi d_b \tau}{A_c E_c} \tag{5.37}$$

A similar equation is obtained from the balance of a small segment of reinforcing bars:

$$dF_s = d(A_s\varepsilon_s E_s) = n\pi d_b \tau dx \tag{5.38}$$

or

$$\frac{d\varepsilon_s}{dx} = \frac{n\pi d_b \tau}{A_s E_s} \tag{5.39}$$

Substituting Eqs. (5.37), (5.39), and (5.35) into Eq. (5.34) gives

$$\frac{d^2s}{dx^2} - \frac{4}{d_b E_s}(1 + \rho n_E) g(s) = 0 \tag{5.40}$$

where $\rho = \frac{A_s}{A_c}$ and $n_E = \frac{E_s}{E_c}$. For convenience, the origin of the x axis is located at the end of a bond stress block, as shown by point o in Figs. 5.19B and 5.20B.

The differential equation, Eq. (5.40), can be resolved when boundary conditions and the bond–slip relationship $g(s)$ are provided. The boundary conditions at the origin of the x-axis for primary cracking in Fig. 5.19B are given by

$$s(0) = 0 \tag{5.41}$$

$$s'(0) = 0 \tag{5.42}$$

The boundary condition for secondary cracking in Fig. 5.20B is given by

$$s(0) = 0 \tag{5.43}$$

$$s'(0) > 0 \tag{5.44}$$

where $s' = \frac{ds}{dx}$. The difference in the preceding two sets of conditions is that in Fig. 5.19B there is a region from section I_1 to J_1 where the slip is zero, which necessitates a zero slope of the slip curve at $x = 0$, while the slope of the slip curve at the midpoint I_1 in Fig. 5.20B is clearly not zero. The solution of the differential equation, Eq. (5.40), is very different for these two cases.

When the solution of $s(x)$ is obtained, the bond stress distribution $\tau(x)$ can be obtained from Eq. (5.35). For the case of Fig. 5.19, the steel force at the crack positions is

$$F_s(S_0) = F \tag{5.45}$$

The steel force at the fully bonded point I_1 or J_1 can be calculated based on the condition that $\varepsilon_s(x) = \varepsilon_c(x)$:

$$F_s(0) + F_c(0) = \varepsilon_s E_s A_s + \varepsilon_s E_c A_c = F \tag{5.46}$$

The transfer length S_0 can then be calculated from

$$F_s(S_0) - F_s(0) = \int_0^{S_0} \tau(x) n\pi d_b dx \tag{5.47}$$

The maximum crack width is two times the slip at $x = S_0$, or

$$w_{max} = 2s(S_0) \tag{5.48}$$

Bálasz (1993) and Oh and Kim (2007) provided a closed-form solution for primary cracking with a bond–slip relationship given by

$$g(s) = \tau_{max} \left(\frac{s}{s_1} \right)^{\alpha} \tag{5.49}$$

where τ_{max} and s_1 are the bond stress and slip at the peak of the bond–slip curve, respectively; and α is a parameter determining the shape of the bond–slip curve. The closed-form solutions of the maximum crack width and transfer length from Oh and Kim (2007) are given by

$$w_{max} = 2\left(\frac{(1+\alpha) {s_1}^{\alpha} d_b f_s^2}{8(1+n_E\rho)\tau_{max} E_s} \right)^{\frac{1}{1+\alpha}} \tag{5.50}$$

$$S_0 = \left(\frac{({s_1}^{\alpha} d_b)^{1/(1-\alpha)} f_s}{(1+n_E\rho)K} \right)^{\frac{1-\alpha}{1+\alpha}} \tag{5.51}$$

where

$$K = \tau_{max}\frac{4(1-\alpha)}{(1+\alpha)}\left(\frac{2(1-\alpha)^2(1+n_E\rho)\tau_{max}}{(1+\alpha)E_s}\right)^{\frac{\alpha}{1+\alpha}} \tag{5.52}$$

The crack spacing can be obtained by substituting Eq. (5.51) into Eq. (5.21).

No closed-form solution of Eq. (5.40) can be obtained for the secondary cracking condition due to the inhomogeneous boundary condition of Eq. (5.44). Nevertheless, a nonclosed form solution for this case was obtained by Tan et al. (2020).

The crack width given by Eq. (5.48) assumes a uniform crack width from the bar-concrete interface to the surface of the concrete. An analytical solution for nonuniform crack width has also been obtained by researchers. A sophisticated analytical solution for nonuniform cracking can be found in Tan et al. (2020).

5.9.5 Other models

Apart from the preceding mechanical models, other typical approaches to crack modeling include empirical models based on statistical analysis (Kaar and Hognestad 1965; Gergely and Lutz 1968) and fracture mechanics-based model (Bazant and Oh 1983). A statistical approach was adopted by Gergely and Lutz (1968) and Frosch (1999) to determine the key factors that affect crack width. Based on test data and statistical analysis, empirical models were developed for estimating the maximum crack width on the surface of RC members. These models have figured prominently in the development of the crack control design guidelines in the ACI codes. A refined model was developed by Beeby (1970, 1971) who observed that cracks near a longitudinal bar are controlled by their distance to the bar surface while cracks distant from a longitudinal bar are governed by the height from the surface of concrete to the neutral axis. Beeby's model has been widely adopted in the BS (British Standards) codes and the codes of former British colonies.

Cracking is also a time-dependent phenomenon. The increase in crack width is mainly due to shrinkage and creep of concrete, breakdown of bond, and other time-dependent variations of concrete structures. Past research has found that the spacing of cracks does not change with time under sustained loading, but the crack width increases. Crack width was observed to double in 2 years; however, the increase of crack width slows down with time (Park and Paulay 1975). Time-dependent crack width models can be developed by adopting time-dependent material properties in conventional crack width models. Fewer works are available in the literature on time-dependent cracking of concrete structures. Extensive studies on time-dependent behavior of cracking can be found in Castel and Gilbert (2014) and Nejadi (2005) who reported variation of crack spacing as well with time.

5.10 Tension stiffening

For calculation of the axial deformation of RC members, the axial rigidity EA is required. Before cracking, one has

$$EA = E_c A_c + E_s A_s \tag{5.53}$$

After cracking and at the section of a crack, Eq. (5.53) becomes

$$EA = E_s A_s \tag{5.54}$$

For a cracked member as shown in Fig. 5.18, it is obvious that the average rigidity of the whole member is between the preceding two cases, i.e., $E_s A_s < EA < E_c A_c + E_s A_s$. This relationship reflects that, although the strength calculation of a cracked member is solely based on a cracked section without concrete, the concrete material between cracks still has an effect of stiffening the RC member. The increase in the rigidity of a cracked RC member, EA, compared with that of the cracked section, $E_s A_s$, is referred to as the tension stiffening. The tension stiffening is closely related to the deformation calculation of RC structures, including EA for axial deformation and EI for flexural deformation. It is also important for crack width calculation, as the term $(\varepsilon_{sm} - \varepsilon_{cm})$ in Eq. (5.25) is directly related to the tension stiffening.

The tension stiffening phenomenon can be well explained through the load vs. strain curve of the RC member in Fig. 5.18, as shown by the solid line in Fig. 5.23 (Wu and Gilbert 2008). Before cracking, the load–deformation path is along Line 0, which is the elastic load–deformation curve of an uncracked member. When the applied load F reaches the crack load F_{cr} where the first crack occurs, there will be a sharp drop of the load and the load–deformation path will touch Line 1 that is the elastic load–deformation curve of a tension member with one crack. Further loading will be along Line 1 until the second crack occurs, where another drop of load

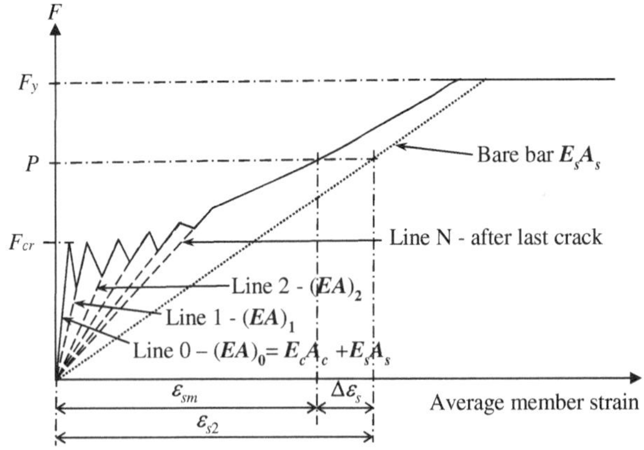

Fig. 5.23 Tension stiffening effect.

moves the curve to Line 2, which is the load–deformation curve of the tension member with two cracks. The drop and rise of the load path continue until the cracking process stabilizes without a new crack. The load path will then move up along a smooth curve, along which the existing cracks open up and debonding continues, until it touches the horizontal line at the load of F_y where the reinforcing steel yields.

The dotted line shows the load path of the "naked" reinforcing steel. It can be seen from Fig. 5.23 that the rigidity EA of the RC tension member decreases when the number of cracks increases, from Line 0, to Line 1, to Line 2, and so on until it reaches Line N, the line after the last crack. The loading curve after formation of the last crack is not linear although steel and concrete are both in their elastic stage, because the bond continues to break down in this process, which is nonlinear.

At a particular load P, the average strain of the member is given by (Fig. 5.23)

$$\varepsilon_{sm} = \varepsilon_{s2} - \Delta\varepsilon_s \tag{5.55}$$

where ε_{sm} is the average strain of the steel bar that is equal to the average strain of the RC member; ε_{s2} is the steel bar strain at a cracked section, which is equal to the strain of the bare steel reinforcement; and $\Delta\varepsilon_s$ is the effect of tension stiffening, i.e., the hardening to the bare steel caused by the cracked concrete. The relationship of Eq. (5.55) is also shown in Fig. 5.19D.

Accurate calculation of the tension stiffening effect is difficult and numerous empirical and semiempirical models have been developed for evaluation of the effect for calculation of flexural deformation. More details of the tension stiffening effect are discussed in Chapter 7.

5.11 Flexural strength calculation without considering slip

It is shown in Section 5.1 that interfacial slip plays a significant role in the flexural strength of composite members. However, it is also shown in Chapters 2 and 3 that RC flexural strength is calculated by assuming the plane section remains plane, which implies no slip between reinforcement and concrete. In other words, the flexural strength is calculated by considering RC structures as full interaction members. In fact, no composite effect, i.e., no slip between reinforcement and concrete, is mentioned in conventional RC flexural theory. This apparently contradicts the experimental observation of significant slips between concrete and rebars in many RC members, as discussed in Sections 5.8–5.10.

The problem can be discussed using Fig. 5.14, for the following typical scenarios:

1. Under pure bending

   It is discussed in Section 5.5 that, under pure bending, the bond force and average bond stress are equal to zero. As shown in Fig. 5.14C, $F_b = 0$ when $T_1 = T_2$. When the bond stress is zero, according to the bond–slip relationship, the slip and slip strain are also zero. Therefore, it is indeed a full interaction member without interfacial slip. The condition of plane section remaining plane is true for this case.

2. At peak of a moment distribution curve

At the peak of a moment distribution curve (bending moment diagram), one has dM/d$x = 0$, where M is the bending moment and x the longitudinal coordinate of the member. d$M = 0$ necessitates $T_1 = T_2$ in Fig. 5.14C. Therefore, this case is identical to pure bending.

3. In a shear span

From dM/d$x = Q$, where Q is the shear force of a section, it can be seen in Fig. 5.14C that when $Q \neq 0$ and hence d$M \neq 0$, $F_b \neq 0$. There must be slip and slip strain between rebar and concrete in this case. If rebar strain is calculated based on a plane section, the strain of rebar will be overestimated, because the slip strain ε_{slp} (Fig. 5.4) is ignored. The error depends on the value of Q of the section and the bonding condition of the longitudinal bars. The omission of slip or slip strain is unlikely to cause significant error in the calculation of flexural strength because the slip strain is usually insignificant for RC members under good bond condition with deformed bars. If the slip is significant, the plane section assumption will overestimate the rebar strain and the tension resistance, and hence potentially will lead to an unconservative design in this case. Therefore, attention should be paid when significant slip of rebar is expected in special cases of RC design.

Nevertheless, ultimate flexural design calculation usually applies to peak moment regions only in practice where there is no slip in theory. The reinforcement in a shear span is usually determined by curtailment rules (rebars calculated at peak moment sections are cut gradually along the span when bending moment decreases) which is usually conservative in determining the required area of reinforcement. Therefore, the current practice of flexural strength design based on full interaction theory will not cause problems in normal practice. On the other hand, for the design of deep beams where shear force dominates, the plane section assumption is not applicable. Design guidelines for the design of deep RC beams are available in Park and Paulay (1975).

5.12 Frictional shear for RC joints

An important type of interfacial shear/bond transfer at defined joints in reinforced concrete structures is referred to as the shear-friction. A joint between two RC elements tends to slide under shear force, as shown by the wall-to-slab joint in Fig. 5.24A. The sliding interface between concrete elements is usually rough. The sliding in the horizontal direction will lead to a movement or separation of the wall at the joint in the vertical direction due to the roughness of the sliding face, as shown in Fig. 5.24A. This separation will mobilize the tensile resistance of the rebars passing through the joint, which induces a normal pressure to the joint. The deformation or movement in the vertical direction is usually significant enough to cause yielding of the rebars.

In the meantime, the frictional coefficient μ of such an interface is quite large due to the roughness of the failure plane. Park and Paulay (1975) reported a μ value of greater than 1.4. ACI-ASCE Committee 455 (1998) recommended $\mu = 1.7$ for monolithic concrete, 1.4 for artificially roughened construction joints, and 0.8 to 1.0 for ordinary construction joints and concrete to steel interfaces. A value of $\mu \approx 1.0$ was reported by Wu et al. (2010) for the interface between FRP and concrete. Therefore, the shear-

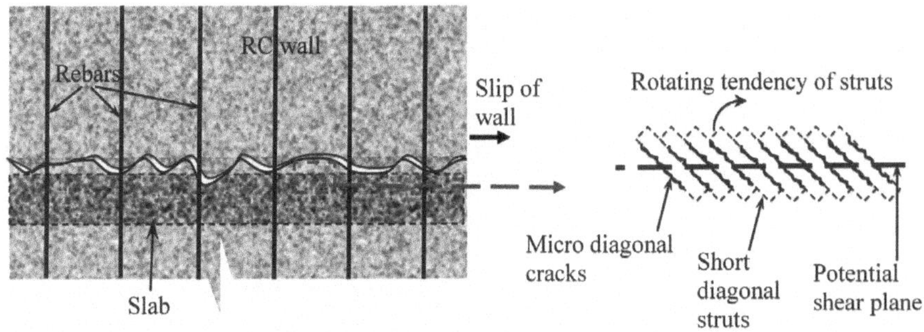

(A) Shear mechanism of cracked plane (B) Shear mechanism of uncracked plane

Fig. 5.24 Shear-friction mechanism.

friction at such a joint can be larger than expected when there is significant reinforcement crossing the joint.

The total frictional shear resistance of the joint is given by

$$S = N \times \mu = f_y A_s \mu \tag{5.56}$$

where S is the frictional shear resistance and N the force perpendicular to the sliding plane provided by all the reinforcing bars at yielding. The average bond stress at sliding of the joint is given by

$$\tau = \frac{S}{A_g} = \mu \frac{f_y A_s}{A_g} = \mu f_y \rho \tag{5.57}$$

where A_g is the gross area of the joint, and ρ is the ratio of reinforcement perpendicular to the sliding plane.

To prevent crushing of the concrete on the edge of the sliding plane caused by the clamping force from the transverse reinforcement before yielding the steel, the transverse reinforcement ratio across the joint is usually limited by design codes, e.g., to 1.5%. An upper limit, e.g., the lesser of $0.2f_{co}$ or 5.5 MPa, is also imposed on the shear stress τ.

Although the mechanism shown in Fig. 5.24A requires the formation of a macro crack followed by sliding movement along the crack, the shear-friction resistance exists before the macro cracking. The tendency of relative shear movement between the RC elements at the joint causes shear deformation of the concrete material at the interface. The shear deformation leads to tensile strain of the concrete at the interface in the perpendicular direction due to the shear expansion mechanism of concrete material (materials under shear action usually expand in the perpendicular direction), which causes tensile stress of the crossing bars. The shear expansion is attributed to the micro diagonal cracking caused by the shear stress, which forms short diagonal struts (Fig. 5.24B). The shear tendency causes the short diagonal struts to rotate, which produces the movement of the concrete in the perpendicular direction and consequent tensile strain of the transverse reinforcement, as shown. This mechanism is referred to as a truss mechanism in Park and Paulay (1975). Therefore, the shear resistance is mobilized before the formation of a macro crack at the joint. In other words, for RC joints with transverse reinforcing bars, the reinforcement induces shear resistance when there is a tendency of shear movement, both before and after the formation of macro cracking along the joint. Hence, the shear-friction resistance can be accounted for without cracking damage occurring at an RC joint. The magnitude of the shear resistance before macro cracking is not less than the frictional resistance of a precracked joint (Park and Paulay 1975). In fact, the shear-friction resistance along an initially uncracked (monolithic) shear plane is usually greater than that along a precracked shear plane, unless the transverse reinforcement ratio is very high (ASCE-ACI Task Committee 426 1973).

Eq. (5.56) only considers the frictional term of the shear-friction mechanism. A cohesion term is included in some other models (Mattock 1974; Duthinh and Carino 1996). In this case, Eq. (5.56) becomes

$$S = cA_c + f_y A_s \mu \tag{5.58}$$

where A_c is the area of concrete interface, and c is the cohesion. ACI-ASCE Committee 455 (1998) recommended $c = 0.3$ MPa for concrete cast against clean hardened concrete, 0.55 MPa for concrete cast against clean hardened concrete intentionally roughened to a full amplitude of 6.25 mm, and 1.10 MPa for concrete cast monolithically.

The shear friction mechanism was first introduced by Birkeland and Birkeland (1966) to deal with the force transfer across joints between precast concrete members. The shear-friction concept permits one to treat problems where failure occurs by sliding along a plane of discontinuity, such as construction joints and discrete cracks. The shear-friction mechanism is widely applied in reinforced concrete. It can be used in structural designs for shear resistance between two structural members, as illustrated in Fig. 5.24. It is applicable in modeling the internal resistance in a structural member, such as the shear transfer in a diagonal crack for the calculation of the shear strength of RC beams. It can also be used for stress–strain modeling of concrete material involving cracking.

References

ACI-ASCE Committee 455. 1998. "Recent approaches to shear design of structural concrete (ACI 445R-99)." *J. Struct. Eng.* 124 (12), 1375–1417.

ASCE-ACI Task Committee 426. 1973. "The shear strength of reinforced concrete members." *J. Struct. Div., Proc. ASCE* 99 (6), 1091–1187.

Bálasz, G. L. 1993. "Cracking analysis based on slip and bond stresses." *ACI Mater. J.* 90 (4), 340–348.

Base, G. D., J. B. Read, A. W. Beeby, and H. P. J. Taylor. 1966. *An Investigation of the crack control characteristics of various types of bar in reinforced concrete beams*, Research Report No. 18, Part I. London: Cement and Concrete Association.

Bazant, A., and B. Oh. 1983. "Spacing of cracks in reinforced concrete." *J. Struct. Eng.* 109 (9), 2066–2085.
Beeby, A. W. 1970. *An investigation of cracking in slabs spanning one way*, Technical Report TRA 433. London: Cement and Concrete Association.
Beeby, A. W. 1971. *Prediction and control of flexural cracking in reinforced concrete members*, Special publication 30, 55–76. Detroit: American Concrete Institute.
Beeby, A. W. 1978. "Corrosion of reinforceing steel in concrete and its relation to cracking." *Struct. Eng.* 56A (3), 77–81.
Beeby, A. W. 1979. "The prediction of crack widths in hardened concrete." *Struct. Eng.* 57A (1), 9–17.
Birkeland, P. W., and H. W. Birkeland. 1966. "Connections in precast concrete construction." *ACI J. Proc.* 63 (3), 345–368.
Borges, J. F. 1966. "Cracking and deformability of reinforced concrete beams." In *Publications*, Vol. 26. Zurich: International Association for Bridge and Structural Engineering, pp. 75–95.
Borosnyói, A., and G. L. Balász. 2005. "Models for flexural cracking in concrete: the state of the art." *Struct. Concr.* 6 (2), 53–62.
Castel, A., and R. I. Gilbert. 2014. "Influence of time-dependent effects on the crack spacing in reinforced concrete beams." *Struct. Concr.* 15 (3), 373–379.
Chen, J. F., and J. G. Teng. 2001. "Anchorage strength models for FRP and steel plates bonded to concrete." *J. Struct. Eng.* 127 (7), 784–791.
Dai, J. G., T. Ueda, and Y. Sato. 2005. "Development of the nonlinear bond stress-slip model of fiber reinforced plastics sheet-concrete interfaces with a simple method." *J. Compos. Constr.* 9 (1), 52–62.
Duthinh, D., and N. J. Carino. 1996. *Shear design of high-strength concrete beams: A review of the state-of-the-art*, US Department of Commerce, Technology Administration, National Institute of Standards and Technology, Office of Applied Economics, Building and Fire Research Laboratory.
fib (federation internationale du beton), and Task Group Bond Models. 2000. *Bond of reinforcement in concrete: state-of-art report*, fib bulletin 10.
Frosch, R. J. 1999. "Another look at cracking and crack control in reinforced concrete." *ACI Struct. J.* 96 (3), 437–442.
Gergely, P., and L. A. Lutz. 1968. *Maximum crack width in reinforced flexural members*, Special Publication 20, 87–117. Detroit: American Concrete Institute.
Harajli, M. H., B. S. Hamad, and A. A. Rteil. 2004. "Effect of confinement of bond strength between steel bars and concrete." *ACI Struct. J.* 101 (5), 595–603.
Kaar, P., and E. Hognestad. 1965. "High strength bars as concrete reinforcement." *J. Portland Cem. Assoc.*, 42–53.
Mattock, A. H. 1974. *Shear transfer in concrete having reinforcement at an angle to the shear plane*, Special Publication 42, 17-42. Farmington Hills, Mich: American Concrete Institute.
Nejadi, S. 2005. "Time-dependent cracking and crack control in reinforced concrete structures." PhD thesis. University of New South Wales.
Oh, B. H., and S. H. Kim. 2007. "Advanced crack width analysis of reinforced concrete beams under repeated loads." *J. Struct. Eng.* 133 (3), 411–420.
Park, R., and T. Paulay. 1975. *Reinforced concrete structures*, New York: John Wiley & Sons.
Piyasena, R. 2002. "Crack spacing, crack width and tension stiffening effect in reinforced concrete beams and one-way slabs." PhD thesis. Australia: Griffith University.

Saliger, R. 1936. "High-grade steel in reinforced concrete." In *Proceedings Second Congress of the International Association for Bridge and Structural Engineering*, pp. 293–315. Berlin-Munich, Germany.

Tan, R. 2019. "Consistent crack width calculation methods for reinforced concrete elements subjected to 1D and 2D stress states." PhD thesis. Norwegian University of Science and Technology.

Tan, R., M. A. N. Hendriks, M. Geiker, and T. Kanstad. 2020. "Analytical calculation model for predicting cracking behavior of reinforced concrete ties." *J. Struct. Eng.* 146 (2), 04019206.

Watstein, D., and D. E. Parsons. 1943. "Width and spacing of tensile cracks in axially reinforced concrete cylinders." *J. Res. Natl. Bur. Stand.* 31 (RP 545), 1–24.

Wu, H. Q., and R. I. Gilbert. 2008. *An experimental study of tension stiffening in reinforced concrete members under short-term and long-term loads*, UNICIV Report. The University of New South Wales.

Wu, Y. F., and X. M. Zhao. 2013. "Unified bond stress–slip model for reinforced concrete." *J. Struct. Eng.* 139, 1951–1962.

Wu, Y. F., J. H. Yan, Y. W. Zhou, and Y. Xiao. 2010. "Ultimate strength of reinforced concrete beams retrofitted with hybrid bonded fiber-reinforced polymer." *ACI Struct. J.* 107 (4), 451–459.

Analytical modeling of composite members

6

Reinforcement other than conventional reinforcing bars can be used to enhance the performance of reinforced concrete structures. The reinforcing materials can be a steel plate or rod, a fiber-reinforced polymer (FRP) laminate or strip, a layer of other additional material, or another structural member. The new reinforcement and the old concrete structure form a composite member. The analytical approaches and theories for the evaluation of the behavior of such composite members are introduced in this chapter.

6.1 Structural rehabilitation

An important application of composite mechanics is the structural rehabilitation of concrete structures where additional reinforcement is attached to an existing structural member to form a composite member. As additional materials are bonded to an existing structure, a critical link of the altered structure is the bond between the existing structure and the new materials, as discussed in Chapter 5. The following terminology is adopted in this chapter:

(a) The term "repair" is defined as reinstatement of the original characteristics of a damaged structure and is confined to dealing with as-built structures.
(b) The term "strengthening" is defined as intervention that leads to enhancement of one or more structural performances (stiffness, strength, ductility, etc.) of existing structures, depending on the requirement.
(c) The term "retrofitting" is defined as intervention that improves the performance of an existing structure that was inadequately designed or constructed.
(d) The term "rehabilitation" is used as a comprehensive term to include all types of repair, retrofitting, and strengthening that lead to improved structural performance and reduced vulnerability to earthquake, fire, and other attacks.

Structural interventions are generally divided into two categories: (1) local intervention where a structural member or members are strengthened without changing the load path of the whole structure, and (2) global intervention where the load path of the whole structure is changed, e.g., by adding additional structural members. Global intervention relates to design of the structural system. This chapter focuses on local interventions.

Various technologies have been developed for rehabilitation of concrete members. Typical ones include:

(1) Shotcrete (Gunite)

Concrete, cement mortar, epoxy, etc. can be sprayed onto the external face of a concrete structure for strengthening. Steel, glass, or other short fibers can be mixed into shotcrete for

Analytical Approaches for Reinforced Concrete. https://doi.org/10.1016/B978-0-12-821164-9.00006-1

better effect. Shotcrete can be used for repairing, strengthening, and retrofitting concrete structures. Reference can be made to the literature on design, construction, and performance of shotcrete rehabilitated concrete structures (Bousias et al. 2007). Shotcrete is sometimes referred to as one kind of RC jacketing, as discussed in the following paragraph.

(2) RC jacketing

Concrete jacketing involves the encasement of an existing RC member with an additional layer of concrete plus additional longitudinal and transverse reinforcement (Frangou et al. 1995), as shown in Fig. 6.1A. Concrete jacketing was the most popular strengthening method in the past that could be effectively used to retrofit (before damage) or repair (after damage) RC columns (Rodriguez and Park 1994). It was widely used for repairing RC structures in Mexico City after the 1985 earthquake. The technique was found to be effective but labor-intensive (Park 2001), as it not only requires the construction of additional reinforced concrete, but it may also involve the roughening of the interface and embedment of dowels/shear keys into the existing member to provide a good bond. Interfacial bond was found by some to have significant influence on the performance of RC jacketed columns (Vandoros and Dritsos 2006), while others believe the effect is insignificant (Julio and Branco 2008). It requires an interruption in the use of the structure while applied. The member size and self-weight are increased significantly. The stiffness of the retrofitted structural system is also significantly increased, which could attract more load in an earthquake. As an alternative, a steel wire net could be wrapped around an RC member and cement mortar applied onto the surface to form a reinforced concrete jacket, which is referred to as ferrocement jacketing in the literature (Kaish et al. 2018). Extensive investigations into the performance, design, and construction of RC jacketing have been reported in the literature (Julio et al. 2003; Vandoros and Dritsos 2006).

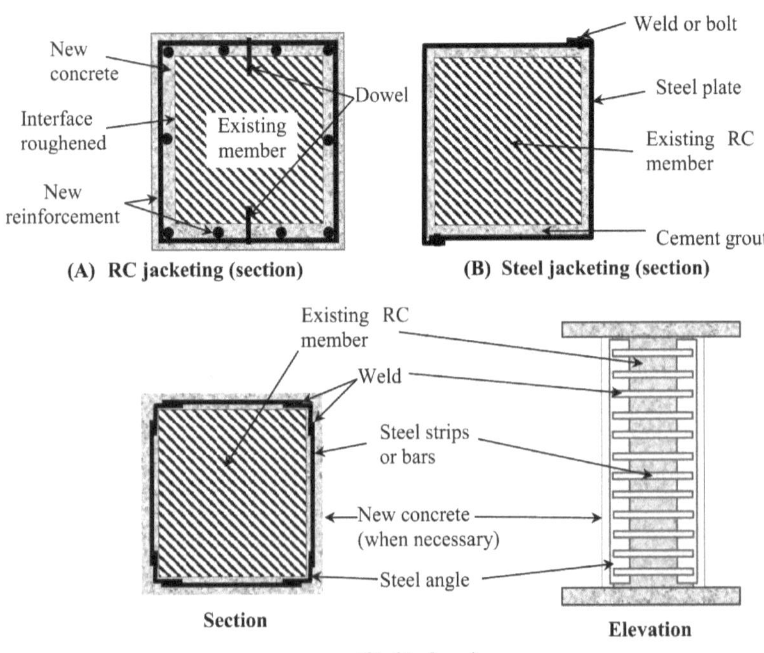

Fig. 6.1 Conventional external jacketing technologies.

(3) Steel jacketing

Steel jacketing is a common rehabilitation technique in practice that involves the total encasement of an RC member, typically an RC column, with thin steel plates, steel strips, or steel bars (Dritsos 1997). It can be divided into two main approaches: steel encasement and steel caging. For steel encasement, an RC member is encased or wrapped by a layer of steel plate around the external face of the member. In construction, two half-circle shells (for circular columns) or two L-shaped plates (for square/rectangular columns) are placed around the column and then site welded or bolted to provide a continuous jacket (Fig. 6.1B). The small gap between the steel plate and column face is normally pressure grouted subsequently with nonshrink grout to ensure a firm contact. Another popular method of steel jacketing involves the use of steel angles placed at the corners of a column. Steel plates are then welded to the angles to form a jacket. Gluing steel plates to the face of a column using epoxy resin and site welding joints is also a common practice.

Steel caging consists of longitudinal angle sections at each corner of a column and transverse steel strips, or bars welded onto the angles to create collars around the concrete column, as shown in Fig. 6.1C. Alternatively, the longitudinal steel angles may be omitted, and thin steel strips or wires can be wrapped in a continuous spiral or in discontinuous rings around the column. The tie strips can be laterally stressed to ensure a tight fit by special wrenches prior to welding or preheated to about 200–400 degrees before welding and left to shrink in cooling after welding to provide prestressing in the lateral direction. A technique involving pretensioning of steel strips around the column using standard strapping machines found in the packing industry and subsequently securing them in place by metal clips has also been used for wrapping steel strips to columns (Dritsos 1997). Prestressing to steel encasement can also be done by pressure grouting the gap between the steel plate and column face with nonshrink or expansive grout.

The external jacket has two effects on an existing member: (1) providing confinement to increase the flexural strength and ductility of the member (see Chapter 8), and (2) as a transverse web reinforcement to increase the shear strength of the member. When both longitudinal stress due to axial force and bending moment and shear stress due to shear action exist in an RC member, the external jacket may act as both a confining material to improve flexural performance and a means of providing additional shear resistance to the member. In such cases, the confining material and shear reinforcement can be designed separately, and the total jacketing material is the summation of the two required quantities of the material, similar to the design of stirrups.

A cover of concrete or shotcrete reinforced with light welded fabric can be applied for corrosion or fire protection when necessary. Compared with RC jacketing, the main advantages of steel jacketing are the minimal increase in cross-sectional dimensions, minimal skilled labor required for application, and speed of installation. The costs of structural intervention and interruption of use are also reduced. Extensive research and design guidelines for steel jacketing have been reported in the literature (Chai et al. 1991; Priestley et al. 1994; Sakino and Sun 2000; Xiao and Wu 2003).

(4) Steel plating

Steel plates can be adhesively or mechanically attached to an RC member (Figs. 6.2 and 6.3), to increase the rigidity and flexural or shear strength of the member. The steel plates can be treated as additional reinforcement of a full interaction RC member when it is adhesively bonded with negligible interfacial slip. The design of such full interaction members is similar to that of normal RC members. RC members with mechanically bonded steel plates are partial interaction composite members. Detailed theory, design, and construction of partial interaction composite members can be found in Oehlers and Bradford (1995, 1999). The fundamental partial interaction theory is introduced in Section 6.2.

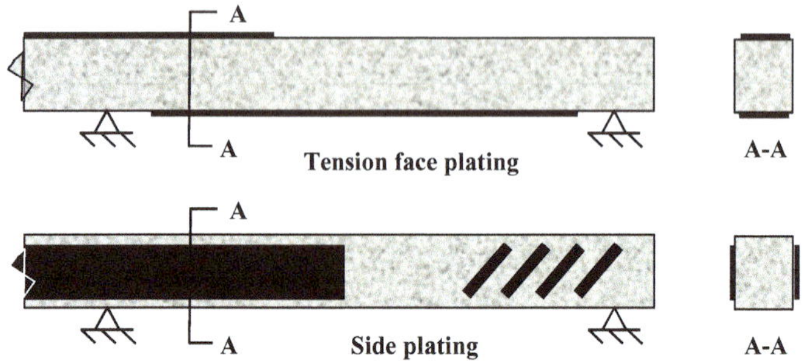

Fig. 6.2 Steel plating.

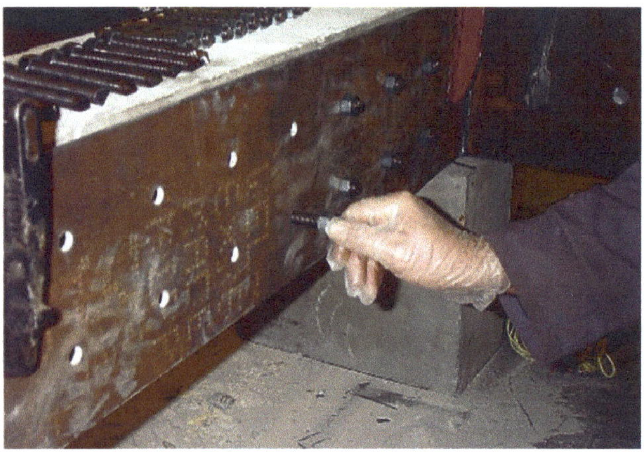

Fig. 6.3 Mechanical fastening of steel plate.

(5) Externally bonded fiber-reinforced polymer

Due to the advantage of being lightweight, high strength, and easy to handle during construction, externally bonded (EB) fiber-reinforced polymer has found more and more applications in structural rehabilitation in recent years. FRP strengthening systems are formed by bonding continuous carbon, glass, aramid, or other synthetic fibers, fiber sheet, or fiber plate to the external face of an RC member with a matrix of epoxy, vinyl ester, or polyester. The strength of FRP is in the fiber direction and its strength perpendicular to the fiber direction is negligible unless fibers are provided in both directions. Similar to externally bonded steel plate, EB-FRP can also be divided into two categories: (1) FRP jacketing where it is wrapped to an RC member to form an enclosed hoop with its fibers in a transverse direction to provide confinement to the concrete (Fig. 6.4B), and (2) FRP plating where FRP plates/laminates/strips are applied along the longitudinal direction for flexural strengthening, or as web reinforcement for shear strengthening (Fig. 6.4C). Prestress can also be applied to FRP, both for jacketing and plating.

Conventional mechanical anchors and bolts usually cannot be used directly to mount thin or nonductile external reinforcement such as FRP materials, which do not have sufficient bearing strength to avoid longitudinal cutting by anchors. Therefore, the current

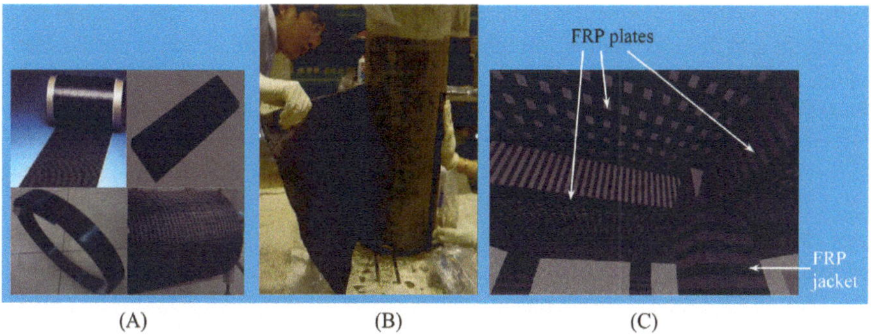

Fig. 6.4 Externally bonded FRP: (A) FRP materials; (B) column jacketing; and (C) FRP strengthened building.
(A and C) From Wu, Z. S., Y. F. Wu, and M. Fahmy. 2020. *Structures strengthened with bonded composites*, Elsevier Woodhead Publishing.

practice in the construction industry is to use adhesive for bonding FRP reinforcement onto the concrete surface. In an adhesively bonded reinforcing system, the interfacial bond between the externally bonded reinforcement and concrete is usually the weakest link, which may debond under significant loading. To improve the bond condition, a technology, namely near-surface mount (NSM), has been developed that embeds a reinforcing strip or rebar into a groove cut on the surface of the concrete structure and bonds the reinforcement with adhesive (Fig. 6.5). There are other technologies that can be used to enhance the bond, such as fiber anchoring and hybrid bonded FRP, which are discussed in Section 6.5.

More details on FRP plating are provided in Sections 6.3 and 6.4. FRP jacketing is examined in Chapter 8. For detailed guidelines on design and construction using these technologies, refer to specialist literature and design codes of practice. This chapter introduces and discusses the theories behind these technologies.

6.2 Mechanically bonded reinforcing systems

Mechanical fastening is suitable for bonding thick and nonbrittle reinforcing materials such as steel plates onto a concrete substrate. It is an effective and reliable method that has been used in the construction industry for decades. In a mechanically fastened

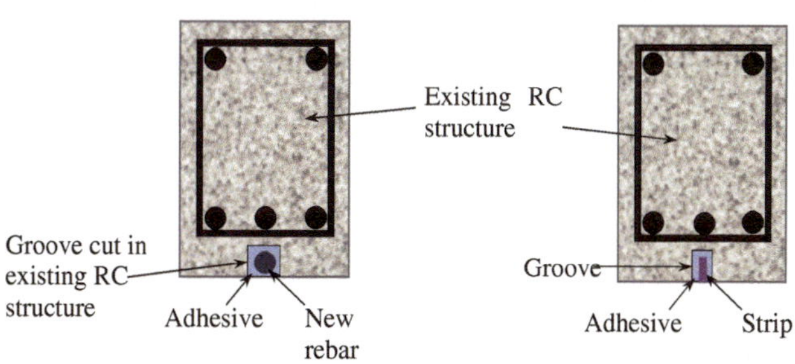

Fig. 6.5 Near-surface mounted reinforcement.

reinforcing system, there will be relative movement or slip between the attached reinforcement and the RC member. In other words, the composite action is partial interaction, as discussed in Chapter 5. The design theory for conventional RC members is not applicable in this case. Additional failure modes such as shear-off of the mechanical anchor and buckling of the attached plate need to be considered and designed for. These failure modes are closely related to the magnitude and distribution of interfacial slip. Composite mechanics is the cornerstone of the design of such composite structures. For engineering design of conventional steel-concrete composite structures, reference can be made to relevant textbooks and specialist literature (Oehlers and Bradford 1999; Liang 2014; Johnson and Wang 2018; Nie et al. 2019). The partial-interaction theory, which lays the foundation for the design of composite members, is introduced in this section.

6.2.1 Classic linear elastic theory

Classic linear elastic partial-interaction theory was first developed for composite steel and concrete beams by Newmark et al. (1951, 1952). Newmark's classic linear-elastic theory for composite beams was extended by Wu et al. (2002) to include axial load and a nonzero slip boundary condition so that the theory is applicable to columns with a plastic hinge. The theory developed by Wu et al. (2002) is introduced in this section.

The basic model under consideration is shown in Fig. 6.6, where the composite member is composed of two elements connected together at the interface by mechanical shear connectors. The origin of the coordinate system is located at the geometric centroid of the cross section of element 1 at the end of the member. The x-axis is in the longitudinal direction of the member. A constant axial load N is applied at the centroid of the cross section of element 1. A constant moment M_0 at the origin and a distributed load $p(x)$ along the length are applied. In the free body diagram of Fig. 6.6A, the

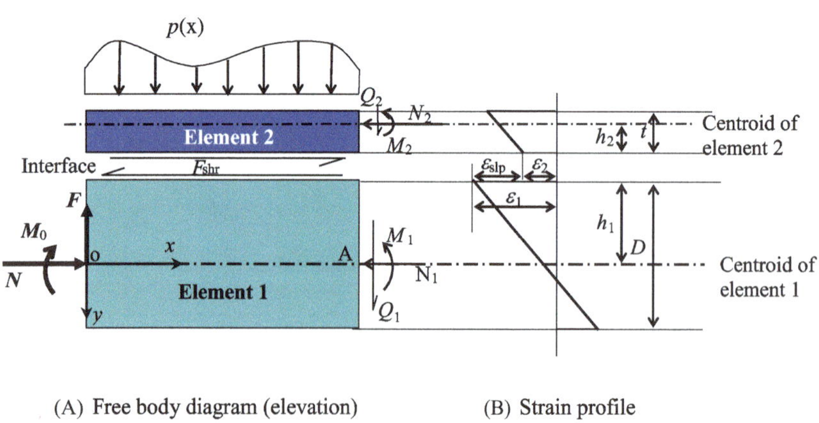

Fig. 6.6 Partial interaction model.

normal (y-direction) stress on the interface is not shown as it has no effect in the following derivations.

The following assumptions are made in the derivations:

(1) All the constituent materials behave linear-elastically;
(2) The cross section is uniform along the length;
(3) Bernoulli's hypothesis that plane sections remain plane applies to individual elements;
(4) The shear connectors between the two elements are continuous and uniformly distributed longitudinally; and
(5) No transverse separation occurs on the contact interface. Therefore, the curvatures are identical for both elements at the same cross section.

6.2.2 *Equilibrium and compatibility*

From the equilibrium of the individual elements in Fig. 6.6, one has

$$N_2 - F_{shr} = 0 \tag{6.1}$$

$$N - F_{shr} - N_1 = 0 \tag{6.2}$$

Taking moments about point A gives

$$M_1 + M_2 + N_2 \cdot (h_2 + h_1) - M(x) = 0 \tag{6.3}$$

where $M(x)$ is the applied moment in the cross section given by $M(x) = M_0 + F \cdot x - \int_0^x p(\xi) \cdot (x - \xi) d\xi$. Substituting Eq. (6.1) into Eq. (6.3) gives

$$M_1 + M_2 + F_{shr} \cdot (h_2 + h_1) - M(x) = 0 \tag{6.4}$$

The strain of element 1 at the interface is given by

$$\varepsilon_1 = \frac{M_1 \cdot h_1}{(EI)_1} + \frac{N_1}{(EA)_1} \tag{6.5}$$

and that at the interface of element 2 is

$$\varepsilon_2 = -\frac{M_2 \cdot h_2}{(EI)_2} + \frac{N_2}{(EA)_2} \tag{6.6}$$

where subscripts 1 and 2 refer to elements 1 and 2, respectively. Assumption 5 requires that

$$\kappa = \frac{M_1}{(EI)_1} = \frac{M_2}{(EI)_2} \tag{6.7}$$

Applying Assumption 4, the longitudinal slip s is given by

$$s = \frac{F_b}{K_b} = \frac{q \cdot L_s}{K_b} \tag{6.8}$$

where F_b is the shear force from a shear connector that has a stiffness of K_b; L_s is the longitudinal spacing of the shear connectors; and q is the longitudinal shear force per unit length, or shear flow, which is given by

$$q = \frac{dF_{shr}}{dx} \tag{6.9}$$

The discontinuous shear forces from shear connectors are transformed to a continuous interfacial shear force through Eq. (6.8). Substituting Eq. (6.9) into Eq. (6.8) gives

$$s = \frac{L_s}{K_b} \cdot \frac{dF_{shr}}{dx} \tag{6.10}$$

The slip strain (Fig. 6.6B) is given by

$$\varepsilon_{slp} = \varepsilon_1 - \varepsilon_2 = -\frac{ds}{dx} \tag{6.11}$$

Differentiating Eq. (6.10) with respect to x gives

$$\frac{ds}{dx} = \frac{L_s}{K_b} \cdot \frac{d^2 F_{shr}}{dx^2} \tag{6.12}$$

By referring to Eqs. (6.1), (6.2), and (6.4)–(6.7), the slip strain in Eq. (6.11) can be transformed to

$$\varepsilon_{slp} = \varepsilon_1 - \varepsilon_2 = -\frac{\overline{EI}}{\overline{EA} \cdot \sum EI} \cdot F_{shr} + \frac{(h_1 + h_2) \cdot M(x)}{\sum EI} + \frac{N}{(EA)_1} \tag{6.13}$$

where

$$\frac{1}{\overline{EA}} = \frac{1}{(EA)_1} + \frac{1}{(EA)_2} \tag{6.14}$$

$$\sum EI = (EI)_1 + (EI)_2 \tag{6.15}$$

$$\overline{EI} = \sum EI + \overline{EA} \cdot (h_1 + h_2)^2 \tag{6.16}$$

where $\overline{EI}$ is the flexural rigidity of the composite member under full interaction. However, $\overline{EA}$ is not the axial rigidity of the composite member.

6.2.3 Governing differential equation

From Eqs. (6.11)–(6.13), the following differential equation is obtained

$$\frac{L_s}{K_b}\cdot\frac{d^2F_{shr}}{dx^2}-\frac{\overline{EI}}{\overline{EA}\cdot\sum EI}\cdot F_{shr}+\frac{(h_1+h_2)\cdot M(x)}{\sum EI}+\frac{N}{(EA)_1}=0 \tag{6.17}$$

Letting

$$a_0=\frac{L_s}{K_b} \tag{6.18}$$

$$a_1=\frac{\overline{EI}}{\overline{EA}\cdot\sum EI} \tag{6.19}$$

$$a_2=\frac{(h_1+h_2)}{\sum EI} \tag{6.20}$$

$$a_3=\frac{1}{(EA)_1} \tag{6.21}$$

Eq. (6.17) simplifies to

$$a_0\cdot\frac{d^2F_{shr}}{dx^2}-a_1\cdot F_{shr}+a_2\cdot M(x)+a_3\cdot N=0 \tag{6.22}$$

This differential equation can be solved for given boundary conditions and an applied moment distribution of $M(x)$. The slip can be obtained from Eq. (6.10). To obtain the deflection of the member, the following relation is used

$$-\frac{d^2v}{dx^2}=\kappa=\frac{M_1}{(EI)_1} \tag{6.23}$$

Combining Eqs. (6.4), (6.7), and (6.17) gives

$$\frac{M_1}{(EI)_1}=\frac{M(x)}{\overline{EI}}-\frac{(h_1+h_2)\cdot\overline{EA}}{\overline{EI}}\cdot\frac{L_s}{K_b}\cdot\frac{d^2F_{shr}}{dx^2}-\frac{(h_1+h_2)\cdot\overline{EA}\cdot N}{\overline{EI}\cdot(EA)_1} \tag{6.24}$$

Substituting Eq. (6.24) into Eq. (6.23) gives

$$\frac{d^2v}{dx^2}=a_4\cdot a_0\frac{d^2F_{shr}}{dx^2}-\frac{M(x)}{\overline{EI}}+a_4\cdot a_3\cdot N \tag{6.25}$$

where

$$a_4 = \frac{(h_1 + h_2) \cdot \overline{EA}}{\overline{EI}} \tag{6.26}$$

6.2.4 Solution for the case of cantilever column

As an example, Eq. (6.22) is solved for the case of a cantilever column shown in Fig. 6.7. In this case, $M(x) = F \cdot x$. The general solution of Eq. (6.22) is given by

$$F_{shr} = c_1 \cdot e^{\sqrt{\frac{a_1}{a_0}} \cdot x} + c_2 \cdot e^{-\sqrt{\frac{a_1}{a_0}} \cdot x} + \frac{a_2 \cdot F}{a_1} \cdot x + \frac{a_3 \cdot N}{a_1} \tag{6.27}$$

where the constants c_1 and c_2 can be determined from the following boundary conditions:

$$F_{shr}\big|_{x=0} = 0 \tag{6.28}$$

$$s\big|_{x=L} = \frac{L_s}{K_b} \cdot \left.\frac{dF_{shr}}{dx}\right|_{x=L} = s_p \tag{6.29}$$

For a fixed-end support as shown in Fig. 6.7, the boundary slip $s_p = 0$ at $x = L$.

For generality, a nonzero value s_p is adopted. This nonzero boundary slip condition is useful for the linear theory to be extended to a nonlinear analysis as discussed later. From Eqs. (6.28) and (6.29), the constants are calculated to be

$$c_1 = \frac{s_p}{2 \cdot \sqrt{a_0 \cdot a_1} \cdot \cos h(\alpha)} - \frac{a_3 \cdot N \cdot e^{-\alpha} + a_2 \cdot F \cdot \sqrt{\frac{a_0}{a_1}}}{2a_1 \cdot \cos h(\alpha)} \tag{6.30}$$

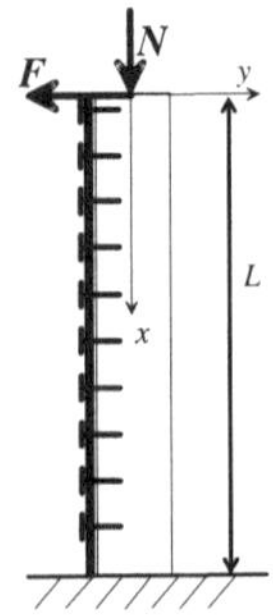

Fig. 6.7 Model of the cantilever column.

$$c_2 = -c_1 - \frac{a_3 \cdot N}{a_1} \tag{6.31}$$

where

$$\alpha = L \cdot \sqrt{\frac{a_1}{a_0}} = L \cdot \sqrt{a_5} = L \cdot \sqrt{\frac{K_b \cdot \overline{EI}}{L_s \cdot \overline{EA} \cdot \sum EI}} \tag{6.32}$$

$$a_5 = \frac{a_1}{a_0} = \frac{K_b \overline{EI}}{L_s \overline{EA} \cdot \sum EI} \tag{6.33}$$

Substituting c_1 and c_2 into Eq. (6.27) and rearranging gives

$$F_{shr} = F_f + F_n + F_{sp} \tag{6.34}$$

where F_f, F_n and F_{sp} are the longitudinal shear forces induced by the lateral force F, the axial load N, and the boundary slip s_p, respectively. They are given by

$$F_f = F \cdot L \cdot (h_1 + h_2) \cdot \frac{\overline{EA}}{\overline{EI}} \cdot \left[\xi - \frac{\sinh(\alpha \cdot \xi)}{\alpha \cdot \cosh(\alpha)} \right] \tag{6.35}$$

$$F_n = \frac{N}{(EA)_1} \cdot \frac{\overline{EA} \cdot \sum EI}{\overline{EI}} \cdot \left[1 - e^{-\alpha \cdot \xi} - \frac{2\sinh(\alpha \cdot \xi)}{e^{2\alpha} + 1} \right] \tag{6.36}$$

$$F_{sp} = s_p \cdot L \cdot \frac{K_b}{L_s} \cdot \frac{\sinh(\alpha \cdot \xi)}{\alpha \cdot \cosh(\alpha)} \tag{6.37}$$

where ξ is the normalized coordinate x, or

$$\xi = \frac{x}{L}, 0 \le \xi \le 1 \tag{6.38}$$

The slip is obtained from Eq. (6.10):

$$s = \frac{L_s}{K_b} \cdot \frac{dF_{shr}}{dx} = s_f + s_n + s_{sp} \tag{6.39}$$

where s_f, s_n and s_{sp} are the slip terms due to F, N, and s_p, respectively, and are given by

$$\begin{aligned} s_f &= F \cdot (h_1 + h_2) \cdot \frac{\overline{EA}}{\overline{EI}} \cdot \frac{L_s}{K_b} \left[1 - \frac{\cosh(\alpha \cdot \xi)}{\cosh(\alpha)} \right] \\ &= F \cdot \frac{a_2}{a_5} \cdot \left[1 - \frac{\cosh\left(L \cdot \sqrt{a_5} \cdot \xi\right)}{\cosh\left(L \cdot \sqrt{a_5}\right)} \right] \end{aligned} \tag{6.40}$$

$$s_n = \frac{N}{(EA)_1} \cdot \sqrt{\frac{L_s}{K_b} \cdot \frac{\overline{EA}}{\overline{EI}} \cdot \sum EI} \cdot \left[e^{-\alpha \cdot \xi} - \frac{2\cosh(\alpha \cdot \xi)}{e^{2\alpha}+1} \right]$$
$$= \frac{N \cdot a_3}{\sqrt{a_5}} \cdot \left[e^{-(L \cdot \sqrt{a_5}) \cdot \xi} - 2\cosh(L \cdot \sqrt{a_5} \cdot \xi) / \left(e^{2(L \cdot \sqrt{a_5})} + 1 \right) \right] \tag{6.41}$$

$$s_{sp} = \frac{s_p \cdot \cosh(\alpha \cdot \xi)}{\cosh(\alpha)} = \frac{s_p \cdot \cosh\left(L \cdot \sqrt{a_5} \cdot \xi\right)}{\cosh\left(L \cdot \sqrt{a_5}\right)} \tag{6.42}$$

To solve for the deflection, Eq. (6.25) is integrated twice to give

$$\nu = v_{full}^F + a_4 a_0 F_{shr} + \frac{a_4 a_3 N}{2} x^2 + c_3 x + c_4 \tag{6.43}$$

where

$$v_{full}^F = \int dx \int \left(-\frac{M}{\overline{EI}} \right) dx = \int dx \int \left(-\frac{F \cdot x}{\overline{EI}} \right) dx$$
$$= -\frac{F \cdot L^3}{6\overline{EI}} \cdot \left(\xi^3 - 3\xi + 2 \right) \tag{6.44}$$

which is the deflection due to the lateral force F under full interaction, i.e., without slip between elements 1 and 2. The constants c_3 and c_4 are determined from the boundary conditions:

$$v|_{x=L} = 0 \tag{6.45}$$

$$\left. \frac{dv}{dx} \right|_{x=L} = 0 \tag{6.46}$$

which gives

$$c_3 = -a_4 \cdot a_3 \cdot L \cdot N - a_4 \cdot a_0 \cdot \frac{s_p \cdot K_b}{L_s} \tag{6.47}$$

$$c_4 = \frac{a_4 \cdot a_3 \cdot L^2 \cdot N}{2}$$
$$- a_4 \cdot a_0 \cdot \left[2c_1 \sinh(\alpha) + \frac{a_3 \cdot N}{a_1} \cdot \left(1 - e^{-\alpha}\right) + \frac{a_3 \cdot F}{a_1} \cdot L - \frac{s_p \cdot L \cdot K_b}{L_s} \right] \tag{6.48}$$

Substituting c_3 and c_4 into Eq. (6.43) and rearranging gives

$$v = v_{full}^F + v_{full}^N + F \cdot g_1(\xi) + N \cdot g_2(\xi) + s_p \cdot g_3(\xi) \tag{6.49}$$

where

$$\begin{aligned} v_{full}^{N} &= \frac{(h_1+h_2)\cdot \overline{EA}\cdot L^2}{2\overline{EI}\cdot (EA)_1}\cdot N\cdot(\xi-1)^2 \\ &= \frac{1}{2}\cdot N\cdot a_4\cdot a_3\cdot L^2\cdot(\xi-1)^2 \end{aligned} \tag{6.50}$$

which is the deflection caused by a constant moment due to the eccentricity of the axial load about the centroid of the combined composite cross section of elements 1 and 2 under full interaction.

The three additional terms in Eq. (6.49), $F\cdot g_1(\xi)$, $N\cdot g_2(\xi)$, and $s_p\cdot g_3(\xi)$, reflect the partial interaction effects of the lateral force, axial force, and boundary slip, respectively, where

$$\begin{aligned} g_1(\xi) &= (h_1+h_2)^2\cdot\left(\frac{\overline{EA}}{\overline{EI}}\right)^2\cdot\frac{L_s}{K_b}\cdot L\cdot\left[\xi-\frac{\sinh(\alpha\cdot\xi)}{\alpha\cdot\cosh(\alpha)}-1+\frac{1}{\alpha}\cdot\tanh(\alpha)\right] \\ &= a_4^2\cdot a_0\cdot L\cdot\left[\xi-\frac{\sinh(\alpha\cdot\xi)}{\alpha\cdot\cosh(\alpha)}-1+\frac{1}{\alpha}\cdot\tanh(\alpha)\right] \end{aligned} \tag{6.51}$$

$$\begin{aligned} g_2(\xi) &= (h_1+h_2)\cdot\left(\frac{\overline{EA}}{\overline{EI}}\right)^2\cdot\frac{\sum EI}{(EA)_1}\cdot\frac{L_s}{K_b}\cdot\left[-e^{-\alpha\cdot\xi}-\frac{2\sinh(\alpha\cdot\xi)}{e^{2\alpha}+1}+\frac{1}{\cosh(\alpha)}\right] \\ &= \frac{a_4\cdot a_3}{a_5}\left[-e^{-\alpha\cdot\xi}-\frac{2\sinh(\alpha\cdot\xi)}{e^{2\alpha}+1}+\frac{1}{\cosh(\alpha)}\right] \end{aligned} \tag{6.52}$$

$$\begin{aligned} g_3(\xi) &= (h_1+h_2)\cdot\frac{\overline{EA}}{\overline{EI}}\cdot L\cdot\left[\frac{\sinh(\alpha\cdot\xi)-\sinh(\alpha)}{\alpha\cdot\cosh(\alpha)}+1-\xi\right] \\ &= a_4\cdot L\cdot\left[\frac{\sinh(\alpha\cdot\xi)-\sinh(\alpha)}{\alpha\cdot\cosh(\alpha)}+1-\xi\right] \end{aligned} \tag{6.53}$$

The functions $g_1(\xi)$ and $g_2(\xi)$ approach zero when the stiffness of the shear connectors approaches infinity, which leads to the result given by the full interaction theory.

6.2.5 Composite parameters

For a composite member, there are many member properties/factors that affect the response. The effect of each property is not obvious. Some of the properties are interrelated, which further complicates the matters. For example, if an engineer would like to reduce the interfacial slip considering the limited-slip capacity of a given shear connector, it is difficult for him/her to find how to achieve that apart from providing more shear connectors. The slip is given by Eq. (6.39), which is very complicated and affected by many factors. For example, the first term in Eq. (6.39) is given by Eq. (6.40) in which the slip appears to be in direct proportion to (h_1+h_2); however,

$\overline{EI}$ and $\overline{EA}$ are also related to $(h_1 + h_2)$. Furthermore, α is also dependent on h_1 and h_2. These complications bring difficulty to the design of composite members.

Careful study of the properties and parameters in the solutions reveals that the response of a composite member is actually determined by only a few composite parameters that are combinations of the basic material and geometric properties, as described in the following sections. The finding of these composite parameters brings convenience to, and sheds light on, design of composite members.

6.2.5.1 Parameters governing longitudinal slip

Interfacial slip governs the behavior of composite members. Therefore, it is important to understand the factors that affect the slip. Eq. (6.13) is rewritten with reference to Eqs. (6.19)–(6.21) as

$$\varepsilon_{slp} = -a_1 \cdot F_{shr} + a_2 \cdot M(x) + a_3 \cdot N \tag{6.54}$$

For a linear system, the superposition law applies. Therefore, the total slip strain of Eq. (6.54) can be considered to be the algebraic sum of the following 3 components:

$$\left(\varepsilon_{slp}\right)_1 = -\frac{ds_1}{dx} = -a_1 \cdot F_{shr} \tag{6.55}$$

$$\left(\varepsilon_{slp}\right)_2 = -\frac{ds_2}{dx} = a_2 \cdot M(x) \tag{6.56}$$

$$\left(\varepsilon_{slp}\right)_3 = -\frac{ds_3}{dx} = a_3 \cdot N \tag{6.57}$$

where $(\varepsilon_{slp})_1$, $(\varepsilon_{slp})_2$ and $(\varepsilon_{slp})_3$ are the slip strain terms caused by (1) the longitudinal shear force F_{shr}, (2) the bending moment $M(x)$, and (3) the axial load N, respectively. The corresponding slip terms are indicated by s_1, s_2 and s_3, respectively. Therefore

$$\varepsilon_{slp} = \left(\varepsilon_{slp}\right)_1 + \left(\varepsilon_{slp}\right)_2 + \left(\varepsilon_{slp}\right)_3 \tag{6.58}$$

$$s = s_1 + s_2 + s_3 \tag{6.59}$$

Integrating Eqs. (6.56) and (6.57) gives

$$s_2 = -a_2 \cdot \int M(x) \cdot dx = a_2 \cdot F_2(x) \tag{6.60}$$

$$s_3 = -a_3 \cdot \int N \cdot dx = a_3 \cdot F_3(x) \tag{6.61}$$

where $F_2(x) = -\int M(x) \cdot dx$ and $F_3(x) = -\int N \cdot dx$. By referring to Eq. (6.10), Eq. (6.55) can be rewritten as

$$-\frac{ds_1}{dx} = -a_1 \cdot \frac{1}{a_0} \int s \cdot dx \tag{6.62}$$

Taking the derivative of both sides and referring to Eqs. (6.33) and (6.59)–(6.61) gives

$$\frac{d^2 s_1}{dx^2} - a_5 \cdot s_1 - a_5 \cdot [a_2 \cdot F_2(x) + a_3 \cdot F_3(x)] = 0 \tag{6.63}$$

The factors that affect the slip can now be clearly seen from Eqs. (6.59)–(6.61) and (6.63). Apart from the loading conditions that determine $F_2(x)$ and $F_3(x)$, there are only three composite parameters that affect the slip: a_2, a_3 and a_5. These three composite parameters are functions of the geometric and material properties of the composite member. Furthermore, these parameters have clear physical meanings as discussed in the following paragraphs. To discuss the parameter a_5, the parameters a_0 and a_1 are also analyzed.

(1) The flexibility of shear connection, a_0
As given by Eq. (6.18), it can be seen that $1/a_0$ is the shear connector stiffness over a unit length of longitudinal interface. In other words, a_0 is the flexibility of the shear connection. Therefore, a larger value of a_0 indicates a smaller resistance to slip from the shear connection, which tends to give a larger value of slip at the interface.

(2) The coefficient of slip sensitivity to shear resistance, a_1
Let $M = N = 0$ and $F_{shr} = 1$ in Eq. (6.54); then $\varepsilon_{slp} = -a_1$, in which a_1 is another function of the cross-sectional properties given by Eq. (6.19). Therefore, the coefficient a_1 is the slip strain of the cross section caused by a unit longitudinal shear force. For a cross section with a larger a_1, the longitudinal shear force will cause a larger longitudinal slip strain and hence slip. As the longitudinal shear force always tries to stop the slip caused by the external forces, a larger a_1 indicates that the longitudinal shear is more effective in resisting the slip caused by external forces.

(3) The coefficient of slip sensitivity to moment, a_2
Letting $F_{shr} = 0$, $N = 0$, and $M = 1$ in Eq. (6.54) gives $\varepsilon_{slp} = a_2$, where a_2 is given by Eq. (6.20). This means that the coefficient a_2 is the slip strain caused by a unit external bending moment at the cross section when no interaction exists at the interface (no shear connectors). A larger value of a_2 means that a certain external bending moment will cause a larger slip strain. In other words, the member with a larger a_2 value is prone to have a larger slip under flexural bending. From Eq. (6.20), it can be seen that a_2 is a cross-sectional property of the individual elements 1 and 2.

(4) The coefficient of slip sensitivity to axial force, a_3
Letting $F_{shr} = 0$, $N = 1$, and $M = 0$, Eq. (6.54) gives $\varepsilon_{slp} = a_3$, where a_3 is given by Eq. (6.21). That is to say, a_3 is the slip strain caused by a unit axial force that is applied at the centroid of element 1. When the axial flexibility a_3 of element 1 is greater, a given axial load will cause a larger axial shortening of element 1, which in turn causes a larger slip between elements 1 and 2.

(5) The coefficient of effective resistance to slip, a_5
As given by Eq. (6.33), this coefficient is the combination of a_0 and a_1 and reflects the effective resistance of the shear connection in restraining the slip at the interface. This can be observed from the physical meanings of a_0 and a_1. When the shear connector stiffness $1/a_0$ is larger, the connectors are stiffer, hence providing greater shear resistance on the interface. When the coefficient of slip sensitivity to shear resistance a_1 is larger, the same shear resistance will produce a larger negative slip against the positive slip caused by loading so that

a smaller slip at the interface is produced. Therefore, for composite members with a larger a_5, the overall effect of shear connection in restraining the interfacial slip is greater, and hence the slip caused by given external loading will be smaller.

These relations can be verified by the slip solutions given by Eqs. (6.40)–(6.42). In all three terms, which are caused by the external moment (Eq. 6.40), the axial load (Eq. 6.41) and the boundary slip (Eq. 6.42), respectively, the coefficient of effective resistance to slip a_5 plays a key role in resisting the slip. Therefore, a_5 is the most important factor of interfacial slip.

6.2.5.2 Parameters affecting flexural deformation

Similar to the slip, the deflection of the composite member is also governed by a few key parameters regardless of the many factors involved. With reference to Eqs. (6.11) and (6.12), Eq. (6.24) can be rewritten as:

$$\begin{aligned}\kappa &= \frac{M(x)}{\overline{EI}} + \frac{(h_1+h_2)\cdot\overline{EA}}{\overline{EI}}\cdot\varepsilon_{slp} - \frac{(h_1+h_2)\cdot\overline{EA}}{\overline{EI}\cdot(EA)_1}\cdot N \\ &= \frac{M(x)}{\overline{EI}} + a_4\cdot\varepsilon_{slp} - a_4\cdot a_3\cdot N\end{aligned} \tag{6.64}$$

where a_4 is given by Eq. (6.26). The first term in the preceding equation is the curvature caused by the external moment under full interaction. The second is the additional curvature of the cross section caused by a given slip strain ε_{slp}. The third term is the curvature caused by the axial load due to the eccentricity between the centroid of element 1 and the centroid of the composite section under full interaction. For beams or columns with the axial force applied at the centroid of full interaction, this term is zero.

Letting $M(x)=0$ and $N=0$ in Eq. (6.64) and considering Eq. (6.54) give

$$\kappa = a_4\cdot\varepsilon_{slp} = -a_4\cdot a_1\cdot F_{shr} \tag{6.65}$$

From Eq. (6.65), it can be seen that a_4 is the curvature when the slip strain $\varepsilon_{slp}=1$, which leads to the following composite parameter:

(6) The coefficient of curvature sensitivity to slip (or slip-induced curvature), a_4

For a given value of slip strain, a larger a_4 means a larger curvature of the member. As curvature is directly related to deformation of a flexural member, this means that a_4 is a key parameter that affects flexural deformation.

Furthermore, a_4 has a geometrical meaning as discussed in the following text. For convenience of study, the strain profile corresponding to Eq. (6.65) is shown in Fig. 6.8.

The strain distribution of element 1 is given by $\varepsilon=\frac{N_1}{(EA)_1}-\kappa\cdot y$. Considering Eq. (6.65) and $N_1 = -F_{shr}$, one has

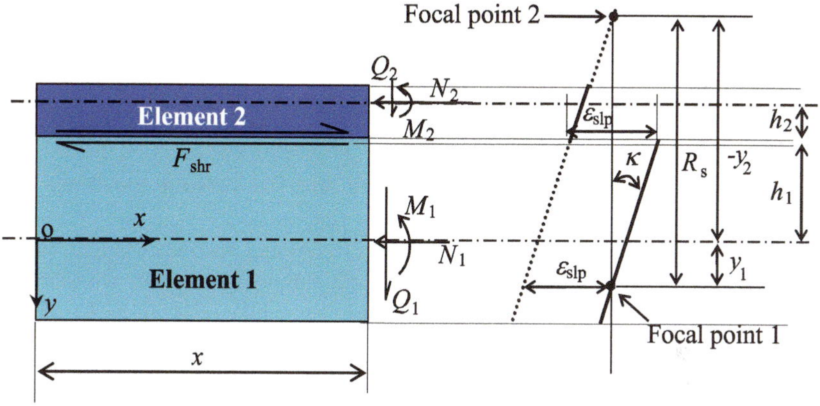

Fig. 6.8 Strain profile due to longitudinal shear force.

$$\varepsilon = -\frac{F_{shr}}{(EA)_1} + a_1 \cdot a_4 \cdot F_{shr} \cdot y = F_{shr} \cdot (a_1 \cdot a_4 \cdot y - a_3) \tag{6.66}$$

It is noted that at the point where $a_1 \cdot a_4 \cdot y - a_3 = 0$ in Eq. (6.66), $\varepsilon \equiv 0$ for any value of F_{shr}. The y coordinate of this point is given by

$$y_1 = \frac{a_3}{a_1 \cdot a_4} \tag{6.67}$$

In other words, the additional strain caused by interface shear (or slip) is always zero at this point no matter how the shear force or slip varies. This means that all the strain profiles, for any design/distribution of shear connections, pass through or "are focused at" this point.

For the same reason, element 2 also has a focal point. The y coordinate of the focal point 2 can be calculated as

$$y_2 = -\frac{1}{a_1 \cdot a_4 \cdot (EA)_2} - h_1 - h_2 \tag{6.68}$$

The distance between these two focal points, R_s (Fig. 6.8), is given by

$$R_s = y_1 - y_2 = \frac{1}{a_1 \cdot a_4 \cdot (EA)_1} + \frac{1}{a_1 \cdot a_4 \cdot (EA)_2} + h_1 + h_2 = \frac{1}{a_4} \tag{6.69}$$

Eq. (6.69) shows that R_s is the inverse of a_4. Since a_4 is the slip-induced curvature, R_s is hence the slip-induced radius of curvature. It is noted that the slip-induced radius of curvature (and slip-induced curvature) is also a function of the cross-sectional properties. When the value of R_s is greater, an interfacial slip strain (or shear force) will cause a smaller curvature, i.e., a smaller flexural displacement.

From the geometrical relationship in Fig. 6.8, it can also be seen that the slip strain is given by

$$\varepsilon_{slp} = R_s \cdot \kappa = \frac{\kappa}{a_4} \tag{6.70}$$

Comparing Eq. (6.70) with Eq. (6.65), it can be seen that they are the same, which further supports the concept of "focal points."

Interestingly, the first and third terms in Eq. (6.64) can also be obtained with the concept of slip-induced radius of curvature. For a composite member without interaction, the curvature due to an external moment is given by $\kappa = M(x)/\sum EI$. The slip strain at the interface is simply

$$\varepsilon_{slp} = \frac{M(x)}{\sum EI} \cdot (h_1 + h_2) \tag{6.71}$$

For a composite member with full interaction, this slip strain must be resisted by the longitudinal shear force so that $\varepsilon_{slp} = 0$. Therefore, the slip strain due to the interfacial shear force must be equal and opposite, i.e.,

$$\varepsilon_{slp} = -\frac{M(x)}{\sum EI} \cdot (h_1 + h_2) \tag{6.72}$$

From Eq. (6.70), the additional curvature due to longitudinal shear is given by

$$a_4 \cdot \varepsilon_{slp} = -a_4 \cdot \frac{M(x)}{\sum EI} \cdot (h_1 + h_2) \tag{6.73}$$

The curvature under full interaction is therefore the summation of the two parts, or

$$\kappa = \frac{M(x)}{\sum EI} - a_4 \cdot \frac{M(x)}{\sum EI} \cdot (h_1 + h_2) = \frac{M(x)}{\overline{EI}} \tag{6.74}$$

which is the same as the first term in Eq. (6.64). The third term of Eq. (6.64) can also be obtained with the same concept.

The "focal point" of the strain profile is an interesting new concept in the study of composite members that was first discovered and introduced by Seracino et al. (2001) from their numerical simulations. They later verified this finding using linear elastic analysis to show that all strain profiles calculated from the elastic theory, with different shear connections, intersected at a common point, but without a mathematical proof for the existence of the focal points.

From the preceding study, it can be seen that the longitudinal shear force F_{shr} causes an additional axial force $N_1 = -F_{shr}$ and an additional moment $M_1 = -a_4 \cdot a_1 \cdot F_{shr} \cdot (EI)_1$ in the cross section of element 1. This additional axial force and moment lead to the additional strain of $-F_{shr}/(EA)_1$ and $a_4 \cdot a_1 \cdot F_{shr} \cdot y$, respectively, in the cross section, which cancel out each other at the focal point regardless of the value of F_{shr}. It is noted that the "focal point" concept is, strictly speaking, a characteristic of linear systems. The linear strain distribution given by Eq. (6.66) is not applicable to a general nonlinear system, which will not have a common y value for $\varepsilon = 0$.

In summary, the key composite parameter that controls the interfacial slip is the effective resistance coefficient to slip, a_5, given by Eq. (6.33). The key composite parameter that controls the flexural deformation is the curvature sensitivity coefficient to slip (or slip-induced curvature), a_4, given by Eq. (6.26), together with a_5.

6.2.6 Slip distribution

Interfacial slip is critical to the design of composite members, as it not only affects the effect of composite action (Eq. 5.2) and member rigidity, hence deflection, but also is directly related to the design of interfacial shear connectors. The analytical solutions in the previous section shed light on the factors that affect the interfacial slip. The slip distribution is important for composite members as it reflects the distribution of interfacial shear stress along with the member. The typical slip distributions of the cantilever column obtained from Section 6.2.4, as given by Eqs. (6.40)–(6.42), are shown in Fig. 6.9A. For a member with a plastic hinge formed, the solutions are only applicable to the elastic zone, as shown. The slip distribution in Fig. 6.9A is typical of a flexural member from the point of contraflexure (zero moment) to the position of maximum elastic moment.

The term of slip due to flexural deformation given by Eq. (6.40) is shown by the curve marked with "By F" in Fig. 6.9A. The slip term caused by the axial load N given by Eq. (6.41) is marked with "By N." When the load and deflection are small, no plastic hinge forms, and the total slip is given by the curve marked with "By $F+N$." The other term given by Eq. (6.42) is induced by the boundary slip s_p at the elastic-zone/plastic-hinge interface when a plastic hinge forms at large deformation, as marked with "By s_p." Summation of these three terms depicts the total slip when large plastic deformation occurs, as shown by the thick line indicated with "Total."

The maximum slip due to either F or N is at the column top in an elastic member. The slip term due to N does not change at a cross section under a constant axial load. The slip term due to F increases when the load increases. As the load F reduces after the formation of a plastic hinge, this term also reduces when F reduces. The analytical and numerical studies show that, after the formation of a plastic hinge, the slip term due to s_p is proportional to the angle of plastic hinge rotation, and hence increases continuously when the deformation of the member increases (Wu 2002). When the slip s_p is big enough, the position of maximum slip will occur at the top of the plastic hinge instead of at the top of the column, as shown in Fig. 6.9A.

The preceding analytical results are verified by the full nonlinear numerical simulations shown in Fig. 6.9B (Wu 2002). The example column is a 1.2 m long cantilever with a concrete cross section of 200 × 200 mm and 4T16 main longitudinal bars under an axial load of 360 kN as shown in Fig. 5.2. A 6 mm thick mild steel plate is attached

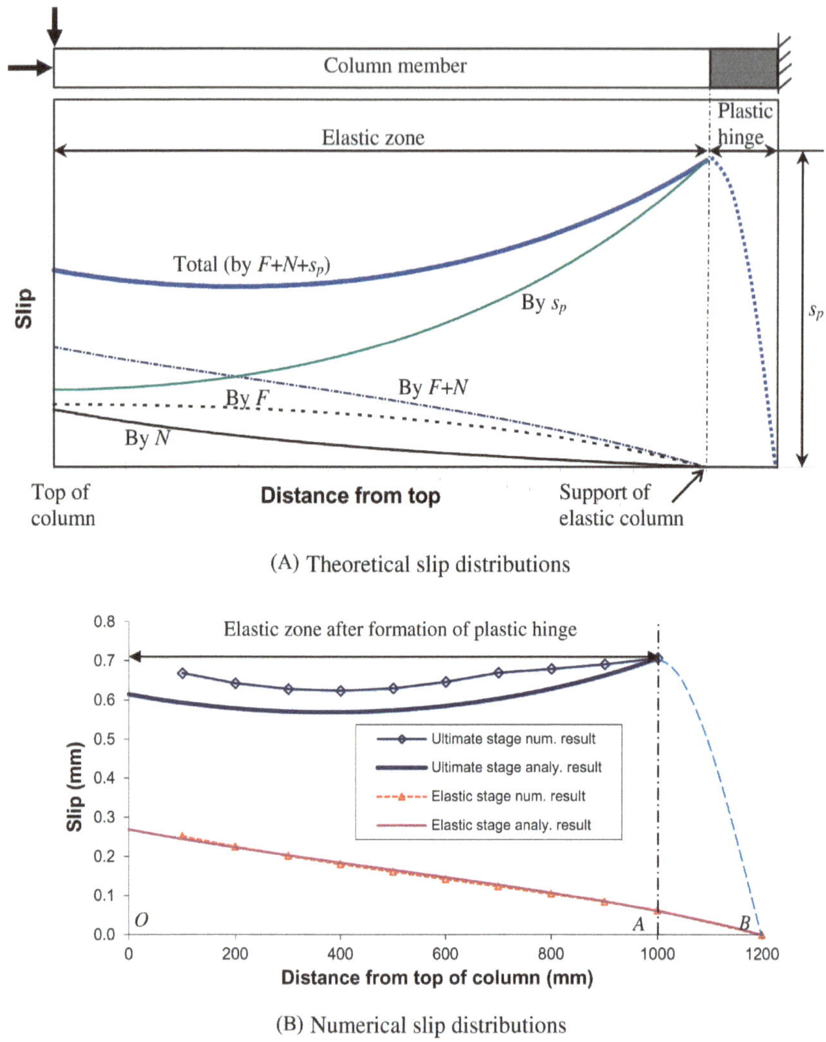

Fig. 6.9 Typical slip distributions.

to the compression face of the RC column with 20 bolts (10 rows, 2 bolts in each row; $K_b = 61$ kN/mm). The slip distribution of the elastic stage in Fig. 6.9B is at $F = 24.2$ kN and the one after the formation of the plastic hinge is at $F = 58.9$ kN. It is noted that, before formation of the plastic hinge, the whole member is elastic, and hence the analytical solutions are applicable to the whole column from point O to B in Fig. 6.9B, with $s_p = 0$ at point B. After formation of the plastic hinge, the elastic solutions are only applicable to the elastic zone from point O to A, with $s_p \neq 0$ at point A. The value of s_p in Fig. 6.9B is obtained from the numerical simulation. It can be seen from the figure that both the elastic result and the result after formation of the plastic hinge agree well between the analytical solutions and the numerical simulations.

In the literature, the elastic slip distribution, where a maximum slip occurs at the position of zero moment (the support of a simply supported beam, the top of a column, the point of contraflexure in a continuous beam), is considered typical in the classic theory of composite members. This classic slip distribution is generally adopted for the design of shear connectors (Oehlers and Bradford 1995). As a result, the failure of a shear connector is considered to most likely occur at the zero moment positions. However, it can be seen from Fig. 6.9A and B that the maximum slip, and hence shear connector failure, may occur at the top of the plastic hinge, when the composite member is loaded to its plastic deformation stage.

The following further conclusions can be drawn from Eqs (6.40)–(6.42):

- All the three terms, s_f, s_n, and s_{sp}, decrease when a_5 increases. It can be seen from the mathematical function that increasing a_5 is very effective in reducing the slips when it is small, but is ineffective when it is larger than a certain value, as shown in Fig. 6.10. In other words, too stiff a shear connection is ineffective and unnecessary in reducing the interfacial slip.
- The slips increase when the external actions F, N, and s_p increase.
- The maximum slip due to F that occurs at the top of the column ($\xi = 0$) (where the moment is zero) is given by

$$s_{f,\max} = F\frac{a_2}{a_5}\left[1 - \frac{1}{\cosh\left(L\cdot\sqrt{a_5}\right)}\right] \tag{6.75}$$

- The maximum slip due to N that occurs at the top of the column (where $M=0$) is given by

$$s_{n,\max} = \frac{N\cdot a_3}{\sqrt{a_5}}\left[1 - 2/\left(e^{2\left(L\cdot\sqrt{a_5}\right)} + 1\right)\right] \tag{6.76}$$

- s_f and s_n decrease monotonically from the maximum at column top to zero at the support.
- The slip due to s_p monotonically reduces from the support (top of plastic hinge) to the column top, with its maximum at the support, $s_{sp,max} = s_p$, and its minimum at the column top, given by

$$s_{sp,\min} = \frac{s_p}{\cosh\left(L\cdot\sqrt{a_5}\right)} \tag{6.77}$$

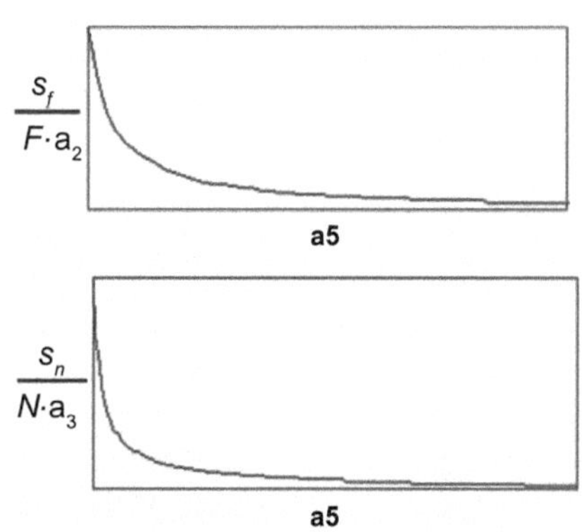

Fig. 6.10 Effect of a_5 on slip distributions.

6.2.7 Other studies on partial-interaction composite members

After Newmark et al. (1951) first introduced the classical linear elastic partial-interaction theory for composite beams, Goodman (1967) independently presented the theory for layered wood systems with interlayer slip. Amana and Booth (1967) studied the problem both theoretically and experimentally. Goodman and Popkov (1968) extended the theory to wood structures, taking into account the nonlinear relation between interface shear force and interlayer slip. Thompson et al. (1975) developed the finite element model of layered wood structures with interlayer slip. Itani and co-workers presented a closed-form solution of the second-order differential equation of the axial force and interlay slips (Itani and Brito 1978) and utilized the finite difference method to analyze the partial-interaction composite beams (Itani and Hiremath 1980). Murakami (1984) developed a theory of partial-interaction composite beams by taking account of the transverse shear deformation and obtained the solution for a beam with two simply supported ends subjected to a point load acting at the midspan. Girhammar and Gopu (1993) analyzed partial-interaction composite beam-columns subjected to transverse and axial loading and provided the analytical solution for simply supported and clamped boundary conditions. Wang (1998) proposed a simplified method for designers to predict the maximum deflection of a composite beam at working load based on the stiffness of its shear connectors. Jasim (1999) obtained an analytical solution of composite beams with shear connector density varying linearly along the span. Gattesco (1999) developed a numerical procedure for the analysis of steel and concrete composite beams accounting for nonlinear behavior of concrete, steel, and shear connectors. Jasim and Atalla (1999), Fabbrocino et al. (2000), and Seracino et al. (2004) extended these theories to continuous beams and emphasized the effect of the negative bending moment. Nguyen et al. (2001) studied side plated beams where both transverse and longitudinal slips between RC beams and bolted side plates are involved. Faella et al. (2002, 2003) and Ranzi et al. (2004) employed the stiffness element method to study the effect of nonlinearity of the shear connection. Kwak and Seo (2002) and Jurkiewiez et al. (2005) took into account the creep and shrinkage characteristics of concrete and studied the long-term time-dependent responses of composite beams. Ranzi and Bradford (2007) studied the behavior of partial-interaction beams under elevated temperature.

Most analytical studies on composite members adopt the original Newmark's assumptions that all constituents (subelements) of the composite beam have identical curvature and deflection, i.e., there is no separation at the interface. This may not always be true, in cases of large space between shear connectors or when one of the subelements is relatively thin. To consider separation at the interface, Adekola (1968) presented an interaction theory of composite beams that takes account of interfacial uplift and friction effects apart from longitudinal slip. The formulated problem is nonlinear, which was solved using a special finite difference scheme. Gara et al. (2006) proposed three novel displacement-based formulations for the analysis of composite beams with longitudinal slip and vertical uplift at the interface, extending the analysis of Adekola (1968) for determinate beams to indeterminate ones.

Different numerical approaches have been reported to analyze the behavior of partial-interaction composite members (Manfredi et al. 1999; Fang et al. 2000; Salari and Spacone 2001; Ranzi et al. 2004; Wu et al. 2004). More advanced theories have also been applied to study the behavior of partial-interaction composite members. Typical ones are introduced here.

(1) State-space solution of partial-interaction members

State-space modeling uses state variables to describe a system by a set of first-order differential equations, rather than by one or more nth-order differential equations. The nth-order governing differential equations of the partial-interaction problem can be represented by a set of first-order differential equations called the state equation, where responses of a structure are the state vector. Solution of the problem is obtained by solving the state equation analytically or numerically. Chen et al. (2007) developed a state space formulation of partial-interaction composite members by introducing appropriate state variables. The state space formulation is able to deal with arbitrary loads readily, including discontinuous point loads, and can be easily used to study inhomogeneous beam-columns with discontinuous interfacial shear. Therefore, it can better reflect the effect of individual mechanical shear connectors. To facilitate the treatment of an arbitrary inhomogeneity, a piecewise homogeneous beam model was adopted, which can model discrete and discontinuous shear stud connections at the interface with nonuniform slip stiffness consisting of rectangular pulses, outside which the shear stiffness is considered to be zero. It was concluded from the piecewise model that an equivalent homogeneous slip stiffness may be insufficient and inaccurate sometimes to capture the actual action of discrete shear connectors in a composite beam. Another superiority of the state space method over classic partial-interaction theory is the direct treatment/description of boundary conditions at end supports as well as continuity conditions at intermediate supports for continuous beams or intermediate nodes for inhomogeneous beams. It is interesting and worthwhile to note that the state space solution found that under point load, a composite beam does show the tendency to separate at the interface, i.e., Newmarks's original assumption of identical deflection of all subelements may not be true in some cases.

(2) Analytical studies using Timoshenko's beam theory

The simplifying assumptions that Newmark et al. (1951) made in their analysis include small elastic deformation, Euler-Bernoulli's hypothesis of plane cross section applicable to each subelement, continuous and uniform shear connection rigidity (slip stiffness) along the interface, and identical deflection/curvature of all subelements. These assumptions have been widely adopted in subsequent studies of composite beams. It is well known that the Euler-Bernoulli beam theory overestimates the stiffness because the transverse shear deformation is neglected. Thus, the deflection of the beams is underestimated, and the error will increase with the increase in the height-to-length aspect ratio. The natural frequency of the beams is also overestimated, and the error will rapidly increase with the increase in the beam depth-to-shear span ratio.

In the early 20th century, the Euler-Bernoulli beam theory was improved by Timoshenko (1941), who took into account the important effect of shear deformation and rotary inertia. This is universally known as Timoshenko's beam theory. Xu and Wu (2007a) applied Timoshenko's beam theory to investigate the static, dynamic, and buckling behavior of partial-interaction composite beams. Analytical solutions were obtained from the governing differential equations for static, free vibration, and buckling behavior under certain boundary conditions. It was found that the deflection due to the shear deformation is significantly dependent on the boundary conditions and is proportional to the square of the height-to-length aspect ratio of the beam, which means that the effect of shear deformation increases rapidly with the increase in the aspect ratio. The shear deformation and rotary inertia also decrease the natural frequency of the beams. The influence of shear deformation and rotary inertia on free vibration frequencies increases with the increase in the order of free vibration mode and the stiffness of the shear connectors. The effect of shear deformation on the frequency is more significant than that of rotary inertia. The shear deformation also decreases the buckling load of the partial-interaction beams; however, the effect is not significant. It was also found that the stiffness of shear connectors has a sensitive range that affects the magnitudes of the frequencies, outside of which it has little effect. Schnabl et al. (2007) also studied the problem by considering the shear deformation and obtained similar conclusions. Berczynski and Wrosblewski (2005) and Ranzi and Zona (2007) analyzed the vibration of steel-concrete composite beams using Timoshenko's beam model.

(3) Two-dimensional theory of partial-interaction composite members

Classical Euler-Bernoulli beam theory, Timoshenko's beam theory and other high-order shear deformation beam theories are all one-dimensional theories that involve assumptions on deformation field. Unlike one-dimensional beam theories, a two-dimensional beam theory does not impose extra restriction or assumption on the deflection and longitudinal displacement along the thickness of the beam. As a result, a two-dimensional beam theory is more precise and realistic than one-dimensional ones, especially for a thick beam. Therefore, the results from a two-dimensional beam theory can be used as a benchmark for other analytical and numerical methods based on one-dimensional beam theories. The two-dimensional beam theory can also be applied to partial-interaction composite members for more precise modeling.

Rao and Ghosh (1980) and Fazio and Hussein (1982) directly adopted elasticity theory to analyze the mechanical behavior of partial-interaction composite beams. Chen et al. (2004) presented an exact analysis of simply supported laminated beams with imperfect bonding based on the two-dimensional theory of elasticity and the state-space approach. Xu and Wu (2007b) presented a two-dimensional theory to investigate the static behavior of partial-interaction composite beams by the state-space method. The exact analytical solution for simply supported beams under arbitrarily distributed transverse loading was obtained, as well as semianalytical solutions for general end conditions. It was found that the rigidity of shear connectors significantly affects the flexural stiffness of the composite beams but has an economical upper bound above which no significant gain on the flexural stiffness of the composite beams can be obtained. In another work (Xu and Wu 2008), the two-dimensional theory in combination with the state-space method was extended to investigate the free vibration and buckling of partial-interaction composite beams. The exact analytical solution of beams with two simply supported ends was obtained through trigonometric function expansion. Moreover, the solution of beams with nonsimply supported ends was obtained by a semianalytical method using the differential quadrature method coupled with the state-space method.

(4) Partial-interaction theory for dynamic response

Introducing the inertia force by d'Alembert's principle, Girhammar and Pan (1993) extended their theory for static analysis of partial-interaction composite members to dynamic problems and presented the governing differential equation of motion without axial force. An improved theory that took the viscous damping into account was then presented by Adam et al. (1997). Biscontin et al. (2000) studied the vibration of the composite beam by modeling the shear connection with a strain energy density function. Wu et al.

(2007) presented a theory for the dynamic analysis of partial-interaction composite members with axial forces that is applicable to composite columns and prestressed composite beams. Design-oriented simplified expressions based on the exact solution for the simply supported case were proposed to predict the fundamental frequency of the partial-interaction composite members. It was found that the interlayer slip decreases the frequencies and a tensile axial force increases the frequencies, while a compression one decreases the frequencies. Berczynski and Wroblewski (2005) considered the free vibration of steel-concrete composite beams based on Timoshenko's beam theory. When compared with the Euler-Bernoulli beam theory, their analysis predicted more accurate frequencies of higher and lower modes of flexural vibrations. More studies on dynamic behavior of partial-interaction composite members, particularly by FEM modeling, have been reported in recent years (Huang and Su 2008; Chakrabarti et al. 2013; Bao and Xu 2015; Zhou et al. 2015; Fu and Yang 2018; Abramowicz et al. 2020).

Studies of forced vibration of partial-interaction composite beams are relatively few. Girhammar and Pan (1993) investigated the dynamic behavior of partial-interaction composite beams without axial force and presented the orthogonality relation of vibration modes. They also obtained the analytical expressions for the transient response of composite beams under impulsive and step loads. Adam et al. (1997) decomposed the dynamic response into two parts: the quasistatic component and the complementary component. The former can be solved by integrating the influence function and can conveniently take account of the singularity due to abrupt variation of load. Meanwhile, the latter was solved by the mode superposition without the necessity to deal with any singularity. Consequently, the calculation of modal shapes converges quickly, and the results were improved when compared with those given by Girhammar and Pan (1993). Girhammar et al. (2009) utilized Hamilton's principle to derive the governing equations for dynamic problems of composite beams under generalized boundary conditions, including the spring-supported end and end with lumped mass. Common boundary conditions (such as simply supported, clamped, and free) were considered in their derivation of the orthogonality relation of modes. At the same time, they studied the forced vibration of composite beams by virtue of mode superposition and proved that the dynamic response was identical with that given in Girhammar and Pan (1993) in the simply supported case. Shen et al. (2011) extended the static analysis in Chen et al. (2007) to a dynamic one without axial force based on the state-space method. In order to obtain the orthogonality relation of vibration modes conveniently, the state variables were chosen slightly differently from those in Chen et al. (2007). The natural frequencies and the corresponding mode shapes were presented based on the state-space formulae. By utilizing the symplectic inner product, the orthogonality relation of vibration modes was derived for generalized boundary conditions at the two ends with spring support and lumped mass. The derivation used a unified matrix form, and hence can be conveniently extended to the proof of mode orthogonality relation for more complex structures. Based on the orthogonality relation, they presented the analytical solution of the inhomogeneous state equation by mode superposition and investigated the dynamic responses of composite beams subject to uniformly distributed impulsive load, step load and moving constant force.

6.3 Adhesively bonded reinforcement

Reinforcing sheets/plates may be fully bonded onto a concrete member, such as by adhesion, with little slip, namely full-interaction plating. In this case, the conventional theory for the design of RC structures is applicable, considering the plates as additional reinforcement. The material strength of the plate may not be achieved at failure of the composite structure, because the adhesive joint is often the weakest link of the composite system and may debond before the occurrence of other conventional failure modes that are designed for in conventional structures. Therefore, the interfacial bond characteristics are critical and usually govern the effectiveness of such rehabilitation work and the performance of the rehabilitated structures.

Prevention of premature debonding or peeling of externally bonded reinforcement (EBR) is a critical aspect of the design and construction of the rehabilitation work, which requires specialist knowledge and expertise. Treatment of bond interface and workmanship is critical to interfacial bond. Extreme care must be exercised in the bonding process and manufacturer's specification and design guidelines (e.g., ACI Committee 440 2017) must be strictly followed.

The adhesive debonding process is usually sudden and catastrophic. Therefore, debonding usually causes the ultimate failure of the structures strengthened by EBR. It is important to develop debonding criteria for a particular debonding failure mode, based on which failure load can be evaluated. There are a few typical and possible debonding failure modes for EBR systems. Each failure mode is a link of the chain in Fig. 1.9. Therefore, debonding design is the identification of the weakest link, which follows the same principle as discussed in Section 1.4.3. Although analytical study and numerical simulation can also point to possible failure modes, failure modes should be identified from observation, either from experimental tests or from practice. Extensive literature on the modeling, design, construction of externally bonded FRP strengthening systems and the performance of FRP strengthened concrete structures can be found in the monograph by Wu et al. (2020).

6.3.1 Failure modes

The typical debonding failure modes of EBR strengthened RC members are shown in Fig. 6.11 (Teng and Chen 2007). The plate-end interfacial debonding is due to the high stress concentration at the plate end caused by sudden cut-off of the plate. The high stress causes the cohesive failure in concrete (see Fig. 5.7, another three debonding failure modes are not allowed to occur by proper treatment of bond interface and use of adequate materials as explained in Section 5.3.1). Debonding failure at a plate end may occur through two paths: (1) interfacial debonding where a thin layer of concrete is peeled off, and (2) concrete cover separation in which the failure plane occurs at the rebar level where the concrete area is reduced by the existence of rebars. Debonding occurs at the weaker link of these two paths.

The other debonding failure modes are caused by cracking. The mechanism of debonding caused by a crack is shown in Fig. 6.12. It can be seen from Fig. 6.12B that it is the relative movement between point A and point B that causes the debonding. This relative moment is essentially the same as the pull of the reinforcement at the loaded end in Fig. 5.8. Therefore, the study of the pull-off test in Fig. 5.8 is an important topic for EBR strengthened RC structures, which will be studied in detail in Section 6.4.

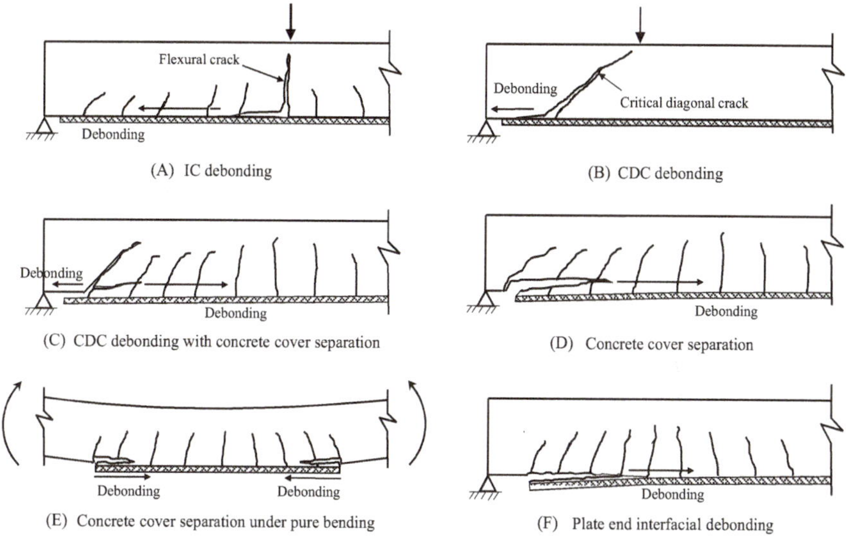

Fig. 6.11 Debonding failure modes in FRP-strengthened RC beams.
Based on Teng, J. G., and J. F. Chen. 2007. "Debonding failures of RC beams strengthened with externally bonded FRP reinforcement: behaviour and modelling." In *Proc. First Asia-Pacific Conf. FRP in Structures (APFIS 2007), 12–14, The University of Hong Kong, Hong Kong, China.*

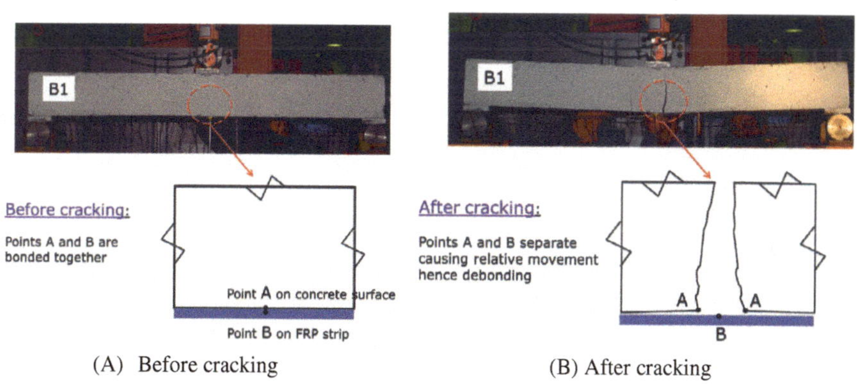

Fig. 6.12 Debonding mechanism due to cracking.

Intermediate crack-induced debonding (IC debonding) is the most common debonding failure mode in EBR-strengthened RC members. It is caused by a flexural crack and spreads over the whole length of the EBR plate quickly. Critical diagonal crack (CDC) induced debonding is caused by shear cracking. As significant shear cracking is not allowed in design, a member with potential CDC must be strengthened for shear strength first before flexural strengthening. Therefore, CDC debonding is not allowed to occur in practice. As a result, the flexural design of an EBR-strengthened member only considers IC debonding and plate-end debonding, as in the following section.

6.3.2 Flexural design approach

The calculation of the flexural strength of EBR-strengthened beams follows the same principles and procedure as those for conventional RC beams, and uses similar strain and stress diagrams, as shown in Fig. 6.13. However, the additional force from the EBR needs to be considered in the calculation. The magnitude of the additional force depends on the failure mode, which complicates the matter. In addition, the concrete stress block will not be the same as that used for the flexural design of conventional RC members in Chapter 2, unless the failure mode happens to be the same as that of conventional RC members which is concrete crushing (softening). It has been shown in Chapter 2 that the shape of the concrete stress block evolves in the deformation process. If a different failure mode from that of conventional RC members occurs, such as debonding or fracture of the additional reinforcement, the concrete stress block will not be the same as that at the failure of conventional RC members. Therefore, the value of the stress block parameters will also be different from that used for the flexural design of conventional RC members.

The following procedure can be adopted for the flexural design of EBR-strengthened concrete beams:

(1) Determine the strain or stress of the EBR for IC debonding failure mode using an appropriate model

Different criteria are used in different design codes for IC debonding. These debonding criteria are either based on fracture mechanics or from an empirical model based on test results. A summary of IC debonding models/criteria can be found in Wu et al. (2020). The IC debonding criteria are usually given in the form of debonding strain of EBR. ACI 440.2R-17 (ACI Committee 440 2017) provides the following debonding criterion:

$$\varepsilon_{pd} = 0.41\sqrt{\frac{f_{co}}{E_f t_f}} \leq 0.9\varepsilon_{fu} \tag{6.78}$$

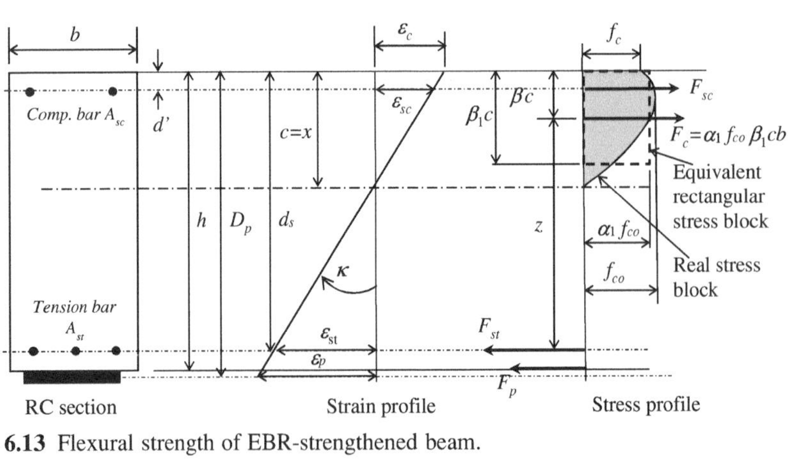

Fig. 6.13 Flexural strength of EBR-strengthened beam.

where ε_{pd} is strain of the FRP plate attached to concrete substrate at the onset of IC debonding; ε_{fu} is the rupture strain of the FRP material. The debonding strain from Eq. (6.78) is smaller than the FRP rupture strain. Based on this model, debonding always occurs before FRP rupture, and hence no FRP rupture will occur.

Calculate the flexural strain of FRP at debonding by ignoring the interfacial slip based on Fig. 6.13. In this calculation, the initial substrate strain ε_{ini} due to the flexure under existing loading at application of the EBR and the prestressing strain of the reinforcement ε_{pre} should be considered, through the following relationship:

$$\varepsilon_p = \varepsilon_{pd} - \varepsilon_{pre} + \varepsilon_{ini} \tag{6.79}$$

where ε_p is the strain at the centroid of the EBR in Fig. 6.13.

Set/assume an initial value of the neutral axis depth x. The correct x will be found by iteration through the next steps. The neutral axis depth of the original unstrengthened beam at the onset of flexural strength can be used as the initial value of x.

(2) Evaluate the stress block parameters α_1 and β based on a code model

With given x and ε_p, the whole strain profile is fixed. The concrete strain at the extreme compression fiber ε_c can be calculated based on the strain profile in Fig. 6.13. When the equivalent stress block is used to calculate the resultant force of concrete, the stress block parameters are required. It is noted that the stress block parameters are different from those for the design of conventional RC structures for the reason given previously. Different codes provide different models for the stress block parameters. The principle in determining these parameters is that it must reflect the shape of the concrete stress block when the failure or the ultimate limit state onsets. Therefore, the stress block parameters are generally a function of the extreme compression strain ε_c. Calculate the stress block parameters α_1 and β for corresponding ε_c.

Since the strain profile is fixed for a given set of ε_p and x values, another way different from the equivalent stress block method is to calculate the concrete stress and its resultant force by directly using a stress-strain relationship of concrete.

(3) Recalculate x value

With the given strain profile, all internal forces can be calculated according to the strain profile and the stress-strain relationship of the materials. The value of the neutral axis x can then be calculated from the equilibrium of forces in the longitudinal direction, i.e., $\sum F - 0$ for all the forces in the stress profile in Fig. 6.13. There is only one unknown x in the equation. If the calculated x value is the same as that used in step (2), go to step (4); otherwise go to step (2) until the calculation converges.

(4) Identify failure mode

If $\varepsilon_c \leq \varepsilon_{cm}$ (see Chapter 3 for ε_{cm}), FRP debonding occurs before concrete crushing (strictly speaking, it is the onset of the optimal concrete stress block rather than concrete crushing; see Chapter 3). Go to step (5).

If $\varepsilon_c > \varepsilon_{cm}$, the IC debonding occurs after the onset of concrete crushing. Therefore, the concrete crushing governs without FRP debonding at the onset of the flexural strength. The member should be designed in the same way as that in Chapters 2 and 3. The strain profile and the tension force from the EBR should be calculated based on $\varepsilon_c = \varepsilon_{cm}$ and the plane section in Fig. 6.13 in this case, with an additional tension force from the FRP given by $A_f E_f \varepsilon_p$, where A_f is the area of the FRP plate, and E_f is the elastic modulus of FRP.

If no valid solution of x can be found from Step (4), it means IC debonding failure will not occur. The EBR is elastic at the onset of the flexural strength. Calculate the flexural strength based on the normal procedure for conventional RC members with an additional tension force from the FRP.

(5) Use the calculated stress and strain profiles to calculate the bending moment according to Section 2.6.1

The previous procedure is suitable when the debonding criterion is given in the form of debonding strain ε_{pd}. If other forms of debonding criterion are given, the procedure needs to be adjusted. In principle, when the stress-strain relationships of the constituent materials are available, there are only two independent unknowns in the stress and strain profiles in Fig. 6.13. There are two equations available: (1) the debonding criterion, and (2) $\sum F = 0$. Therefore, a fixed solution can be obtained for the problem as long as an appropriate procedure is adopted for the calculation.

(6) After obtaining the maximum loading of the member from the preceding calculation, check the plate-end debonding

For the plate-end debonding, the relevant debonding criterion is given in a design code that can be used to check whether it occurs at the maximum loading condition. Different criteria or models should apply for the two different paths of plate-end debonding, i.e., plate-end interfacial debonding and concrete cover separation (Fig. 6.11), because the two different paths have different failure resistances. The plate-end debonding is usually related to both the shear force and bending moment at the plate-end section. If the calculation shows that it occurs, a mechanical anchorage device can be utilized to suppress the plate-end debonding failure.

The shear strength of the strengthened member should also be checked according to relevant provisions of a design code. If the shear strength of the member is insufficient to take the maximum load calculated from the flexural design, shear strengthening should be carried out. Reference can be made to Wu et al. (2020) for shear strengthening of RC members.

6.3.3 Measures to suppress interfacial debonding

Apart from mechanical fastening, another effective way to reduce the stress concentration at the plate end, and hence suppress the plate-end interfacial debonding, is by tapering the plate toward the plate end, as shown in Fig. 6.14. This approach can significantly reduce the stress concentration at the plate-end interface due to the gradual reduction in plate thickness (Xu and Wu 2009). The reduction in stress concentration depends on the rate of reduction in plate thickness. Theoretically, the plate-end debonding can always be avoided by adopting a sufficient tapering to reduce the stress concentration to a sufficiently small value.

IC debonding can also be suppressed by a technology called debonding localization (Wu and Lu 2013). It is discussed in Section 6.3.2 that IC debonding occurs when the strain of the EBR reaches a certain value. Through local weakening of the externally

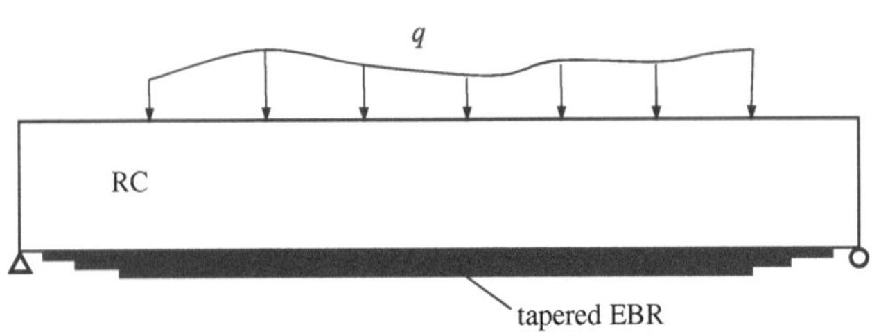

Fig. 6.14 Plate-end tapering.

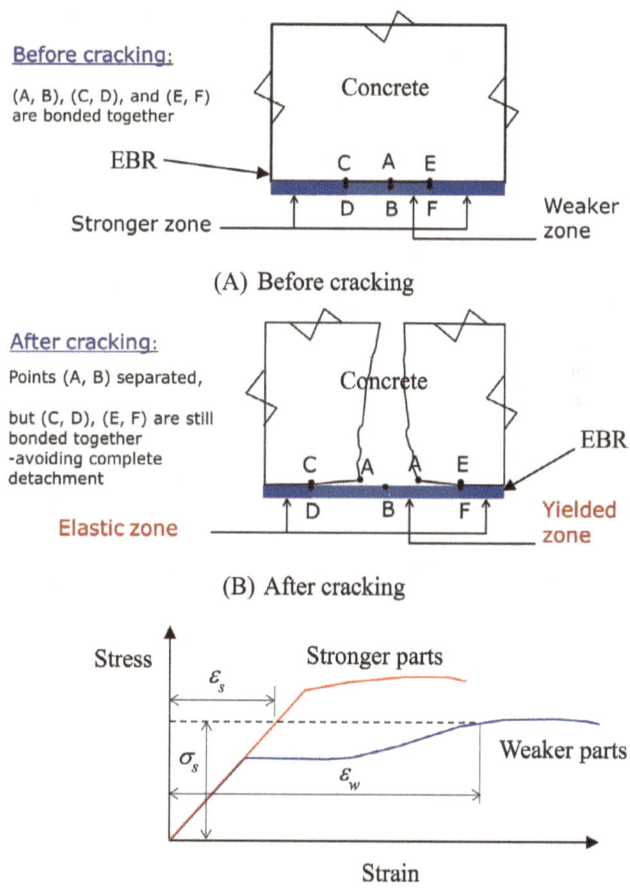

(C) Stress-strain relationship of reinforcing plate

Fig. 6.15 Accommodating cracking by strain localization.

bonded reinforcing material, as shown in Fig. 6.15, larger strain of EBR can be confined into a small region, while the strain in other areas is kept below a certain value so that debonding is confined inside the weaker zone.

The EBR material is weaker in the region from point C to E. At a given tension stress σ_s in the EBR, the strain of the weaker parts (ε_w) can be much larger than that in the stronger parts (ε_s), as shown in Fig. 6.15C, which causes the strain localization. This scheme can be utilized to design the EBR so that the strain in the stronger parts is always smaller than ε_{pd} (see Eq. 6.78) to avoid debonding in the stronger regions. In other words, debonding is localized in the strain localization system, without detachment of the whole plate from the RC member during debonding that happens in a normal EBR system.

As the cross section between C and E is weaker than the cross section outside the region, cracking will occur within the weaker zone rather than in the stronger section as shown in Fig. 6.15B, as long as regularly spaced weaker and stronger zones are

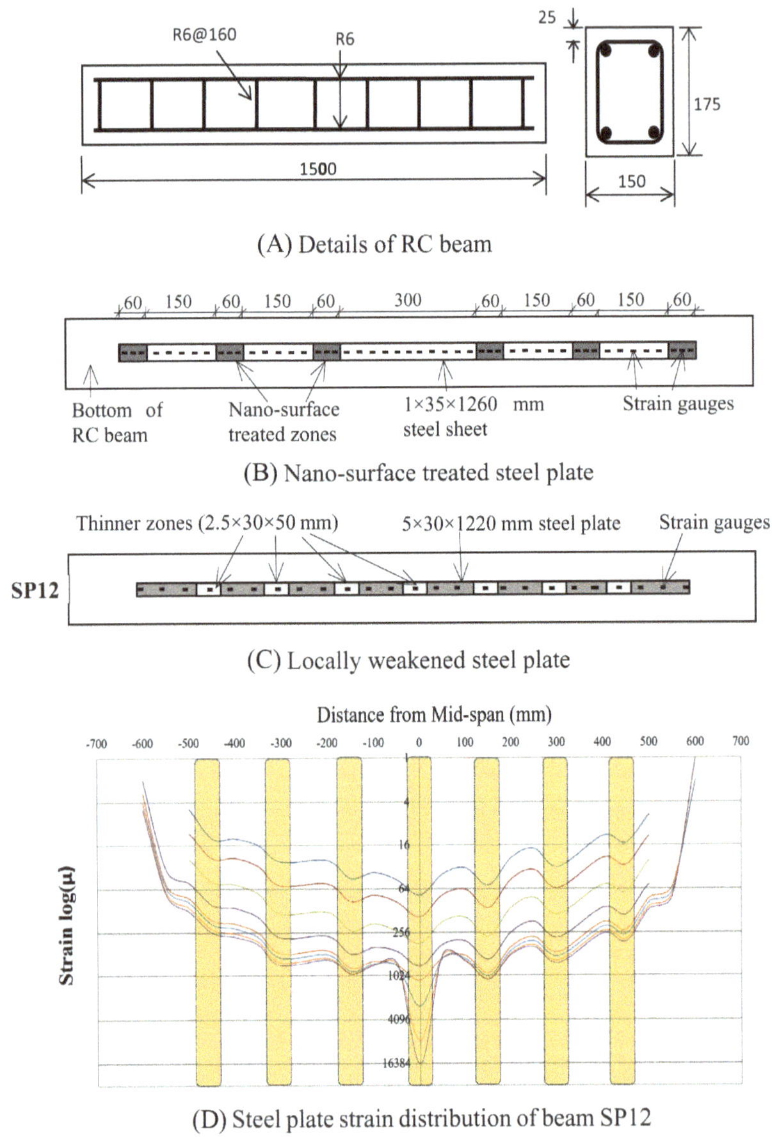

Fig. 6.16 Strain localized beam tests.

placed along the length of the member, as shown in Fig. 6.16B and C. When a steel plate is used as the EBR, strain localization can be achieved by two means: (1) by local reduction of plate width or thickness in the weaker zones, as shown in Fig. 6.16C, and (2) by using a technology called surface nanocrystallization (Wu and Lu 2013) to strengthen the steel material locally in the stronger parts. Surface nanocrystallization is a surface treatment technology that can be used to increase the yield strength of stainless steel plate. By nano-treating some parts while keeping other parts untreated, stronger and weaker zones can be produced along an EBR, as shown in Fig. 6.16B. Experimental tests were carried out by Wu and Lu (2013) that evaluated the strain localization system and demonstrated the effectiveness of the system in avoiding debonding. One of the test results (beam SP12) is shown in Fig. 6.16D, where the logarithm scaled longitudinal strain (in microstrain) clearly shows the strain localization in the weakened parts of the steel plate.

There are other bond enhancement systems that make use of anchors. These systems are introduced in Section 6.5.

6.4 Analytical solution of EBR pull-off test

The interfacial bond is usually studied through the simple pull-off (or shear) tests, as shown in Fig. 5.8B, experimentally, analytically, and numerically. Given the heterogeneous properties of the constituent materials and the high nonlinearity involved in the debonding process, experimental results are usually highly scattered, and analytical modeling and numerical simulations are often used to study the problem and develop models for the bond properties and interfacial behavior.

Numerous empirical and semiempirical models have been developed for the bond strength and effective bond length of the EBR systems. A summary of these models can be found in Wu and Jiang (2013) and Wu and He (2019). Finite element models of the EBR systems have been developed to simulate the debonding process and facilitate better understanding of the interfacial failure mechanisms (Lu et al. 2006; Lin and Wu 2016; Lin et al. 2017).

The key in studying an EBR system and the behavior of EBR strengthened structures is the identification of the interfacial bond-slip relationship. When a bond-slip relationship is known, the behavior of an EBR system and the response of its strengthened structure can be evaluated either by mathematical solutions or numerical simulations. Experimental testing is the main means to obtain the bond-slip curves of EBR systems. The bond stress and slip at different locations of a bond interface can be measured directly in a test to give a local bond-slip ($\tau - s$) response curve. Due to nonuniformity of materials and the highly nonlinear nature of the local fracture process, the experimentally obtained local $\tau - s$ curves are largely scattered. To overcome the problem, an indirect approach was introduced by Dai et al. (2005), which derives average bond parameters of an EBR system from the global load vs. slip response curves (Fig. 5.8E). Zhou et al. (2010) further extended the approach by using slip distribution curves (Fig. 5.8C) to derive the bond parameters. These approaches provide average values of local interfacial properties, and hence the derived results are usually more stable and consistent. Details of the experimental methods for identification of bond parameters can be found in Wu et al. (2016).

Mathematical modeling is an important means in the study of the interfacial bond problem. Theoretical solutions have been derived for the interfacial response, including bond strength and effective bond length, by different approaches (Taljsten 1997; Dai et al. 2005; Wu et al. 2009b; Zhou et al. 2010; Liu and Wu 2012). With a nonlinear bond-slip relationship, these early solutions were only applicable to the EBR systems with an infinite bond length, where the bond length L approaches infinity in Fig. 5.8B. The mathematical solution for the EBR system in Fig. 5.8B with a limited bond length was first obtained by Wu et al. (2012). A more sophisticated solution was later provided by He et al. (2015). The solution of Wu et al. (2012) is introduced in detail in this section as an example of mathematical modeling of the bond problem.

6.4.1 Governing equations

Considering elastic adherents (concrete and reinforcing plate) with a nonlinear interfacial bond-slip relationship for the EBR joint as shown in Fig. 5.8B, its mechanical behavior is governed by the following equations:

The constitutive relationship,

$$\sigma(x) = E_f \varepsilon(x) = \frac{f(x)}{b_f t_f} \tag{6.80}$$

the compatibility condition,

$$\varepsilon(x) = \frac{ds}{dx} = s' \tag{6.81}$$

and the equilibrium,

$$\tau = t_f \frac{d\sigma}{dx}, \text{or } \tau = E_f t_f \frac{d^2 s}{dx^2} = E_f t_f s'' \tag{6.82}$$

where σ and ε are the stress and strain of the reinforcing plate, respectively; $f(x)$ is the tensile force in the plate at x; t_f, b_f, and E_f are the thickness, breadth, and Young's modulus of the plate, respectively; and τ and s are the interfacial bond stress (shear stress) and the relative slip between the plate and the substrate concrete, respectively. The boundary conditions are given by:

$$\begin{cases} x=0: & \sigma=0 \\ x=l: & \sigma=f(l)/t_f b_f \end{cases}, \text{or} \begin{cases} \left.\dfrac{ds}{dx}\right|_{x=0} = 0 \\ \left.\dfrac{ds}{dx}\right|_{x=l} = \dfrac{f(l)}{E_f t_f b_f} \end{cases} \tag{6.83}$$

Integration of the interfacial bond stress gives the tensile force in the plate:

$$b_f \int_0^l \tau(x) dx = E_f t_f b_f \int_0^l s'' dx = f(l) \tag{6.84}$$

where l is the length of the bond interface. The bond-slip relationship of the EBR system can be represented by (Dai et al. 2005; Zhou et al. 2010):

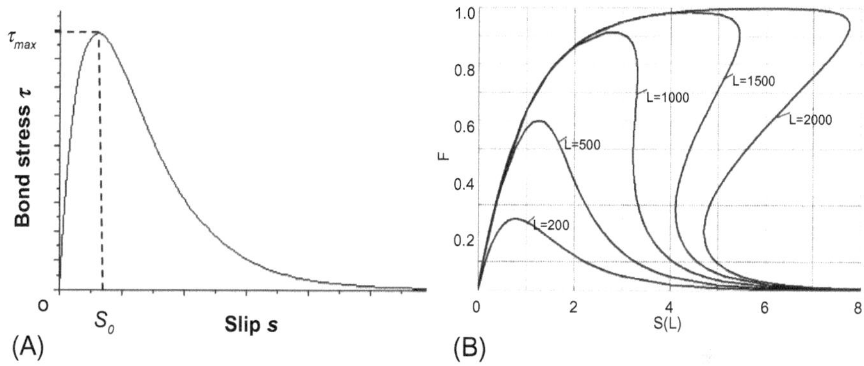

Fig. 6.17 Analytical solution of EBR pull-off test: (A) Bond-slip relationship; and (B) load vs. loaded-end slip response.
From Wu, Y. F., X. S. Xu, J. B. Sun, and C. Jiang. 2012. "Analytical solution for the bond strength of externally bonded reinforcement." *Compos. Struct.* 94 (11), 3232–3239.

$$s'' = \frac{\alpha}{\beta^2} e^{-s/\alpha}\left(1 - e^{-s/\alpha}\right) \tag{6.85}$$

where $s'' = \tau/(E_f t_f)$ (Eq. 6.82) and α and β are the bond parameters that control the amplitude and shape of the bond-slip relationship. The general shape of Eq. (6.85) is shown in Fig. 6.17A. This set of equations has been mathematically solved by different approaches when l approaches infinity, as mentioned earlier. The solution to the problem with a limited bond length l is given here.

Introducing the dimensionless variables $X = x/\alpha, S = s/\alpha, L = l/\alpha, F = f\beta/(\alpha E_f t_f b_f)$, $T = \tau/E_f$, $\lambda = \alpha/\beta$, $B_f = b_f/\alpha$, and $T_f = t_f/\alpha$, the governing differential equation, Eq. (6.85), is transformed to a dimensionless form:

$$\frac{d^2S}{dX^2} - \lambda^2 e^{-S}\left(1 - e^{-S}\right) = 0 \tag{6.86}$$

6.4.2 Analytical solutions

Integrating Eq. (6.86) gives

$$\left(\frac{dS}{dX}\right)^2 = \lambda^2\left(1 - e^{-S}\right)^2 + C_1 \tag{6.87}$$

Substituting the boundary conditions, Eq. (6.83), into Eq. (6.87) leads to

$$C_1 = -\lambda^2\left(1 - e^{-S(0)}\right)^2 \tag{6.88}$$

$$\left(1 - e^{-S(L)}\right)^2 = F^2 + \left(1 - e^{-S(0)}\right)^2 \tag{6.89}$$

As the plate is always under tension, or $\partial S(X)/\partial X \geq 0$ and $S(X) \geq S(0) \geq 0$, therefore $(1 - e^{-S(X)})^2 \geq (1 - e^{-S(0)})^2$. From Eq. (6.87), one has

$$\frac{dS}{dX} = \lambda\sqrt{(1-e^{-S})^2 - (1-e^{-S(0)})^2} \tag{6.90}$$

Let $y = e^{-S(X)}$ and $y_0^2 = 1 - (1 - e^{-S(0)})^2$, and considering Eq. (6.89), one has

$$y(0) = 1 - \sqrt{1-y_0^2} \tag{6.91}$$

$$y(L) = e^{-S(L)} = 1 - \sqrt{F^2 + 1 - y_0^2} \tag{6.92}$$

Also, from $y = e^{-S(X)}$, one has $dS = -(1/y)dy$. Eq. (6.90) can be written as

$$-\frac{1}{y}\frac{dy}{dX} = \lambda\sqrt{(1-y)^2 - \left(1-y_0^2\right)} \tag{6.93}$$

Integrating Eq. (6.93) gives

$$\int \frac{dy}{y\sqrt{y^2 - 2y + y_0^2}} = -\lambda X + C_2 \tag{6.94}$$

The solution of Eq. (6.94) is given by

$$y = \frac{4y_0^2 e^{-\lambda X y_0 + C_2 y_0}}{e^{-2\lambda X y_0 + 2C_2 y_0} + 4e^{-\lambda X y_0 + C_2 y_0} + 4\left(1-y_0^2\right)} \tag{6.95}$$

From Eq. (6.95) and considering Eq. (6.91), one has

$$y(0) = \frac{4y_0^2 e^{+C_2 y_0}}{e^{2C_2 y_0} + 4e^{C_2 y_0} + 4\left(1-y_0^2\right)} = 1 - \sqrt{1-y_0^2} \tag{6.96}$$

Solving Eq. (6.96) gives $e^{C_2 y_0} = 2\sqrt{1-y_0^2}$. Substituting this solution into Eq. (6.95) gives

$$y = \frac{2y_0^2 e^{-\lambda X y_0}}{e^{-2\lambda X y_0}\sqrt{1-y_0^2} + 2e^{-\lambda X y_0} + \sqrt{1-y_0^2}} \tag{6.97}$$

The solution of Eq. (6.97) can be simplified to

$$S(X) = \ln \frac{\eta \cosh\left(\lambda\sqrt{1-\eta^2}X\right) + 1}{1-\eta^2} \tag{6.98}$$

Let $\eta \equiv 1 - e^{-S(0)}$, $y_0^2 = 1 - (1 - e^{-S(0)})^2 = 1 - \eta^2$. By evaluating Eq. (6.97) at $X = L$ and equating it to Eq. (6.92), one has

$$\eta^2 - \sqrt{F^2+\eta^2} + \eta\left(1 - \sqrt{F^2+\eta^2}\right)\cosh\left(\lambda L\sqrt{1-\eta^2}\right) = 0 \tag{6.99}$$

Eq. (6.99) can be rewritten into the following form to give the explicit expression for the dimensionless pull force:

$$F = \frac{\eta\sqrt{1-\eta^2}\sinh\left(\lambda L\sqrt{1-\eta^2}\right)}{\left[1+\eta\cosh\left(\lambda L\sqrt{1-\eta^2}\right)\right]} \tag{6.100}$$

It can be seen from Eqs. (6.99) or (6.100) that the parameter η is a function of the load F, or $\eta = \phi(F)$, and can be obtained by solving one of the two equations. By substituting $\eta = \phi(F)$ into Eq. (6.98), the load F vs. slip S relationship can be determined.

Typical load vs. slip curves are shown in Fig. 6.17B for a normal λ value of 0.005 (Zhou et al. 2010). It can be seen from the figure that the peak load or bond strength increases with the increase in bond length. However, the increase in bond strength slows down when the bond length is longer than a certain value, after which further increase in bond length will increase the ductility of bond only, as discussed in Section 5.3.1. Furthermore, there is an interesting "snapback" phenomenon when L is greater than a certain value, which will be discussed in the next section.

6.4.3 Snapback problem

Substituting $X = L$ into Eq. (6.98) gives

$$1 - \eta^2 - \left[1+\eta\cosh\left(\lambda L\sqrt{1-\eta^2}\right)\right]e^{-S(L)} = 0 \tag{6.101}$$

from which η can be resolved as a function of $S(L)$, or $\eta = \eta(S(L)) \equiv \eta_{S(L)}$. From Eqs. (6.98), (6.100), one has

$$S(0) = -\ln\left(1 - \eta_{S(L)}\right) \tag{6.102}$$

$$F = e^{-S(L)}\eta_{S(L)}\sinh\left(\lambda L\sqrt{1-\eta_{S(L)}^2}\right)/\sqrt{1-\eta_{S(L)}^2} \tag{6.103}$$

Substituting $\eta = 1 - e^{-S(0)}$ into Eqs. (6.98), (6.100) and letting $X = L$ gives

$$S(L) = \ln\frac{1+\left(1-e^{-S(0)}\right)\cosh\left(\lambda L e^{-S(0)}\sqrt{2e^{S(0)}-1}\right)}{e^{-S(0)}\left(2-e^{-S(0)}\right)} \tag{6.104}$$

$$F = \frac{e^{-S(0)}\left(1-e^{-S(0)}\right)\sqrt{2e^{S(0)}-1}\sinh\left(\lambda L e^{-S(0)}\sqrt{2e^{S(0)}-1}\right)}{1+\left(1-e^{-S(0)}\right)\cosh\left(\lambda L e^{-S(0)}\sqrt{2e^{S(0)}-1}\right)} \tag{6.105}$$

Eqs. (6.102) and (6.103) show that the free-end slip $S(0)$ and applied load F are functions of the loaded-end slip $S(L)$, while Eqs. (6.104), (6.105) give the loaded-end slip $S(L)$ and corresponding force F as functions of $S(0)$. Now let us look at the relationship between $S(L)$ and $S(0)$. Letting $X=0$ and $X=L$, respectively, in Eq. (6.98) leads to

$$\begin{cases} S(0) = -\ln(1-\eta) \\ S(L) = \ln\dfrac{\eta\cosh\left(\lambda L\sqrt{1-\eta^2}\right)+1}{1-\eta^2} \end{cases} \tag{6.106}$$

Eliminating η gives the $S(0)$ vs. $S(L)$ relationship:

$$e^{-S(0)}\left(2-e^{-S(0)}\right) = e^{-S(L)} + \left(1-e^{-S(0)}\right)e^{-S(L)}\cosh\left(\lambda L\sqrt{2e^{-S(0)}-e^{-2S(0)}}\right) \tag{6.107}$$

For $\lambda = 0.005$, the $S(L)$ vs. F, $S(0)$ vs. F, and $S(0)$ vs. $S(L)$ curves for different L values are shown in Figs. 6.17, 6.18, and 6.19, respectively. From Fig. 6.18, it can be seen that if $S(0)$ is used as the horizontal axis to replace $S(L)$, snapback does not appear. Mathematically, Fig. 6.17 shows that for a given $S(L)$ value, Eq. (6.103) may have multiple solutions of F, while Eq. (6.105) has only a single solution for a fixed $S(0)$. From Fig. 6.19 it can be seen that when L is small the relationship between $S(L)$ and $S(0)$

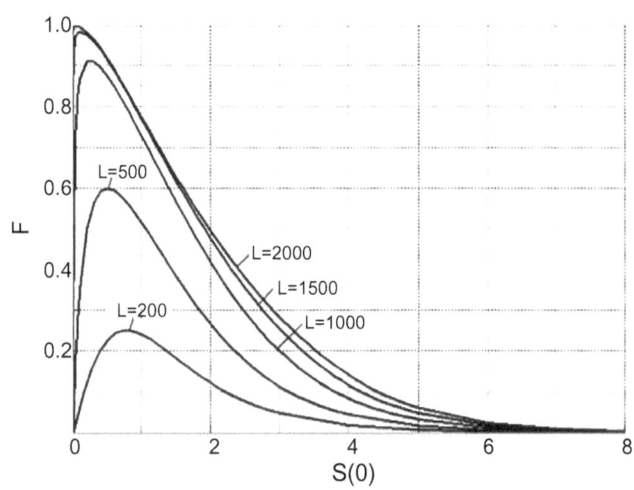

Fig. 6.18 Load vs. free-end slip response curves.
From Wu, Y. F., X. S. Xu, J. B. Sun, and C. Jiang. 2012. "Analytical solution for the bond strength of externally bonded reinforcement." *Compos. Struct.* 94 (11), 3232–3239.

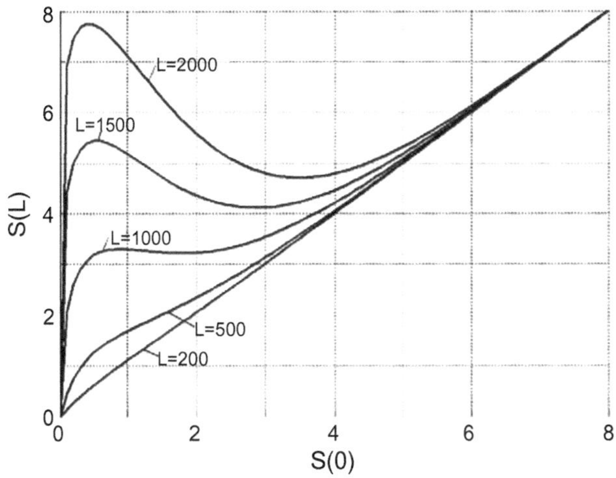

Fig. 6.19 Free-end slip vs. loaded-end slip response curves.
From Wu, Y. F., X. S. Xu, J. B. Sun, and C. Jiang. 2012. "Analytical solution for the bond strength of externally bonded reinforcement." *Compos. Struct.* 94 (11), 3232–3239.

is monotonic. In other words, an increase in the loaded-end slip will always cause an increase in the free-end slip. When L is larger than a critical value L_c, the monotonic relationship will be lost, and an increase in the free-end slip may correspond to a reduction in the loaded-end slip.

Snapback occurs when L is greater than L_c. The critical length L_c is derived in the following text. From Eq. (6.89), $S(L)$ can be written as

$$S(L) = -\ln\left(1 - \sqrt{F^2 + \eta^2}\right) \tag{6.108}$$

where η can be seen as a function of F, or $\eta = \eta(F)$ (see Eq. 6.99). Letting $dS_L(F)/dF = 0$ and $d^2S_L(F)/dF^2 = 0$ leads to

$$2\sqrt{1-\eta^2}\left[1+\eta\cosh\left(\lambda L\sqrt{1-\eta^2}\right)\right] + \left[\begin{array}{c}\eta^3\sqrt{1-\eta^2}(\lambda L)^2\cosh\left(\lambda L\sqrt{1-\eta^2}\right)+ \\ (2\eta^3-3\eta)\lambda L\sinh\left(\lambda L\sqrt{1-\eta^2}\right)\end{array}\right] = 0 \tag{6.109a}$$

$$2\eta\sqrt{1-\eta^2}\left[1+\eta\cosh\left(\lambda L\sqrt{1-\eta^2}\right)\right] + \left(1-\eta^2\right)\left[\begin{array}{c}\sqrt{1-\eta^2}\cosh\left(\lambda L\sqrt{1-\eta^2}\right) \\ -\eta^2\lambda L\sinh\left(\lambda L\sqrt{1-\eta^2}\right)\end{array}\right] = 0 \tag{6.109b}$$

Eq. (6.109) gives the conditions for snapback to occur. The solution of Eq. (6.109) is $\lambda L = 4.57067$ and $\eta = 0.714242$. As $\lambda L = l/\beta$, the critical bond length is

$$L_c = \frac{4.57067}{\lambda} \quad \text{or} \quad l_c = 4.57067\beta \tag{6.110}$$

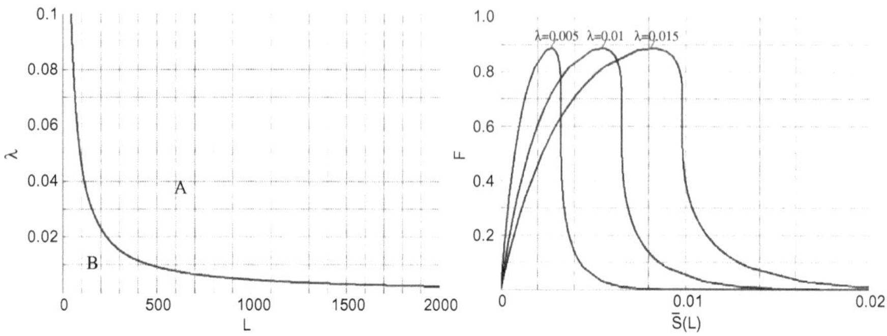

(A) Parameter space (B) Boundary cases

Fig. 6.20 Critical length for snapback.
From Wu, Y. F., X. S. Xu, J. B. Sun, and C. Jiang. 2012. "Analytical solution for the bond strength of externally bonded reinforcement." *Compos. Struct.* 94 (11), 3232–3239.

When the bond length is greater than the critical bond length, the snapback will occur. Eq. (6.110) divides the parameter space into two zones (Fig. 6.20A). Snapback will occur in zone A but will not occur in zone B. The $S(L)$ vs. F curves for three critical cases on the boundary line are shown in Fig. 6.20B, where $\bar{S}(L) = S(L)/L$.

From $\eta = 1 - e^{-S(0)} = 0.714242$, $S(0) = 1.25261$, or in dimensional form, $s(0) = 1.25261\alpha$. This gives the slip value at the free end of the plate at which the snapback starts to occur.

6.4.4 Control of pull-off test

The pull-off tests are usually controlled by pulling at the loaded end. An example with $\lambda = 0.005$ and $L = 2000$ is used to study the problem, as shown in Fig. 6.21. When the test is controlled by the loaded-end slip and $S(L)$ increases from 0 to point

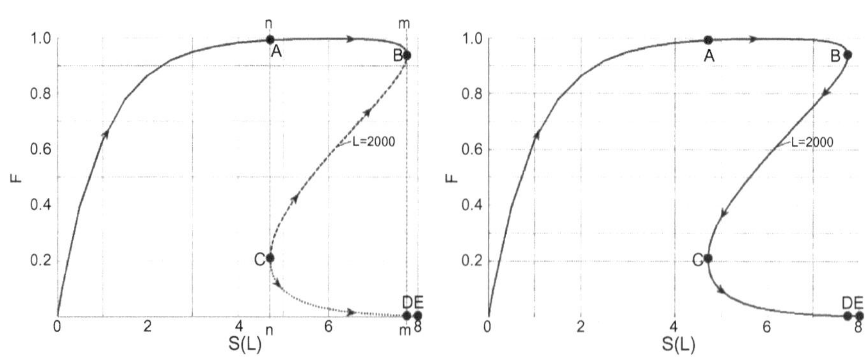

(A) Loaded-end slip control (B) Free-end slip control

Fig. 6.21 Control of pull-off tests.
From Wu, Y. F., X. S. Xu, J. B. Sun, and C. Jiang. 2012. "Analytical solution for the bond strength of externally bonded reinforcement." *Compos. Struct.* 94 (11), 3232–3239.

A, F and $S(0)$ increase monotonically (Fig. 6.21A). Therefore, the pull can be under either load control or displacement control at this stage. At point A, Eq. (6.103) has two roots, one at point A and another at point C. When $S(L)$ further increases, Eq. (6.103) has three roots. Further pull from point A to point B will lead to a drop of the load in the later stage and should thus be controlled by displacement (with an increasing loaded-end slip rather than increasing F). At point B, the snapback branch can be obtained only by reducing $S(L)$ until point C, and increasing it again after point C. As points B and C are unstable bifurcation points, obtaining the whole path 0-A-B-C-D-E by controlling $S(L)$ is practically impossible. In a real test controlled by the loaded-end slip $S(L)$, the curve will jump from point B to point D directly, producing the path 0-A-B-D-E.

Figs. 6.18 and 6.19 show that for a certain free-end slip $S(0)$, there is only one corresponding point for both F and $S(L)$. From a mathematical point of view, for a certain $S(0)$ value, both $S(L)$ and F have a unique and single solution. When $S(0)$ increases monotonically, the point on the curve in Fig. 6.21 moves along the path 0-A-B-C-D-E. The whole process is clear and stable without a bifurcation point. Therefore, by controlling $S(0)$ the whole $S(L)$ vs. F curve can be obtained. Carrara et al. (2011) adopted the free-end slip control method in their experimental tests, and successfully obtained the whole $S(L)$ vs. F curve with snapback. They pointed out that as the free-end slip is very small (particularly for long joints) and thus very difficult to measure, the free-end control method is applicable only to joints with a small L value.

6.4.5 Peak strength

Eq. (6.89) can be rewritten as $F^2 + \eta^2 = (1 - e^{-S(L)})^2$. Substituting it into Eq. (6.99) gives

$$\eta^2 - \left(1 - e^{-S(L)}\right) + \eta\left[1 - \left(1 - e^{-S(L)}\right)\right]\cosh\left(\lambda L\sqrt{1-\eta^2}\right) = 0 \tag{6.111}$$

Eq. (6.111) gives the relationship between η and $S(L)$, or $\eta = \eta(S_L)$. The peak point of an $S(L)$ vs. F curve (Fig. 6.17) can be obtained by letting $dF(S_L)/dS_L = 0$. Taking the derivative of $F^2 + \eta^2 = (1 - e^{-S(L)})^2$ with respect to $S(L)$ and considering Eq. (6.111), the peak point is obtained as

$$S(L) = \ln\left[\frac{1+\eta\cosh\left(\lambda L\sqrt{1-\eta^2}\right)}{1-\eta^2}\right] \tag{6.112}$$

$$F_{max}(L) = \frac{\eta\sqrt{1-\eta^2}\sinh\left(\lambda L\sqrt{1-\eta^2}\right)}{1+\eta\cosh\left(\lambda L\sqrt{1-\eta^2}\right)} \tag{6.113}$$

where η is given by the following equation:

$$2\eta^2\sqrt{1-\eta^2}\lambda L\cosh\left(\lambda L\sqrt{1-\eta^2}\right)-2\left(1-2\eta^2\right)\sinh\left(\lambda L\sqrt{1-\eta^2}\right)$$
$$+2\lambda L\eta^3\sqrt{1-\eta^2}+\eta^3\sinh\left(2\lambda L\sqrt{1-\eta^2}\right)=0 \quad (6.114)$$

For a certain parameter λ and length L, Eq. (6.114) gives a certain value of η. Substituting this η into Eq. (6.113) produces the peak strength F_{max}. The F_{max} vs. λL relationship is shown in Fig. 6.22A. Three curves of $F_{\max}$ for the λ values of 0.005, 0.010, 0.015 are shown in Fig. 6.22B. The load capacity F_{max} increases monotonically when the bond length L increases and approaches the limit value of 1 asymptotically. When $\lambda=\alpha/\beta$ is larger, the curve approaches its limit faster.

As Eq. (6.114) is a transcendental equation, there is no closed-form solution. Through regression to Eq. (6.114), an approximate solution of η is obtained, as given by Eq. (6.115). For engineering calculation, Eq. (6.115) is sufficiently accurate.

$$\eta=-3.61e^{-0.4454\lambda L}+4.11e^{-0.3835\lambda L} \quad (6.115)$$

Eqs. (6.113), (6.115) can be expressed with their original dimensional parameters as

$$f_{max}(l)=\frac{\alpha E_f t_f b_f \eta\sqrt{1-\eta^2}\sinh\left(\sqrt{1-\eta^2}\,l/\beta\right)}{\beta\left[1+\eta\cosh\left(\sqrt{1-\eta^2}\,l/\beta\right)\right]} \quad (6.116)$$

$$\eta=-3.61e^{-0.4454\frac{l}{\beta}}+4.11e^{-0.3835\frac{l}{\beta}} \quad (6.117)$$

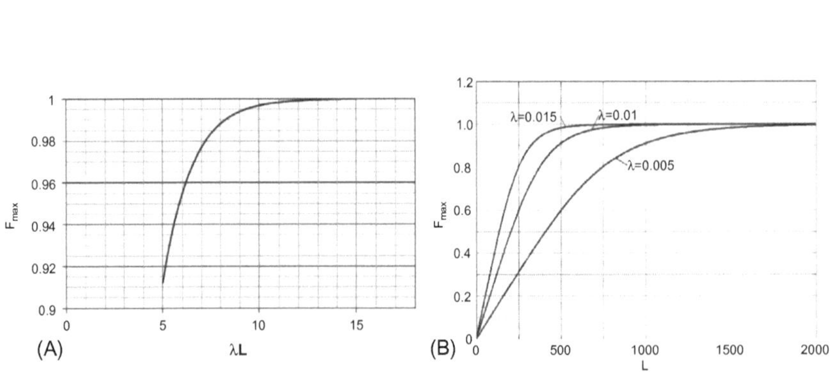

Fig. 6.22 Bond strength vs. bond length.
From Wu, Y. F., X. S. Xu, J. B. Sun, and C. Jiang. 2012. "Analytical solution for the bond strength of externally bonded reinforcement." *Compos. Struct.* 94 (11), 3232–3239.

From Eq. (6.114), it can be mathematically proved that when $L \to \infty$, $\eta \to 0$, or

$$\eta^{\infty} = \underset{L \to \infty}{\text{Lim}} [\eta(L)] = 0 \tag{6.118}$$

and

$$F_{max}^{\infty} = \underset{L \to \infty}{\text{Lim}} [F_{max}(L)] = \underset{L \to \infty}{\text{Lim}} \frac{\eta\sqrt{1-\eta^2}\sinh\left(\lambda L\sqrt{1-\eta^2}\right)}{1+\eta\cosh\left(\lambda L\sqrt{1-\eta^2}\right)} = 1 \tag{6.119}$$

Since $F = f\beta/(\alpha E_f t_f b_f)$,

$$f_{max}^{\infty} = E_f t_f b_f \frac{\alpha}{\beta} \tag{6.120}$$

Eq. (6.120) is identical to the solution for joints with an infinite bond length, which has been derived by many others.

The interfacial fracture energy is defined as $G_f = \int_0^{\infty} \tau(s) ds$. Substituting Eq. (6.85) into the equation gives

$$G_f = \int_0^{\infty} \tau(s) ds = \frac{1}{2} E_f t_f \left(\frac{\alpha}{\beta}\right)^2 \tag{6.121}$$

Considering Eq. (6.121), Eq. (6.120) can be rewritten

$$f_{max}^{\infty} = b_f \sqrt{2 G_f E_f t_f} \tag{6.122}$$

Eq. (6.122) is also identical to the fracture mechanics based ultimate bond strength model that has been derived by others for EB joints with an infinite bond length (Dai et al. 2005; Wu et al. 2009b), which is only a special case of the preceding solution. For the general case in Fig. 5.8 with a finite bond length, Eq. (6.116) can be rewritten as

$$f_{max}(l) = f_{max}^{\infty} k_l \tag{6.123}$$

where f_{max}^{∞} is given by Eq. (6.120) and k_l is a length factor given by

$$k_l = F_{max}(L) = \frac{\eta\sqrt{1-\eta^2}\sinh\left(\sqrt{1-\eta^2}\, l/\beta\right)}{1+\eta\cosh\left(\sqrt{1-\eta^2}\, l/\beta\right)} \tag{6.124}$$

An approximate but sufficiently accurate form of Eq. (6.124) for engineering calculation is obtained by He et al. (2015), given by

$$k_l = \tanh\left(\frac{0.3l}{\beta}\right) \tag{6.125}$$

It is noted that the preceding mathematical solutions are for a two-dimensional problem that involves dimensions x and z only in Fig. 5.8B. In other words, it is only applicable when $b_c = b_f$ (Fig. 5.8B) where variation in the y-direction is not involved. However, in practice $b_c \neq b_f$ in most cases where the variation in the y-direction is involved, so that it becomes a three-dimensional problem. In order to apply the two-dimensional solution to the three-dimensional problem, people introduced the concept of width effect, i.e., multiplying the two-dimensional solution by a width factor. Apparently, the width factor is closely related to the ratio of b_f/b_c. The width factor will be involved in the solutions discussed in the following section.

6.4.6 Parameter identification

The analytical solutions derived in the previous sections require two parameters: α and β. These two parameters are usually considered as material properties of a bonded joint. In principle, they should be obtained from experimental testing. Extensive experimental works on the EBR joints have been reported in the literature. Numerous empirical models are available to evaluate the responses (such as bond strength) of the joints. These models can be used to determine α and β.

Fig. 6.22 shows that the ultimate bond strength F_{max} increases with an increase in the bond length L, and approaches but never reaches its asymptotic value of 1.0. This is a result of the bond-slip relationship of Eq. (6.85) that approaches but never drops to zero when the slip increases, as shown in Fig. 6.17A. If the bond stress drops to zero at a slip value of s_u, F_{max} will reach its peak of 1.0 when the loaded-end slip is equal to s_u, as shown in Fig. 5.8D and E. For the bond-slip relationship of Eq. (6.85), a value of F_{max} that is sufficiently close to the asymptotic value, e.g., $F_{max} = k_l = 0.96, 0.98$, or 0.99, is considered as its maximum bond strength, and its corresponding bond length is defined as the effective bond length, L_e. When $L > L_e$, the bond strength is considered to be a constant. Eq. (6.125) can also be used for the calculation of effective bond length:

$$l_e = \beta \frac{a\tanh(F_{max})}{0.3} = \beta\lambda_0 \tag{6.126}$$

For example, substituting $F_{max} = 0.96$ into Eq. (6.126) gives $\lambda_0 = 6.4864$ or $l_e = 6.4864\beta$.

Based on a linear fracture model, Wu et al. (2009b) provided an effective bond length equation as

$$l_e = \frac{2\sqrt{2G_f E_f t_f}}{\tau_{max}} \tag{6.127}$$

where τ_{max} is the peak bond strength (Fig. 6.17A). From the bond-slip relationship, Eq. (6.85), the peak bond strength can be easily derived (by letting $d\tau/ds = 0$) as

$$\tau_{max} = \frac{1}{4} E_f t_f \frac{\alpha}{\beta^2} \tag{6.128}$$

Substituting Eqs. (6.121), (6.128) into Eq. (6.127) gives

$$l_e = 8\beta \text{ or } \lambda L_e = 8 \tag{6.129}$$

Comparing this value with Eq. (6.110), it can be seen that the minimum bond length for the snapback to occur is smaller than the effective bond length. In other words, for a joint with a bond length greater than the effective bond length, the snapback phenomenon will certainly occur (but may not be observed if it is under loaded-end control).

Substituting $\lambda L_e = 8$ into Eq. (6.114), one has $\eta = 0.0849485$. Further, substituting these two values into Eq. (6.113), the maximum bond strength is found to be $F_{max}(L_e) = 0.9887$. This value is very close to the asymptotic value of 1.0 (directly substituting Eq. 6.129 into Eq. 6.125 gives $F_{max}(L_e) = 0.984$). Therefore, Eq. (6.127) is reasonable. If a different value of $F_{\max}$ is selected, the corresponding l_e value can be determined using Eq. (6.126). The relationship between $F_{\max}$ and λL is graphically shown in Fig. 6.22A.

In an empirical bond strength model, the empirical equations for l_e or f_{max}^{∞} are given. Therefore, the material parameters α and β can be determined from the empirical equations. An example is given here. The empirical model of Chen and Teng (2001) is as follows:

$$f_{max}^{\infty} = 0.427 k_w b_f l_e \sqrt{f_{co}} \text{ (in N, mm)} \tag{6.130}$$

$$l_e = \sqrt{\frac{E_f t_f}{\sqrt{f_{co}}}} \text{ (in N, mm)} \tag{6.131}$$

where k_w is the width factor given by

$$k_w = \sqrt{(2 - b_f/b_c)/(1 + b_f/b_c)} \tag{6.132}$$

By substituting Eqs. (6.130), Eq. (6.131) into Eqs. (6.120), (6.126), respectively, the two parameters can be derived as

$$\begin{cases} \alpha = 0.427 k_w/\lambda_0 \\ \beta = \dfrac{1}{\lambda_0}\sqrt{\dfrac{E_f t_f}{\sqrt{f_{co}}}} \end{cases} \text{ (in N, mm)} \tag{6.133}$$

As Eqs. (6.130), (6.131) are empirical equations that involve specific dimensions, Eq. (6.133) should have dimensions that are identical to those in the original equations.

The model proposed by Wu et al. (2009b) is given by

$$\begin{cases} l_e = 0.395 (E_f t_f)^{0.54} (f_{co})^{-0.09} \\ f_{max}^{\infty} = 0.595 b_f (E_f t_f)^{0.54} (f_{co})^{0.1} \end{cases} \text{ (in N, mm)} \tag{6.134}$$

Through a similar process as previously, the bond parameters are derived as

$$\begin{cases} \alpha = 0.0294\left(E_f t_f\right)^{0.08}\left(f_{co}\right)^{0.01} \\ \beta = 0.0494\left(E_f t_f\right)^{0.54}\left(f_{co}\right)^{-0.09} \end{cases} \text{(in N, mm)} \tag{6.135}$$

The bond parameters can also be evaluated from bond test results directly. For example, one way is to use experimental bond-slip curves measured from tests. Regression analysis can then be used to best fit Eq. (6.85) to a measured bond-slip curve to derive one pair of α and β. A test database can be formed by producing a large number of α and β pairs, which can be used to develop empirical models of the bond parameters. Instead of using an experimental bond-slip curve directly, Wu and Jiang (2013) used an indirect method for identification of bond parameters from bond test results. They used the bond strength or the peak point of a load-slip curve rather than the whole response curve, for identification of bond parameters. The general procedure of their approach is as follows:

(1) Factors that affect the bond parameters, and the rational form of the bond parameter models, are identified through analytical studies;
(2) Using a bond test with a bond length significantly larger than l_e and Eq. (6.120) to calculate the value of α/β from one bond test. A database of α/β is formed by using a sufficient number of tests;
(3) Based on the results of step (1) and step (2), regression analysis is used to form the model of α/β;
(4) From Eqs. (6.120), (6.123), (6.125), one has $f_{max}(l) = E_f t_f b_f \frac{\alpha}{\beta} \tanh\left(\frac{0.3l}{\beta}\right)$. Using the bond test results with a bond length less than l_e, and incorporating the model from step (3) into this equation, a database of β can be formed;
(5) Based on the results from step (1) and step (4), the model of β can be produced; and
(6) Combining the models from step (3) and step (5) gives the model of α.

Wu and Jiang (2013) proposed the following model of the bond parameters based on this approach:

$$\alpha = 0.094 f_{co}^{0.026} \text{ (in N, mm)} \tag{6.136}$$

$$\beta = 0.134 \frac{\left(E_f t_f\right)^{0.5}}{k_w f_{co}^{0.082}} \text{ (in N, mm)} \tag{6.137}$$

$$k_w = 1 + 0.222\left(f_{co}\right)^{0.304}\left(1 - \frac{b_f}{b_c}\right) \text{ (in N, mm)} \tag{6.138}$$

Wu and He (2019) used a more advanced measuring technique, digital image correlation (DIC), in their experimental tests to identify bond parameters (Fig. 6.23). The DIC technology can obtain $\tau - s$ curves at different points of a bond interface directly. It was found from their tests that there are two different zones in a bond interface. In zone 1, a certain part in the middle region of a bond surface, the longitudinal strain is

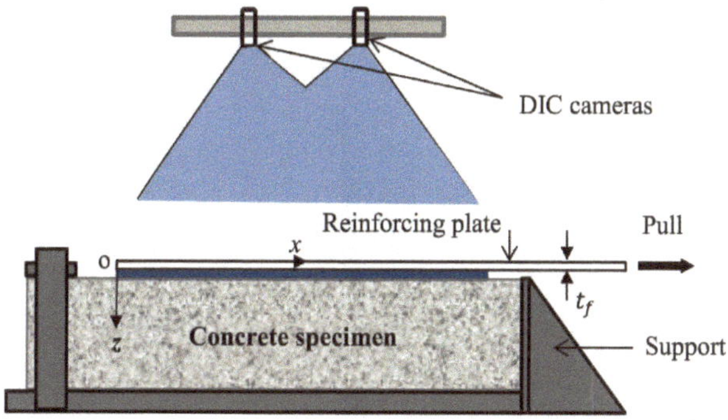

(A) Pull-off test

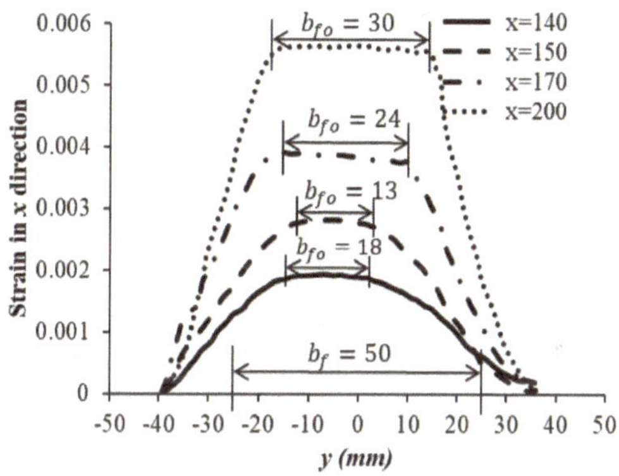

(B) ε_{xx} at different cross-sections

Fig. 6.23 Bond test for identification of bond parameters.
Based on Wu, Y. F., and L. He. 2019. "Width effect of interfacial bond characteristics." *Constr. Build. Mater.* 220, 712–726.

uniformly distributed (the region marked by b_{fo} in Fig. 6.23B) where it is under pure Mode II fracture (shear stress acting parallel to the plane of the crack and perpendicular to the crack front). In zone 2, the bond area outside zone 1, it is under a mixed fracture of Mode II and Mode III (shear stress acting parallel to the plane of the crack and parallel to the crack front). The area in zone 1 reflects a pure two-dimensional bond problem, and hence it does not involve the width effect. The width effect is caused by the Mode III fracture mode in zone 2. Therefore, the $\tau - s$ curves obtained in zone 1 can be used to derive bond parameters without width effect. The $\tau - s$ curves obtained in this way are depicted in Fig. 6.24A, in comparison with those obtained based on the conventional definition in Fig. 6.24B, which is the average $\tau - s$ curve for a cross section. It can be seen from the comparison in Fig. 6.24 that the scattering of bond properties reported in the literature was partly due to the mixing up of the two different aspects of the bond problem: the two-dimensional solution and the width effect. Therefore, Wu and He (2019) proposed to separate the modeling of the bond into two parts: (1) use the $\tau - s$ curves from zone 1 for modeling of bond parameters

Fig. 6.24 Bond-slip curves of specimen C30-400-2 at different cross sections.
From Wu, Y. F., and L. He. 2019. "Width effect of interfacial bond characteristics." *Constr. Build. Mater.* 220, 712–726.

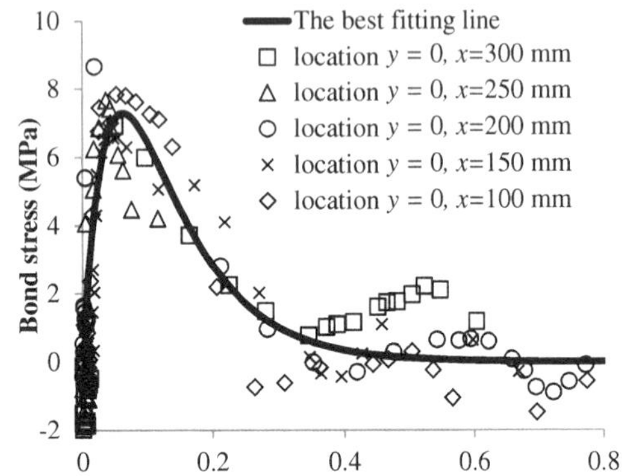

(A) Bond-slip curves without width effect

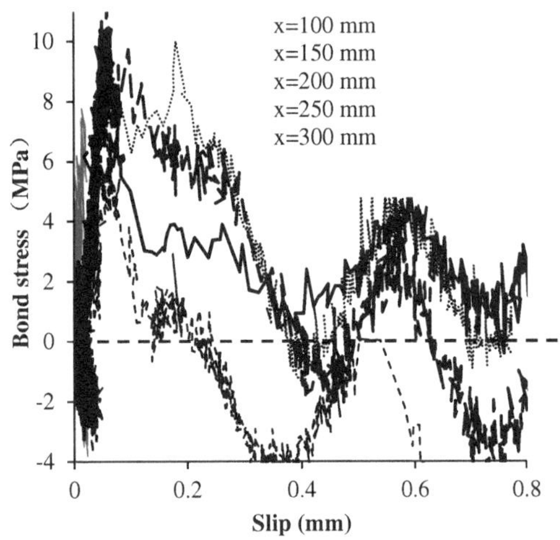

(B) Bond-slip curves with width effect

α and β, and (2) considering the width effect separately by factoring Eq. (6.123) with a width factor k_w, i.e.,

$$f_{max}(l) = \left(f_{max}^{\infty} k_l\right)_{2D} k_w \tag{6.139}$$

where $(f_{max}^{\infty} k_l)_{2D}$ is the bond strength of the joint when $b_c = b_f$. The following model was derived from their study:

$$\alpha = 0.124 f_{co}^{-0.103} (\text{in N, mm}) \tag{6.140}$$

$$\beta = 0.174 (f_{co})^{-0.205} \left(E_f t_f\right)^{0.5} / k_w (\text{in N, mm}) \tag{6.141}$$

$$\kappa_w = 1 + f_{co}^{0.385}\left[8\left(E_f t_f\right)^{-0.25} + 0.001\right]\left(1 - b_f/b_c\right)^{0.61}/\left(1 + 0.01 b_f^{1.5}\right)(\text{in N, mm}) \tag{6.142}$$

It is noted that k_w takes the value of 1 in calculating the term $(f_{max}^{\infty} k_l)_{2D}$ in Eq. (6.139), i.e., $\beta = \beta_{2D} = 0.174(f_{co})^{-0.205}(E_f t_f)^{0.5}$. Eq. (6.123) is equivalent to Eq. (6.139), because $f_{max}(l) = f_{max}^{\infty} k_l = E_f t_f b_f \frac{\alpha}{\beta} k_l = E_f t_f b_f \frac{\alpha}{\left(\frac{\beta_{2D}}{k_w}\right)} k_l = \left(f_{max}^{\infty} k_l\right)_{2D} k_w$.

Comparing the previous models proposed by different people, significant differences are noted. The differences are mainly caused by the following factors:

(1) Different understandings of the physical problem;
(2) Different test results or databases used for modeling;
(3) Different techniques used in experimental tests and analytical studies;
(4) Relatively higher scattering of test results in bond tests; and
(5) Different workmanship, materials, and lab conditions in construction of bond test specimens and bond testing.

It can be seen from the preceding analyses that modeling the bond problem is highly complicated. Apart from the previous factors, there are other ones that are ignored in the studies of the bond problem, such as the assumption of an invariable bond-slip relationship for the whole bond interface. Zhou et al. (2010) found that a uniform bond-slip relationship is suitable only for long joints with a sufficient bond length. It is inappropriate for short joints. Therefore, accurate modeling of the EBR systems is still a challenge to the research community.

6.5 Hybrid bonded reinforcement

6.5.1 Bond enhancement systems

As discussed in Section 5.3.1, the adhesive bond is weak and unreliable. A practical EBR system should have a bond enhancement system such as mechanical anchoring to provide a sufficiently strong and durable bond. However, a high-strength reinforcing plate such as FRP plate/laminate and high-strength steel plate are usually thin, and hence normal dowel action provided by mechanical anchors will cause splitting and bearing failure of the plate, as shown in Fig. 5.10A. Therefore, the dowel mechanism is not suitable for thin reinforcing plates, especially FRP materials. To resolve the critical debonding problem, various bond enhancement technologies have been developed (Kalfat et al. 2013), such as MF-FRP (Fig. 5.10A), end anchorage, U-jacketing (Fig. 6.25), and fiber anchoring (or anchor spikes) (Fig. 5.10B).

Mechanically fastened FRP (MF-FRP) uses a special type of FRP strip, the SafStrip that is made by using nonbrittle matrix and multidirectional fibers, and hence has a significant resistance against the bearing failure (Lamanna et al. 2004). This special FRP strip can be mounted onto a concrete surface by mechanical fasteners in the same way as mounting a thick steel plate, and the fastening pins can be conveniently

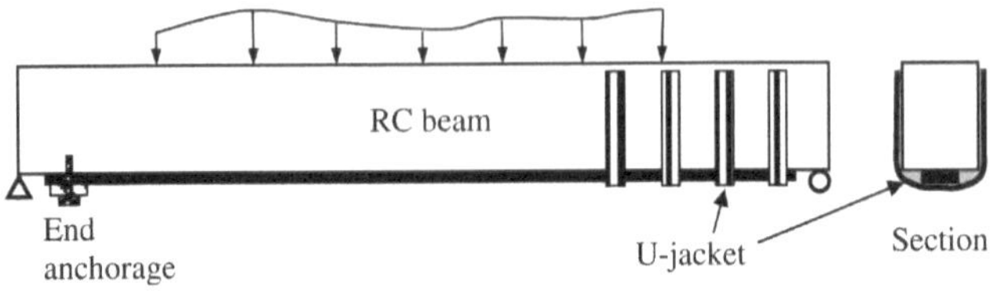

Fig. 6.25 End anchorage and U-jacket.

inserted into the concrete substrate using a powder-actuated fastening "gun." However, progressive splitting of the FRP strip at sustained bearing still occurs (Fig. 5.10A). Due to the progressive bearing failure, MF-FRP has a limited lifetime, and is only intended for temporary strengthening works or structures that are to be replaced soon.

End anchorage is usually made through clamping the plate end, which relies on a strong frictional bond to avoid a bearing failure. It uses large mechanical fasteners and is effective for restraining debonding at the two ends of the reinforcement plate. However, it cannot block debonding elsewhere and hence is usually ineffective for the common and important IC debonding (Fig. 6.11A).

U-jacketing uses numerous U-shaped FRP or steel strips to wrap the longitudinal reinforcing plate. The U-jackets provide certain restraint or resistance against debonding. Due to the flexibility of the restraint and considering the very brittle nature of debonding failure, a large increase in bond strength cannot be relied upon in this system. This method has been reported to increase the IC debonding strength by about 30% (Ye et al. 2005).

In a fiber anchoring system, fiber anchors made from bundles of fibers are inserted into epoxy-filled holes in the concrete substrate (Fig. 5.10B). The free ends of the fibers are then splayed and bonded to the surface of the reinforcing laminate with adhesive. It was reported that a single fiber anchor increased bonding strength by 73% and a second anchor behind the first further increased bonding strength by 86% (Zhang and Smith 2012). The further increase by the second anchor fell to 0% when the distance between the two anchors was increased from 75 to 125 mm. Apparently successive shearing-off of the anchors one by one occurred. Therefore, a large increase in total bond strength is difficult, as one anchor has a limited bond resistance.

Near-surface mounting is a technique whereby reinforcement is embedded into a groove cut into the cover concrete layer (Fig. 6.5). This technology can improve the external bonding condition of reinforcing strip/bar, and hence increase bond strength. The embedment depth of NSM is between internal rebar and EBR, and hence the bonding condition is better than EBR but not as good as that of internal rebar. Details of the NSM technology can be found in De Lorenzis and Teng (2007). In construction, saw cutting is used for groove making. Therefore, it is possible that the existing reinforcement bars, especially transverse bars, be cut during the groove-making process, which causes a critical problem with the technology. This technical difficulty hinders the application of this technology in practice.

Another bond enhancement system is the hybrid-bonded (HB) FRP system developed by Wu and Huang (2008), which can be used to mount a reinforcing plate made

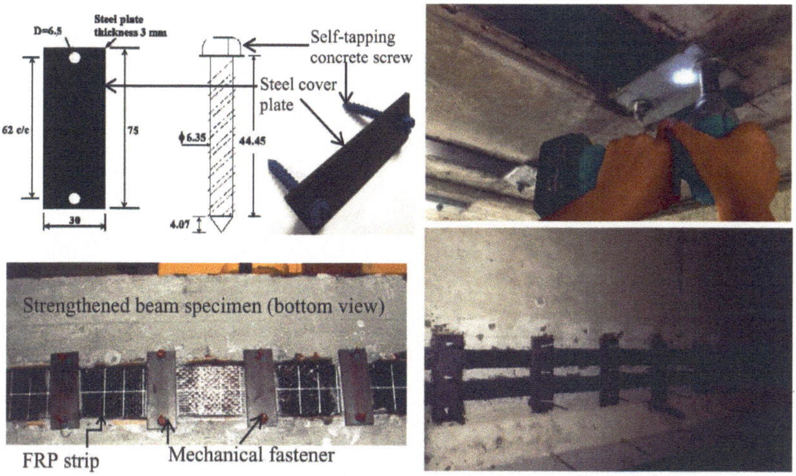

(A) (B)

Fig. 6.26 HB-FRP system: (A) Lab test; and (B) bridge strengthening project.
(A) From Wu, Y. F., and J. Lu. 2013. "Preventing debonding at the steel to concrete interface through strain localization." *Compos. Part B Eng.* 45 (1), 1061–1070; (B) Courtesy F. Zhang.

of any material such as steel and FRP, as shown in Fig. 6.26. The mechanical fastener is composed of a thin steel capping plate that presses against the reinforcing plate and two small bolts/anchors that are used to fasten the capping plate onto the substrate concrete. The major difference between the mechanical fasteners used in the HB system and those used in the MF-FRP system is that no bearing resistance of the reinforcing plate is required in the HB system. Experimental tests and theoretical studies showed that the use of the HB system can result in a several-fold increase in the bond strength, which is generally sufficient to prevent the critical IC debonding (Wu et al. 2010, 2011). The HB-FRP system is introduced and studied in detail in the following sections.

The technologies for increasing bond strength usually come with a significant increase in construction time, costs, and labor. Therefore, convenience and costs are important considerations in selecting an appropriate bond enhancement system. EBR construction requires temporary supports of the reinforcing plate before the adhesive sets. One of the temporary supporting methods is shown in Fig. 6.27A. The HB-FRP system can make use of the temporary brackets to reduce the construction work (Fig. 6.27B) while using stronger, closer spaced mechanical fasteners.

6.5.2 Mechanisms of the HB system

The higher bond strength of the hybrid bonded system comes from its special bonding mechanism. When adequate adhesive bond between the reinforcing plate and concrete is ensured, debonding of the plate usually occurs in concrete with a rough surface (cohesive failure in concrete; see Section 5.3.1), as shown in Fig. 6.28. This kind

(A) Temporary brackets

(B) HB-FRP fasteners

Fig. 6.27 Temporary and permanent fasteners.
Courtesy F. Zhang.

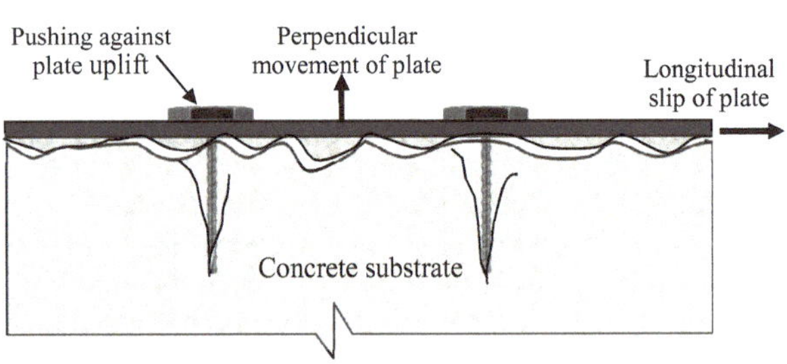

Fig. 6.28 Mechanism of HB-FRP system.

of debonding failure plane will induce the frictional bond mechanism both before interfacial cracking and after interfacial cracking, as explained in Section 5.12.

It is noted that, although frictional bond is a major mechanism of the HB system, appropriate adhesive bond is essential to mobilize the friction. This is because shear dilation is the condition (see Section 5.12) to mobilize the passive normal pressure from the anchors that pushes against the uplift of the plate during the shear deformation (Fig. 6.28). If the plate and the substrate concrete are not properly bonded, the

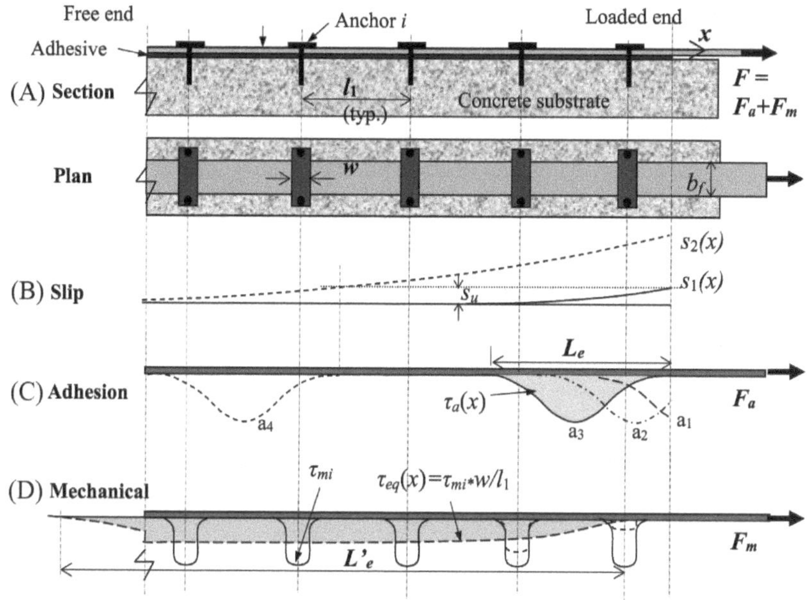

Fig. 6.29 Development of bond in HB system.
From Wu, Y. F., and K. Liu. 2013. "Characterization of mechanically enhanced FRP bonding system." *J. Compos. Constr.* 17 (1), 34–49.

plate will slip along a smooth plane, which will not cause significant uplift of the plate, and hence cannot mobilize the passive normal pressure from the anchors against the plate uplift. As friction is equal to normal pressure multiplied by frictional coefficient, an insignificant normal pressure does not produce a significant frictional resistance.

In fact, all the three mechanisms, adhesion, friction, and dowel action, exist in the HB system, as discussed through Fig. 6.29, which shows a pull-off test of a plate bonded by the HB technology. When the slip at the loaded end increases, the adhesive bond stress develops from a_1 to a_2, a_3, and then to a_4, as discussed in Section 5.3.1 (see Fig. 5.8). The dowel force from an anchor is transmitted through two paths as shown in Fig. 5.9A: (1) the bearing force F_b, and (2) the adhesion and friction T_1 at the interface 1, which is transmitted to the capping plate and then to the bolt shaft. For the HB system, there is no direct contact between the anchor shafts and the reinforcing plate. Therefore, the path 1 does not exist. The dowel action through the path 2 is marginal and relatively small compared with the frictional shear mechanism on the interface 2 in the HB system, because the interface 1 is smooth compared with interface 2 (Fig. 5.9A and Fig. 6.28).

As discussed earlier, the longitudinal slip of the FRP against the concrete substrate induces a vertical movement of the reinforcing plate. This perpendicular movement of the plate pushes against the fasteners and causes a passive normal pressure on the plate to produce frictional resistance at the interface. Since friction is essentially constant to any slip (Fig. 5.6C), the frictional bond contributed by different individual fasteners is similar, as depicted by the solid lines in Fig. 6.29D, although the slip reduces monotonically from the loaded end to the free end (see $s_2(x)$ in Fig. 6.29B). The maximum bond strength is attained when the bond resistance provided by the first mechanical fastener reduces to zero (see the dotted line in Fig. 6.29D) if the anchor loses its function when the slip exceeds a certain limit. In this case, a further pull at the loaded end can only shift the zero-bond position backward in a manner similar to the shifting of the adhesive bond stress block. The distance between the two zero frictional bond positions between the loaded end and the free end is the effective bond length, as shown by L_e' in Fig. 6.29D. This is the case when small concrete nails or screws are used as anchors, as observed in tests where nails were pushed out at a certain slip (Wu and Huang 2008). However, pushing out of anchors was not observed when larger concrete screws were used. A residual bond stress is present in this case even at a large slip, similar to τ_f in Fig. 5.11. In this case, an effective bond length does not exist in the HB system; and the bond strength always increases with the increase in bond length, similar to the bond of a rebar discussed in Section 5.4.

6.5.3 Experimental tests

Flexural tests of RC beams strengthened with HB-FRP were conducted to evaluate the system (Wu et al. 2010). The test program and results are shown in Fig. 6.30.

The test specimen EB8 in the figure represents the beam strengthened by normal EB-FRP without a mechanical fastener. HB6, HB7, and HB9 are specimens strengthened by HB-FRP with different numbers of fasteners. The test results in Fig. 6.30B and C show the load-carrying capacity of the beams and the maximum strain (force) in the FRP plate, respectively. In Fig. 6.30C and D, P, P_a, and P_f represent the total bond strength and the bond strength contributions from adhesion and mechanical fasteners, respectively. Fig. 6.30D shows the theoretical relationship between the bond strengths and the number of fasteners. For a bond-slip relationship that approaches zero at a large slip (Fig. 6.29D), the total bond strength approaches an asymptotic value when the number of fasteners increases. For a bond-slip relationship with a residual bond, the total bond increases continuously when the number of fasteners increases.

Pull-off tests were conducted to characterize the bond of the HB-FRP system (Wu and Liu 2013). The test specimens are shown in Fig. 6.31A. The test results are illustrated in Fig. 6.31B. For these particular test specimens, the effective bond length of the HB-FRP system was identified to be 2.5 m (Fig. 6.31B). This is significantly larger than that of normal EB-FRP systems, which are around 50–200 mm. The bond-slip relationship of the HB-FRP interface was developed from the pull-off tests, as discussed in the following section.

6.5.4 Bond modeling

Based on the experimental results and analyses of the failure mechanisms, the total bond strength P provided by an HB system is the summation of the three bond mechanisms, i.e., adhesion, friction, and dowel action:

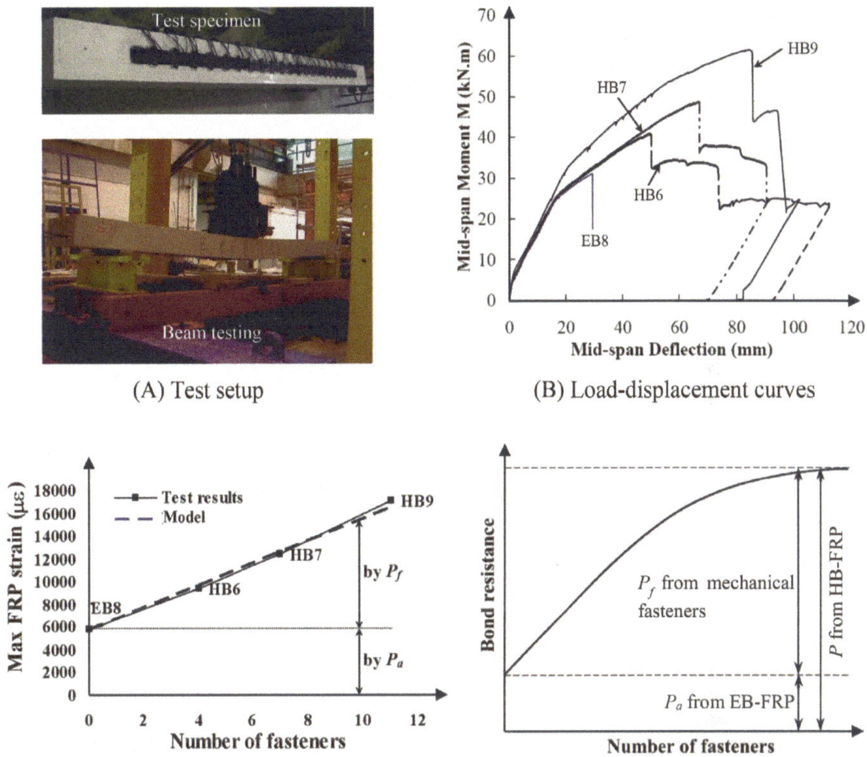

(A) Test setup (B) Load-displacement curves

(C) Enhanced bond (D) Total bond strength

Fig. 6.30 HB-FRP strengthened beam tests.
Based on Wu, Y. F., J. H. Yan, Y. W. Zhou, and Y. Xiao. 2010. "Ultimate strength of reinforced concrete beams retrofitted with hybrid bonded fiber-reinforced polymer." *ACI Struct. J.* 107 (4), 451–459.

$$P = P_a + P_f + P_d \tag{6.143}$$

where P_d is the bond strength contribution from dowel action, which can be neglected compared with the other two terms in the HB-FRP system. The contribution from adhesive bond P_a is similar to that in normal EB-FRP systems and can be calculated using an available bond model, as discussed in Sections 6.4.5 and 6.4.6. P_f is the frictional bond strength, given by

$$P_f = \mu \cdot N_a \cdot \frac{L_f}{s} \tag{6.144}$$

where μ is the frictional coefficient between two rough concrete surfaces; N_a is the normal pressure applied on the FRP strip from one mechanical fastener, which is

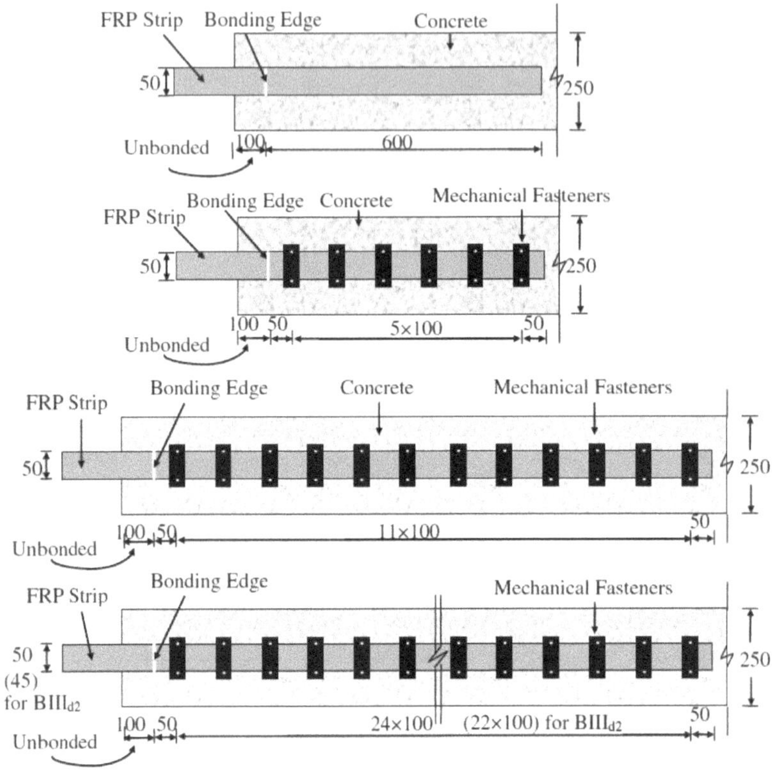

(A) Test specimens

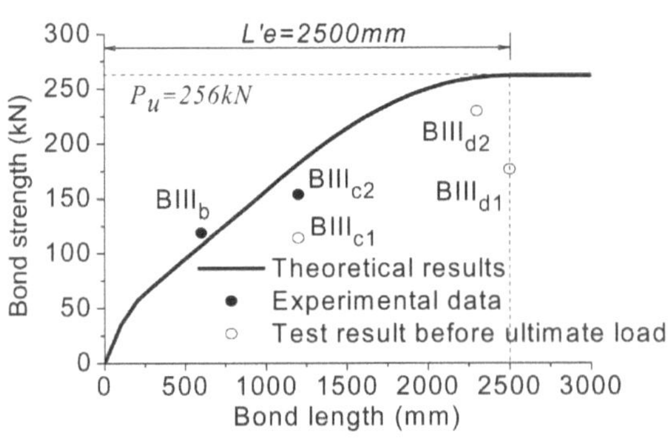

(B) Test results

Fig. 6.31 Pull-off tests of HB-FRP bond system.
From Wu, Y. F., and K. Liu. 2013. "Characterization of mechanically enhanced FRP bonding system." *J. Compos. Constr.* 17 (1), 34–49.

found equal to the pull-out resistance of the fasteners with the push-out failure and known for a given concrete strength and particular mechanical fastener. For anchors that do not fail even at large slip, tests should be undertaken to evaluate the maximum normal pressure N_a during slip; s is the longitudinal spacing between the mechanical fasteners; and L_f is the bond length within which mechanical fasteners are provided, which should not be larger than the effective bond length, i.e., $L_f \leq L'_e$. L_f/s gives the number of mechanical fasteners within the bond length. When the number of fasteners is equal to zero, $P_f = P_d = 0$, the total bond strength is equal to the adhesive bond strength or $P = P_a$, as shown in Fig. 6.30D. The frictional coefficient μ for HB-FRP joints was identified to be 0.96 in the tests. The pull-out resistance for the small fastener shown in Fig. 6.26 was tested to be $N_a = 6$ kN. When different fasteners are used, tests should be conducted to evaluate N_a.

The bond-slip relationship for the local areas under fasteners is different from that outside it. The model shown in Fig. 6.32 and given by Eq. (6.145) was developed for simulating the bond stress due to the mechanical fastening that includes friction and dowel action:

$$\tau_m(s) = \begin{cases} \tau_{max} \cdot \left(\dfrac{s}{s_1}\right)^{\gamma_1}, & s \leq s_1 \\ \tau_{max}, & s_1 < s \leq s_2 \\ (\tau_{max} - \tau_0) \cdot \left(\dfrac{s - s_3}{s_2 - s_3}\right)^{\gamma_2} + \tau_0, & s_2 < s \leq s_3 \\ \tau_0, & s > s_3 \end{cases} \tag{6.145}$$

where τ_{max}, τ_0, γ_1, γ_2, s_1, s_2, and s_3 are parameters of the model, as shown in Fig. 6.32. The overall bond-slip relationship of the HB-FRP system is described as

$$\tau(x) = E_f \cdot t_f \cdot \frac{d^2 s}{dx^2} = \begin{cases} \tau_a, & \text{for } x \text{ outside fasteners} \\ \tau_a + \tau_m, & \text{for } x \text{ under fasteners} \end{cases} \tag{6.146}$$

For the particular anchor shown in Fig. 6.26A, γ_1 and γ_2 were experimentally identified to be 1.0. Other parameters were identified to be $\tau_{max} = 4.4$ MPa, $\tau_0 = 0$, $s_1 = 0.2$ mm, $s_2 = 8.0$ mm, and $s_3 = 30$ mm. More details of the model refer to Wu

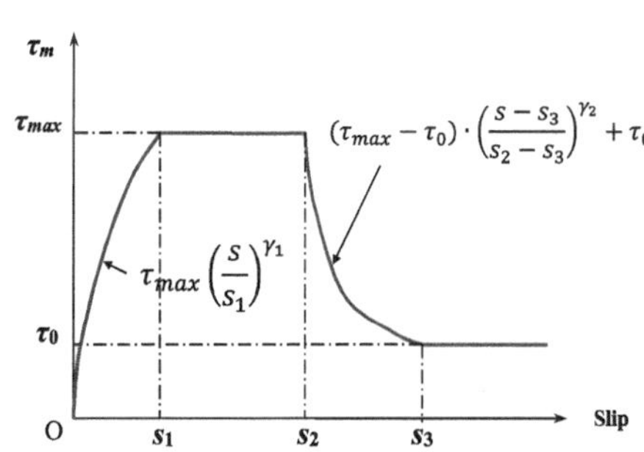

Fig. 6.32 Bond-slip relationship of HB-FRP interface.

and Liu (2013). When different anchors are used, tests should be conducted to determine the parameters for Eq. (6.145). The test method can be found in Wu et al. (2016).

The differential equation, Eq. (6.146), can be solved numerically with appropriate boundary conditions to evaluate the interfacial behavior of the HB joints. It can also be applied in the numerical simulation of HB-FRP strengthened members to study the response of the composite members. The response of HB-FRP strengthened members can also be evaluated by the finite element method (Wu et al. 2009a).

References

Abramowicz, M., S. Berczyński, and T. Wróblewski. 2020. "Modelling and parameter identification of steel–concrete composite beams in 3D rigid finite element method." *Archiv. Civ. Mech. Eng.* 20, 103.

ACI Committee 440. 2017. *Guide for the design and construction of externally bonded FRP systems for strengthening concrete structures, ACI 440.2R17*. American Concrete Institute.

Adam, C., R. Heuer, and A. Jeschko. 1997. "Flexural vibrations of elastic composite beams with interlayer slip." *Acta Mech.* 125, 17–30.

Adekola, A. O. 1968. "Partial interaction between elastically connected elements of a composite beam." *Int. J. Solids Struct.* 4, 1125–1135.

Amana, E. J., and L. G. Booth. 1967. "Theoretical and experimental studies of nailed and glued stressed-skin components." *J. Inst. Wood Sci.* 4 (1), 43–69.

Bao, G., and R. Xu. 2015. "Dynamic stiffness matrix of partial-interaction composite beams." *Adv. Mech. Eng.* 7 (3). 1687814015575990.

Berczyński, S., and T. Wróblewski. 2005. "Vibration of steel-concrete composite beams using the Timoshenko beam model." *J. Vib. Control* 11, 829–848.

Biscontin, G., A. Morassi, and P. Wendel. 2000. "Vibrations of steel–concrete composite beams." *J. Vib. Control* 6, 691–714.

Bousias, S. N., D. Biskinis, M. N. Fardis, and A. L. Spathis. 2007. "Strength, stiffness, and cyclic deformation capacity of concrete jacketed members." *ACI Struct. J.* 104 (5), 521.

Carrara, P., D. Ferretti, F. Freddi, and G. Rosati. 2011. "Shear tests of carbon fiber plates bonded to concrete with control of snap-back." *Eng. Fract. Mech.* 78, 2663–2678.

Chai, Y. H., M. J. N. Priestley, and F. Seible. 1991. "Seismic retrofit of circular bridge columns for enhanced flexural performance." *ACI Struct. J.* 88 (5), 572–584.

Chakrabarti, A., A. H. Sheikh, M. C. Griffith, and D. J. Oehlers. 2013. "Dynamic response of composite beams with partial shear interaction using a higher-order beam theory." *J. Struct. Eng.* 139 (1), 47–56.

Chen, J. F., and J. G. Teng. 2001. "Anchorage strength models for FRP and steel plates bonded to concrete." *J. Struct. Eng.* 127 (7), 784–791.

Chen, W. Q., J. Ying, J. B. Cai, and G. R. Ye. 2004. "Benchmark solution of laminated beams with bonding imperfections." *AIAA J.* 42, 426–429.

Chen, W. Q., Y. F. Wu, and R. Q. Xu. 2007. "State space formulation for composite beam-columns with partial interaction." *Compos. Sci. Technol.* 67 (11-12), 2500–2512.

Dai, J. G., T. Ueda, and Y. Sato. 2005. "Development of the nonlinear bond stress-slip model of fiber reinforced plastics sheet-concrete interfaces with a simple method." *J. Compos. Constr.* 9 (1), 52–62.

De Lorenzis, L., and J. G. Teng. 2007. "Near-surface mounted FRP reinforcement: an emerging technique for strengthening structures." *Compos. Part B Eng.* 38 (2), 119–143.

Dritsos, S. E. 1997. "Jacket retrofitting of reinforced concrete columns." *Constr. Repair* 11 (4), 35–40.

Fabbrocino, G., G. Manfredi, and E. Cosenza. 2000. "Analysis of continuous composite beams including partial interaction and bond." *J. Struct. Eng.* 126 (11), 1288–1294.

Faella, C., E. Martinelli, and E. Nigro. 2002. "Steel and concrete composite beams with flexible shear connection: "exact" analytical expression of the stiffness matrix and applications." *Comput. Struct.* 80, 1001–1009.

Faella, C., E. Martinelli, and E. Nigro. 2003. "Shear connection nonlinearity and deflections of steel-concrete composite beams: a simplified method." *J. Struct. Eng.* 129, 12–20.

Fang, L. X., S. L. Chan, and Y. L. Wong. 2000. "Numerical analysis of composite frames with partial shear–stud interaction by one element per member." *Eng. Struct.* 22, 1285–1300.

Fazio, P., and R. Hussein. 1982. "Sandwich beam-columns with interlayer slips." *J. Eng. Mech. Div.* 108 (EM2), 354–366.

Frangou, M., K. Pilakoutas, and S. Dritsos. 1995. "Structural repair/strengthening of RC columns." *Constr. Build. Mater.* 9 (5), 259–266.

Fu, C., and X. Yang. 2018. "Dynamic analysis of partial-interaction Kant composite beams by weak-form quadrature element method." *Arch. Appl. Mech.* 88, 2179–2198.

Gara, F., G. Ranzi, and G. Leoni. 2006. "Displacement-based formulations for composite beams with longitudinal slip and vertical uplift." *Int. J. Numer. Methods Eng.* 65, 1197–1220.

Gattesco, N. 1999. "Analytical modeling of nonlinear behavior of composite beams with deformable connection." *J. Constr. Steel Res.* 52, 195–218.

Girhammar, U. A., and V. K. A. Gopu. 1993. "Composite beam-columns with interlayer slip-exact analysis." *J. Struct. Eng.* 119 (4), 1265–1282.

Girhammar, U. A., and D. H. Pan. 1993. "Dynamic analysis of composite members with interlayer slip." *Int. J. Solids Struct.* 30 (6), 797–823.

Girhammar, U. A., D. H. Pan, and A. Gustafsson. 2009. "Exact dynamic analysis of composite beams with partial interaction." *Int. J. Mech. Sci.* 51 (8), 565–582.

Goodman, J. R. 1967. "Layered wood Systems with interlayer slip." Ph.D. thesis. Berkeley, CA: University of California.

Goodman, J. R., and E. P. Popkov. 1968. "Layered beam systems with interlayer slip." *J. Struct. Div.* 24 (ST11), 2535–2547.

He, L., Y. F. Wu, and Y. Xiao. 2015. "Analytical solution for externally bonded joints considering snap-back." *J. Compos. Constr.* 19 (5), 04014077.

Huang, C. W., and Y. H. Su. 2008. "Dynamic characteristics of partial composite beams." *Int. J. Struct. Stab. Dyn.* 8 (4), 665–685.

Itani, R. Y., and F. A. Brito. 1978. "Elastometric bonded wood beams with transverse gaps." *J. Struct. Div.* 104 (ST10), 1595–1609.

Itani, R. Y., and G. S. Hiremath. 1980. "Partial composite action in diaphragms." *J. Eng. Mech. Div.* 106 (EM4), 739–751.

Jasim, N. A. 1999. "Deflections of partially composite beams with linear connector density." *J. Constr. Steel Res.* 49, 241–254.

Jasim, N. A., and A. Atalla. 1999. "Deflections of partially composite continuous beams: a simple approach." *J. Constr. Steel Res.* 49, 291–301.

Johnson, R. P., and Y. C. Wang. 2018. *Composite structures of steel and concrete: beams, slabs, columns and frames for buildings*, John Wiley and Sons.

Julio, E. N. B. S., and F. Branco. 2008. "Reinforced concrete jacketing—interface influence on cyclic loading response." *ACI Struct. J.* 105 (4), 471.

Julio, E. N. B. S., F. Branco, and V. D. Silva. 2003. "Structural rehabilitation of columns with reinforced concrete jacketing." *Prog. Struct. Eng. Mater.* 5 (1), 29–37.

Jurkiewiez, B., S. Buzon, and J. G. Sieffert. 2005. "Incremental viscoelastic analysis of composite beams with partial interaction." *Comput. Struct.* 83, 1780–1791.

Kaish, A. B. M. A., M. Jamil, S. N. Raman, M. F. M. Zain, and L. Nahar. 2018. "Ferrocement composites for strengthening of concrete columns: a review." *Constr. Build. Mater.* 160, 326–340.

Kalfat, R., R. Al-Mahaidi, and S. T. Smith. 2013. "Anchorage devices used to improve the performance of reinforced concrete beams retrofitted with FRP composites: state-of-the-art review." *J. Compos. Constr.* 17 (1), 14–33.

Kwak, H. G., and Y. J. Seo. 2002. "Time-dependent behavior of composite beams with flexible connectors." *Comput. Methods. Appl. Mech. Eng.* 191, 3751–3772.

Lamanna, A. J., L. C. Bank, and D. T. Borowicz. 2004. "Mechanically fastened FRP strengthening of large scale RC bridge T beams." *Adv. Struct. Eng.* 7 (6), 525–538.

Liang, Q. Q. 2014. *Analysis and design of steel and composite structures*, London: Taylor & Francis Ltd.

Lin, J. P., and Y. F. Wu. 2016. "Numerical analysis of interfacial bond behavior of externally bonded FRP-to-concrete joints." *J. Compos. Constr.* 20 (5), 04016028.

Lin, J. P., Y. F. Wu, and S. T. Smith. 2017. "Width factor for externally bonded FRP-to-concrete joints." *Constr. Build. Mater.* 155, 818–829.

Liu, K., and Y. F. Wu. 2012. "Analytical identification of bond-slip relationship of EB-FRP joints." *Compos. Part B Eng.* 43 (4), 1955–1963.

Lu, X. Z., J. J. Jiang, J. G. Teng, and L. P. Ye. 2006. "Finite element simulation of debonding in FRP-to-concrete bonded joints." *Constr. Build. Mater.* 20 (6), 412–424.

Manfredi, G., G. Fabbrocino, and E. Cosenza. 1999. "Modeling of steel-concrete composite beams under negative bending." *J. Eng. Mech.* 125, 654–662.

Murakami, H. 1984. "A laminated beam theory with interlayer slip." *J. Appl. Mech.* 51 (9), 551–559.

Newmark, N. M., C. P. Siess, and I. M. Viest. 1951. "Test and analysis of composite beams with incomplete interaction." *Proc. Soc. Exper. Stress Analysis* 9 (1), 75–92.

Newmark, N. M., C. P. Siess, and I. M. Viest. 1952. *Studies of slab and beam highway bridges, part III—small scale tests of shear connectors and composite T-beams*, Bulletin 396. Urbana, IL: University of Illinios.

Nguyen, N. T., D. J. Oehlers, and M. A. Bradford. 2001. "An analytical model for reinforced concrete beams with bolted side plates accounting for longitudinal and transverse partial interaction." *Int. J. Solids Struct.* 38, 6985–6996.

Nie, J., J. Wang, S. Gou, Y. Zhu, and J. Fan. 2019. "Technological development and engineering applications of novel steel-concrete composite structures." *Front. Struct. Civ. Eng.* 13 (1), 1–14.

Oehlers, D. J., and M. A. Bradford. 1995. *Composite steel and concrete structural members-fundamental behaviour*, Oxford: Elsevier Science Ltd.

Oehlers, D. J., and M. A. Bradford. 1999. *Elementary behaviour of composite steel and concrete structural members*, London: CRC Press.

Park, R. 2001. "Improving the resistance of structures to earthquakes." *Bull. N. Z. Soc. Earthq. Eng.* 34 (1), 1–39.

Priestley, M. J. N., F. Seible, Y. Xiao, and R. Verma. 1994. "Steel jacket retrofitting of reinforced concrete bridge columns for enhanced shear strength-part 1: theoretical consideration and test design." *ACI Struct. J.* 91 (4), 394–405.

Ranzi, G., and M. A. Bradford. 2007. "Analytical solutions for elevated-temperature behavior of composite beams with partial interaction." *J. Struct. Eng.* 133, 788–799.

Ranzi, G., and A. Zona. 2007. "A composite beam model including the shear deformability of the steel component." *Eng. Struct.* 29, 3026–3041.
Ranzi, G., M. A. Bradford, and B. Uy. 2004. "A direct stiffness analysis of a composite beam with partial interaction." *Int. J. Numer. Methods Eng.* 61, 657–672.
Rao, K. M., and B. G. Ghosh. 1980. "Imperfectly bonded unsymmetric laminated beam." *J. Eng. Mech. Div.* 106 (EM4), 685–697.
Rodriguez, M., and R. Park. 1994. "Seismic load tests on reinforced concrete columns strengthened by jacketing." *ACI Struct. J.* 91 (2), 150–159.
Sakino, K., and Sun, Y.P. (2000). "Steel jacketing for improvement of column strength and ductility." *Proc. Twelth World Conf. on Earthquake Engineering*, 2525, New Zealand.
Salari, M. R., and E. Spacone. 2001. "Finite element formulations of one-dimensional elements with bond-slip." *Eng. Struct.* 23, 815–826.
Schnabl, S., M. Saje, G. Turk, and I. Planinc. 2007. "Analytical solution of two-layer beam taking into account interlayer slip and shear deformation." *J. Struct. Eng.* 133, 886–894.
Seracino, R., D. J. Oehlers, and M. F. Yeo. 2001. "Partial-interaction strain distributions in composite bridge beams." *Eng. Struct.* 23 (9), 1186–1193.
Seracino, R., C. T. Lee, T. C. Lim, and J. Y. Lim. 2004. "Partial interaction stresses in continuous composite beams under serviceability loads." *J. Constr. Steel Res.* 60 (10), 1525–1543.
Shen, X., W. Chen, Y. F. Wu, and R. Xu. 2011. "Dynamic analysis of partial-interaction composite beams." *Compos. Sci. Technol.* 71 (10), 1286–1294.
Taljsten, B. 1997. "Defining anchor lengths of steel and CFRP plates bonded to concrete." *Int. J. Adhesion Adhesives* 17 (4), 319–327.
Teng, J. G., and J. F. Chen. 2007. "Debonding failures of RC beams strengthened with externally bonded FRP reinforcement: behaviour and modelling." In *Proc. First Asia-Pacific Conf. FRP in Structures (APFIS 2007), 12-14*, Hong Kong, China: The University of Hong Kong.
Thompson, E. G., J. R. Goodman, and M. D. Vanderbilt. 1975. "Finite element analysis of layered wood systems." *J. Struct. Div.* 101 (ST12), 2659–2671.
Timoshenko, S. P. 1941. "On the correction for shear of the differential equation for transverse vibration of prismatic bars." *Philos. Mag.* 41, 744.
Vandoros, K. G., and S. E. Dritsos. 2006. "Concrete jacket construction detail effectiveness when strengthening RC columns." *Constr. Build. Mater.* 22 (3), 264–276.
Wang, Y. C. 1998. "Deflection of steel-concrete composite beams with partial shear interaction." *J. Struct. Eng.* 124 (10), 1159–1165.
Wu, Y. F. 2002. "Seismic retrofitting of rectangular reinforced concrete columns with partial interaction plating." Ph.D. thesis. Australia: The University of Adelaide.
Wu, Y. F., and L. He. 2019. "Width effect of interfacial bond characteristics." *Constr. Build. Mater.* 220, 712–726.
Wu, Y. F., and Y. Huang. 2008. "Hybrid bonding of FRP to reinforced concrete structures." *J. Compos. Constr.* 12 (3), 266–273.
Wu, Y. F., and C. Jiang. 2013. "Quantification of bond-slip relationship for externally bonded FRP–to-concrete joints." *J. Compos. Constr.* 17 (5), 673–686.
Wu, Y. F., and K. Liu. 2013. "Characterization of mechanically enhanced FRP bonding system." *J. Compos. Constr.* 17 (1), 34–49.
Wu, Y. F., and J. Lu. 2013. "Preventing debonding at the steel to concrete interface through strain localization." *Compos. Part B Eng.* 45 (1), 1061–1070.
Wu, Y. F., D. J. Oehlers, and M. C. Griffith. 2002. "Partial interaction analysis of composite beam/column members." *Mech. Struct. Mach.* 30 (3), 309–332.

Wu, Y. F., M. C. Griffith, and D. J. Oehlers. 2004. "Numerical simulation of steel plated RC columns." *Comput. Struct.* 82, 359–371.

Wu, Y. F., R. Xu, and W. Q. Chen. 2007. "Free vibrations of the partial-interaction composite members with axial force." *J. Sound Vib.* 299, 1074–1093.

Wu, Y. F., Z. Wang, K. Liu, and W. He. 2009a. "Numerical analyses of hybrid-bonded FRP strengthened concrete beams." *Comput.-Aided Civ. Infrastruct. Eng.* 24 (5), 371–384.

Wu, Z. S., S. M. Islam, and H. Said. 2009b. "A three-parameter bond strength model for FRP–concrete interface." *J. Reinf. Plastics Compos.* 28 (19), 2309–2323.

Wu, Y. F., J. H. Yan, Y. W. Zhou, and Y. Xiao. 2010. "Ultimate strength of reinforced concrete beams retrofitted with hybrid bonded fiber-reinforced polymer." *ACI Struct. J.* 107 (4), 451–459.

Wu, Z. M., C. H. Hu, Y. F. Wu, and J. J. Zheng. 2011. "Application of improved hybrid bonded FRP technique to FRP debonding prevention." *Constr. Build. Mater.* 25 (6), 2898–2905.

Wu, Y. F., X. S. Xu, J. B. Sun, and C. Jiang. 2012. "Analytical solution for the bond strength of externally bonded reinforcement." *Compos. Struct.* 94 (11), 3232–3239.

Wu, Y. F., L. He, and L. C. Bank. 2016. "Bond test protocol for plate-to-concrete interface involving all mechanisms." *J. Compos. Constr.* 20 (1), 04015022.

Wu, Z. S., Y. F. Wu, and M. Fahmy. 2020. *Structures strengthened with bonded composites*, Elsevier Woodhead Publishing.

Xiao, Y., and H. Wu. 2003. "Retrofit of reinforced concrete columns using partially stiffened steel jackets." *J. Struct. Eng.* 129 (6), 725–732.

Xu, R. Q., and Y. F. Wu. 2007a. "Static, dynamic, and buckling analysis of partial interaction composite members using Timoshenko's beam theory." *Int. J. Mech. Sci.* 49 (10), 1139–1155.

Xu, R. Q., and Y. F. Wu. 2007b. "Two-dimensional analytical solutions of simply supported composite beams with interlayer slips." *Int. J. Solids Struct.* 44, 165–175.

Xu, R. Q., and Y. F. Wu. 2008. "Free vibration and buckling of composite beams with interlayer slip by two-dimensional theory." *J. Sound Vib.* 313 (3-5), 875–890.

Xu, R. Q., and Y. F. Wu. 2009. "Analytical study of beams strengthened by adhesively bonded reinforcement with variable properties using state space method." *Compos. Sci. Technol.* 69 (11-12), 1912–1918.

Ye, L. P., X. Z. Lu, and J. F. Chen. 2005. "Design proposals for the debonding strengths of FRP strengthened RC beams in the Chinese design code." In *Proc. Int. sympo. bond behaviour FRP Structures (BBFS 2005), 45-54, Hong Kong.*

Zhang, H. W., and S. T. Smith. 2012. "Influence of FRP anchor fan configuration and dowel angle on anchoring FRP plates." *Compos. Part B Eng.* 43 (8), 3516–3527.

Zhou, Y. W., Y. F. Wu, and Y. Yun. 2010. "Analytical modeling of the bond-slip relationship at FRP-concrete interfaces for adhesively-bonded joints." *Compos. Part B Eng.* 41 (6), 423–433.

Zhou, W., S. Li, L. Jiang, and S. Qin. 2015. "Vibration analysis of steel-concrete composite box beams considering shear lag and slip." *Math. Probl. Eng.* 2015, 601757.

Flexural deflection

7

7.1 Introduction

Deflection calculation is required for both serviceability limit states to ensure satisfactory working conditions of a member under service loads and ultimate limit states to check for deformation capacity or ductility under ultimate loads. Under service loads, structures are considered as elastic and linear theories from statics are generally applicable. However, concrete structures crack, which causes complications in the calculation of flexural rigidity of a member. In the meantime, the strain of concrete varies with time under sustained loadings due to shrinkage and creep that further complicate the deflection calculation. Deformation under ultimate loads further involves plastic deformation that is usually concentrated in a plastic hinge zone. Therefore, it involves the evaluation of plastic hinge length and ultimate curvature, which is a complicated issue that has not been well resolved so far. Therefore, the evaluation of flexural deformation is not a simple matter for reinforced concrete.

The deflection of reinforced concrete members is usually assessed by one of the following three approaches:

(1) By deemed-to-satisfy rules through some simple evaluations such as the ratio of effective span-to-effective depth of a member. If the evaluated results are within the limits given in a code of practice/design guidelines, it is deemed that the deflection of the member satisfies the design requirement without calculating the value of deflection. This is the simplest and the most widely used approach in current engineering practice.

(2) Calculation of deflection with a relatively simple method. Although much more involved than the first approach, the deflection calculation can usually be done manually. The calculated results are then compared with allowable deflections specified in a design code to check whether the deflections meet design requirements.

(3) Evaluation by a more complicated or refined approach, where more complex issues such as shrinkage, creep, loading history, cracking, tension stiffening, construction procedure, etc. are evaluated more accurately. This approach usually needs calculations by a sophisticated computer program. Such a refined calculation is included in modern design codes but seldom used in practice. It may be required for special structures such as excessively long span structures under complicated loading and environments.

This chapter introduces the basic theories for the deflection calculation of reinforced concrete members and discusses the complex issues involved in the calculations.

7.2 Deflection under serviceability limit states

7.2.1 General

Loading under service conditions is relatively small, and hence elastic theory is usually applicable to the calculations. The following assumptions are usually made for the calculation of flexural deflection of RC members in service conditions:

Analytical Approaches for Reinforced Concrete. https://doi.org/10.1016/B978-0-12-821164-9.00007-3

- The stresses of reinforcement are in the linear elastic range.
- Before cracking concrete is linear elastic.
- After cracking concrete has no tensile stress in the tension zone and is linear elastic in the compression zone.
- Plane sections before loading remain plane after loading.

Under these assumptions, the elastic theory of flexural members is applicable.

From the elastic theory of beam, the governing differential equation for the deflection of a beam is given by

$$\frac{d^2y}{dx^2}=\frac{M(x)}{EI}=\kappa \tag{7.1}$$

where the coordinates x and y are shown in Fig. 7.1; EI is the flexural rigidity of the member; $M(x)$ is the bending moment of the section at x; and κ is the curvature of the section. Eq. (7.1) can be solved for a particular beam with given boundary conditions and loading.

For example, for a simply supported beam under a uniformly distributed loading w, the bending moment is given by

$$M(x)=-\frac{wL}{2}x+w\frac{x^2}{2} \tag{7.2}$$

The boundary conditions are

$$\begin{cases} y|_{x=0}=0 \\ y|_{x=L}=0 \end{cases} \tag{7.3}$$

Substituting Eq. (7.2) into Eq. (7.1) and integrating both sides twice give

$$EIy=-\frac{wL}{12}x^3+w\frac{x^4}{24}+c_1x+c_2 \tag{7.4}$$

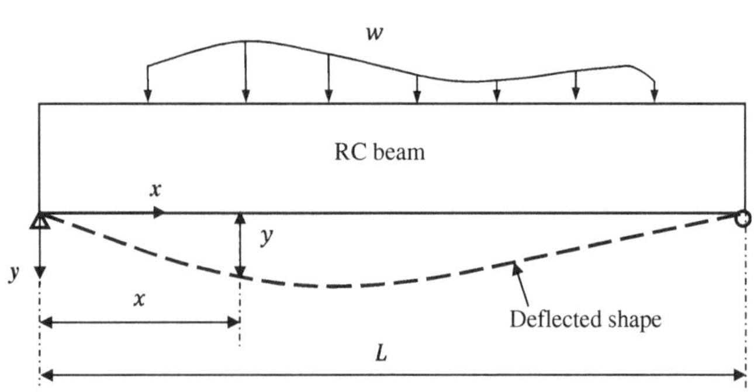

Fig. 7.1 Deflection of RC beam.

Substituting the boundary conditions of Eq. (7.3) into Eq. (7.4) gives

$$\begin{cases} c_2 = 0 \\ c_1 = \dfrac{w}{24}L^3 \end{cases} \tag{7.5}$$

Hence, the deflection is given by

$$y = \frac{1}{EI}\left(-\frac{wL}{12}x^3 + w\frac{x^4}{24} + \frac{w}{24}L^3 x\right) \tag{7.6}$$

The maximum deflection at the mid-span is given by

$$y_{mid} = \frac{5wL^4}{384EI} \tag{7.7}$$

For convenience, y_{mid} is usually provided in design manuals in the following form:

$$y_{mid} = C_\Delta \frac{M_{max}L^2}{EI} = C_\Delta \kappa_b L^2 \tag{7.8}$$

where M_{max} is the maximum bending moment within the considered span; C_Δ is a coefficient that depends on loading and boundary conditions; $\kappa_b = \frac{M_{max}}{EI}$, the curvature of the member at the section of maximum moment. C_Δ and M_{max} are given in design manuals for typical loads and boundary conditions. Typical ones are illustrated in Table 7.1.

The calculation of deflection using Eq. (7.7) is not a simple matter for reinforced concrete due to cracking, shrinkage, and creep. Although the member can still be considered as elastic after cracking, the flexural rigidity before and after cracking is significantly different, which means that loading sequence affects final deflection. These effects are discussed in the following sections.

Another common method for the deflection calculation that is different from the preceding procedure is by using a few curvatures at critical sections rather than one curvature in Eq. (7.8). Gilbert (2001) proposed the following equation to calculate the displacement at the mid-span of a beam by using three curvatures and assuming a parabolic variation of curvature along the span:

$$y_{mid} = \frac{L^2}{96}(\kappa_L + 10\kappa_M + \kappa_R) \tag{7.9}$$

where κ_L, κ_M, and κ_R are the curvature at the left end, mid-span, and right end of a beam, respectively. Eq. (7.9) is obtained by double integration of the curvature along the span (see Eq. (7.1)). The advantage of Eq. (7.9) is that C_Δ is not needed and the disadvantage is that three curvatures are required to calculate the deflection.

Table 7.1 Coefficients for calculation of deflection of beam.

Beam and loading	M_{max}	C_Δ
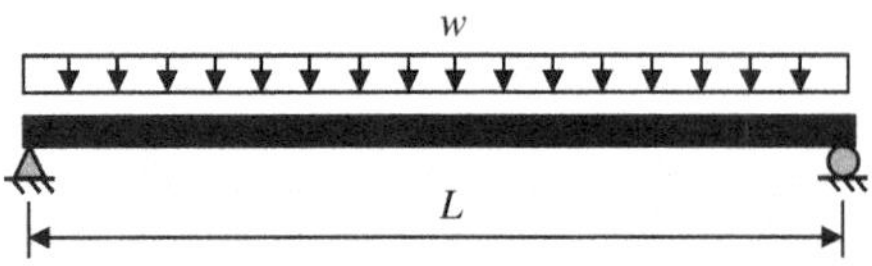	$\frac{wL^2}{8}$	$\frac{5}{48}$
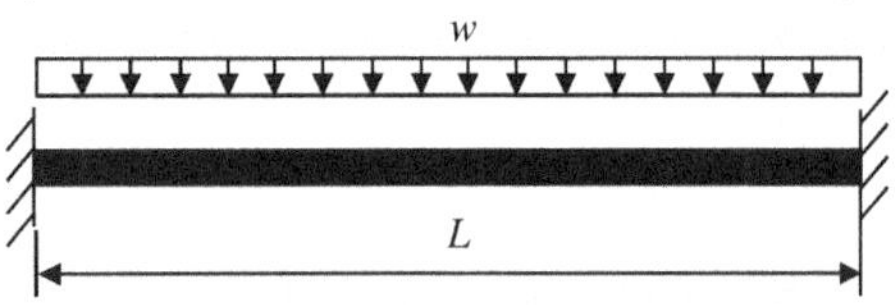	$\frac{wL^2}{12}$	$\frac{1}{32}$
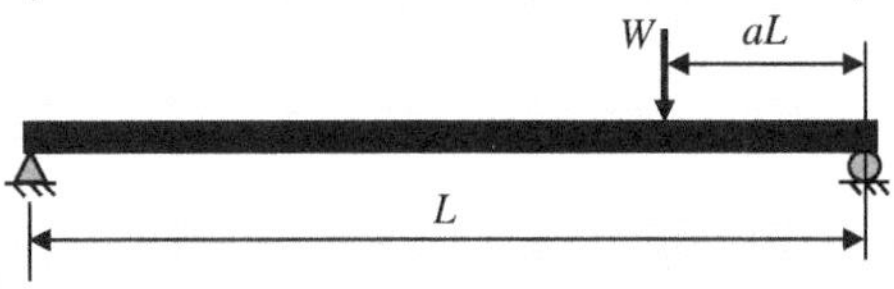	$W(1-a)aL$	$\frac{3-4a^2}{48(1-a)}$
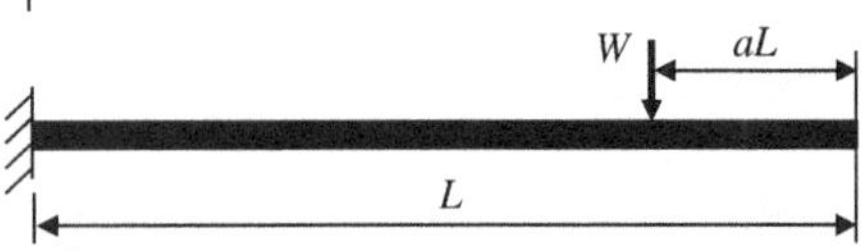	$W(1-a)L$	$\frac{(1-a)(2+a)}{6}$
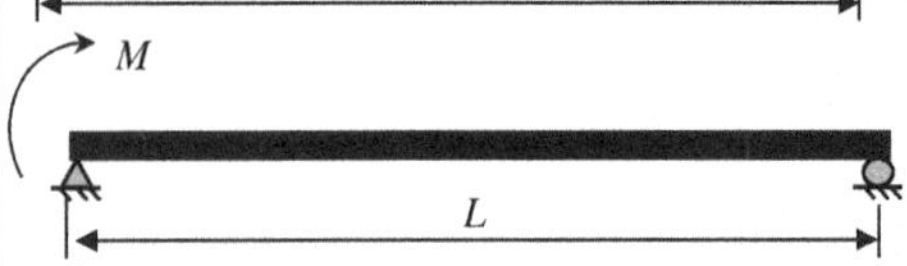	M	$\frac{1}{16}$
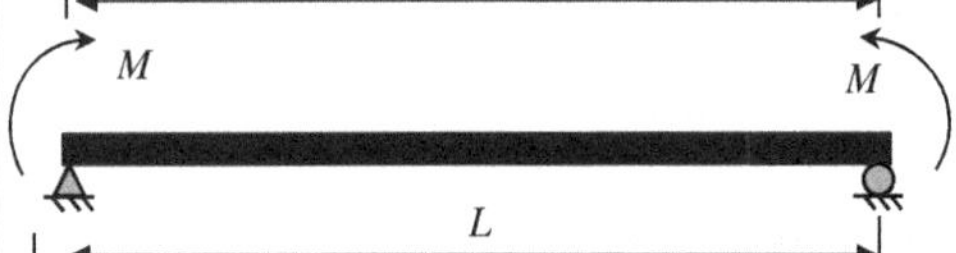	M	$\frac{1}{8}$
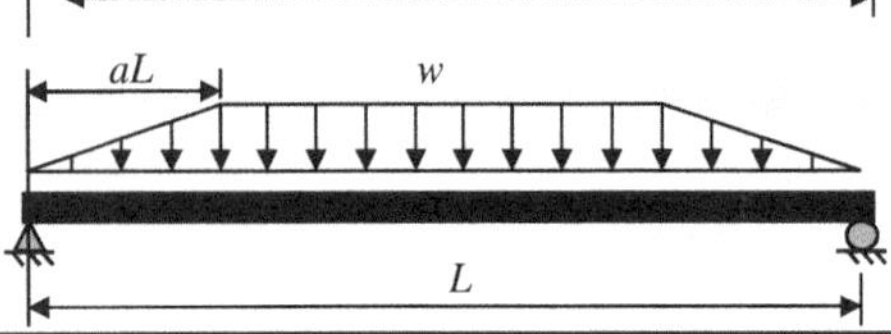	$\frac{(3-4a^2)wL^2}{24}$	$\frac{(5-4a^2)^2}{80(3-4a^2)}$

7.2.2 Short-term flexural rigidity and tension stiffening

Cracking must be considered in the calculation of deflection for reinforced concrete members because (1) it is normal for reinforced concrete members even under small loading and (2) it has a significant effect on the flexural rigidity of RC members. A typical section of cracked RC member is shown in Fig. 7.2, where it can be seen that the flexural rigidity EI of an uncracked section is significantly larger than that

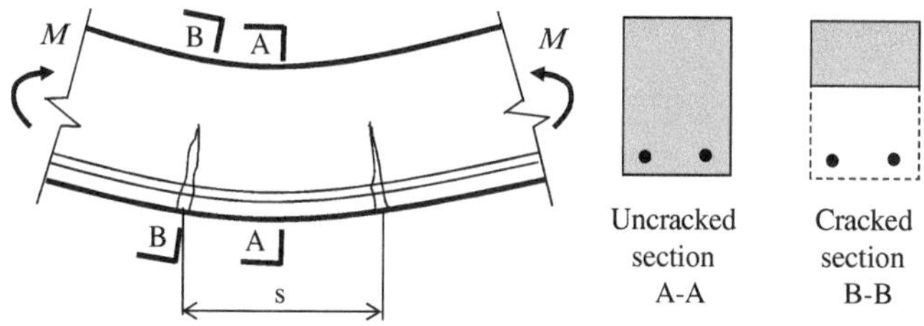

Fig. 7.2 Cracked and uncracked sections.

of a cracked one. Therefore, cracking causes significant variation of flexural rigidity *EI* along the length of the member and must be considered in the calculation of flexural deflection of RC members.

Although the concrete in the tension zone of a cracked section has no contribution to the flexural strength, as shown by section B-B in Fig. 7.2, the concrete in the cracked zone between two adjacent cracks still applies tensile stress to the RC member due to the bond, as discussed in Chapter 5. This tensile stress of concrete in the tension zone stiffens the member and reduces deflection. As discussed in Section 5.10, the reduction of deformation caused by the concrete under tension between cracks is called "tension stiffening." Due to the complexity of cracking, accurate analytical modeling of the tension stiffening effect for flexural deflection calculation of RC members is difficult, if not impossible. Numerous semiempirical approaches for evaluation of the tension stiffening effect have been developed. The following sections describe the common methods for the evaluation of flexural deflection of RC members considering the tension stiffening effect. Nevertheless, with the development of numerical tools such as finite element analysis, more accurate calculation of the flexural deflection can be made through numerical simulations (Piyasena 2002).

7.2.2.1 Transformed section

An RC section is composed of two materials: reinforcing bars and concrete. Therefore, both materials contribute to the overall flexural rigidity *EI* that is given by

$$EI = E_c I_c + E_s I_s \tag{7.10}$$

where E_c and E_s are the Young's modulus of the concrete and steel bar, respectively; and I_c and I_s are the moments of inertia (second moment of area) of concrete and steel bar with respect to the cross-sectional centroid, respectively. Eq. (7.10) can be transformed as follows:

$$\begin{aligned} EI &= E_c I_c + E_s I_s = E_c\left(I_c + \frac{E_s}{E_c} I_s\right) = E_c(I_c + nI_s) \\ &= E_c[(I_c + I_s) + (n-1)I_s)] = E_c[I + (n-1)I_s)] = E_c I_t \end{aligned} \tag{7.11}$$

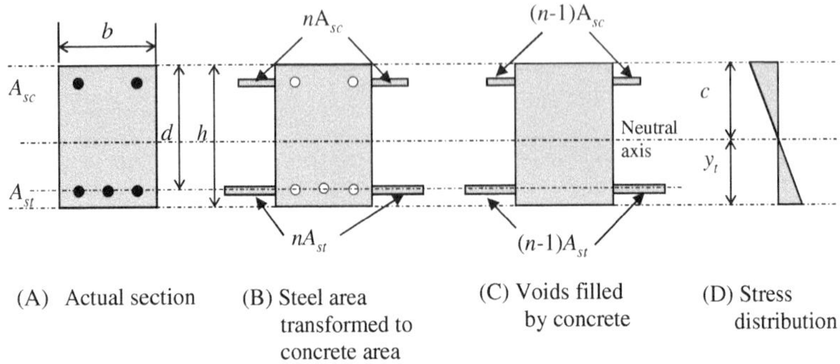

Fig. 7.3 Transformed section of uncracked RC section.

where $n = E_s/E_c$; $I = I_c + I_s$; and $I_t = I + (n-1)I_s$. Eq. (7.11) gives the expression for calculation of the moment of inertia of the composite RC section by the so-called transformed section approach, where the steel bars are transformed to concrete and the only material in the transformed section is concrete, as shown in Fig. 7.3.

With the transformed section approach, the actual RC section in Fig. 7.3A is transformed to the pure concrete section in Fig. 7.3B where the steel areas are multiplied by a factor of n (see Eq. 7.11). When the steel bars are replaced by the additional concrete area nA_{sc} and nA_{st} in Fig. 7.3B, the original locations of the steel bars become voids. If the holes of the steel bars are filled back by concrete, the additional concrete area is reduced to $(n-1)A_{sc}$ and $(n-1)A_{st}$, as shown in Fig. 7.3C. Therefore, the moment of inertia of the RC section can be calculated based on a pure concrete section as E_cI_t, where I_t is the moment of inertia calculated from the cross section as shown in Fig. 7.3C.

It is noted that, for a cracked section, there is no concrete, and hence no void, in the tension zone. Therefore, the additional concrete area in the tension zone is nA_{st}, as shown in Fig. 7.4C. The moment of inertia I_t for an uncracked section is referred to as I_g, and for a cracked one as I_{cr} in the following sections. The moment of inertia discussed in the following section refers to I_t of the transformed section without showing the subscript "t."

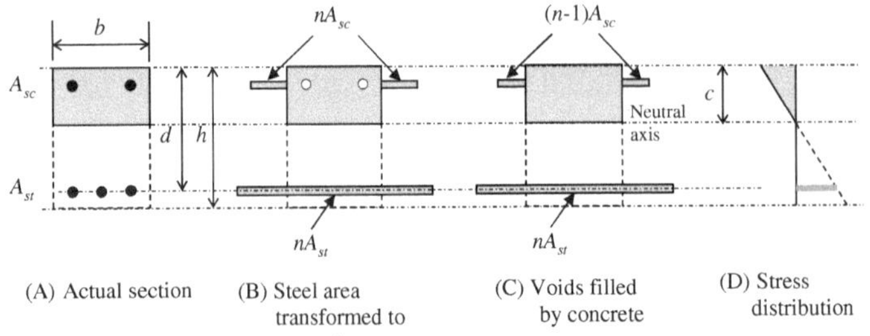

Fig. 7.4 Transformed section of cracked RC section.

The moment of inertia of the pure concrete sections in Figs. 7.3C and 7.4C can be obtained following the method introduced in statics. The neutral axis position must be determined before the moment of inertia can be calculated. For the cracked section in Fig. 7.4C, the neutral axis position affects the concrete area of the section. It is fixed and remains unchanged in the elastic deformation stage for the reason given here. Once cracking occurs at the extreme tension fiber of a cross section, the crack front propagates immediately to a fixed position, because when the tensile stress of concrete is insignificant, there is only one possible position of the neutral axis for the cross-sectional forces to achieve a balance, i.e., $\sum F = 0$. In reality, there is small tensile stress in the tension zone close to the neutral axis after a crack occurs. The neutral axis position calculated by ignoring the small tensile stress in the tension zone is sufficiently accurate and close enough to the real one for engineering design purposes. Therefore, for a cracked section, the neutral axis can be calculated first from $\sum F = 0$ (or using the first moment of area method in statics, which gives the same result). With a given neutral axis, the cross section is fixed and the moment of inertia of a cracked section can then be calculated following the method introduced in statics for a given cross section.

7.2.2.2 Effective moment of inertia

Although EI in Eq. (7.1) significantly varies along the span due to cracking, an average or "effective" value of the moment of inertia, I_e, can be used to calculate the deflection of a cracked RC beam. The only criterion for deriving I_e is that, when $E_c I_e$ is used as EI in Eq. (7.1), the calculated deflection is the same as that of the actual cracked RC member under a given loading. For a cracked RC member, I_e must be something between the moment of inertia of the uncracked section, I_g, and that of the cracked section, I_{cr}, or $I_g > I_e > I_{cr}$. As cracking depends on loading, I_e should be a function of loading. Different approaches have been developed for modeling I_e. These approaches are briefly introduced here.

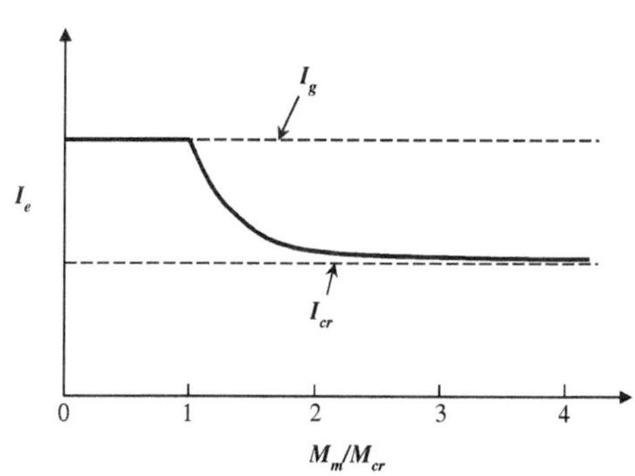

Fig. 7.5 Variation of effective moment of inertia.

Based on the experimental results illustrated in Fig. 7.5, Branson (1963) developed the following equation for calculating the effective moment of inertia of cracked reinforced concrete beams and one-way slabs:

$$I_e = I_{cr} + (I_g - I_{cr})\left(\frac{M_{cr}}{M_m}\right)^\beta \le I_g \tag{7.12}$$

where β is an exponential coefficient determined from tests, which was evaluated to be 3; M_m is the maximum bending moment of the beam; and M_{cr} is the moment at the onset of cracking of the RC section, which can be calculated using the stress-moment relationship from statics, $\sigma = \frac{M}{I}y$, when the tensile stress at the extreme tension fiber just reaches the modulus of rupture (cracking strength of concrete) f_{to}:

$$M_{cr} = f_{to}\frac{I_g}{y_t} \tag{7.13}$$

where y_t is the distance from the neutral axis to the extreme tension fiber as shown in Fig. 7.3D.

Branson's formula has been widely adopted in RC design codes including ACI codes. It was developed from test results under uniformly distributed loads. Others modified the formula to improve the accuracy for different cases. Al-Zaid et al. (1991) suggested a 20% increase of I_e calculated from Eq. (7.12) for RC beams loaded by a point load at mid-span. Al-Shaikh and Al-Zaid (1993) found that the exponential coefficient β in Eq. (7.12) is related to the reinforcement ratio ρ and proposed the following relationship:

$$\beta = 3 - 0.8\rho \tag{7.14}$$

Rangan (1982) proposed the following equations based on Branson's formula for normal service loads when the tensile steel stress reaches 60% of yield strength:

$$\frac{I_e}{bd^3} = 0.1599\sqrt{n\rho},\ if\ n\rho > 0.045 \tag{7.15a}$$

$$\frac{I_e}{bd^3} = 0.0019/n\rho,\ if\ n\rho \le 0.045 \tag{7.15b}$$

where b and d are the breadth and effective depth of beam, respectively, and $n = E_s/E_c$.

7.2.2.3 Smeared crack approach

The curvature for deflection calculation in Eq. (7.8) can be calculated from the stress distribution in Figs. 7.3D and 7.4D. The "average" curvature for the deflection calculation must be between the curvature calculated from Fig. 7.3D and that from Fig. 7.4D. The difference between the stress profile in Fig. 7.3D and that in Fig. 7.4D is the tensile stress of concrete in the tension zone. This gives an idea that

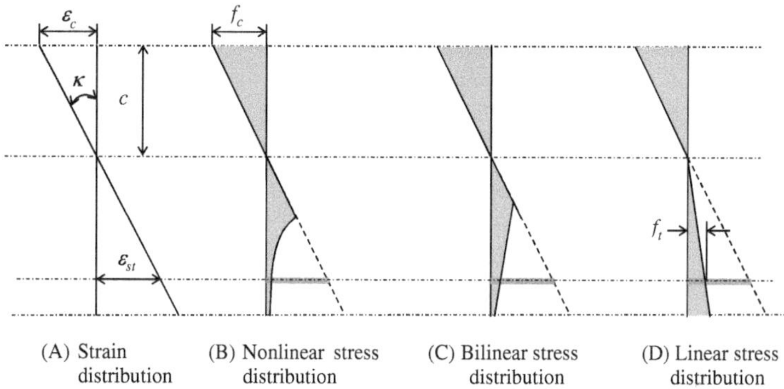

Fig. 7.6 Smeared crack models.

if an appropriate tensile stress of concrete in the tension zone is taken, which is between the one in Fig. 7.3D and that in Fig. 7.4D, the deflection can be calculated correctly. This approach essentially smears the cracks so that there is no discontinuity (discrete crack) along the length.

Different stress profiles of the concrete in the tension zone have been proposed for the smeared crack approach, as shown in Fig. 7.6. The tensile stress distribution in Fig. 7.6B is determined from the stress-strain relationship given by Eq. (7.16) that is shown in Fig. 7.7:

$$\sigma_t = f_{to}\frac{\varepsilon_t}{\varepsilon_{to}}, \text{ if } \frac{\varepsilon_t}{\varepsilon_{to}} \leq 1 \tag{7.16a}$$

$$\sigma_t = \frac{\beta f_{to}\left(\dfrac{\varepsilon_t}{\varepsilon_{to}}\right)}{\beta - 1 + \left(\dfrac{\varepsilon_t}{\varepsilon_{to}}\right)^{\beta}}, \text{ if } \frac{\varepsilon_t}{\varepsilon_{to}} > 1 \tag{7.16b}$$

where σ_t and ε_t are the tensile stress and strain of concrete, respectively; f_{to} and ε_{to} are the cracking stress and strain at the peak point of the tensile stress-strain curve, respectively (Fig. 7.7); and β is an empirical parameter related to beam properties. A larger

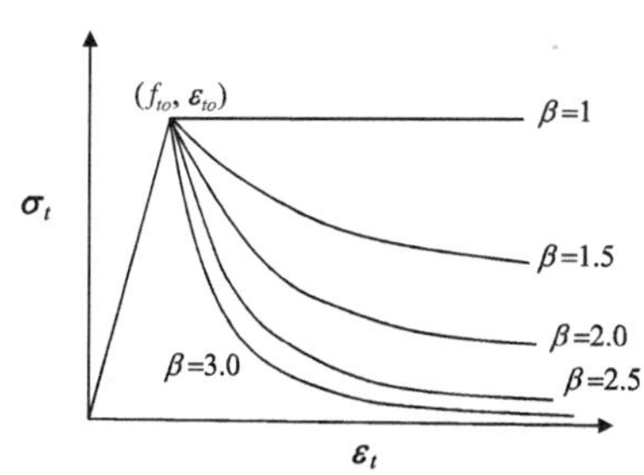

Fig. 7.7 Nonlinear tensile stress model for smeared crack approach.

value of β indicates a steeper descending part of the curve, which means less tensile resistance or smaller tension stiffening effect.

When the stress-strain relationships are given, the strain and stress profiles only have two independent unknowns, which can be solved through the balance of force and moment in the cross section, i.e., $\sum F = 0$ and $\sum M = 0$. The curvature can then be calculated from

$$\kappa = \frac{\varepsilon_c}{c} = \frac{f_c}{c \cdot E_c} \tag{7.17}$$

The curvature calculated from Eq. (7.17) can be substituted into Eq. (7.8) to obtain the deflection.

The empirical parameter β in Eq. (7.16) is determined by matching calculated deflections to test results. Research found that β is related to reinforcement ratio, diameter, and spacing of the main bars, neutral axis depth, and clear concrete cover thickness (Prakhya and Morley 1990).

Some researchers suggested a straight line rather than the nonlinear curve for the descending part of the stress-strain curve in Fig. 7.7 (Bazant and Oh 1984; Kaklauskas and Ghaboussi 2001). In this case, the stress distribution in the cross section is shown in Fig. 7.6C. The former British code BS8110-1997 (BSI 1997) further simplified the bilinear stress distribution in the tension zone to a straight line, as shown in Fig. 7.6D. It was specified in the BS code that the tensile concrete stress at the rebar level be $f_t = 1$ MPa for the calculation of short-term deflection, and $f_t = 0.55$ MPa for long-term deflection.

It is noted that none of the three stress distributions in Fig. 7.6 reflects the actual stress distribution of concrete in the tension zone of cracked RC beams. It is simply a modeling approach, and the aim of the models is to match the calculated deflection to the actual deflection of beams. From this point of view, there is no advantage for using a complicated model of the tensile stress-strain profile in the tension zone of the concrete for this particular problem, as long as a simple one can serve the purpose.

7.2.2.4 Curvature modeling considering slip

Another approach to consider the tension stiffening effect in the flexural rigidity is by considering the bond-slip relationship directly in an analytical solution. The free-body diagram of a small segment of RC beam is shown in Fig. 7.8. The following equations can be obtained from the equilibrium and compatibility conditions:

$$\sum F = 0: \sigma_{sc}A_{sc} + \sigma_{st}A_{st} + \int_{A_c} \sigma_c dA_c = 0 \tag{7.18}$$

$$\sum M = 0: \sigma_{sc}A_{sc}y_c + \sigma_{st}A_{st}y_t + \int_{A_c} \sigma_c y dA_c = M(x) \tag{7.19}$$

$$\text{Balance of the rebar}: A_{st}\mathrm{d}\sigma_{st} = \tau(x)p_s\mathrm{d}x \tag{7.20}$$

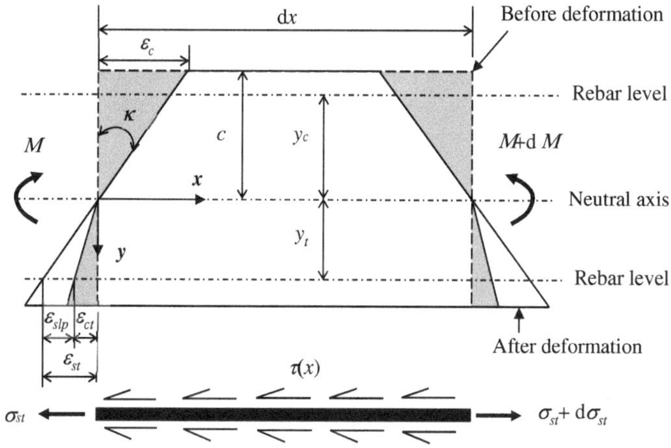

Fig. 7.8 Curvature modeling considering slip.

$$\text{Strain compatibility}: \frac{ds}{dx} = \varepsilon_{slp} = \varepsilon_{st} - \varepsilon_{ct} \tag{7.21}$$

where p_s is the total perimeter length of the rebars and ε_{slp} is the slip strain shown in Fig. 7.8.

Fantilli et al. (1998) used the preceding equations to produce the moment-curvature relationship of an RC section by making two plane section assumptions: (1) the compression concrete together with the rebars forming a plane section and (2) the tension concrete itself forming a separate linear strain distribution, as shown in Fig. 7.8. When all constitutive relationships are given, i.e., the stress-strain relationship of concrete in both compression and tension, the stress-strain relationship of steel bar, the bond-slip relationship of the concrete-to-rebar interface, the preceding equation set, Eqs. (7.18)–(7.21) that form the so-called "two-point boundary value problem," can be resolved numerically for given boundary conditions and loadings. Using this approach, Fantilli et al. (1998) obtained average curvature and the "local" curvature between two adjacent cracks for a beam subjected to constant bending moment and another one subjected to variable bending moment. The calculated results found good agreement with the experimentally measured curvatures. With the curvature known, the flexural deflection can be calculated by Eqs. (7.8) or (7.9).

7.2.3 Long-term deflection

The deflection of RC members increases with time under sustained loading due to shrinkage of concrete and creep of materials and bond, where the creep of concrete plays a major role. The long-term deflection can be a few times that of a short-term deflection. Therefore, the long-term effect of deflection must be considered in the design of RC members. Creep of concrete is related to the stress-to-strength ratio of concrete, age of concrete at initial loading, loading type, and environment such as relative humidity. Provision of compression steel can significantly reduce the creep of concrete and reduce long-term deflection. The mechanism of long-term flexural deflection and different approaches for the calculation of long-term deflection are introduced in the following sections.

7.2.3.1 Long-term effective modulus of concrete

The time-dependent deformation of concrete is illustrated in Fig. 7.9, where ε_e is the instantaneous strain of concrete immediately after a load is applied and ε_{cr} is called the creep strain, which is an inelastic strain due to creep of concrete material under sustained loading. The creep coefficient is defined as

$$\phi_c = \frac{\varepsilon_{cr}}{\varepsilon_e} \tag{7.22}$$

which may take a value greater than 2. The elastic strain ε_e is recovered immediately when the stress is released, while the creep strain ε_{cr} is not. However, some part of creep strain may recover with time. If a modulus of deformation is defined as $E = \sigma/\varepsilon$, the long-term effective modulus of concrete E_{eff} allowing for creep is given by

$$E_{eff} = \frac{\sigma}{\varepsilon_e + \varepsilon_{cr}} = \frac{\sigma}{\varepsilon_e + \phi_c \varepsilon_e} = \frac{E_c}{1+\phi_c} \tag{7.23}$$

By considering variation of sustained loading, the following model of the long-term effective modulus of concrete was proposed by Bazant (1972):

$$E_{eff}(t, t_0) = \frac{E_c(t_0)}{1 + \chi(t, t_0) \cdot \phi_c(t, t_0)} \tag{7.24}$$

where t = time when the deflection is evaluated; t_0 = age of concrete at initial loading; $E_{eff}(t, t_0)$ is the time-dependent effective modulus of concrete, which is a function of t and t_0; $E_c(t_0)$ is the Young's modulus of concrete at initial loading; $\chi(t, t_0)$ is the aging coefficient depending on loading type, which is taken as 1.0 for constant loading

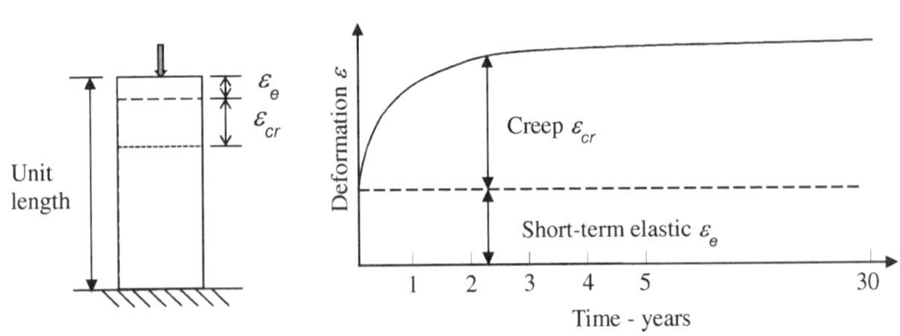

Fig. 7.9 Creep of concrete material.

and reduced down to 0.8 depending on the duration of loading for periodical loading; and $\phi_c(t, t_0)$ is the creep coefficient of concrete material. The creep coefficient is affected by many factors, including

- cement and aggregate properties,
- mix design,
- curing condition,
- stress-to-strength ratio,
- age of first loading,
- duration of loading,
- section dimensions,
- relative humidity, and
- temperature.

Design tables and charts are given in design codes to evaluate the creep coefficient that is obtained from experimental tests and engineering practice.

By using the long-term effective modulus of concrete, the long-term deformation of RC structures may sometimes be calculated with the equations in Section 7.2.2 in which the Young's modulus E_c is replaced by the effective modulus of concrete E_{eff}. In this case, the structure is considered to be made of a concrete that is softer, and consequently the structure will have a larger deflection under the same loading.

However, the creep coefficient is related to stress level and concrete stress is not uniformly distributed in a flexural member. Therefore, the calculation of long-term deflection may not be as simple as replacing the instantaneous Young's modulus with a long-term effective modulus. More detailed analysis of this case is discussed in the next section.

7.2.3.2 Long-term curvature and deflection due to creep

The effect of long-term creep on flexural deformation can be seen in Fig. 7.10 (Nilson et al. 2003). For an RC section as shown in Fig. 7.10, the solid line in Fig. 7.10B and C indicates the instantaneous strain and stress profiles, respectively. Concrete will creep under long-term stress while steel bars essentially do not. Since the strain of steel bar does not increase with time, the strain profile will rotate about the point of the rebar after a long period of time under sustained loading, as shown by the dashed line in Fig. 7.10B. As the value of $\varepsilon_{cr}/\varepsilon_e$ is different at different y values (Fig. 7.10B), the

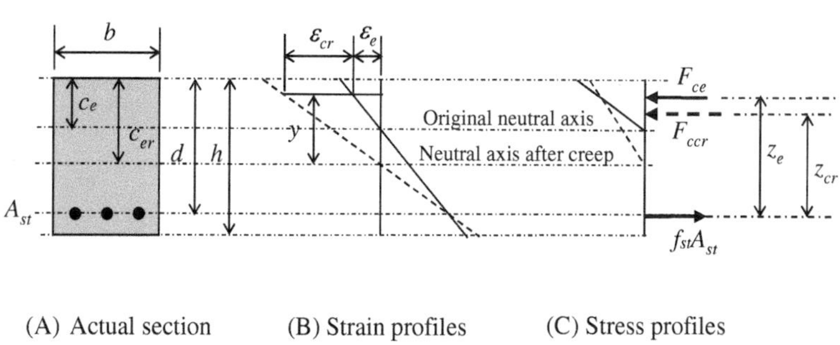

Fig. 7.10 Effect of creep on flexure.

creep coefficient is different at different levels according to Eq. (7.22). Based on Eq. (7.23), E_{eff} is different across the cross section.

The actual stress distribution is slightly more complicated than that discussed earlier. Assuming that the bending moment remains constant during creep, then the following balances must be maintained during creep:

$$(f_{st})_e A_{st} = F_{ce} \tag{7.25}$$

$$(f_{st})_{cr} A_{st} = F_{ccr} \tag{7.26}$$

$$F_{ce} \cdot z_e = F_{ccr} \cdot z_{cr} \tag{7.27}$$

where F_{ce} and F_{ccr} are the concrete resultant force in the compression zone before and after creep, respectively; $(f_{st})_e$ and $(f_{st})_{cr}$ are the tension steel stress before and after creep, respectively; and z_e and z_{cr} are the lever arm of F_{ce} and F_{ccr}, respectively. Since $z_e > z_{cr}$ (Fig. 7.10C), to maintain balance, one has $F_{ccr} > F_{ce}$, and hence $(f_{st})_{cr} > (f_{st})_e$. In other words, the steel stress actually slightly increases during creep.

The provision of compression bars will further complicate the case. Clearly, the existence of compression bars can significantly reduce the creep strain in the compression zone of concrete, and hence reduce the long-term deflection due to creep. Nevertheless, the earlier discussion on nonuniform distribution of E_{eff} is still valid.

As a result of these complexities, the approach of using a long-term effective modulus of elasticity E_{eff} to replace the short-term Young's modulus of concrete E_c for calculation of deflection is generally inapplicable for RC flexural members, unless another definition of E_{eff} in an "average" sense, which is different from Eqs. (7.22), (7.23) is adopted. A simple approach is used in practice to calculate the long-term curvature or deflection, as given by the following equation:

$$\kappa_{cr} = \lambda \cdot \kappa_e \tag{7.28}$$

where κ_e and κ_{cr} are the initial elastic curvature and the additional long-term curvature due to creep, respectively; and λ is an empirical coefficient determined from experimental data and practical experience, which is provided in design codes/guidelines based on duration of loading, compression and tension reinforcement ratios, support condition, concrete strength, etc. More details on evaluation of λ can be found in Paulson et al. (1991). In ACI codes, the calculation of Eq. (7.28) includes both the effects of creep and shrinkage unless for a separate and more accurate estimation of the deflection due to shrinkage is required. The shrinkage deflection is discussed in the next section.

7.2.3.3 Deflection due to shrinkage and temperature change

The effect of shrinkage on flexural deflection is shown in Fig. 7.11. When the rebars and concrete are fully bonded, shrinkage of concrete causes compression of the rebars, which in return produces tensile stress in the concrete as a reaction. If concrete and

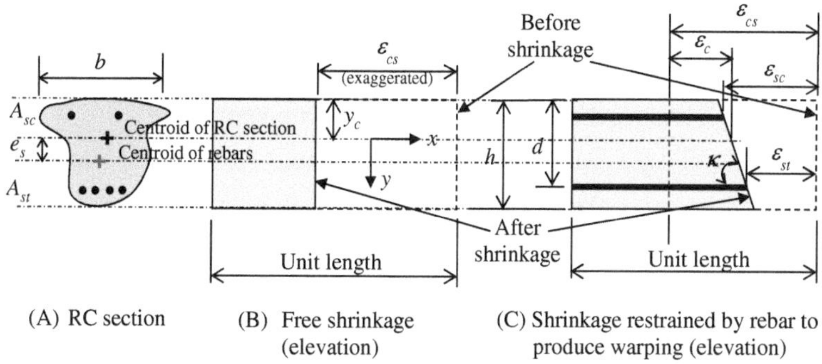

Fig. 7.11 Effect of shrinkage of concrete on flexure.

rebars are not symmetrical, the shrinkage-induced deformation will cause warping and hence flexure of the RC member. Warping or flexure is produced due to the non-uniform restraint of the rebar to shrinkage of concrete. The free shrinkage of concrete tends to have a uniform shortening with a free shrinkage strain of concrete ε_{cs} (Fig. 7.11B). However, shrinkage of concrete is not free in an RC member, as it is resisted by the rebars that do not shrink. The equilibrium of internal forces, i.e., the compression forces in the rebars and the tension force in the concrete, produces the balanced position as shown in Fig. 7.11C. In the balanced position, the rebars have the compression strains of ε_{st} and ε_{sc}, and the concrete section has an axial tensile strain of ε_c at the centroid and a curvature κ as shown. Assuming the plane section remains plane, then there are only two unknowns for the problem, ε_c and κ, which can be resolved from the two equations: $\sum F = 0$ and $\sum M = 0$.

A simpler procedure based on the principle of superposition can be used to calculate the stress and curvature caused by shrinkage of concrete (Park and Paulay 1975). If there is no bond between the rebars and concrete, free shrinkage will occur in the concrete and the top of the concrete block will be at the position shown by the blue dashed lines in Fig. 7.11C. For the RC section where the rebars and concrete are fully bonded together, the concrete block will shorten if a compression force is applied at the top of the block. To push the top of concrete block to the position of the blue dashed line, the only possible scenario is to apply a force P at the centroid location of the rebars (the red cross in Fig. 7.11A) which causes the rebar to have a compression strain of ε_{cs}. Therefore, this force must be equal to

$$P = A_s E_s \varepsilon_{cs} \tag{7.29}$$

where $A_s = (A_{st} + A_{sc})$.To remove the artificial external force P, another force equal and opposite to P is applied at the same position. This opposite tension force will cause axial tension and bending of the RC member to generate the following tensile stress in concrete:

$$\sigma_c = \frac{P}{A} + y\frac{P \cdot e_s}{I} \tag{7.30}$$

where σ_c is the concrete stress at a distance y from the centroid (shown by the black cross in Fig. 7.11A) of the RC section; A and I are the area and moment of inertia of the transformed section, respectively; and e_s is the distance between the two centroids as shown in Fig. 7.11A. The curvature of the section is given by

$$\kappa_{cs} = \frac{M}{EI} = \frac{P \cdot e_s}{E_c I} = \frac{nA_s \cdot e_s}{I} \varepsilon_{cs} \tag{7.31}$$

where κ_{cs} is the curvature of the RC section due to shrinkage and $n = E_s/E_c$.

It can also be seen from the preceding analysis that the compression bars play a significant role in offsetting the warping effect, and hence reducing the shrinkage induced flexural deflection. It is noted that concrete may crack if the tensile stress of concrete due to shrinkage exceeds the tensile strength of concrete. In practice, the following equation is usually provided in RC design codes for the curvature caused by shrinkage:

$$\kappa_{cs} = \frac{\lambda}{d} \varepsilon_{cs} \tag{7.32}$$

where λ is a factor determined by tension and compression reinforcement ratios, concrete strength, and geometrical properties of the section, which is provided in design codes.

Differential temperature also causes flexure of RC members due to the nonuniform strain across the cross section of an RC member. However, the thermal-induced flexural deflection is relatively simpler than that caused by shrinkage and creep, because the thermal coefficient of steel is very close to that of concrete. In other words, thermal change does not cause significant internal action between rebar and the surrounding concrete. Therefore, thermal-induced curvature can simply be calculated from the differential strain gradient caused by the temperature gradient.

With the curvature known, the flexural deflection can be calculated by Eqs. (7.8) or (7.9).

7.2.4 Superposition of flexural deflections

Although the constituent materials are assumed to be elastic in serviceability limit states, the flexural rigidity significantly changes after cracking, as shown by the turning point of the curves in Fig. 1.5 and Fig. 2.5. Therefore, the behavior of RC flexural members is not linear even in serviceability limit states. Consequently, the sequence of loading affects the final deflection of the member, as explained in Fig. 7.12. In reality, live loads are applied after permanent/dead loads. The short-term curvature caused by the moment due to live load, M_{live}, is shown as $\kappa_{s,live}$ in red in the figure. If we ignore the loading sequence and directly calculate the curvature due to M_{live} as shown in blue in the figure, the result is shown by $\kappa_{s,live}$ in blue in the figure, which is much smaller than the actual one.

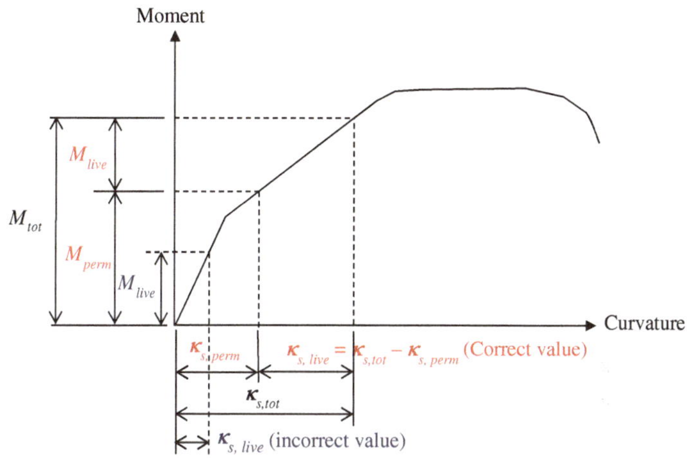

Fig. 7.12 Effect of loading sequence on final deflection.

To avoid this problem, the curvature due to short-term live load should be calculated as

$$\kappa_{s,live} = \kappa_{s,tot} - \kappa_{s,perm} \tag{7.33}$$

where $\kappa_{s,\,tot}$ is the short-term curvature due to total loads (dead/permanent loads + live loads); and $\kappa_{s,\,perm}$ is the short-term curvature due to permanent loads. For this reason, the total long-term curvature κ_l for calculation of long-term deflection should be based on the following equation:

$$\kappa_l = \kappa_{s,perm} + \kappa_{s,live} + \kappa_{cr} + \kappa_{cs} \tag{7.34}$$

where $\kappa_{s,\,perm}$ is the short-term curvature due to relevant permanent loads; κ_{cs} is given by Eq. (7.32); and κ_{cr} is the additional long-term curvature due to creep calculated by Eq. (7.28). It is noted that the loads for calculation of κ_e in Eq. (7.28) should be long-term loads, which are usually permanent loads; however, due consideration should be allowed to include some live loads that are applied on the structure for a long period of time and that also produce significant long-term effects of deflection.

Deflections under different loads should be checked for different conditions of use according to design codes. Design codes usually require a check of the deflections of structures: (1) due to total load relative to the casting line to maintain design shape and space of structure, (2) due to live load to ensure its functionality, and (3) after fixing floor finishes to avoid damage of the nonstructural constructions. The loads used for different deflection calculations are therefore different. For example, for the calculation in (3), $\kappa_{s,\,perm}$ in Eq. (7.33) is the total dead load while that in Eq. (7.34) is the dead load that is applied after the construction of floor finishes. Another important factor to be considered in deflection calculation is the loading sequence. It can be seen from Fig. 7.12 that the same load will produce different deflections when it is applied at different stages and, similarly, unloading and further loading at different stages will also produce a significant effect on the deflection.

7.3 Deflection under ultimate limit states

The importance of ductility is discussed in Chapter 1. From a structural safety point of view, adequate ductility is often more important than strength. This is because the requirement on strength is determined by loading that usually cannot be controlled directly and accurately, while sufficient ductility can be controlled in design, which can control/cap loading that is applicable onto a structure in a passive and safe manner. The calculation of ductility requires the evaluation of ultimate deformation of a structure.

As discussed in Section 7.2, deflection is usually calculated through integration of curvature. However, the approach of curvature integration will encounter difficulty after a plastic hinge forms in an RC member. This is explained through the RC column shown in Fig. 7.13. The RC column is assumed to have a uniform cross section that has a typical moment-curvature relationship shown in Fig. 7.14. The RC column in Fig. 7.13 can also represent a typical part of a flexural member, such as a continuous beam, from the point of contraflexure with a zero moment (column top) to the point of maximum moment (column base).

There is no problem to calculate the column deflection using the typical moment-curvature curve through the approach described in Section 7.2, before the maximum moment reaches the peak point A in Fig. 7.14A. Once the section of maximum moment at the bottom of the column passes the peak point A, only the bottom cross section can pass the peak and proceed onto the descending part of the

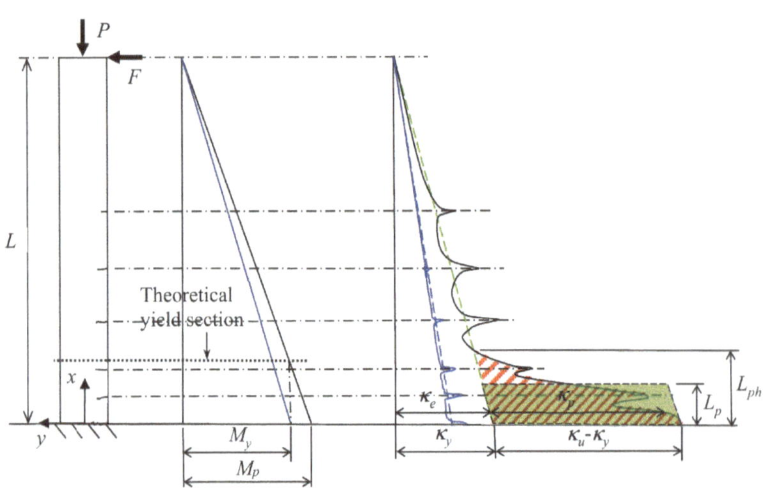

(A) RC column (B) Moment distribution (C) Curvature distribution

Fig. 7.13 RC column with a plastic hinge.

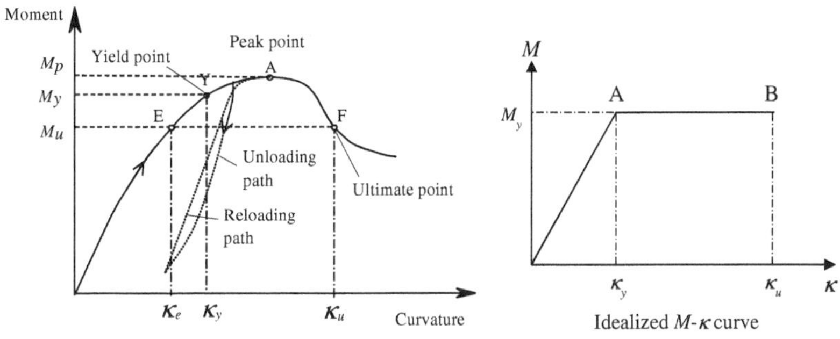

(A) Real curve (B) Idealized curve

Fig. 7.14 Typical curvature-moment relation of an RC section.

moment-curvature curve where the curvature further increases while the moment decreases. All other cross sections above the bottom section will move along an unloading path as shown in Fig. 7.14A when the moment of the bottom cross section decreases. This is because the bending moments of all cross sections are corelated by the linear distribution shown in Fig. 7.13B. In other words, the curvature of all cross sections reduces except the one at the bottom cross section when the bottom section passes the peak point of the moment-curvature curve. If Eq. (7.1) is used directly to integrate curvature for deflection calculation, the total deflection of the column reduces after the onset of the peak moment at the bottom section, because the bottom section has a zero length and has no effect on the integration of curvature from a mathematical point of view, while the curvature of all other cross sections above the base reduces. This is against the real behavior of the column.

This problem is caused by the use of the typical moment-curvature curve for all cross sections. Strictly speaking, the application of one typical moment-curvature curve to all cross sections is suitable only when the bending moment does not vary along the span, i.e., under a constant bending moment. When the moment varies along span, difference in stress at adjacent cross sections occurs, which tends to move the material in between from one side toward the other. This is illustrated in Fig. 5.14, where the steel bar tends to move or slip when the forces at the two ends are different, i.e., $T_1 \neq T_2$. The movement or slip will cause slip strain ε_{slp} as shown in Fig. 5.4, which invalidates the plane section assumption. In other words, the section is no longer a plane. Therefore, the moment-curvature curve, which is based on the plane section assumption, is no longer applicable. The error caused by the plane section assumption to curvature is relatively small before a plastic hinge forms in a flexural member but is significant after that. Therefore, this problem should, and can only, be dealt with properly by two- or three-dimensional (2D or 3D) modeling, such as finite element analysis, rather than the one-dimensional beam theory. More discussion of the problem is given in Sections 7.3.2 and 7.3.4.

However, 2D or 3D modeling/calculation is too complicated and not suitable for engineering design. Research works found that the moment-curvature relationship for one-dimensional beams/columns can still be applied to engineering design if the concept of plastic hinge is introduced. This approach is introduced in detail in the following sections.

7.3.1 Physical plastic hinge length and equivalent plastic hinge length

The typical curvature distribution of the cantilever column is shown in Fig. 7.13C. Before yielding of the member, the curvature distribution can be illustrated by the blue solid line that reflects cracking in Fig. 7.13C, while the blue dashed line shows the "average" curvature distribution. It is discussed in Section 2.8.2 that the plane section assumption can be used to produce the moment-curvature relationship for three cases: (1) before the column cracks using E_cI_g, (2) at cracked sections using E_cI_{cr}, and (3) the effective or "average" curvature at critical sections for the calculation of deflection after cracking using E_cI_e. For a section between two cracks, some tensile stress exists but it cannot be calculated from the plane section assumption. Therefore, the typical moment-curvature curve that is calculated from the plane section assumption is only applicable to sections where concrete stress is zero after cracking, i.e., to the places where the column cracks.

The curvature at a cracked section is given by

$$\kappa=\frac{M}{E_cI_{cr}} \tag{7.35}$$

where I_{cr} is the moment of inertia of the cracked section. The effective curvature at the critical section for deflection calculation is given by

$$\kappa=\frac{M}{E_cI_e} \tag{7.36}$$

where I_e is the effective moment of inertia given by Eq. (7.12).

After yielding of the bottom section, plastic deformation will occur in the member. Experiments and practical observation find that the plastic deformation only concentrates in a small zone, namely the plastic hinge zone. The deformation beyond the plastic hinge zone remains elastic. Assume the moment-curvature relationship is an idealized curve as shown in Fig. 7.14B. When the moment reaches the yield moment M_y at point A, the curvature at the bottom of the column reaches the yield curvature κ_y as shown in Fig. 7.13C. When the curvature further increases, the curvature near the bottom of the column will further increase, as shown by the red shaded area. The curvature curve above the shaded area remains unchanged because the bending moment remains unchanged. The column length corresponding to the red shaded area is called physical plastic hinge length, L_{ph}. For simplicity of calculation, the red shaded area is represented by an equivalent rectangular area in the literature, as shown in green, which has an equal area to that of the red shaded area. The column length corresponding to the equivalent area is called equivalent plastic hinge length, L_p, as shown in the figure.

The deflection can be obtained by double integration of curvature. The actual curvature κ (shown by the black solid curve in Fig. 7.13C) can be separated into two parts: the elastic part κ_e represented by the green dashed line and the plastic part κ_p from the green dashed line to the black solid line, i.e., $\kappa = \kappa_e + \kappa_p$, as shown in the figure. With double integration of Eq. (7.1), and considering the boundary conditions at the fixed end, the deflection of the column at the top is given by

$$y = \int_0^L dx \int_0^x (\kappa_e + \kappa_p) dx = \int_0^L dx \int_0^x \kappa_e dx + \int_0^L dx \int_0^x \kappa_p dx \quad (7.37)$$

where

$$\kappa_e = \kappa_y \left(1 - \frac{x}{L}\right) \quad (7.38)$$

and

$$\kappa_p = \kappa - \kappa_e \quad (7.39)$$

Using Eq. (7.38), the first part of the integration in Eq. (7.37) gives the elastic deformation at the top of the column

$$y_e = \int_0^L dx \int_0^x \kappa_e dx = \frac{\kappa_y}{3} L^2 \quad (7.40)$$

The integration of the second part of Eq. (7.37) cannot be obtained, because κ in Eq. (7.39) is unknown. To resolve this problem, the integration is taken over the equivalent rectangular area shown by the green area. In this case, the plastic curvature is given by

$$\kappa_p = \kappa_u - \kappa_y \quad (7.41)$$

The integration of the second part of Eq. (7.37) using Eq. (7.41) gives the plastic deflection at the top of the column

$$y_p = \int_0^L dx \int_0^x \kappa_p dx = (\kappa_u - \kappa_y) L_p \left(L - \frac{L_p}{2}\right) \quad (7.42)$$

where L_p is the equivalent plastic hinge length. Empirical models of L_p are given in RC design codes/guidelines for calculation of the plastic deflection. The term $(\kappa_u - \kappa_y)L_p$ is the area of the equivalent rectangular block and $\left(L - \frac{L_p}{2}\right)$ is the distance from the centroid of the equivalent rectangular block to the top of the column. Therefore, Eq. (7.42) means a rotation that is equal to the area of the plastic curvature and is concentrated at the center of the equivalent plastic hinge, as shown in Fig. 7.15C.

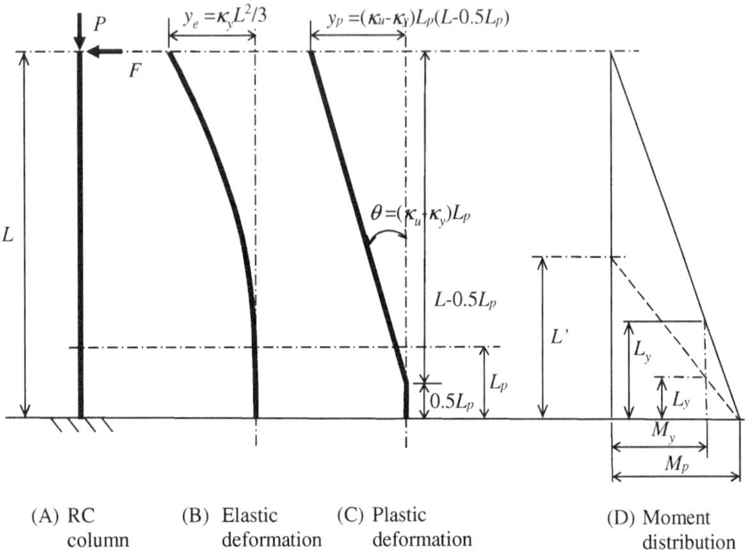

Fig. 7.15 Plastic hinge and deflection.

It should be noted that, although the area of the actual plastic curvature (the red shaded area in Fig. 7.13C) is equal to the area of the equivalent plastic curvature (the green rectangular area), the result of Eq. (7.37) is different when Eq. (7.39) is replaced by Eq. (7.41). A full equivalence can be achieved by introducing another parameter α so that

$$y_p = \int_0^L dx \int_0^x (\kappa - \kappa_e)dx = \alpha(\kappa_u - \kappa_y)L_p\left(L - \frac{L_p}{2}\right) \tag{7.43}$$

This is because the effect of plastic curvature on the total deflection is equivalent to a rotation hinged at the plastic centroid. The location of the hinge is related to the distribution of the plastic curvature. For the equivalent rectangular plastic curvature distribution, its centroid is fixed at $0.5L_{\mathrm{p}}$ from the fixed end (Fig. 7.15C) and L_p is only determined by the area of the plastic curvature, which is irrelevant to its distribution. The introduction of α can adjust the width of the equivalent plastic curvature from $(\kappa_u - \kappa_y)$ (Fig. 7.13C) to $\alpha(\kappa_u - \kappa_y)$ so that its centroid can be located properly while keeping its area unchanged. In this case, another empirical model for α is required for the calculation, which would increase the complexity of the calculation. In developing engineering design equations, one should always be conscious of a good balance between accuracy and convenience.

The total deflection at the ultimate limit state, y_u, is the summation of the two parts, i.e.,

$$y_u = y_e + y_p \tag{7.44}$$

In the preceding equations, the idealized curve in Fig. 7.14B is assumed where $M_y = M_u$. For real columns, the yield moment M_y and ultimate M_u are determined using the method discussed in Section 1.5. Subsequently, κ_y is calculated from Eq. (7.36) with $M = M_y$ (Section 7.2.2.2), and κ_u from Eq. (7.35) with $M = M_u$ (actual curvature of the cracked section). The previous calculations may cause a significant error if M_u is significantly different from M_y. When the load passes the peak point and moves along the descending branch, M_u decreases, which also reduces the load and elastic bending moment of the whole column. In other words, the elastic moment shown by the green dashed line in Fig. 7.13C changes when the ultimate curvature increases. In this case, the elastic deflection given by Eq. (7.40) would have a significant error if M_u is significantly different from M_y. The actual load for the elastic deformation is given by $F = M_u/L$ (when the P-Δ effect is ignored). Therefore, the correct ultimate deflection is given by

$$y_u = y_e + y_p = \frac{\kappa_e}{3}L^2 + (\kappa_u - \kappa_e)L_p\left(L - \frac{L_p}{2}\right) \tag{7.45}$$

where κ_e is given by

$$\kappa_e = \frac{M_u}{E_c I_e} \tag{7.46}$$

7.3.2 *Factors affecting plastic hinge length*

In Eq. (7.45), the equivalent plastic hinge length is required to calculate the ultimate deflection. Empirical models of the equivalent plastic hinge length have been developed for the calculation, which are summarized in Section 7.3.3. The mechanisms and factors that affect the plastic hinge length are discussed herein.

(1) Strength hardening
It can be seen in Fig. 7.13 that the first yield section occurs at the theoretical yielding section (Fig. 7.13B) where the bending moment is equal to M_y. Yielding of rebar and plastic deformation occurs in the region where the bending moment exceeds M_y. If the increase of moment from M_y to M_p in Fig. 7.14A is larger, the extent of yielding from the theoretical yielding section to the bottom of the column is greater, which provides a larger yielding zone and hence a larger plastic hinge length. The difference between M_y and M_p is usually governed by the strain hardening of the steel bar from the yield strength f_y to the ultimate strength f_{su} (Fig. 1.8A) for tension-controlled members (tension rebars yield first followed by concrete crushing). For compression-controlled members (concrete crushes without tension rebar yielding), the concrete strength plays a significant role in the shape of the moment-curvature curve and hence the difference between M_y and M_p.

(2) Moment distribution
The effect of moment distribution is illustrated in Fig. 7.15D where two moment distributions are shown, one for a column with a length of L and another with a length of L'. For columns with an identical RC cross section, i.e., columns having the same values of M_y and M_p, it can be seen from the figure that the rebar yielding zone L_y is different when the

column length is different. The extent from the theoretical yielding section to the bottom of the column is greater when the column length is greater. This means that the moment gradient in a longitudinal direction affects the column yielding zone, hence plastic hinge length.

(3) Base rotation

The column is assumed to be fixed at the base. However, the rebar is pulled out of the base slightly in the tension side in reality, particularly after rebar yielding. On the compression side, the concrete and rebar are pushed into the base slightly. The pull-out and push-in at the column base form a rotation of the column base as shown in Fig. 7.16. The base rotation is referred to as yield penetration into the base in the literature (Paulay and Priestley 1992). It should be noted that the base rotation is caused not only by the penetration of yielding into the base but also by elastic deformation of the base. The additional curvature that causes the support rotation can be represented by the red area in Fig. 7.17A, which represents the yield penetration (plastic deformation of the base) plus the yellow area that represents the elastic deformation of the support. The green area represents the equivalent plastic curvature of the column above the base, which provides an equivalent deflection of the member to the base rotation.

The calculation of the additional deformation of the column caused by the base rotation is simplified by adding an extra length to the equivalent plastic hinge length, as shown in Fig. 7.17B. Such simplification is generally acceptable from an engineering design point of view. However, it increases the error of the calculation as the plastic deformation is shifted up (moving up the hinge position in Fig. 7.15C), unless an additional parameter α for the equivalent plastic curvature model is used as given by Eq. (7.43). In this case, the integration in Eq. (7.43) is extended into the base and the curvature κ in Eq. (7.43) should include the red and yellow areas in Fig. 7.17A.

(4) Tension shift effect

Shear cracking in a flexural member can extend the length of the yielding zone, and hence the plastic hinge length, as explained with Fig. 7.18. For a cantilever beam without transverse reinforcement, the internal forces at the base are shown in Fig. 7.18A before diagonal cracking, and in Fig. 7.18B after diagonal shear cracking. Before diagonal cracking, the tension force of the bar is given by $T = M/z$, where z is the lever arm from the tension resultant to the compression resultant, as shown in the figure. As $M_A > M_B$, hence $T_A > T_B$ (regardless of z changing), where the subscripts A and B indicate the sections A and B as shown in the figure. After diagonal cracking, the internal forces are shown in Fig. 7.18B where the stress on the crack face is neglected. Taking the moment about the position of the compression resultant force F_A at section A, one has $T_B \cdot z_A = F \cdot L$, or

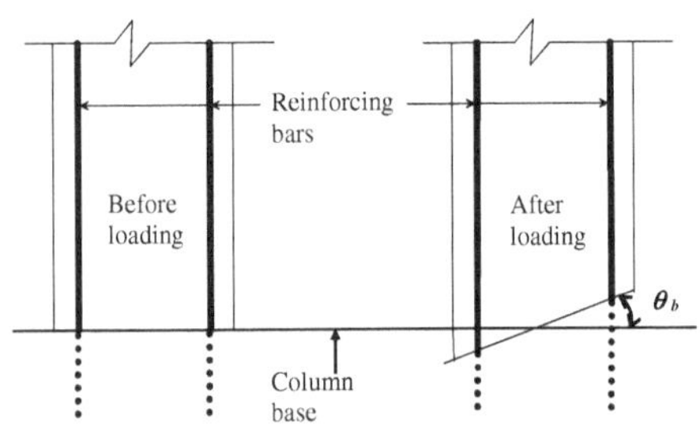

Fig. 7.16 Base rotation.

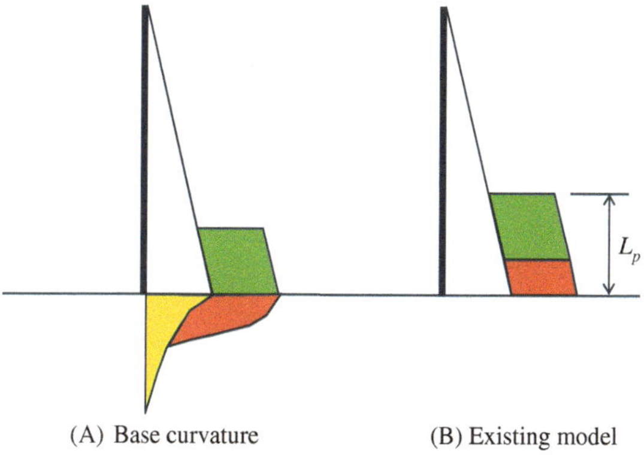

Fig. 7.17 Additional curvature causing base rotation

$$T_B = \frac{F \cdot L}{z_A} = \frac{M_A}{z_A} = T_A \tag{7.47}$$

which means that the tension force of the bar at section B is equal to that at section A, or the tension force of section A is shifted up to section B. This phenomenon is called tension shift in the literature (Park and Paulay 1975). If the rebar is yielded, the yielding of rebar at section A is also shifted to section B. As the diagonal crack is about 45 degrees, the yielding at section A is shifted for a distance about the effective depth *d*. This is probably the reason that the plastic hinge length is considered to be about the depth of the section in some existing plastic hinge models (see Table 7.2).

If transverse bars (links) exist as shown in Fig. 7.18C, the preceding equation becomes

$$M_A = T'_B \cdot z'_A + T_i \cdot s_i \tag{7.48}$$

where T_i and s_i are stirrup force and its lever arm shown in Fig. 7.18C. From Eq. (7.48), one has

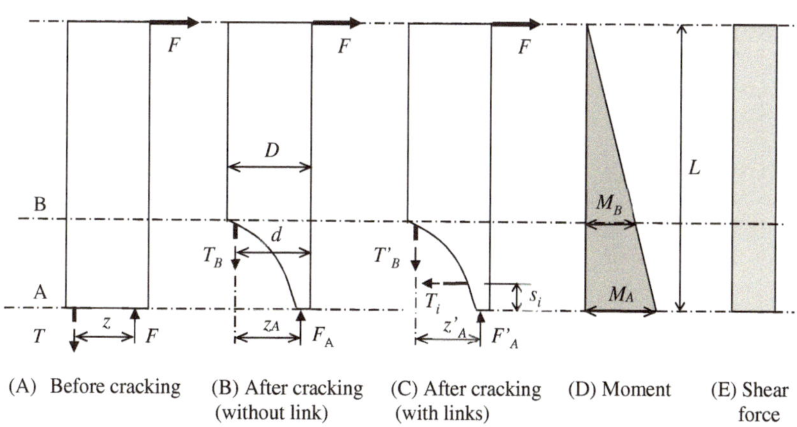

Fig. 7.18 Tension shift effect.

Table 7.2 Existing plastic hinge length models.

No.	Reference	Model
1 2	Baker (1956) Bate et al. (1962)	$L_p = k_1 k_2 k_3 \left(\frac{z}{d}\right)^{0.25} d$ $k_1 = \begin{cases} 0.7 & \text{for mild steel} \\ 0.9 & \text{for cold-worked steel} \end{cases}, k_2 = 1 + 0.5\frac{P}{P_u},$ $k_3 = \begin{cases} 0.6 & \text{for } f_c' = 42\text{MPa} \\ 0.9 & \text{for } f_c' = 14\text{MPa} \end{cases}$
3	Baker and Amarakone (1965)	$L_p = 0.8 k_1 k_3 (z/d) c$ k_1 and k_3 same as in Model 2, c = neutral axis depth
4	Sawyer (1964)	$L_p = 0.25d + 0.075z$
5	Mattock (1965)	$L_p = \frac{d}{2}\left[1 + \left(1.14\sqrt{\frac{z}{d}} - 1\right)\left(1 - \frac{q-q'}{q_b}\sqrt{\frac{d}{16.2}}\right)\right]$ $q = \frac{A_{st} f_y}{bdf_c'}, q' = \frac{A_{sc} f_y}{bdf_c'}, q_b = \frac{A_b f_y}{bdf_c'}$
6	Corley (1966)	$L_p = d/2 + 0.2z/\sqrt{d}$ (in inches)
7	Mattock (1967)	$L_p = d/2 + 0.05z$
8	ACI-ASCE (1968)	$L_p = \begin{cases} R_\varepsilon\left(\frac{d}{4} + 0.03 z_0 R_m\right) & \text{or} R_\varepsilon d \text{ lower bound} \\ R_\varepsilon\left(\frac{d}{4} + 0.1 z_0 R_m\right) & \text{upper bound} \end{cases}$ $R_\varepsilon = \frac{0.004 - \varepsilon_{cue}}{\varepsilon_{cuo} - \varepsilon_{cue}}, z_0 = \frac{4M_m}{4V_z + \sqrt{wM_m R_m}}, R_m = \frac{M_m - M_e}{M_u - M_e}$
9	Park et al. (1982)	$L_p = 0.4d$
10	Mander (1984)	$L_p = 32\sqrt{d_b} + 0.06z$
11	Zahn (1985)	$L_p = (0.08z + 6d_b)(0.5 + 1.67n)$, $n = 0.3$ when $n \geq 0.3$
12	Priestley and Park (1987)	$L_p = 0.08z + 6d_b$
13	Paulay and Priestley (1992)	$L_p = 0.08z + 0.022 d_b f_y$
14	Sheikh et al. (1994)	$L_p = 1.0d$ (for high axial load level)
15	Panagiotakos and Fardis (2001)	$L_p = \begin{cases} 0.12z + 0.014\alpha_{01} d_b f_y & \text{for cyclic loading} \\ 0.18z + 0.021\alpha_{01} d_b f_y & \text{for monotonic loading} \end{cases}$
16	Ho (2003)	$L_p = [20(P/P_u)^{0.5}(f_c'/f_y)^{1.5}(\rho_s/\rho_{sv})^{0.5} + 0.6]d$
17	Ruangrassamee and Kawashima (2003)	$L_p = 0.2z - 0.1d$
18	Bae and Bayrak (2008)	$L_p/d = [0.3(P/P_u) + 3(A_{sv}/A_g) - 0.1](z/h) + 0.25 \geq 0.25$ $P_u = 0.85 f_c'(A_g - A_s) + f_y A_s$

Table 7.2 Continued

No.	Reference	Model
19	Berry (2006)	$L_p = 0.05z + 0.1f_y d_b/\sqrt{f'_c} \le z/4$
20	Berry et al. (2008)	$L_p = \begin{cases} 0.0375z + 0.12f_y d_b/\sqrt{f'_c} & \text{optimal} \\ 0.05z + 0.1f_y d_b/\sqrt{f'_c} & \text{recommended} \end{cases}$
21	Kolias et al. (2008)	$L_p = 0.1z + 0.015f_y d_b$
22	Alemdar (2010)	$L_p = \left[\frac{1}{4} + \frac{3f_y d_b}{10,000\sqrt{f'_c}} + \frac{z}{25,000}\right] d$ based on deformation $L_p = \left[0.3 + \frac{3f_y d_b}{10000\sqrt{f'_c}} + \frac{z}{5000}\right] d$ based on curvature
23	Biskinis and Fardis (2010)	$L_p = \begin{cases} 0.2d\left\{1 + \left[\min\left(9; \frac{z}{d}\right)\right]/3\right\}, \text{cyc loading} \\ d\left[1.1 + 0.04\min\left(9; \frac{z}{d}\right)\right], \text{mono loading} \end{cases}$
24	Gu et al. (2012)	$L_p = (0.59 - 2.30\lambda_f + 2.28\lambda_f{}^2)z + 0.022f_y d_b,$ $0.1 \le \lambda_f(\,confinement\,ratio) < 0.5$
25	Köroğlu et al. (2014)	$L_p = -218.7 + 0.642f_{yv} + 0.201s$ $-0.07A_{sv} + 0.001A_g + 2.544f_{co}$

Note: Definitions of the model parameters are given in notations. Detailed explanations of the parameters refer to relevant references.

$$T'_B = \frac{M_A - T_i \cdot s_i}{z'_A} \tag{7.49}$$

Comparing Eqs. (7.47) and (7.49), the transverse reinforcement would reduce the tension shift effect, or reduce the plastic hinge length. However, the stirrups' effect is more complicated than that, because of other mechanisms involved. For example, the confinement provided by the transverse reinforcement will change the moment capacity of section A and its internal lever arm, i.e., M_A and z'_A in Eq. (7.49). More discussions on the effect of these factors on the plastic hinge length can be found in the numerical studies in Section 7.3.4.

(5) Bond effect

It is shown in Fig. 5.14C that the tension forces of two adjacent sections T_1 and T_2 are equal if the bond force F_b in between is zero. In other words, the bond force affects the spread of yielding into adjacent area. Yielding of rebar can be extended further when bond is smaller. Researchers have made use of this mechanism by purposely debonding the rebars in the plastic hinge region to increase the plastic hinge length and hence the ductility of the RC members (Wang et al. 2019).

(6) Other factors

Apart from the preceding factors or mechanisms that affect the extent of plastic deformation in the maximum moment region, there are many other factors that play a significant role in determining the plastic hinge length. These factors are further discussed in the numerical studies in Section 7.3.4.

7.3.3 Existing plastic hinge length models

Baker (1956) was the first who proposed an empirical plastic hinge model. Numerous other models have since been developed. Typical ones are provided in Table 7.2. Almost all existing plastic hinge models refer to the equivalent plastic hinge length. Existing empirical plastic hinge models are usually regressed from experimentally measured column test results. Most researchers including the renowned research group at the University of Canterbury used Eq. (7.45) (see Eq. (7.54) in Section 7.3.4.6) to calculate the equivalent plastic hinge length from test results. In this case, the determination of the ultimate and yield (or elastic) displacements (y_u and y_e) and the corresponding curvatures (κ_u and κ_e) are critical. The ultimate failure is often defined at the point on the postpeak load-deformation curve where the strength drops a certain percent, generally in the range of 10%–30%. When different strength drops are used, the ultimate displacement y_u can be significantly different. The definition of the yield (or elastic) point, i.e., y_e and κ_e could also be different in different works. As a result, the equivalent plastic hinge length calculated from Eq. (7.45) could be significantly different.

On the other hand, some models such as that by Biskinis and Fardis (2010) do not allow for the effect of strain penetration into the support. Also, some researchers developed an equivalent plastic hinge length model by measuring the physical plastic hinge region (severely damaged region) from tests (Bae and Bayrak 2008). Because of these differences, the empirical models reported in the literature can be significantly different in both the form and the results, as shown in Table 7.2. Even the factors that were considered in a model are very different from one model to another. In fact, no model has included all the factors that are discussed in Section 7.3.2. The factors that are considered in the existing plastic hinge models include: (1) cross-sectional depth h or effective depth d, (2) shear span length z or span-depth ratio z/d, (3) the type or yield strength of steel bar f_y, (4) concrete strength f_{co}, (5) axial load level $n = P/P_u$, (6) confinement effect by stirrups or external jacket, (7) diameter of tension bar d_b, (8) reinforcement ratios, (9) loading types (monotonic or cyclic), etc.

The previous discussion shows that the modeling of plastic hinge length is essentially empirical and irrational so far. Furthermore, the methods for model development adopted by different researchers were inconsistent and the database used for model regression was not only small but also different. To date, the plastic hinge has not been well understood and modeled. In fact, the key factors that affect the plastic hinge length have not been correctly identified (Zhao et al. 2012). The evaluation of the existing models by Wu et al. (2015) using a test database showed that the existing models give very different results and provide inadequate accuracy of prediction. Therefore, further research works are required.

7.3.4 Numerical and analytical studies on plastic hinge length

It is mentioned at the beginning of Section 7.3 that the plastic hinge problem is a 2D or 3D problem. Therefore, an appropriate 2D/3D analysis or model can be used to evaluate the plastic hinge length of an RC member. An early study on the plastic hinge problem by 2D finite element modeling was reported by Zhao et al. (2012). The same approach was subsequently used by Yuan et al. (2017) and Yuan and Wu (2017) to further study the problem. Details of the FEM models refer to the respective works. Their findings are summarized in this section.

It is shown in Fig. 7.13 that the equivalent plastic hinge length L_p is different from the physical plastic hinge length L_{ph}. The physical plastic hinge length is closely related to the plastic deformation zone in the RC column. Different plastic deformation zones such as rebar yielding zone and concrete crushing zone can be studied by FEM modeling. The numerical results for the cantilever RC columns shown in Fig. 7.13 with a constant axial load applied at the top of the column from the FEM modeling are reported and discussed in this section in combination with other relevant results.

One of the typical RC columns studied by numerical modeling is shown in Fig. 7.19. The columns have a constant axial force of $N = 0.38N_c$, where $N_c = f_{co}A_g$ in which A_g is the gross cross-sectional area of the column. The column sizes are 400 × 400 × 1700 mm that is connected to a 600 × 800 × 1400 mm stub. The effective cantilever length is 1600 mm from the center of the lateral force to the column base. The compressive strength of the unconfined concrete is $f_{co} = 30$ MPa. The longitudinal bars have a diameter of $\phi = 20$ mm and the links consist of diameter $\phi = 10$ mm bars with a spacing of 100 mm. The steel bars have a yield strength of $f_y = 460$ MPa, an elastic modulus of $E_s = 200$ GPa, and a hardening modulus of $E_h = 3$ GPa. Details of the numerical model refer to Yuan and Wu (2017).

7.3.4.1 Rebar yielding zone

The extent of an RC member where the strain of the tension bar exceeds the yield strain of the steel reinforcement is referred to as the rebar yielding zone in this section. The length of this zone is defined as L_{sy}. Fig. 7.20 shows the variation of L_{sy} of the column when its deflection increases. It can be seen from the figure that the extent

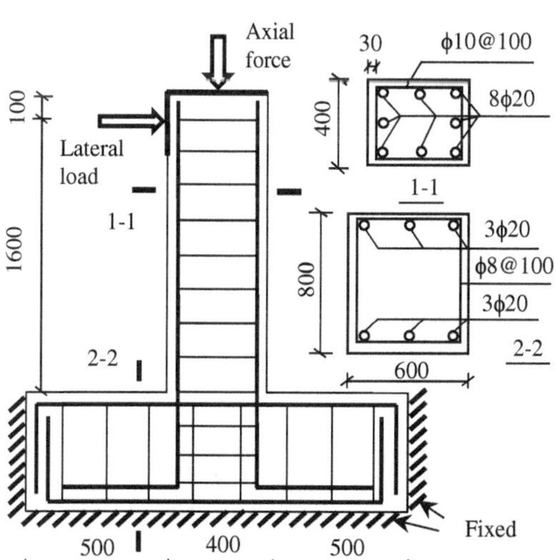

Fig. 7.19 RC reference column (unit: mm).
From Yuan, F., and Y. F. Wu. 2017. "Effect of load cycling on plastic hinge length of RC columns." *Eng. Struct.* 147, 90–102.

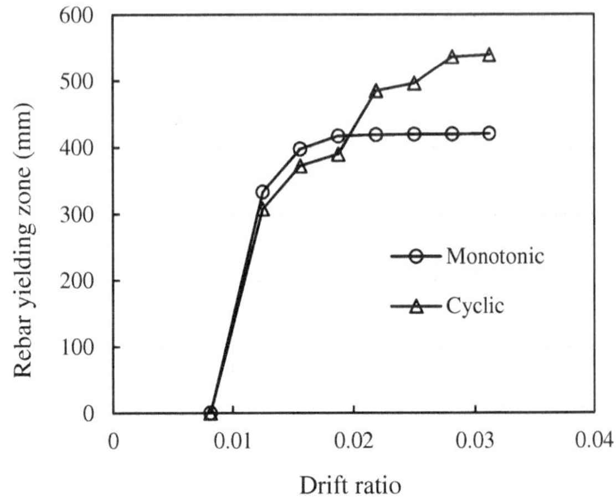

Fig. 7.20 Variations of rebar yielding zones.
From Yuan, F., and Y. F. Wu. 2017. "Effect of load cycling on plastic hinge length of RC columns." *Eng. Struct.* 147, 90–102.

of the rebar yielding zone changes when the column drift ratio, which is defined as the top displacement divided by column length, increases. However, the rebar yielding zone approaches an upper limit under both monotonic loading and cyclic loading. This finding is generally consistent with test results reported in the literature (Goodnight et al. 2016). It is noted from the figure that L_{sy} under cyclic loading is larger than that under monotonic loading.

7.3.4.2 Concrete softening zones

The lengths of the two compression regions of concrete are important to the plastic hinge analysis: (1) the length L_{cs} of the compression zone of the concrete where the compressive strain is larger than the peak strain of unconfined concrete ε_{co} that usually takes the value of 0.002, and (2) the length L_{cc} of the region where the compressive strain is greater than the crushing strain of unconfined concrete ε_{cu} that usually takes the value of 0.006. The numerical results in Fig. 7.21 show that L_{cs} stays almost constant after the drift ratio of 1.5% while L_{cc} keeps increasing with the increase in deflection. It can also be found from the figure that both L_{cs} and L_{cc} for cyclically loaded columns are significantly larger than those of monotonically loaded ones. These numerical results are consistent with the test observations that more concrete cover spalling and a larger extent of severe concrete crushing were experienced by cyclically loaded columns than by monotonically loaded ones (Jiang et al. 2014).

7.3.4.3 Curvature localization zone

Fig. 7.22 shows the curvature distributions of the columns at different drift ratios. The curvature of the columns distributes more evenly along the column length before yielding (the solid line). After yielding, the curvatures around the column base increase quickly and the plastic curvature localizes in a certain region. A dividing point is observed (the solid circle) from which, to the column base, the curvatures

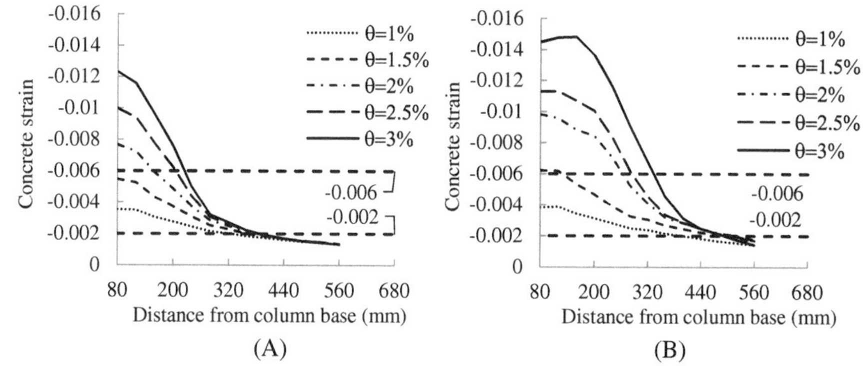

Fig. 7.21 Concrete compressive strain distribution on extreme compression surface: (A) monotonic loading (B) cyclic loading.
From Yuan, F., and Y. F. Wu. 2017. "Effect of load cycling on plastic hinge length of RC columns." *Eng. Struct.* 147, 90–102.

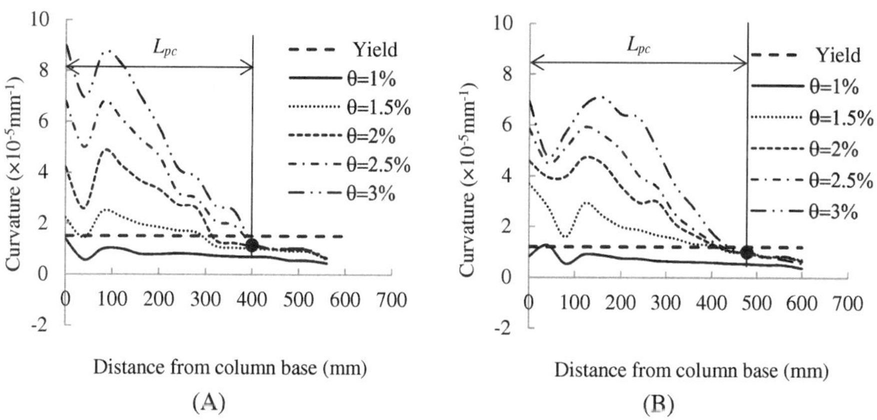

Fig. 7.22 Curvature distribution of columns: (A) monotonic loading (B) cyclic loading.
From Yuan, F., and Y. F. Wu. 2017. "Effect of load cycling on plastic hinge length of RC columns." *Eng. Struct.* 147, 90–102.

increase rapidly after column yielding, while those outside the region remain almost unchanged. The distance from the column base to the dividing point is defined as the length of significant curvature localization zone L_{pc}.

It can be seen from the figures that the dividing point of the column under cyclic loading is further away from the column base compared with that of the column under monotonic loading, which means a larger L_{pc} under cyclic loading than that under monotonic loading. In addition, the maximum curvature value decreases from $9.0 \times 10^{-5}\ \mathrm{mm}^{-1}$ for monotonic loading to $7.1 \times 10^{-5}\ \mathrm{mm}^{-1}$ for cyclic loading, indicating that the column deformation distributes more evenly under cyclic loading. As both tensile strain and compressive strain contribute to the curvature, the curvature concentration is caused not only by rebar yielding but also by concrete plastic deformation. Since both L_{sy} and L_{cs} of the cyclically loaded columns are larger than those of the monotonically loaded columns, L_{pc} under cyclic loading is consequently larger than that under monotonic loading.

7.3.4.4 Parametric studies

The effect of various factors or parameters on the plastic hinge lengths is discussed in this section. The key factors that are believed to have a significant effect on the plastic hinge length in the literature are numerically studied by Yuan and Wu (2017) and Yuan et al. (2017). The major findings are summarized and discussed in this section. When one factor is varied in a parametric study, other factors are kept the same as those of the reference column.

(1) Effect of loading schemes
The effect of load cycling on the plastic hinge lengths, L_{sy}, L_{cs}, and L_{pc}, of RC columns is studied by considering different load schemes as shown in Fig. 7.23A. The cyclic displacement is applied at the column top at various displacement increments and cycles up to a maximum displacement of 50 mm. The numerical results in Fig. 7.23B show that the number of load cycles N has a significant influence on the plastic hinge zones. It can be found that all the three plastic hinge lengths of the column under monotonic loading are evidently smaller than those of the column under cyclic loading. The three lengths increase rapidly when the load cycles increase from 2 to 4. However, they approach an asymptotical value and keep almost constant when N is greater than 5, which indicates that there is an upper bound to the plastic hinge lengths when N further increases.

(2) Effect of aspect ratio
To study the influence of the moment gradient that is reflected by the aspect ratio L/d, the parametric study was carried out by changing the column length L from 1200 to 3200 mm while fixing the column depth h to be 400 mm. Fig. 7.24A shows the influence of L/h on the three plastic hinge lengths, where L_{sy}, L_{cs}, and L_{pc} represent the three lengths for columns under cyclic loading while L'_{sy}, L'_{cs}, and L'_{pc} indicate those under monotonic loading. It is clearly seen from Fig. 7.24A that L_{sy} and L_{pc} increase almost linearly with L/h while L_{cs} shows a much smaller increasing rate. As the increase of L/h is equivalent to the increase in L, it is clear that the column length L is a key factor that controls the extent of plastic hinge regions, as discussed in Section 7.3.2. This effect is well reflected in existing models of plastic hinge length (see Table 7.2).

A much smaller increasing rate of L_{cs} can be explained by the experimental phenomenon that the postpeak damage of concrete columns under compression is always localized (Vonk 1991; Wu and Wei 2015; Wei and Wu 2016). It can also be found from Fig. 7.24A that L'_{sy}, L'_{cs}, and L'_{pc} are generally smaller than L_{sy}, L_{cs}, and L_{pc}, respectively, while their trends of variation are almost identical for the two different loading schemes.

(3) Effect of axial load level
The numerical results depicted in Fig. 7.24B show the effect of axial load level $n = N/N_c$ on the lengths of the plastic hinge zones. The axial load level varies from 0.1 to 0.5 in the parametric study. It can be found that, for both monotonic and cyclic loading, L_{sy} decreases while L_{cs} increases with the increase of axial load level. With increasing axial force, the compression zone becomes larger and the tensile zone is less stressed, leading to a larger L_{cs} and a smaller L_{sy}. This numerical result is consistent with the test observations that a

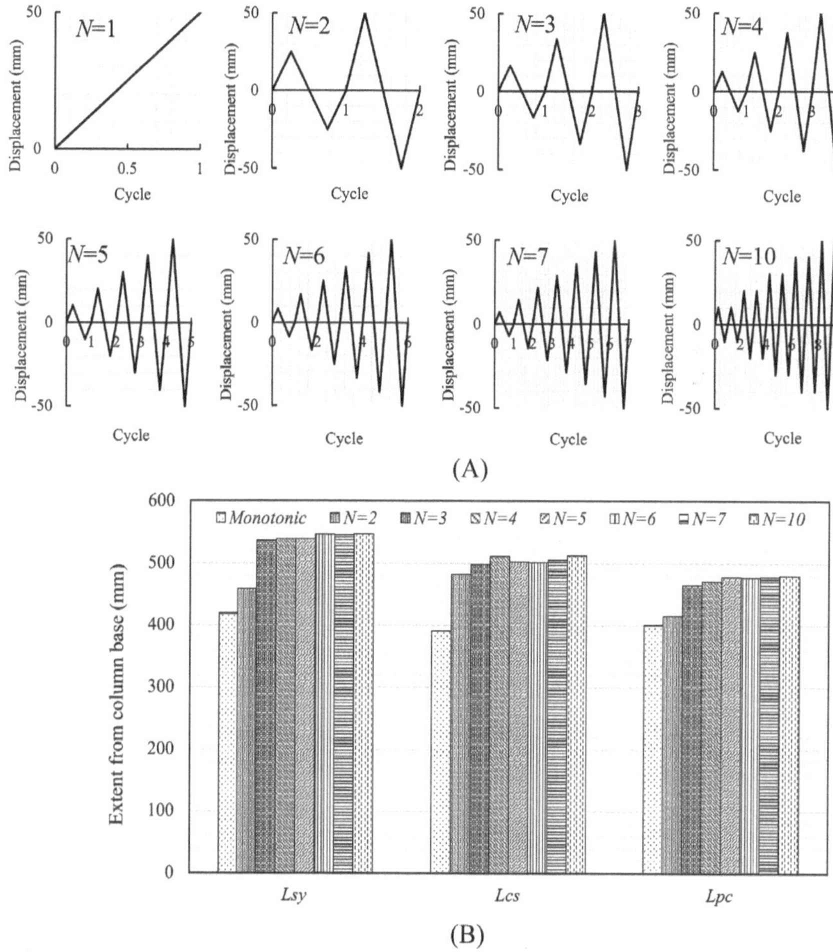

Fig. 7.23 Effect of load cycles on plastic hinge zones: (A) loading scheme and (B) length of plastic hinge zones.
From Yuan, F., and Y. F. Wu. 2017. "Effect of load cycling on plastic hinge length of RC columns." *Eng. Struct.* 147, 90–102.

higher axial load level resulted in a larger concrete damage zone (Bae and Bayrak 2008). L_{pc} has a similar trend to L_{sy} when the axial load level is lower than 0.3. When the axial load level further increases, the influence of L_{cs} on L_{pc} increases, but the increasing trend of L_{cs} and decreasing trend of L_{sy} cancel out each other, resulting in an almost constant trend of L_{pc}. These numerical results show that the axial load level is important to the plastic hinge lengths of RC columns. The trends of variation of the plastic hinge lengths for the two different loading schemes are similar.

An increase in the equivalent plastic hinge length with the increase of axial load level was proposed in the empirical models 16 and 18 in Table 7.2, which is consistent with the

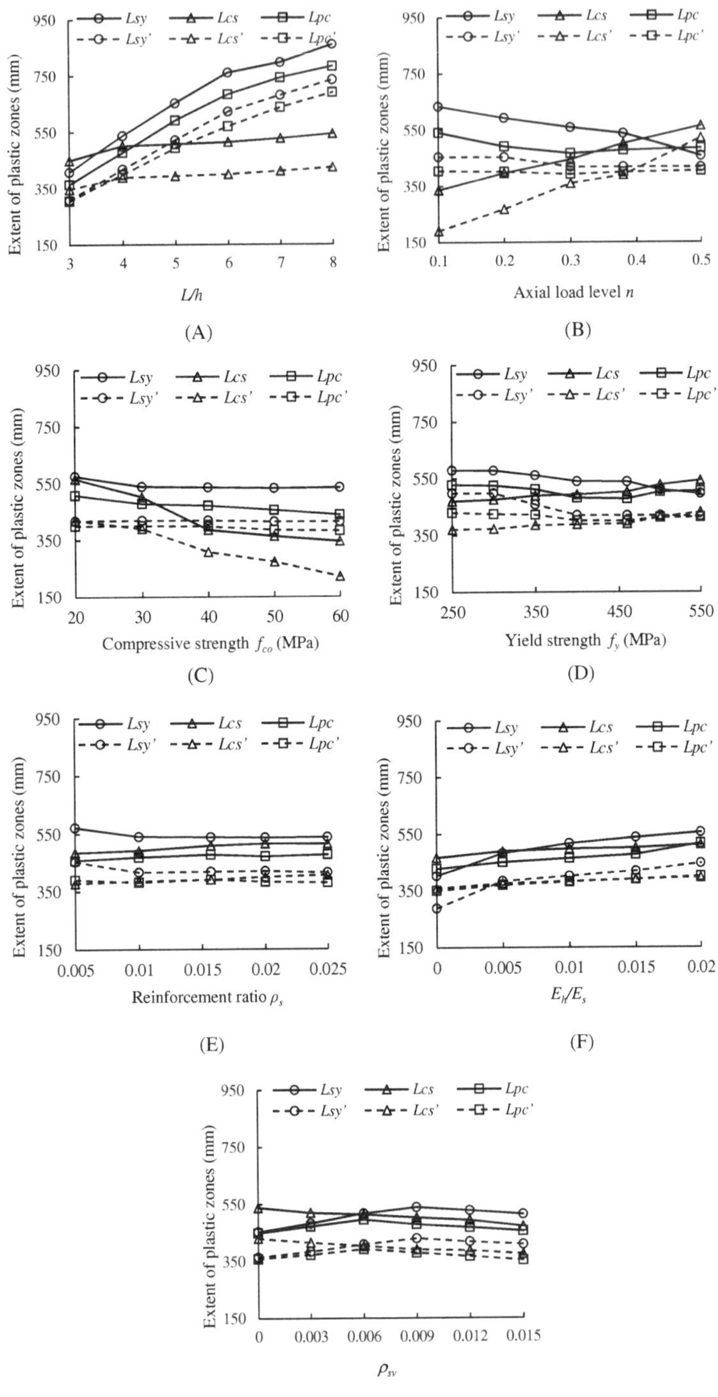

Fig. 7.24 Variation of plastic hinge zones: (A) effect of aspect ratio; (B) effect of axial load level; (C) effect of compressive strength of concrete; (D) effect of yield strength of longitudinal bars; (E) effect of reinforcement ratio; (F) effect of hardening modulus of steel; and (G) effect of volume ratio of stirrups.

From Yuan, F., and Y. F. Wu. 2017. "Effect of load cycling on plastic hinge length of RC columns." *Eng. Struct.* 147, 90–102.

numerical results of L_{cs} but not L_{sy} and L_{pc}. In fact, the empirical models 16 and 18 were based on the observation of the concrete crushing zone, and therefore are for L_{cs} rather than L_{sy} and L_{pc}.

(4) Effect of concrete grade
The compressive strength of unconfined concrete f_{co} is believed to influence the plastic hinge length of RC members and is included in a few empirical models (Table 7.2). However, results reported in the literature are inconsistent and no consensus has been reached on the influence of f_{co} on plastic hinge length. Ho (2003) showed that an increase in plastic hinge length occurs as f_{co} increases while Coleman and Spacone (2001) and Berry (2006) reported the opposite trend. Fig. 7.24C shows the variation of the plastic hinge regions as f_{co} changes from 20 to 60 MPa. It is seen that a decrease in L_{cs} and L_{cs}' is observed as f_{co} increases. The neutral axis moves toward the compression face as f_{co} increases, leading to a decrease in the area of the compression zone and L_{cs}. It is also seen that L_{sy} first decreases and then stays constant as f_{co} increases and L_{sy}' stays almost constant. Careful study of the numerical results shows the mechanism: an increase in f_{co} results in the increase of moment capacity of the RC column, which has a positive effect on L_{sy} or L_{sy}'. However, the bond stress between steel and concrete also increases with the increase of f_{co}, which has an adverse effect on L_{sy} or L_{sy}'. These two effects coexist and result in the variation of L_{sy} and L_{sy}' described earlier.

(5) Effect of yield strength of longitudinal reinforcement
The effect of yield strength of longitudinal reinforcement f_y on the equivalent plastic hinge length is included in the most widely used empirical model, proposed by Paulay and Priestley (1992). This parameter is used to account for yield penetration of the longitudinal bar into the column base. The numerical results in Fig. 7.24D show the variations of the plastic hinge lengths as f_y varies from 250 to 550 MPa. An ascending trend of L_{cs} and a descending trend of L_{sy} are observed as f_y increases. The tensile zone is strengthened as f_y increases, resulting in a larger compression zone and L_{cs} to balance the larger tensile resistance. On the other hand, concrete crushing happens relatively earlier for columns with a larger f_y, leading to a smaller part of the longitudinal bar yielding, and thus a smaller L_{sy}. L_{pc} first decreases and then increases with the increase of f_y, indicating that L_{pc} is determined by L_{sy} when f_y is small and is governed by L_{cs} when f_y is large. The trends of variation of the three plastic hinge lengths for columns under cyclic loading are identical to those under monotonic loading, but the values of L_{sy}, L_{cs}, and L_{pc} are larger than those of L_{sy}', L_{cs}', and L_{pc}', respectively. It is noted that the yield penetration into the column base is not reflected by the preceding three lengths of plastic zones.

(6) Effect of reinforcement ratio
Extensive studies have been carried out to examine the influence of reinforcement ratio ρ_s on plastic hinge length (Mattock 1965; Mendis 2001; Lopes and Bernardo 2003; Kheyroddin and Naderpour 2007; Ko et al. 2001). However, there has been no consensus on the influence of this factor until recently. Mattock (1965) reported that the plastic hinge length decreases with the increase of ρ_s while Mendis (2001) reported the opposite trend. Other researchers suggested that there was no direct relation between the plastic hinge length and ρ_s (Ko et al. 2001). Fig. 7.24E shows the numerical results of the effect of ρ_s on plastic hinge zones. It is seen that an increase in L_{cs} is observed as ρ_s increases. It is due to the fact that the tensile zone is strengthened as ρ_s increases and a larger area of compression zone is needed in order to balance the force, leading to a larger L_{cs}. L_{sy} first decreases and then stays almost constant with the increase of ρ_s. Due to the complex variations of L_{cs} and L_{sy}, no clear trend in L_{pc} is observed. The plastic hinge lengths for columns under monotonic loading are approximately 80% of those under cyclic loading.

(7) Effect of hardening modulus of steel
The strain hardening modulus E_h has not been considered in most of the previous empirical models. Few systematic studies on the influence of E_h on the extent of plastic hinge regions can be found in the extant literature. Zhao et al. (2012) studied this parameter by numerical simulations of RC beams and concluded that the plastic hinge lengths significantly increase as E_h increases, and suggested that E_h should not be ignored in the plastic hinge model. Priestley et al. (2007) replaced the term 0.08 in the model no. 13 (Table 7.2) with the parameter $k = 0.2(f_u/f_y - 1)$ to account for the influence of steel strain hardening on the spread of plasticity.
The effect of E_h on plastic hinge zones is analyzed by changing E_h from 0 to $0.02E_s$ in Fig. 7.24F, which shows that all the six lengths of plastic hinge zones increase with the increase of E_h. The mechanism is discussed in Section 7.3.2(1). A significant increase in plastic hinge lengths is observed in Fig. 7.24F when the loading scheme changes from monotonic loading to cyclic one.

(8) Effect of volumetric ratio of stirrups
It was reported that the external confinement has a great influence on the lengths of the plastic hinge regions of RC columns (Yuan et al. 2017). However, the volumetric ratio of stirrups ρ_{sv} is not included in the existing models of L_p for RC members. In Fig. 7.24G, the effect of ρ_{sv} on the plastic hinge zones is analyzed by changing ρ_{sv} from 0 to 1.5%. It is clearly seen that a decrease in L_{cs} is observed as ρ_{sv} increases. It is due to the fact that the compression zone is strengthened as ρ_{sv} increases and a smaller area of the compression zone is needed to balance the forces in the cross section, leading to a smaller L_{cs}. L_{sy} first increases and then decreases as ρ_{sv} increases. On one hand, the increase of ρ_{sv} leads to an increase in moment resistance that increases the rebar yielding zone due to the strength hardening effect (Section 7.3.2(1)). On the other hand, the frictional bond between longitudinal tensile reinforcement and concrete increases linearly with the increase in ρ_{sv} due to confinement, which opposes the spreading of rebar stress and has an adverse effect on the rebar strain and hence L_{sy}. These two effects coexist and lead to first an increase and then a decrease of L_{sy} with the increase in the volumetric ratio of stirrups. This effect is further discussed in Section 7.3.4.5. It is also found from Fig. 7.24G that the values of L_{sy}, L_{cs}, and L_{pc} are larger than those of L_{sy}', L_{cs}', and L_{pc}', respectively.

(9) Size effects
The effect of member size on plastic rotational capacity of RC members has been investigated previously by experimental study (Bigaj and Walraven 2002) and numerical modeling (Carpinteri et al. 2009). It was found that the rotational capacity of the plastic hinge is size dependent and is inversely proportional to member size. Yuan et al. (2017) studied the effect of cross-sectional depth on the plastic hinge zones of FRP-confined RC circular columns. The columns had one identical slenderness ratio ($L/D = 3.65$) but four different cross-sectional diameters of $D = 200$, 400, 600, and 800 mm. Other factors, including the reinforcement ratio, axial load level, and confinement ratio, are kept constant. The numerical results in Fig. 7.25A show that the member size has a significant influence on the lengths of the plastic hinge zones. L_{sy} shows a proportional increase with the increase in member size while L_{cs} and L_{pc} exhibit a slower increasing rate. It can also be found from Fig. 7.25B that the rotation capacity and ductility decrease as the column depth increases from 200 to 800 mm. The rotation capacity is defined as the rotation of the column top at the onset of FRP rupture in Fig. 7.25B, while that in Fig. 7.25C is defined as the rotation of the column top when the load drops to 80% of its peak load. The ductility is defined as the ratio of the rotation capacity to the yielding rotation. The numerical results indicate

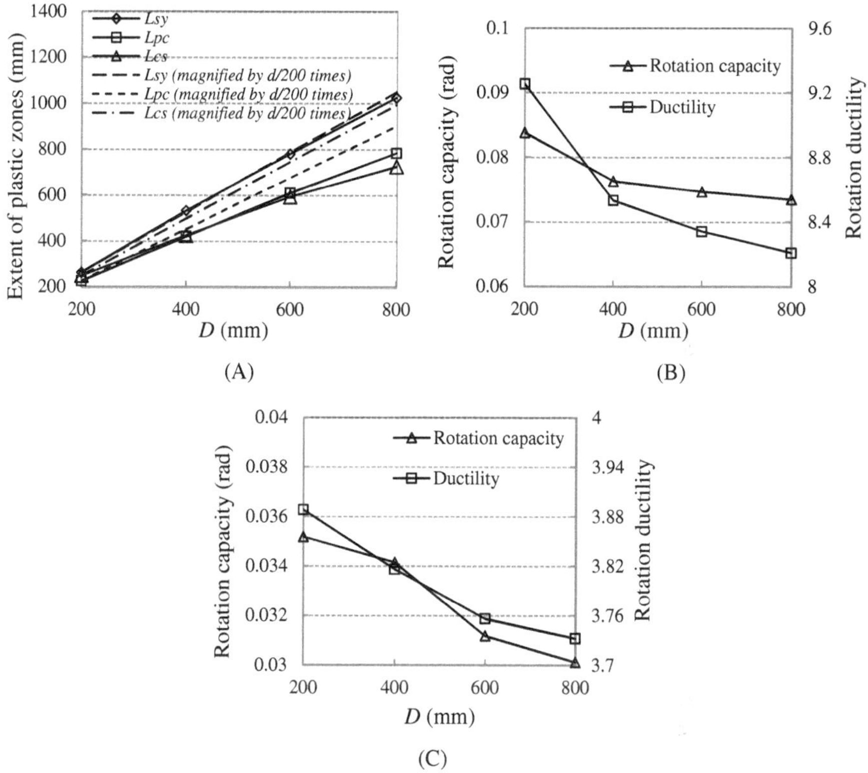

Fig. 7.25 Size effects on: (A) extent of plastic zones; (B) deformation capacity at FRP rupture; and (C) deformation capacity at 20% load drop from peak.
From Yuan, F., Y. F. Wu, and C. Q. Li. 2017. "Modelling plastic hinge of FRP-confined RC columns." *Eng. Struct.* 131, 651–668.

that members with smaller dimensions show a more ductile response than those with larger dimensions. This conclusion is consistent with the conclusion made from the analytical study in Section 3.4.1 (Eq. 3.25) and the experimental finding in Bigaj and Walraven (2002).

7.3.4.5 Effect of confinement and minimum jacket length

Due to the importance of confinement to RC columns, the effect of external confinement on the plastic hinge is studied in a separate section. When confined, concrete exhibits a higher strength and ductility. External jacketing can effectively provide confinement to the concrete in an RC column to increase not only its strength but also its ductility through increasing its ultimate curvature as well as the plastic hinge length. The drop of postpeak moment resistance can be delayed, and hence the ultimate curvature significantly increased, when confinement is provided to slow down the softening of concrete material. This can be clearly seen in Eq. (3.25), where the curvature ductility is increased when ε_{cu} and N_c are significantly increased by confinement. The effect of confinement on the stress-strain behavior of concrete is studied in detail in Chapter 8. This section discusses the effect of confinement on the plastic hinge length.

The effect of confinement on the plastic hinge length was not well understood before Gu et al. (2012), as the literature gave contradictory conclusions on the plastic hinge length of confined columns. Numerous researchers suggested that the plastic hinge length of an FRP-confined RC column is smaller than that of a normal RC column (Priestley et al. 1996; Elsanadedy and Haroun 2005). This conclusion was derived from the test observation of steel-jacketed RC columns in which the jacket restricted the spread of plastic yielding (Chai et al. 1991). However, some people believed that the plastic hinge length of FRP-confined columns is generally larger than that of normal RC columns based on their experimental results (Ozbakkaloglu and Saatcioglu 2006). In the meantime, there were others who proposed that the plastic hinge length of an FRP-confined column should be equal to that of a normal RC column for simplicity (Monti et al. 2001; Binici 2008). Gu et al. (2012) resolved the conflicting issue, as discussed in Section 7.3.4.6.

When the confinement due to transverse reinforcement to an RC column increases, the rebar yielding zone first increases and then reduces, as discussed in Section 7.3.4.4 (8). Yuan et al. (2017) studied the effect of confinement due to FRP jacketing on the plastic hinge length and the ultimate deflection of RC columns through FEM simulations. The results are shown in Fig. 7.26, which is consistent with the results in Section 7.3.4.4(8), where λ_f is the confinement ratio provided by the FRP jacket, defined as

$$\lambda_f = \frac{2f_f \cdot t}{b \cdot f_{co}} \tag{7.50}$$

where t and f_f are the nominal thickness and strength of the FRP jacket, respectively. For more discussions on the confinement ratio, see Chapter 8.

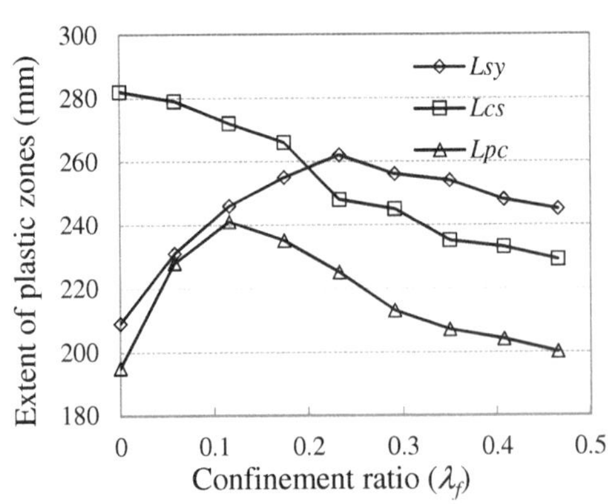

Fig. 7.26 Effect of confinement ratio on plastic hinge length. From Yuan, F., Y. F. Wu, and C. Q. Li. 2017. "Modelling plastic hinge of FRP-confined RC columns." *Eng. Struct.* 131, 651–668.

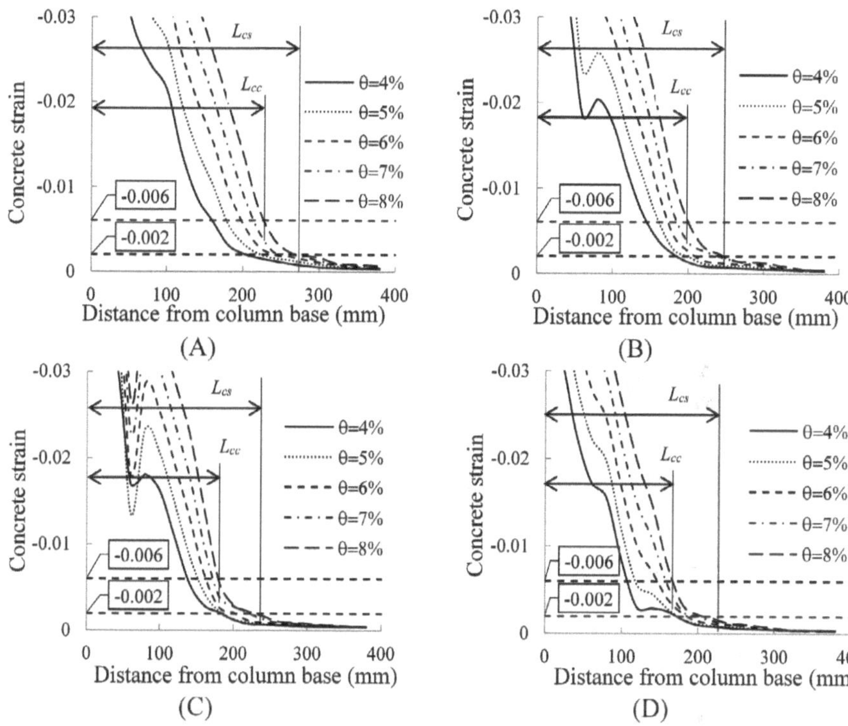

Fig. 7.27 Concrete compressive strain distribution: (A) $\lambda_f = 0.117$; (B) $\lambda_f = 0.233$; (C) $\lambda_f = 0.350$; and (D) $\lambda_f = 0.466$.
From Yuan, F., Y. F. Wu, and C. Q. Li. 2017. "Modelling plastic hinge of FRP-confined RC columns." *Eng. Struct.* 131, 651–668.

An important purpose of jacketing RC columns is to restrain concrete crushing at a large deflection. Fig. 7.27 shows the effect of confinement on the distribution of concrete strain on the column surface at the compression side. It can be seen from the figure that the concrete compressive strain increases during plastic hinge rotation but the area where concrete strain exceeds ε_{co} and ε_{cu} continues to decrease when the confinement increases. Fig. 7.28 shows the variation of the concrete compression zone of the columns with different confinement ratios. It can be seen that the concrete softening area decreases when confinement increases. In other words, a higher level of confinement results in a smaller area of damage zone.

The effects of confinement ratio on L_{sy}, L_{cs}, and L_{pc} are shown together in Fig. 7.26. L_{sy} exhibits an increasing and then decreasing trend when the confinement ratio increases, while L_{cs} shows a descending trend. L_{pc} is determined by L_{sy} when the confinement is low, resulting in the ascending trend of L_{pc}. Afterward, both L_{sy} and L_{cs} decrease with further increase in the confinement ratio, leading to a reducing trend of L_{pc}. When the confinement ratio increases to a certain large value, the FRP jacket essentially becomes a rigid hoop and hence L_{pc} should remain constant after that.

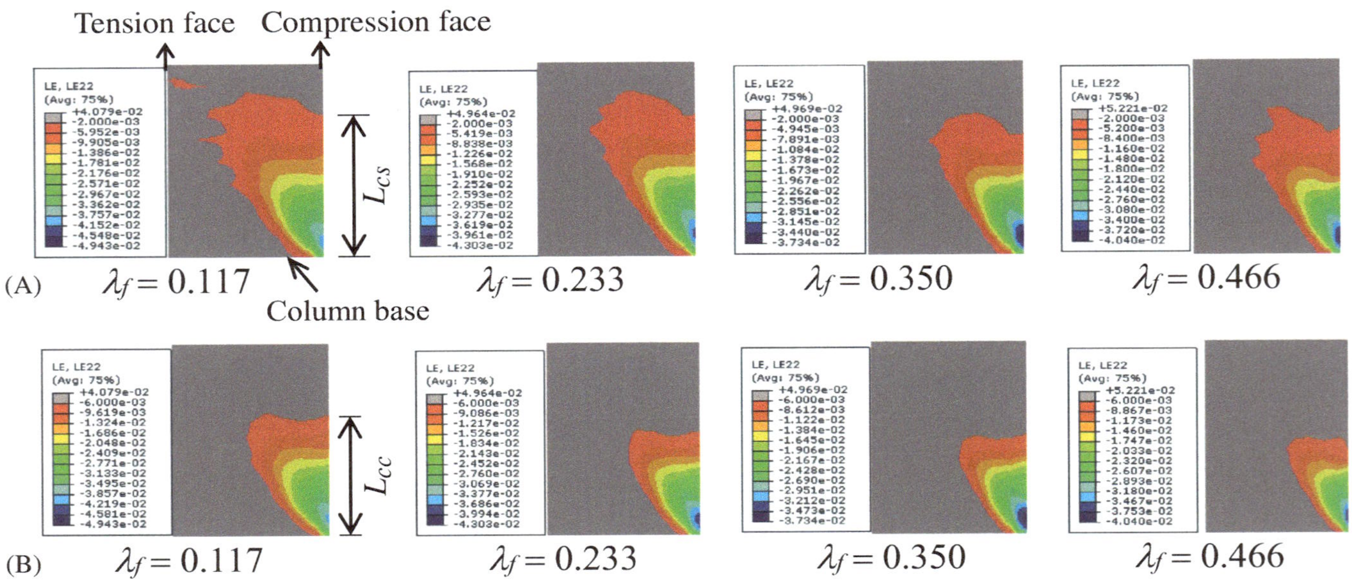

Fig. 7.28 Concrete compressive strain contour at 8% drift: (A) area with $\varepsilon_c > 0.002$ and (B) area with $\varepsilon_c > 0.006$.
From Yuan, F., Y. F. Wu, and C. Q. Li. 2017. "Modelling plastic hinge of FRP-confined RC columns." *Eng. Struct.* 131, 651–668.

For retrofitting RC columns by FRP jacketing, an important problem is to determine the minimum length of FRP jacket to save on construction work and costs. Since the function of an FRP jacket is to confine concrete dilation and prevent concrete crushing, the FRP jacket only needs to cover the concrete softening area in principle. Therefore, the minimum jacket length can be based on L_{cs} as shown in Fig. 7.28. FRP is applied on the whole column length for the results in Figs. 7.26–7.28. When the jacket length changes, the value of L_{cs} may change. Fig. 7.29 shows the effect of the length of the FRP jacket on L_{cs}. The length of the FRP jacket is calculated from the column base. The dashed line is used to determine the minimum length of FRP jacket, which is referred to as the critical L_{cs}. A point on the dashed line means that the FRP jacket length exactly equals L_{cs}. The points on the left side of the intersecting point between the dashed line and a curve mean insufficient FRP jacket length, while the points on the right side indicate an excessive jacket length. The effect of confinement ratio on the critical L_{cs} is shown in Fig. 7.29.

Based on the numerical parametric study, the influencing factors, including confinement ratio λ_f, axial load level n, and shape factor $2r/b$, are found to have a significant effect on the critical L_{cs}. By analyzing the variations of the critical L_{cs}, a power relationship between the critical L_{cs} and n and $2r/b$ is observed, while an exponential relationship is found between the critical L_{cs} and λ_f. Thus, the following model of the critical L_{cs} was proposed by Yuan et al. (2017) through regression analysis:

$$L_{cs} = \left[1.07e^{-0.6\lambda_f} n^{0.16} \left(\frac{2r}{b} + 0.2\right)^{0.1} + 0.6\right] d \tag{7.51}$$

The plastic hinge lengths including L_{cs} under cyclic loading are larger than those under monotonic loading (Fig. 7.24). The minimum jacketing length given by Eq. (7.51) should be increased by a factor of $5^{0.14}$ (see Eq. 7.61) for seismic retrofitting of

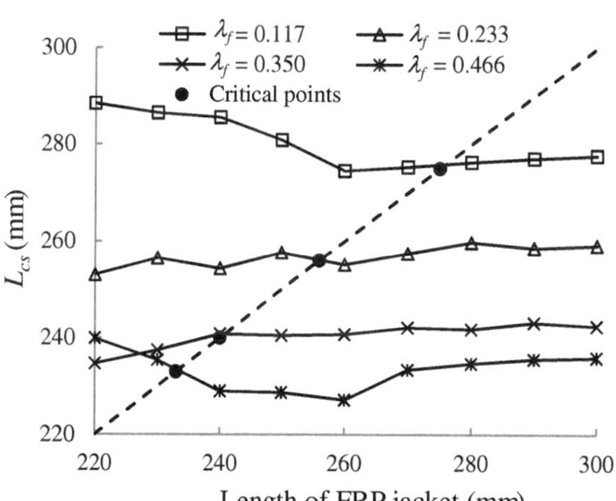

Fig. 7.29 Effect of FRP jacket length on L_{cs}.
From Yuan, F., Y. F. Wu, and C. Q. Li. 2017. "Modelling plastic hinge of FRP-confined RC columns." *Eng. Struct.* 131, 651–668.

RC column that would experience many load cycles under an earthquake. Therefore, the minimum jacket length for seismic retrofitting of RC columns is given by

$$L_{min} = 5^{0.14} L_{cs} = 1.25 \left[1.07 e^{-0.6\lambda_f} n^{0.16} \left(\frac{2r}{b} + 0.2 \right)^{0.1} + 0.6 \right] d \tag{7.52}$$

It is noted that this minimum jacketing length is only applicable for flexural strengthening. Jacketing design needs to consider other requirements sometimes, such as shear strengthening, where the shear strength of an RC column needs to be increased for a higher shear strength demand. For shear strengthening of RC columns, a full-length jacket is usually required.

7.3.4.6 Equivalent plastic hinge length

As shown in Fig. 7.13 and discussed in the previous sections, the physical plastic hinge lengths L_{sy}, L_{cs}, and L_{pc} for RC columns are different from the equivalent plastic hinge length L_p. It should be noted that a plastic hinge length in the literature usually refers to the equivalent plastic hinge length L_p that is used for calculation of ultimate displacement and member ductility. It is not the length of the physical plastic hinge zones L_{sy}, L_{cs}, L_{cc}, and L_{pc} studied in the previous sections. However, the length of the plastic curvature concentration zone L_{pc} is closely related to plastic hinge rotation. Therefore, L_p has a close relation with L_{pc}.

Numerous empirical models of L_p have been proposed for RC columns (Table 7.2). A well-known and widely used model of L_p for cantilever columns is in the form of

$$L_p = \alpha L + \beta f_y d_b \tag{7.53}$$

where α and β are model coefficients. The first term in Eq. (7.53) allows for the influence of bending moment gradient along the column length and the second term accounts for yield penetration of the longitudinal reinforcement into the column base (Paulay and Priestley 1992). Different values of α and β have been proposed by different researchers. Priestley and Park (1987) proposed $\alpha = 0.08$ and $\beta = 6/f_y$, while Paulay and Priestley (1992) revised α and β to 0.08 and 0.022, respectively. Panagiotakos and Fardis (2001) suggested the values of 0.18 and 0.021, respectively, and Berry (2006) proposed 0.05 and $0.1/\sqrt{f_{co}}$, respectively.

Yuan and Wu (2017) calculated L_p for RC columns under various loading schemes from their FEM simulation results, through the following procedure. Eq. (7.45) can be rewritten as

$$L_p = L - \sqrt{L^2 - 2\frac{y_u - y_e}{\kappa_u - \kappa_e}} \tag{7.54}$$

Since y_u, y_e, κ_u, and κ_e can be obtained directly from the FEM results, the equivalent plastic hinge length L_p can be calculated from the results of numerical simulations using Eq. (7.54). The values of y_u and κ_u are evaluated at the ultimate displacement, defined at the point where the lateral resistance of the column drops to 80% of its peak value. The elastic curvature is given by $\kappa_e = \kappa_y \frac{M_u}{M_y}$, where M_y and κ_y are the moment and curvature at first yield of the longitudinal reinforcement, respectively, and M_u is the bending moment of the base section at the ultimate displacement. The displacement corresponding to κ_e is $y_e = \frac{\kappa_e L^2}{3}$ (Eq. 7.45). It is noted that Eqs. (7.41), (7.42) are typically used in the literature, in which case $\kappa_e = \kappa_y$. This assumption is usually fine unless the ultimate moment is significantly different from the yielding moment, as discussed in Section 7.3.1.

Fig. 7.30 shows the results of L_p calculated from Eq. (7.54) with the assumption of $\kappa_e = \kappa_y$, in comparison with those from the four empirical models. It is noted that "MR" in the figure indicates the mean value of the ratio of L_p predicted by the empirical model to the simulated L_p by FEM, and "SD" represents the standard deviation of the ratios. It can be found from the figure that the L_p values of cyclically loaded columns are significantly larger than those of monotonically loaded columns. This is consistent with the variation of the three physical plastic hinge lengths discussed previously. The simulation results of L_p are relatively close to the predictions by the empirical model of Paulay and Priestley (1992) for monotonic loading. However, the errors are large under cyclic loading for all the empirical models. Therefore, the existing empirical models are not suitable for cyclically loaded columns.

The difference in the results from the two approaches of (1) assuming $\kappa_e = \kappa_y$ and (2) using $\kappa_e = \kappa_y \frac{M_u}{M_y}$ is insignificant. The ratios of L_p calculated by the second approach to that by the first approach have a mean value of 0.99 and an SD of 3.5%, indicating a minor difference in L_p calculated from the two approaches. The results of L_p calculated by the second approach are shown in Fig. 7.31 in comparison with those predicted by the four empirical models.

As shown in Figs. 7.30 and 7.31, the errors of the existing plastic hinge models are significant. Through experimental, numerical, and analytical studies of test results, Gu et al. (2012) proposed an equivalent plastic hinge model (Eq. 7.55a) for circular RC columns, which includes the confinement effect. The model was extended to include square columns by Jiang et al. (2014) (Eq. 7.55b):

$$L_p = L_{p0} + L_{pf} \tag{7.55a}$$

$$L_p = L_{p0} + \left(\frac{2r}{b}\right)^{0.72} \cdot L_{pf} \tag{7.55b}$$

where L_{p0} is the plastic hinge length without confinement. The model proposed by Paulay and Priestley (1992) was adopted for L_{p0} in Eq. (7.55), i.e.,

$$L_{p0} = 0.08L + 0.022 f_y d_b = 0.08L + L_{pb} \tag{7.56}$$

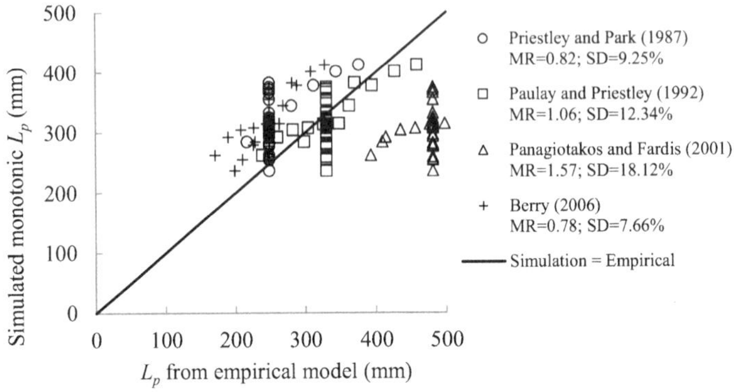

(A)

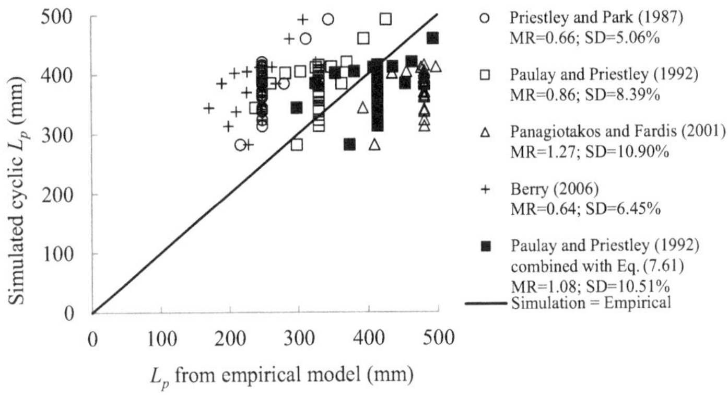

(B)

Fig. 7.30 Performance of the empirical equivalent plastic hinge models: (A) monotonic loading and (B) cyclic loading.
From Yuan, F., and Y. F. Wu. 2017. "Effect of load cycling on plastic hinge length of RC columns." *Eng. Struct.* 147, 90–102.

where $L_{pb} = 0.022 f_y d_b$, the additional plastic hinge length caused by yield penetration into the column base. The second term in Eq. (7.55) allows for the effect of confinement on the plastic hinge length, where L_{pf} is the additional plastic hinge length due to confinement for circular columns:

$$L_{pf} = \begin{cases} 3.028\lambda_f L \text{ when } 0 \le \lambda_f < 0.1 \\ (0.51 - 2.30\lambda_f + 2.28\lambda_f^{\,2})L \text{ when } 0.1 \le \lambda_f < 0.5 \end{cases} \tag{7.57}$$

The term $\frac{2r}{b}$ in Eq. (7.55b) accounts for the reduction of confinement effectiveness caused by the square shape of the column cross section, where b and r are the breadth and corner radius of the column cross section, respectively. For detailed discussions on the confinement effectiveness, see Chapter 8.

The confinement effect on the plastic hinge length is reflected by the term $(L_p - L_{p0})/L$, as shown in Fig. 7.32. The ascending part for $0 < \lambda_f < 0.1$ is caused

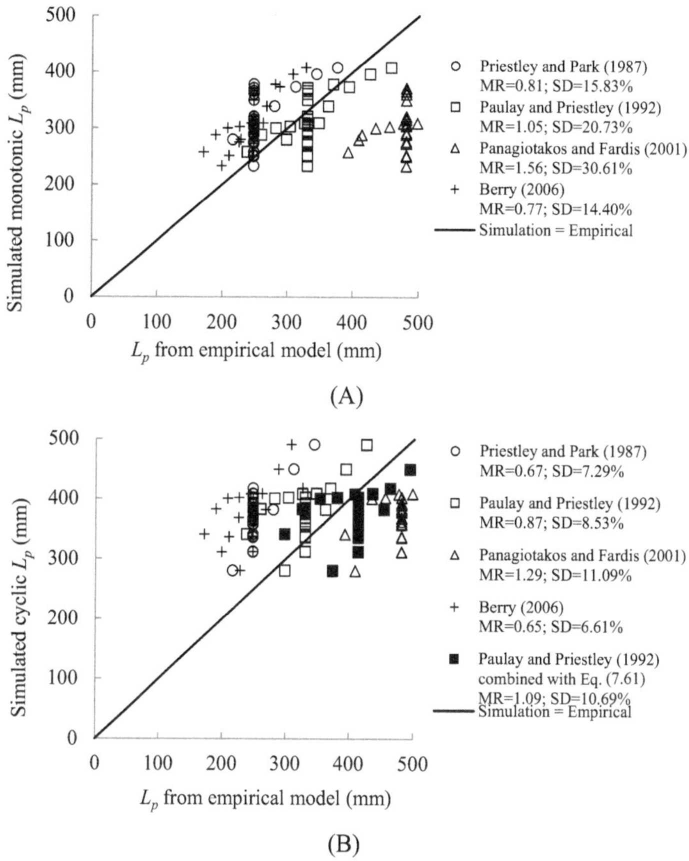

Fig. 7.31 Equivalent plastic hinge length by approach 2 compared with the empirical models: (A) monotonic loading and (B) cyclic loading.
From Yuan, F., and Y. F. Wu. 2017. "Effect of load cycling on plastic hinge length of RC columns." *Eng. Struct.* 147, 90–102.

by the first mechanism: the strength hardening effect, as discussed in Section 7.3.2(1). Confinement significantly increases M_p while M_y is not much affected, thereby increasing the plastic hinge length. The increase of M_p is related to the confinement ratio λ_f. As a result, the plastic hinge length increases with an increase in the confinement ratio.

In the meantime, there is a second mechanism that simultaneously affects the plastic hinge length: the strain reduction of the longitudinal bars due to confinement at the plastic hinge zone. The confinement from an external jacket applies normal pressure to the rebar, which produces an additional frictional bond that is in direct proportion to the confinement pressure. The additional bond causes a larger difference in the longitudinal steel stress between the column base and a section above it (see Fig. 5.14; a large F_b causes a larger difference between T_1 and T_2). Even a small increase in the bond stress will produce a significant reduction in the strain of the longitudinal bar at a

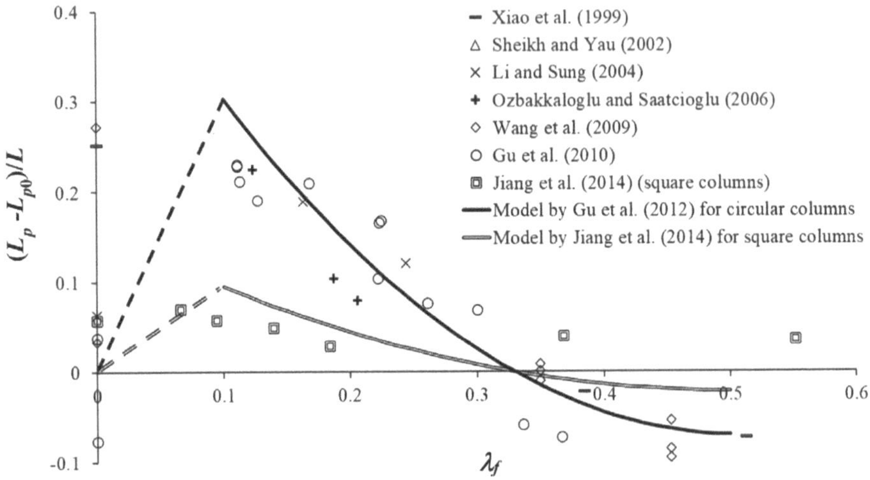

Fig. 7.32 Effect of confinement on plastic hinge length.

cross section above the base section after the longitudinal bar yields, because a large change of strain is required to produce a small change of steel stress after rebar yielding (see the stress-strain curve in Fig. 1.8A). A larger confinement causes a larger difference in the strain of the longitudinal bar between the base and an upper section, and consequently a smaller plastic strain region, leading to a smaller plastic hinge length.

In the ascending part of Fig. 7.32, the first mechanism dominates and the strain variation of the longitudinal bars due to the second mechanism is small due to the small confinement. When the confinement ratio increases beyond $\lambda_f = 0.1$, the increase in moment capacity due to the first mechanism slows down because further increase in M_p at higher confinement is small, and hence the increase in the plastic hinge length due to the increase in moment capacity is relatively smaller than the reducing effect caused by the second mechanism. Thus, the second mechanism dominates at higher confinement, which causes a reducing trend of the plastic hinge length. In general, confinement is positive in increasing the plastic hinge length when $0 < \lambda_f < 0.33$, but it reduces the plastic hinge length when $\lambda_f > 0.33$ as shown in Fig. 7.32.

The foregoing analysis is consistent with the FEM results in Fig. 7.26, and also the experimental observation of the effect of confinement on the curvature distribution near the column base for steel-jacketed columns by Chai et al. (1991) that steel jacketing can restrict the spreading of plastic yielding. It also explains why the curvature of confined columns with a lap splice of reinforcement bars in the plastic hinge region is concentrated in a smaller length than that of columns with continuous bars (Saadatmanesh et al. 1996): apart from reducing the rebar stress due to the additional rebar area in the overlapped zone, the existence of more reinforcement bars in the plastic hinge region enhances the bond stress between the concrete and the longitudinal bars.

By assuming a linear distribution of the physical plastic curvature (assume a triangular red-shaded area in Fig. 7.13C) and according to the principle of area equivalence, a relationship between the physical plastic hinge length L_{ph} and the equivalent plastic hinge length L_p was proposed by Hines et al. (2004):

$$L_p = \frac{L_{ph}}{2} + L_{pb} \tag{7.58}$$

where L_{pb} is the additional plastic hinge length due to yield penetration of longitudinal bars into the column base, which can be evaluated by $L_{pb} = 0.022 f_y d_b$ (Paulay and Priestley 1992). The relationship of Eq. (7.58) was evaluated by Yuan and Wu (2017) using FEM simulation results. It was found that the calculated results of L_p by Eq. (7.58) are larger than those calculated from Eq. (7.54). The ratios of the L_p values evaluated from Eq. (7.58) to those from Eq. (7.54) have a mean of 1.23 and an SD of 14.0%. This error could be attributed to the fact that the plastic curvature is closer to exponential distribution in the plastic hinge region (Ho 2003; Ang et al. 1981) rather than a linear one. If an exponential distribution for the plastic curvature distribution is assumed, the distance from the centroid of the plastic curvature diagram to the column base will be smaller than $L_{ph}/2$. In this case, the calculated L_p will be smaller than that from Eq. (7.58) and closer to that calculated from Eq. (7.54).

Through extensive FEM simulations of the plastic hinge of RC columns and using Eq. (7.54), Yuan et al. (2017) proposed the following model:

$$\begin{aligned} L_p &= L_{p0} + L_{pf} \\ &= (0.08L + 0.022 f_y d_b) + 0.13\left(\frac{2r}{b} + 0.2\right)^{0.1}\left(e^{-1.5\lambda_f} - e^{-40\lambda_f}\right)L \end{aligned} \tag{7.59}$$

A model of the ultimate curvature of RC columns at rupture of the FRP jacket was also proposed:

$$\kappa_u = \left(0.55 + 0.45\left(\frac{2r}{b}\right)\right)\frac{0.012 + 0.05\lambda_f}{(0.14 + n(0.65 - 0.67\lambda_f))d} \tag{7.60}$$

Using Eqs. (7.45), (7.59), and (7.60), the ultimate deflections of columns tested by others were calculated and compared with those predicted by Gu et al. (2012) and Youssf et al. (2015), as shown in Fig. 7.33. Improved accuracy of predictions is shown in the figure by using Eqs. (7.59), (7.60). In general, there is still room for further improvement in the accuracy of the model.

7.3.4.7 Plastic hinge under cyclic loading

By analyzing the variation of the equivalent plastic hinge length under an increasing number of load cycles, Yuan and Wu (2017) observed that the equivalent plastic hinge length at a given number of load cycles, L_{p-c}, increases when load cycle number increases but approaches an upper bound. Regression analysis shows that the trend of increase follows a power relationship with loading cycle number N when N varies from 1 to 5, beyond which an asymptotic upper limit is reached. This relationship can be represented by the following equation:

$$L_{p-c} = [\min(5; N)]^{0.14} L_{p-m} \tag{7.61}$$

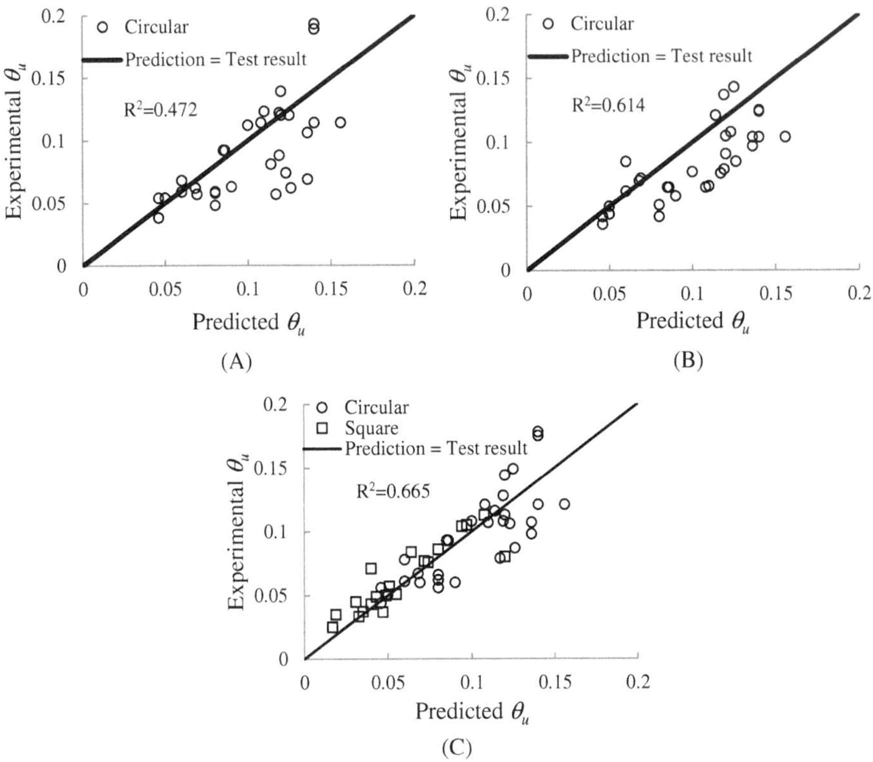

Fig. 7.33 Performance of models: (A) Gu et al. (2012); (B) Youssf et al. (2015); and (C) Eqs. (7.45), (7.59), (7.60).
From Yuan, F., Y. F. Wu, and C. Q. Li. 2017. "Modelling plastic hinge of FRP-confined RC columns." *Eng. Struct.* 131, 651–668.

where L_{p-m} is the equivalent plastic hinge length under monotonic loading. An existing empirical model can be used for L_{p-m} in Eq. (7.61). When Eq. (7.61) is combined with the empirical model by Paulay and Priestley (1992) (Model 13 in Table 7.2), the ratios of the predicted values of L_{p-c} to the FEM simulated ones have a mean value of 1.08, compared with 0.86 by the model from Paulay and Priestley (1992), 0.66 by the model from Priestley and Park (1987), 1.27 by the model from Panagiotakos and Fardis (2001), and 0.64 by the model from Berry (2006).

It was proposed by Panagiotakos and Fardis (2001) that L_p for RC columns under cyclic loading is 50% smaller than that under monotonic loading (Model 15 in Table 7.2). This conclusion is inconsistent with Eq. (7.61) where L_p under cyclic loading is approximately 25% larger than that under monotonic loading when N is larger than 5. The derivation of the equivalent plastic hinge model by Panagiotakos and Fardis (2001) was based on the assumption that load cycling has no influence on the ultimate curvature of RC columns, which caused the problem. L_p is not only related to ultimate displacement but also closely related to ultimate curvature at the column base, as reflected by Eq. (7.54). In fact, load cycling results in a reduction in the ultimate curvature, as explained shortly.

It is found from the numerical results discussed in the previous sections that larger physical plastic zones are produced under cyclic loading than under monotonic loading. This finding is supported by the experimental observations that more concrete cover spalling and a larger extent of severe concrete crushing were experienced by cyclically loaded columns than by monotonically loaded ones (Wu et al. 2003; Fardis 2009). It has been experimentally observed that the ultimate displacement of RC columns under monotonic loading corresponding to the point when the lateral resistance drops to 80% of its peak load was always slightly larger than that under cyclic loading (Park and Paulay 1975; Wu et al. 2003; Fardis 2009; Gu et al. 2010). If the maximum curvature of the cyclically loaded column is larger than that of the monotonically loaded one ($\kappa_{u-c} > \kappa_{u-m}$), the ultimate displacement of the column under cyclic loading will be larger than that of the column under monotonic loading based on Eq. (7.54), i.e., $y_{u-c} > y_{u-m}$. As a result, κ_{u-c} must be smaller than κ_{u-m} in order to satisfy $y_{u-c} < y_{u-m}$. This means that the actual plastic curvature distributes more uniformly along the column length for cyclically loaded columns, as shown in Fig. 7.34. This conclusion is consistent with the observation by Grammatikou et al. (2016) from test results that the ultimate strain of tension reinforcement under cyclic loading is smaller than that under monotonic loading, while the ultimate strain of confined concrete is similar, which results in a smaller ultimate curvature under cyclic loading. The decrease of ultimate curvature due to load cycling leads to a larger L_p, which is well reflected in the model of Eq. (7.61).

If the ultimate curvatures of monotonically loaded columns are used to derive L_p in Eq. (7.54), the average L_{p-c} is smaller than the average L_{p-m}, which will be consistent with Panagiotakos and Fardis (2001). This observation shows that the assumption made by Panagiotakos and Fardis (2001) that load cycling has no influence on the ultimate curvature of RC columns causes a significant error in evaluating the equivalent plastic hinge length.

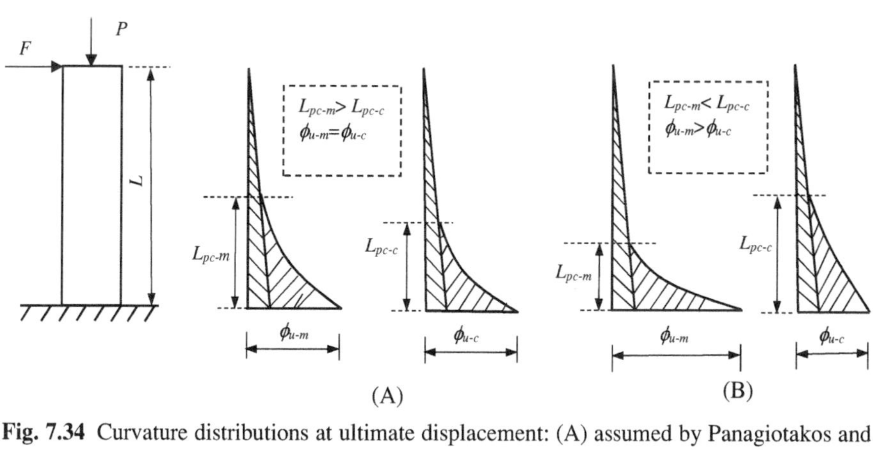

Fig. 7.34 Curvature distributions at ultimate displacement: (A) assumed by Panagiotakos and Fardis (2001) and (B) actual.
From Yuan, F., and Y. F. Wu. 2017. "Effect of load cycling on plastic hinge length of RC columns." *Eng. Struct.* 147, 90–102.

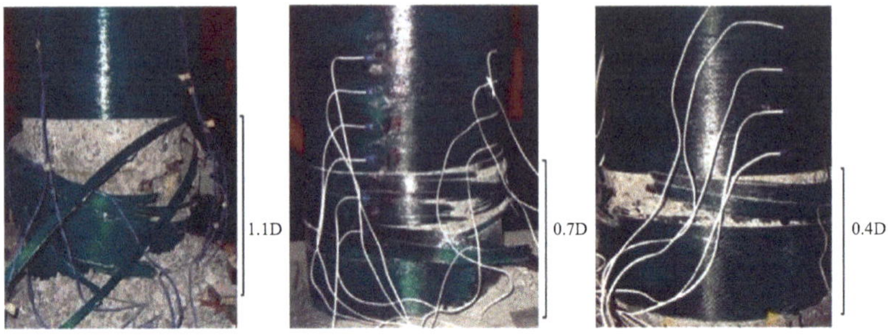

(A) Column CH3 (B) Column CL2 (C) Column CL3

Fig. 7.35 Decreasing damage zone of columns at increasing confinement.
From Gu, D. S., Y. F. Wu, G. Wu, and Z. S. Wu. 2012. "Plastic hinge analysis of FRP confined circular concrete columns." *Constr. Build. Mater.* 27, 223–233.

7.3.4.8 Effect of FRP-to-concrete interfacial bond on plastic hinge

Fig. 7.35 shows the reducing trend of the damage zone of FRP-confined columns when confinement increases (Gu et al. 2012). This shows a reduction of the plastic hinge length when the confinement provided by FRP jacket increases. Although confinement can significantly increase the ultimate curvature of a column, the positive effect of confinement in increasing the ductility of RC columns decreases due to the significant reduction in plastic hinge length. In fact, it was experimentally observed and theoretically predicted that confining RC columns by jacketing may lead to a reduction in the deformation capacity and ductility of RC columns, as shown in Fig. 7.36 where θ_u is the ultimate drift ratio of column (Gu et al. 2012). This is undesirable and against the purpose of jacketing columns.

It is discussed in Section 7.3.4.6 that the negative effect on the plastic hinge length is caused by the second mechanism, while the first mechanism is positive to both the plastic hinge length and the ultimate curvature. Therefore, the column ductility can be

Fig. 7.36 Relationship between confinement ratio and ultimate drift ratio.
From Gu, D. S., Y. F. Wu, G. Wu, and Z. S. Wu. 2012. "Plastic hinge analysis of FRP confined circular concrete columns." *Constr. Build. Mater.* 27, 223–233.

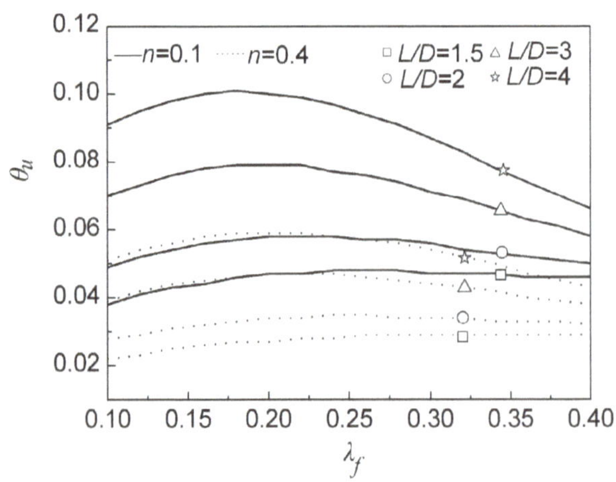

increased if the second mechanism is inhibited or reduced while the first mechanism is retained. Based on this idea, Jiang et al. (2019) proposed a new jacketing scheme that provides effective confinement to the concrete in the compression zone while restraining its confinement to the tension rebars. The technology is shown in Fig. 7.37.

The mechanism of the jacketing scheme is as follows: normal jackets confine an RC column in a passive way, i.e., the compressive stress causes dilation of the concrete, which pushes against the jacket, causing the hoop resistance and passive pressure from the jacket. For columns under bending, the FRP in the tension zone is pulled toward the compression side by the dilation of concrete in the compression zone. Although an adhesive bond is usually provided at the FRP-to-concrete interface, the bond is usually insufficient to prevent the interfacial slip of the FRP laminate, particularly when bonding is poor or under extensive concrete dilation. Different degrees of interfacial slip cause different redistributions of the confining pressure. The redistribution of the confinement pressure has two effects: (1) relaxation of the confining resistance on the compression side, which reduces the first mechanism and (2) increase of the confining pressure on the tension side, which increases the second mechanism. By fixing the jacket to the column side (Fig. 7.37), the first mechanism can be enhanced while the second mechanism is restrained.

To study the effect of the interfacial slip between FRP jacket and column surface on the plastic hinge length and ultimate deformation capacity of the columns, Jiang et al. (2019) investigated the four cases as shown in Fig. 7.38, through experimental testing and numerical simulations. Specimen S1 was the original RC column without FRP jacketing. S2 was retrofitted by normal FRP jacketing where the FRP was adhesively bonded to the column surface. S3 was also jacketed with the same FRP but the contact face between the FRP jacket and column surface was lubricated to remove the bond. S4 was identical to S2 except that anchors were provided on the column side to prevent slipping of the FRP jacket from the tension side to the compression side. The test results showed that the curvature localization zone L_{pc} of the four cases are: 220 mm for S1, 360 mm for S2, 315 mm for S3, and 380 mm for S4. The lengths of rebar yielding zone are consistent with the curvature localization zone. These results demonstrated that the physical plastic hinge lengths of the FRP-confined RC columns are larger than that of the unconfined column (positive value in Fig. 7.32). The interfacial debonding (Specimen S3) decreases L_{pc}, whereas the mechanical anchorage (Specimen S4) increases L_{pc}. This finding matches the expected effects discussed earlier.

The test results were further verified by the numerical simulation results as shown in Fig. 7.39. The numerical results validate the discussion earlier that confinement first increases the plastic zone due to the first mechanism but further increase in confinement will reduce the yielding zone due to the second mechanism. In the meantime, the interface debonded columns have the smallest rebar yielding zone among the three types of interfaces due to the free slip of the jacket toward the compression side, which induces a higher confinement pressure on the tension bars and hence an enhanced second mechanism. The FRP-anchored columns produce the largest rebar yielding zone as expected due to the restraint to the second mechanism. The curvature localization zone L_{pc} for the columns with adhesively bonded FRP is larger than that with

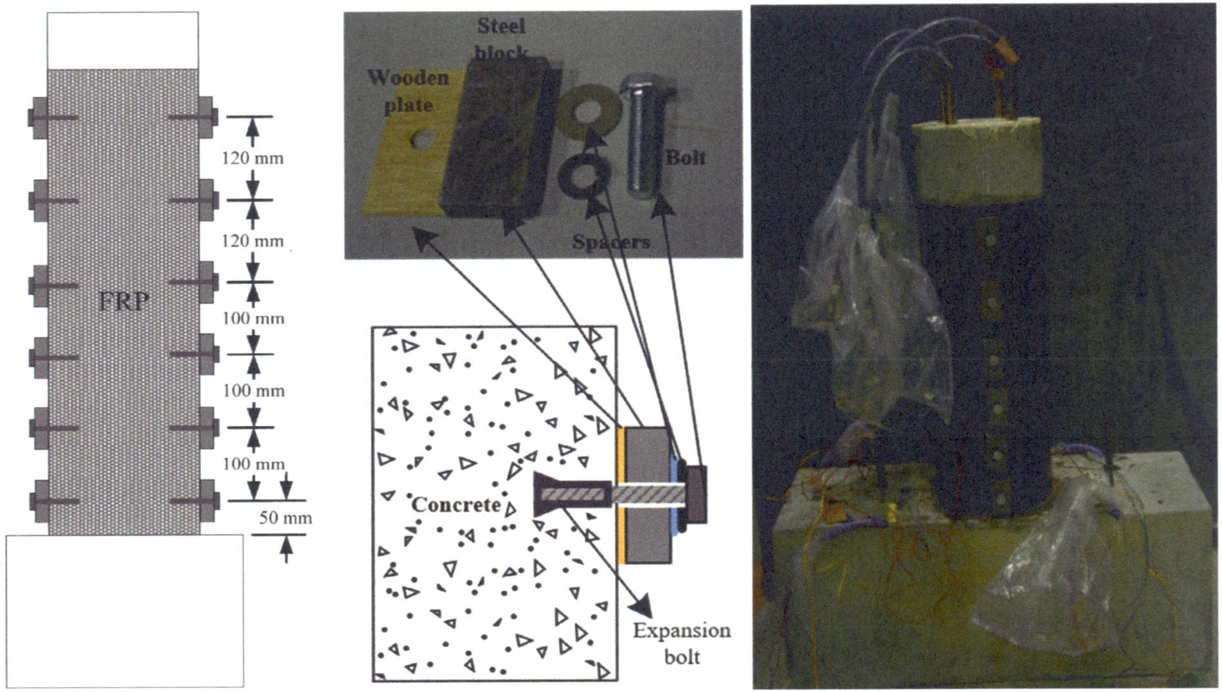

(A) Column side view (B) Details of anchor (C) Test specimen

Fig. 7.37 Anchoring of FRP jacket to reduce frictional bond.
From Jiang, C., F. Yuan, Y. F. Wu, and X. M. Zhao. 2019. "Effect of interfacial bond on plastic hinge length of FRP-confined RC columns." *J. Compos. Constr.* 23 (2), 04019007.

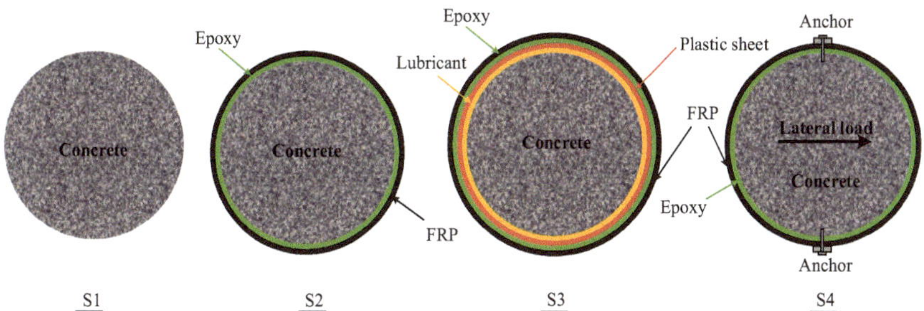

Fig. 7.38 FRP-wrapping schemes.
From Jiang, C., F. Yuan, Y. F. Wu, and X. M. Zhao. 2019. "Effect of interfacial bond on plastic hinge length of FRP-confined RC columns." *J. Compos. Constr.* 23 (2), 04019007.

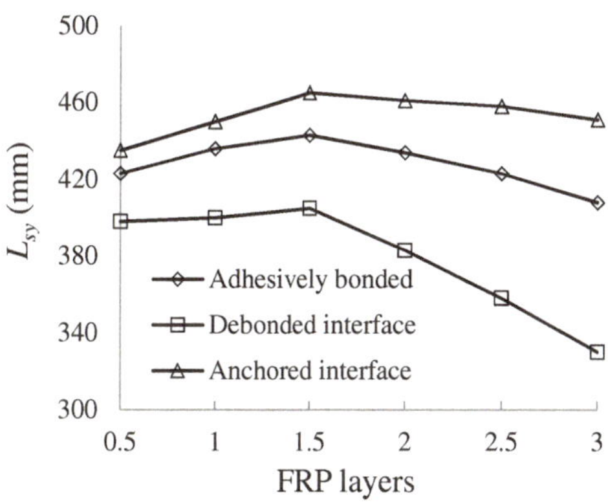

Fig. 7.39 Variation of rebar yielding zone.
From Jiang, C., F. Yuan, Y. F. Wu, and X. M. Zhao. 2019. "Effect of interfacial bond on plastic hinge length of FRP-confined RC columns." *J. Compos. Constr.* 23 (2), 04019007.

debonded interface but smaller than that with anchored FRP. The variation in L_{pc} is identical to the variation in L_{sy}. Therefore, for the same ultimate curvature, the column with the FRP jacket anchored at its side will have a larger ultimate deflection and ductility. This conclusion is validated by the numerical simulation results shown in Fig. 7.40, where the ultimate deflections of the columns at FRP rupture are shown.

In conclusion, the measure of fixing the FRP jacket to reduce the confinement to tension reinforcement can effectively increase the deformation capacity and ductility of FRP-confined RC columns.

7.3.5 The P-Δ effect

The load-deflection curve can be produced by the methods introduced in the previous sections. It can also be produced by using an analytical or numerical approach. However, the produced load-deflection curve usually does not include the P-Δ effect, which may significantly overestimate the load resistance. This is shown by the numerically simulated load-deflection curve in Fig. 7.41 for the RC cantilever column in Fig. 5.2 with a 6 mm steel plate and $K_b = 23$ kN/mm (Wu 2002). It can be seen in

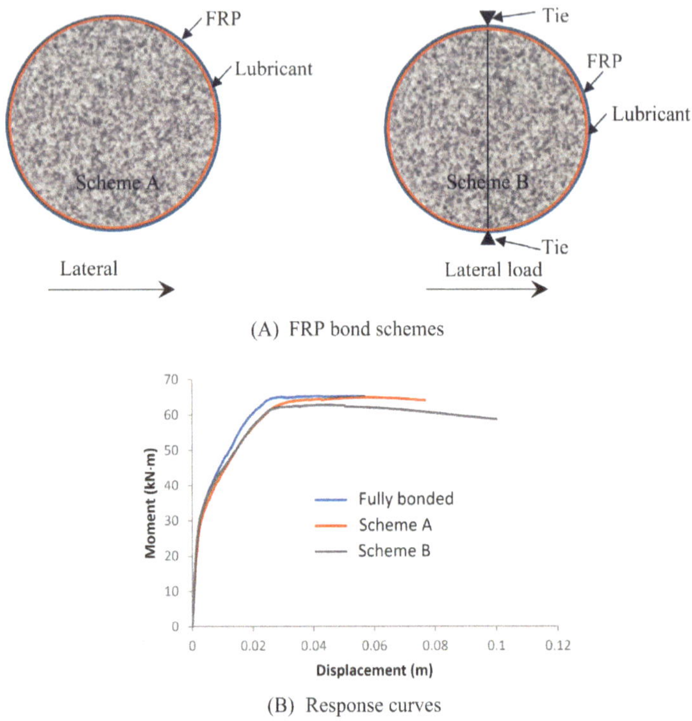

Fig. 7.40 Ultimate deflection of FRP-jacketed RC columns.

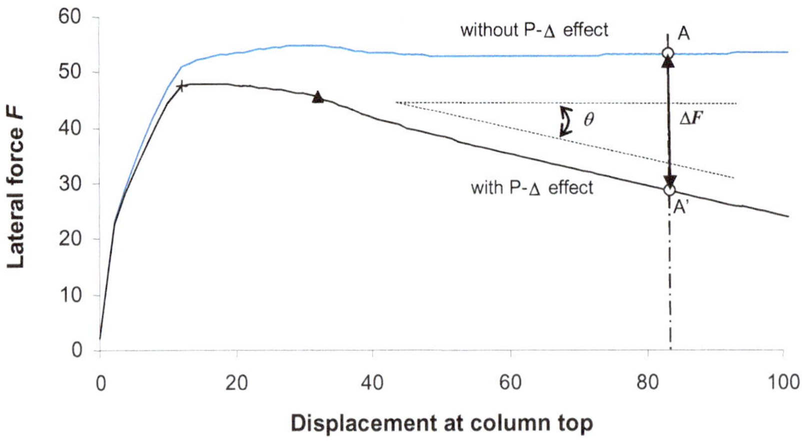

Fig. 7.41 The P-Δ effect.

the figure that the P-Δ effect can significantly reduce the lateral load resistance of the member.

The maximum bending moment is at the bottom of the column in Fig. 7.42 and is given by

$$M_m = F \cdot L + P \cdot \Delta_0 \tag{7.62}$$

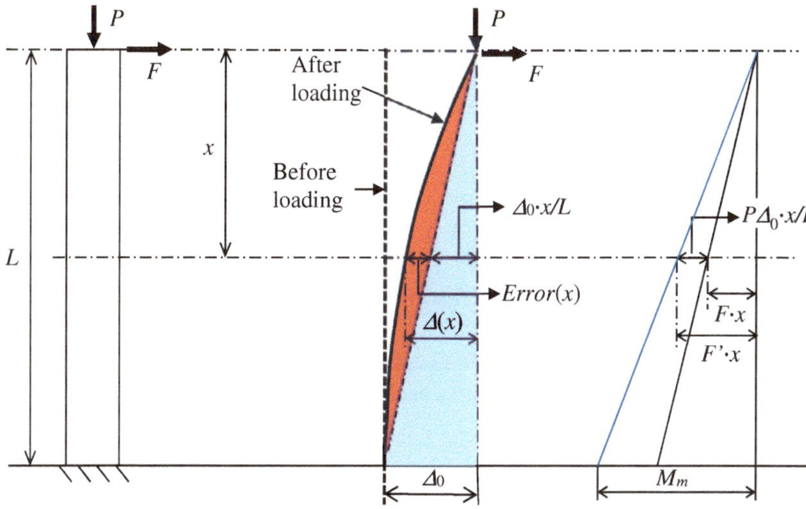

(A) RC column (B) Deformed shape (C) Moment distribution

Fig. 7.42 The simplified approach of the P-Δ effect.

The additional bending moment caused by the deflection of the member, i.e., the second term in Eq. (7.62), is called the P-Δ effect. The second term in Eq. (7.62) is relatively small compared with the first term for short columns and can usually be ignored. But for slender columns, the second term is significant and cannot be ignored. For a slender column, the blue line in Fig. 7.41 is the response curve when the P-Δ effect is ignored. The black line is the actual response curve that considers the P-Δ effect. It can be seen from Fig. 7.41 that the P-Δ effect should be properly accounted for slender columns.

However, the calculation of the P-Δ effect is not straightforward, which requires a tedious iteration process. It can be seen from Eq. (7.62) that, for a given F, the deflection Δ_0 increases the bending moment of the beam, while the increase of bending moment further increases the deflection. For a given moment M_m at the bottom of the member, an iteration process is required to evaluate the deflection Δ_0 and lateral load F by trial and error until it converges. The iteration actually involves all cross sections as the deflections at all cross sections along the column length affect the longitudinal moment distribution and are unknown. Therefore, the iteration process is calculation-intensive and time-consuming. Reference can be made to Wu et al. (2004) for a detailed iteration process.

A simplified method was proposed by Wu (2002) to evaluate the P-Δ effect. The method involves two steps:

(1) Produce the load-deflection curve of the column without considering the P-Δ effect, as shown by the blue line in Fig. 7.41. This process does not involve iterations and can be done using the approach introduced in the previous sections or by a simple numerical process.

(2) Draw the response curve that includes the P-Δ effect by adjusting the vertical coordinates of all the points on the blue curve by a value of

$$\Delta F = -\frac{P}{L} \cdot \Delta_0 \tag{7.63}$$

where Δ_0 is the displacement of a point on the original curve without considering the P-Δ effect. For example, the point A on the blue curve in Fig. 7.41 is changed (dropped) to a new point A′. The old *y* coordinate at point A is dropped by a value of $\frac{P}{L} \bullet \Delta_0$ to the new one at A′. The adjustment of coordinates by Eq. (7.63) is similar to "rotating" the blue curve without considering the P-Δ effect by an angle θ clockwise about the origin, where

$$\tan(\theta) = P/L \tag{7.64}$$

The curve produced by the simplified method is very close to the black line in Fig. 7.41 which considered the P-Δ effect exactly. This simplified approach is not mathematically exact but with an acceptably small error in an engineering sense. A detailed derivation and error evaluation of the approach are given in Section 7.3.5.2.

7.3.5.1 *Implications of the P-Δ effect*

It can be seen from Eq. (7.64) that columns with a larger axial load *P* will have a larger rotation or a greater P-Δ effect. Similarly, the reduction of the cantilever length will increase the slope of the descending branch. The effect of *L* seems counterintuitive in a sense that the P-Δ effect should be more significant for a longer member. However, a short column will have a much smaller lateral response than a slender column. For a certain additional lateral displacement or certain increment of horizontal axis, the shorter column will have a larger drop in lateral strength compared with the longer one, which causes a steeper descending branch. This can be clearly seen from the numerical simulation results shown in Fig. 7.43 where the P-Δ effect is strictly calculated (Wu 2002).

The P-Δ effect causes an increase in the descending slope of the response curve, and hence a reduction of the ultimate deflection based on the definition given in Fig. 1.11 and Eq. (1.1), as can be observed clearly in Fig. 7.41. It should be noted that

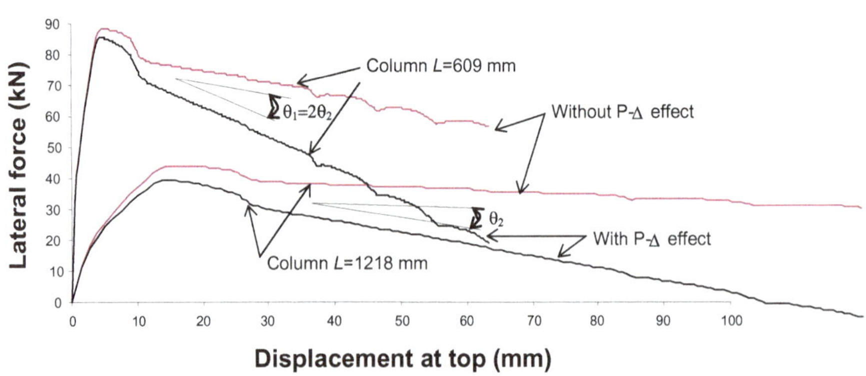

Fig. 7.43 Effect of column length to the P-Δ effect.
From Wu, Y.F. 2002. "Seismic retrofitting of rectangular reinforced concrete columns with partial interaction plating." Ph.D. thesis, Australia: The University of Adelaide.

the smaller calculated ultimate displacement and ductility due to the P-Δ effect does not mean a more significant physical deterioration of the member. The point A on the curve without the P-Δ effect and point A′ with P-Δ effect in Fig. 7.41 actually have the same moment resistance and hence an identical physical condition at the bottom section. The drop of the lateral force is caused by the additional moment term due to the P-Δ effect (the second term in Eq. 7.62), not the physical deterioration of the member.

As discussed in Section 2.2.1, structural failure depends on loading type. If the structural failure of an RC column is under load control where the lateral resistance F is required to be maintained within a smaller drop than a certain limit, the response curve including the P-Δ effect should be considered for the ductility calculation. However, if the member failure is under a displacement control, which is often the case in seismic design, designers should not be misled by the significant drop in the lateral load resistance F due to the P-Δ effect in assessing the condition of the member. In other words, an intact member may have a small ductility ratio if the P-Δ effect is considered. In the design of RC buildings, the lateral resistance is often provided by shear walls and RC columns are often designed mainly to carry vertical loads. In this case, the drop of the lateral load resistance of a column due to the P-Δ effect does not affect the integrity of the whole building, because the lateral resistance of the building relies on the shear walls. The RC columns should be considered as intact as long as the cross-sectional moment resistance under design axial loads does not deteriorate significantly, i.e., the blue curve in Fig. 7.41 without considering the P-Δ effect does not drop much. Therefore, care should be exercised in calculating the ductility of columns by selecting a right curve for structural evaluation. In general, the curve including the P-Δ effect (the black curve in Fig. 7.41) should be used for ductility calculation when the column is designed to carry lateral force. Otherwise, the curve without considering the P-Δ effect should be used when the column is mainly designed to carry axial loads.

7.3.5.2 Evaluation of the simplified method for the P-Δ effect

The simplified method for the P-Δ effect developed by Wu (2002) is derived and evaluated in this section. Based on Fig. 7.42B, the actual bending moment including the P-Δ effect is given by

$$\begin{aligned} M(x) &= F\cdot x + P\Delta(x) = F\cdot x + \left(\frac{x}{L}\cdot\Delta_0\right)\cdot P + Error(x)\cdot P \\ &\approx F\cdot x + \left(\frac{x}{L}\cdot\Delta_0\right)\cdot P \end{aligned} \tag{7.65}$$

where $Error(x)$ is the difference between the actual deflection and the assumed linearly distributed deflection at section x, as shown in Fig. 7.42B. When the term $Error(x)$ is neglected, Eq. (7.65) or $M(x)$ becomes a linear moment distribution as shown by the blue line in Fig. 7.42C. Dividing both sides of Eq. (7.65) by x, one has

$$F \approx F' - \frac{P}{L}\cdot\Delta_0 \tag{7.66}$$

where $F' = \frac{M(x)}{x} = \frac{M_m}{L}$. The following observations can be made from Eqs. (7.65) and (7.66):

(1) For a given value of the cross-sectional bending moment M_m at the bottom of the column, the moment distribution without considering the P-Δ effect is a straight line, which corresponds to a lateral force of $F' = \frac{M_m}{L}$. The displacement Δ_0 corresponding to this F' value can be obtained by double integration of the curvature distribution along the column length caused by this moment distribution.

(2) The actual moment distribution due to the P-Δ effect along the length of the column is a nonlinear distribution as shown by the shaded area in Fig. 7.42B (red area + blue area). When the nonlinear part of the deflected shape (the red area) is neglected, the additional moment due to the P-Δ effect is a linear distribution (the blue area).

(3) When only the blue area in Fig. 7.42B is considered for the P-Δ effect, the bending moment in (1) above can be separated into two parts (Fig. 7.42C). The first part is the additional bending moment caused by the P-Δ effect corresponding to the deflection produced in (1), i.e., Δ_0. The remaining part corresponds to the moment distribution caused by a reduced lateral force F given by Eq. (7.66).

(4) The deflection from Step (1) is identical to that from Step (3) because the total moment distribution from these two steps is the same.

(5) As a result, the P-Δ effect to a column is a drop of the lateral force given by Eq. (7.63).

(6) Eq. (7.66) is an approximation. Its error is caused by the neglecting of the red area in Fig. 7.42B. This error is insignificant for the reason discussed in the following text.

It can be seen from Fig. 7.42B that the error term is given by

$$Error(x) = \Delta(x) - \frac{x}{L} \cdot \Delta_0 \tag{7.67}$$

This error term is evaluated as follows:

(a) When the deflection of the column is small, the P-Δ effect is small. The error of the total moment is therefore small and negligible compared with the moment caused by the lateral force F.

(b) When the deflection is large and the plastic hinge forms at the column base, the deflection of the column is mostly due to the plastic hinge rotation at the bottom, and the deflected shape above the hinge base of the column is relatively straight, as shown in Fig. 7.15C. In this case, $Error(x)$ given by Eq. (7.67) is relatively small compared with the linear distribution part $\frac{x}{L} \cdot \Delta_0$. Therefore, the error introduced by Eq. (7.66) is also small.

(c) For a deflection in between the preceding two cases (between small elastic deflection and large plastic deflection), based on Eq. (7.65), the relative error can be estimated by

$$\frac{M(x)}{P \cdot Error(x)} = \frac{F \cdot x + P\Delta(x)}{P \cdot Error(x)} = \frac{F}{P} \cdot \frac{x}{Error(x)} + \frac{\Delta(x)}{Error(x)} \tag{7.68}$$

If Eq. (7.68) gives a value much greater than 1, then the error of Eq. (7.66) is relatively small. In the first part of Eq. (7.68), i.e., $\frac{F}{P} \cdot \frac{x}{Error(x)}$, $x \gg Error(x)$. In the meantime, this case is in between (a) and (b) or between the small and large deflections, and F is not very small. Hence F/P is usually not very small. Therefore, the term $\frac{F}{P} \cdot \frac{x}{Error(x)}$ usually gives a large value.

From Fig. 7.42B, it can be seen that the second part of Eq. (7.68), i.e., $\frac{\Delta(x)}{Error(x)}$, is also a large value near the bottom of the column where the moment contributes more to the deflection of the column. In the middle and top parts of the column, this second part is relatively smaller than that in the bottom of the column; however, its contribution to the deflection of the whole column is also smaller than that from the bottom part.

The summation of these two parts gives an even bigger value. Therefore, Eq. (7.68) is usually much greater than 1, unless the column is an extremely slender one that is not allowed for reinforced concrete members.

For all the cross sections above the bottom section, the moment-curvature relation is in the ascending part of the curve (on the left of the peak point A in Fig. 7.14A). A small error in moment will only cause a small error in curvature value, hence a small error in the deflection. Although a small error in moment may cause a large error in curvature for the cross section at the column base when the descending part of the curve is relatively flat, the error in moment from the simplified method (the red area in Fig. 7.42B) is very small at the column base. In other words, the error in the calculated curvature caused by the simplified method, hence the calculated deflection, is small.

The error given by Eq. (7.67) will cause an underestimation of the deflection because the moment shown by the red area in Fig. 7.42B is omitted. That is to say, the actual deflection should be slightly larger than Δ_0, which is obtained from double integration of the curvature corresponding to the linear moment distribution in Fig. 7.42C, as shown by the point B in Fig. 7.44 at the actual deflection Δ_0'. The small increment in displacement from the simplified method compared with the actual response curve where the P-Δ effect is strictly considered is illustrated in the numerical examples in Fig. 7.45. Details of the columns and the numerical simulations of the column responses in Fig. 7.45 refer to Wu (2002). It can be seen from Fig. 7.45 that the error caused by the simplified method is usually very small and negligible for normal RC columns.

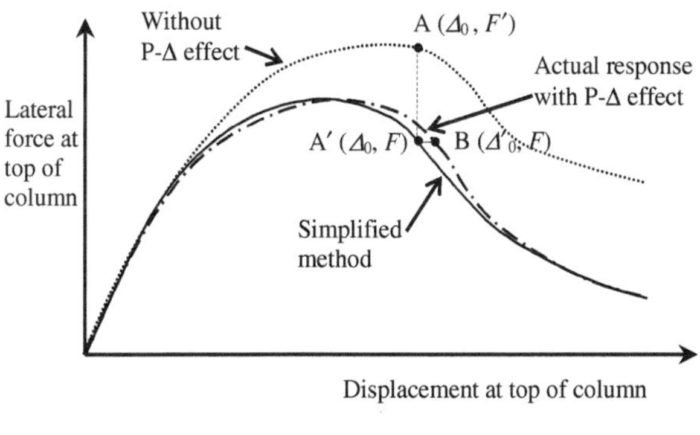

Fig. 7.44 Error due to the simplified method for P-Δ effect.
From Wu, Y. F., M. C. Griffith, and D. J. Oehlers. 2004. "Numerical simulation of steel plated RC columns." *Comput. Struct.* 82 (4–5), 359–371.

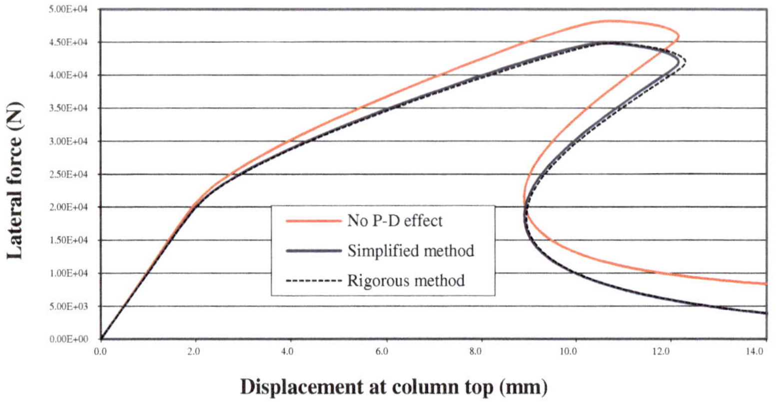

(A) Column without stirrups, L=1115 mm, L_p=200 mm

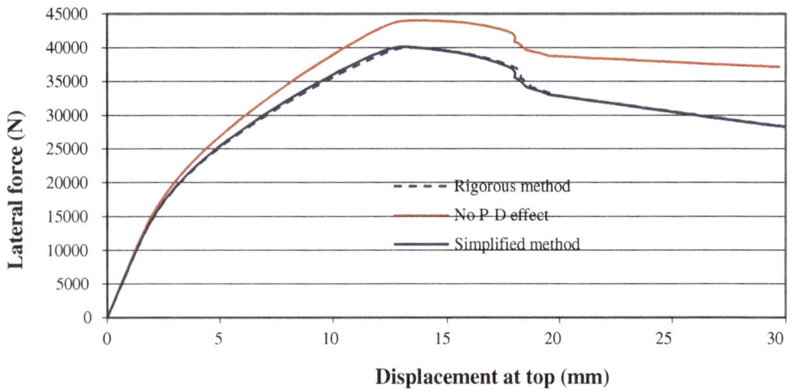

(B) Column with T6-100 stirrups, L=1218 mm, L_p=100 mm

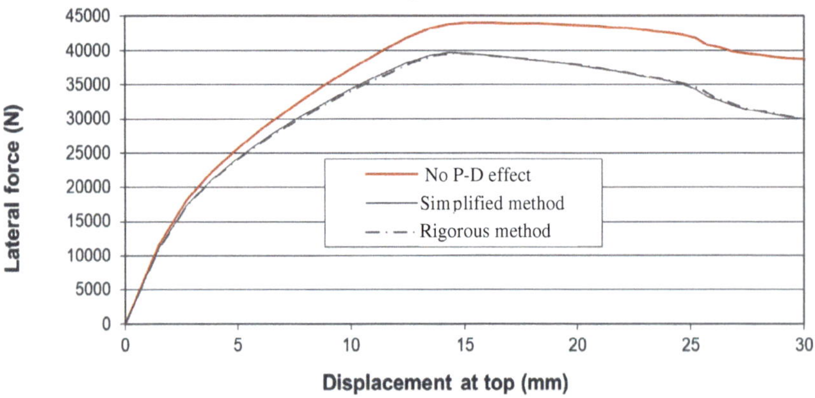

(C) Column with T6-100 stirrups, L=1218 mm, L_p=200 mm

Fig. 7.45 Numerical illustrations of the error caused by the simplified method for P-Δ effect. From Wu, Y.F. 2002. "Seismic retrofitting of rectangular reinforced concrete columns with partial interaction plating." Ph.D. thesis, Australia: The University of Adelaide.

References

ACI-ASCE Committee 428. 1968. "Progress report on code clauses for limit design." *ACI J.* 65, 713–715.

Alemdar, F. 2010. "Plastic hinging behavior of reinforced concrete bridge columns." Ph.D. thesis. University of Kansas.

Al-Shaikh, A. H., and R. Z. Al-Zaid. 1993. "Effect of reinforcement ratio on the effective moment of inertia of reinforced concrete beams." *ACI Struct. J.* 90 (2), 144–149.

Al-Zaid, R. Z., A. H. Al-Shaikh, and M. M. Abu-Hussein. 1991. "Effect of loading type on the effective moment of inertia of reinforced concrete beams." *ACI Struct. J.* 88 (2), 184–190.

Ang, B. G., M. J. N. Priestley, and R. Park. 1981. *Ductility of reinforced concrete bridge piers under seismic loading*, Research Report 81-3. Christchurch, New Zealand: Department of Civil Engineering, University of Canterbury.

Bae, S., and O. Bayrak. 2008. "Plastic hinge length of reinforced concrete columns." *ACI Struct. J.* 105 (3), 290–300.

Baker, A. L. L. 1956. *The ultimate load theory applied to the design of reinforced & prestressed concrete frames*, Concrete Publ. Ltd.

Baker, A., and A. Amarakone. 1965. "Inelastic hyperstatic frames analysis." *ACI Spec. Publ.* 12, 85–142.

Bate, S. C. C., M. R. Horne, W. T. Marshall, C. W. Yu, K. W. Johansen, A. L. L. Baker, W. W. L. Chan, K. H. Konyi, P. B. Morice, and R. Gartner. 1962. "Ultimate load design of concrete structures." *ICE Proc.* 21 (2), 399–442.

Bazant, Z. P. 1972. "Prediction of concrete creep effects using age-adjusted effective modulus method." *ACI J. Proc.* 69 (4), 212–217.

Bazant, Z. P., and B. H. Oh. 1984. "Deformation of progressively cracking reinforced concrete beams." *ACI Struct. J.* 81 (3), 268–278.

Berry, M. P. 2006. "Performance modeling strategies for modern reinforced concrete bridge columns." Ph.D. thesis. Seattle, WA: University of Washington.

Berry, M. P., D. E. Lehman, and L. N. Lowes. 2008. "Lumped-plasticity models for performance simulation of bridge columns." *ACI Struct. J.* 105, 270–279.

Bigaj, A. J., and J. C. Walraven. 2002. "Size effects in plastic hinges of reinforced concrete members." *Heron* 47, 53–75.

Binici, B. 2008. "Design of FRPs in circular bridge column retrofits for ductility enhancement." *Eng. Struct.* 30, 766–776.

Biskinis, D., and M. N. Fardis. 2010. "Flexure-controlled ultimate deformations of members with continuous or lap-spliced bars." *Struct. Concr.* 11, 93–108.

Branson, D. E. 1963. *Instantaneous and time-dependent deflections of simple and continuous reinforced concrete beams*, HPR Report No. 7, Part 1. Alabama Highway Department, Bureau of Public Roads, pp. 1–78.

British Standards Institution (BSI). 1997. *Structural use of concrete, parts 1–3*, BS8110, London.

Carpinteri, A., M. Corrado, G. Mancini, and M. Paggi. 2009. "Size-scale effects on plastic rotational capacity of reinforced concrete beams." *ACI Struct. J.* 106 (6), 887–896.

Chai, Y. H., M. J. N. Priestley, and F. Seible. 1991. "Seismic retrofit of circular bridge columns for enhanced flexural performance." *ACI Struct. J.* 88 (5), 572–584.

Coleman, J., and E. Spacone. 2001. "Localization issues in force-based frame elements." *J. Struct. Eng.* 127 (11), 1257–1265.

Corley, W. 1966. "Rotational capacity of reinforced concrete beams." *Proc. ASCE Struct. Div.* 92, 121–146.

Elsanadedy, H. M., and M. A. Haroun. 2005. "Seismic design guidelines for squat composite-jacketed circular and rectangular reinforced concrete bridge columns." *ACI Struct. J.* 102 (4), 505–514.

Fantilli, A. P., D. Ferretti, I. Iori, and P. Vallini. 1998. "Flexural deformation of reinforced concrete beams." *J. Struct. Eng.* 124 (9), 1041–1049.

Fardis, M. N. 2009. *Seismic design, assessment and retrofitting of concrete buildings: based on EN-Eurocode 8*, Springer Science & Business Media.

Gilbert, R. I. 2001. "Shrinkage, cracking and deflection-the serviceability of concrete structures." *Electron. J. Struct. Eng.* 1 (1), 15–37.

Goodnight, J. C., M. J. Kowalsky, and J. M. Nau. 2016. "Modified plastic hinge method for circular RC bridge columns." *J. Struct. Eng.*, 04016103.

Grammatikou, S., D. Biskinis, and M. N. Fardis. 2016. "Ultimate strain criteria for RC members in monotonic or cyclic flexure." *J. Struct. Eng.* 142 (9), 04016046.

Gu, D. S., G. Wu, Z. S. Wu, and Y. F. Wu. 2010. "Confinement effectiveness of FRP in retrofitting circular concrete columns under simulated seismic load." *J. Compos. Constr.* 14 (5), 531–540.

Gu, D. S., Y. F. Wu, G. Wu, and Z. S. Wu. 2012. "Plastic hinge analysis of FRP confined circular concrete columns." *Constr. Build. Mater.* 27, 223–233.

Hines, E. M., J. I. Restrepo, and F. Seible. 2004. "Force-displacement characterization of well-confined bridge piers." *ACI Struct. J.* 101 (4), 537–548.

Ho, C. M. 2003. "Inelastic design of reinforced concrete beams and limited ductile high-strength concrete columns." Ph.D. thesis. Hong Kong: University of Hong Kong.

Jiang, C., Y. F. Wu, and G. Wu. 2014. "Plastic hinge length of FRP-confined square RC columns." *J. Compos. Constr.* 18 (4), 04014003.

Jiang, C., F. Yuan, Y. F. Wu, and X. M. Zhao. 2019. "Effect of interfacial bond on plastic hinge length of FRP-confined RC columns." *J. Compos. Constr.* 23 (2), 04019007.

Kaklauskas, G., and J. Ghaboussi. 2001. "Stress-strain relations for cracked tensile concrete from RC beam tests." *J. Struct. Eng.* 127 (1), 64–73.

Kheyroddin, A., and H. Naderpour. 2007. "Plastic hinge rotation capacity of reinforced concrete beams." *Int. J. Civil Eng.* 5 (1), 30–47.

Ko, M. Y., S. W. Kim, and J. K. Kim. 2001. "Experimental study on the plastic rotation capacity of reinforced high strength concrete beams." *Mater. Struct.* 34 (5), 302–311.

Kolias, B., D. Sa, and K. Ave. 2008. "Eurocode 8–part 2. Seismic design of bridges." In *Eurocodes: Background and Applications Workshop, Brussels, February 2008*.

Köroğlu, M. A., M. H. Arslan, and M. K. Körez. 2014. "Use of regression analysis in determining the length of plastic hinge in reinforced concrete columns." *Int. J. Civil Archit. Sci. Eng.* 8 (4), 395–400.

Lopes, S. M. R., and L. F. A. Bernardo. 2003. "Plastic rotation capacity of high-strength concrete beams." *Mater. Struct.* 36 (1), 22–31.

Mander, J. B. 1984. "Seismic design of bridge piers." Ph.D. thesis. University of Canterbury.

Mattock, A. H. 1965. *Rotational capacity of hinging regions in reinforced concrete beams*, Special Publication 12. pp. 143–181.

Mattock, A. H. 1967. "Discussion of 'rotational capacity of reinforced concrete beams' by W. G. Corley." *J. Struct. Div.* 93, 519–522.

Mendis, P. 2001. "Plastic hinge lengths of normal and high-strength concrete in flexure." *Adv. Struct. Eng.* 4 (4), 189–195.

Monti, G., N. Nisticò, and S. Santini. 2001. "Design of FRP jackets for upgrade of circular bridge piers." *J. Compos. Constr.* 5 (2), 94–101.
Nilson, A. H., D. Darwin, and C. W. Dolan. 2003. *Design of concrete structures*, New York: McGraw-Hill.
Ozbakkaloglu, T., and M. Saatcioglu. 2006. "Seismic behavior of high-strength concrete columns confined by fiber-reinforced polymer tubes." *J. Compos. Constr.* 10 (6), 538–549.
Panagiotakos, T. B., and M. N. Fardis. 2001. "Deformations of reinforced concrete members at yielding and ultimate." *ACI Struct. J.* 98 (2), 135–148.
Park, R., and T. Paulay. 1975. *Reinforced concrete structures*, New York: John Wiley & Sons.
Park, R., M. J. N. Priestley, and W. D. Gill. 1982. "Ductility of square-confined concrete columns." *J. Struct. Div.* 108, 929–950.
Paulay, T., and M. J. N. Priestley. 1992. *Seismic design of reinforced concrete and masonry buildings*, New York: John Wiley & Sons.
Paulson, K., A. H. Nilson, and K. C. Hover. 1991. "Long-term deflection of high-strength concrete beams." *ACI Mater. J.* 88 (2), 197–206.
Piyasena, R. 2002. "Crack spacing, crack width and tension stiffening effect in reinforced concrete beams and one-way slabs." Ph.D. thesis. Australia: Griffith University.
Prakhya, G. K. V., and C. T. Morley. 1990. "Tension-stiffening and moment-curvature relations of reinforced concrete elements." *ACI Struct. J.* 87 (5), 597–605.
Priestley, M. J. N., and R. Park. 1987. "Strength and ductility of concrete bridge columns under seismic loading." *ACI Struct. J.* 84 (1), 61–76.
Priestley, M. J. N., F. Seible, and M. Calvi. 1996. *Seismic design and retrofit of bridges*, New York: Wiley.
Priestley, M. J. N., G. M. Calvi, and M. J. Kowalsky. 2007. *Displacement-based seismic design of structures*, Pavia, Italy: IUSS Press.
Rangan, B. V. 1982. "Control of beam deflections by allowable span-depth ratios." *ACI J. Proc.* 79 (5), 372–377.
Ruangrassamee, A., and K. Kawashima. 2003. "Control of nonlinear bridge response with pounding effect by variable dampers." *Eng. Struct.* 25, 593–606.
Saadatmanesh, H., M. R. Ehsani, and L. M. Jin. 1996. "Seismic strengthening of circular bridge pier models with fiber composites." *ACI Struct. J.* 93 (6), 639–647.
Sawyer, H. A. 1964. "Design of concrete frames for two failure states." In *Int. Symp. on the Flexural Mechanics of Reinforced Concrete, Miami, Florida, November 1964*, pp. 405–431.
Sheikh, S. A., D. V. Shah, and S. S. Khoury. 1994. "Confinement of high-strength concrete columns." *ACI Struct. J.* 91, 100–111.
Vonk, R. A. 1991. "Softening of concrete loaded in compression." Ph.D. thesis. Eindhoven: Eindhoven University of Technology.
Wang, J., H. Zhao, and J. He. 2019. "Seismic behaviors and resilient capacity of CFRP-confined concrete columns with partially debonded high-strength steel rebars." *Compos. Struct.* 222, 110912.
Wei, Y., and Y. F. Wu. 2016. "Experimental study of concrete columns with localized failure." *J. Compos. Constr.*, 04016032.
Wu, Y. F. 2002. "Seismic retrofitting of rectangular reinforced concrete columns with partial interaction plating." Ph.D. thesis. Australia: The University of Adelaide.
Wu, Y. F., and Y. Wei. 2015. "Stress-strain modeling of concrete columns with localized failure: an analytical study." *J. Compos. Constr.* 20 (3), 04015071.

Wu, Y. F., M. C. Griffith, and D. J. Oehlers. 2003. "Improving the strength and ductility of rectangular reinforced concrete columns through composite partial interaction: tests." *J. Struct. Eng.* 129 (9), 1183–1190.

Wu, Y. F., M. C. Griffith, and D. J. Oehlers. 2004. "Numerical simulation of steel plated RC columns." *Comput. Struct.* 82 (4–5), 359–371.

Wu, Y. F., C. Jiang, X. M. Zhao, and J. S. Wang. 2015. "Plastic hinge of reinforced concrete columns." *Innov. Constr.* 2, 69–77.

Youssf, O., M. A. ElGawady, and J. E. Mills. 2015. "Displacement and plastic hinge length of FRP-confined circular reinforced concrete columns." *Eng. Struct.* 101, 465–476.

Yuan, F., and Y. F. Wu. 2017. "Effect of load cycling on plastic hinge length of RC columns." *Eng. Struct.* 147, 90–102.

Yuan, F., Y. F. Wu, and C. Q. Li. 2017. "Modelling plastic hinge of FRP-confined RC columns." *Eng. Struct.* 131, 651–668.

Zahn, F. A. 1985. "Design of reinforced concrete bridge columns for strength and ductility." Ph.D. thesis. University of Canterbury.

Zhao, X. M., Y. F. Wu, and A. Y. Leung. 2012. "Analyses of plastic hinge regions in reinforced concrete beams under monotonic loading." *Eng. Struct.* 34 (1), 466–482.

Confined concrete

8

8.1 Introduction

Concrete is the most popular construction material in the world, but it has certain drawbacks. Low tensile strength and brittleness are two of the major problems. The problem of low tensile strength is overcome by using reinforcement. The problem of low ductility can be resolved by confinement. Confinement increases both the compressive strength and ductility of concrete material. When a reinforced concrete structure is well confined, it can be visualized as consisting of two ductile materials: ductile reinforcing bars and ductile concrete. As a result, confined RC structures can be much more ductile than normal RC structures. The importance and effectiveness of confinement for RC columns were pointed out by Shah and Ahmad (1994), who stated that "high deformability requirements in the first-story columns of multistory buildings can only be achieved through confinement of the core concrete." The properties of confined concrete and relevant concrete members are discussed in this chapter.

8.2 Confinement effects on compression failure of concrete

The typical failure modes of concrete cylinders/prisms under compression are shown in Fig. 8.1. Concrete dilates under compression and failure occurs when the lateral dilation strain exceeds a certain limit. Restraint of the lateral dilation of concrete delays the failure and hence increases the strength and ductility of concrete. Such effects can be seen from the standard compression tests of concrete, as illustrated in Fig. 8.2. For the same concrete material and cross-sectional sizes, the concrete strength of the prism in Fig. 8.2A is lower than that of the cube in Fig. 8.2B (about 80% of the cube strength). The reason is that the loading platen incurs friction on the top and bottom faces of the test specimens, which imposes restraint on the dilation of the concrete. For the cube, the restraint is closer to the whole body of the concrete material (Fig. 8.2B) and hence is more effective in restraining dilation, while the restraining effect is smaller on the prism due to the longer side length. Further reduction of the restraining effect can be made by adding grease or other lubricant materials to reduce the friction at the concrete-to-platen contact face. In this case, the concrete failure mode will change from the one shown in Figs. 8.1A and 8.2A to that shown in Figs. 8.1B and 8.2C, and the concrete strength of the prism in Figs. 8.1B and 8.2C is lower than that in Figs. 8.1A and 8.2A.

The effect of friction at the contact faces between the test platens and the concrete specimen is shown by the compression tests in Fig. 8.2D and E (Mohammadi and Wu 2017). In the tests, the friction at the contact face was reduced by using Teflon sheets. Grease was applied between Teflon layers to further reduce friction. The friction was

Analytical Approaches for Reinforced Concrete. https://doi.org/10.1016/B978-0-12-821164-9.00008-5

(A) Failure with friction at loading platen (B) Failure without friction at loading platen

Fig. 8.1 Typical compression failure modes of concrete.

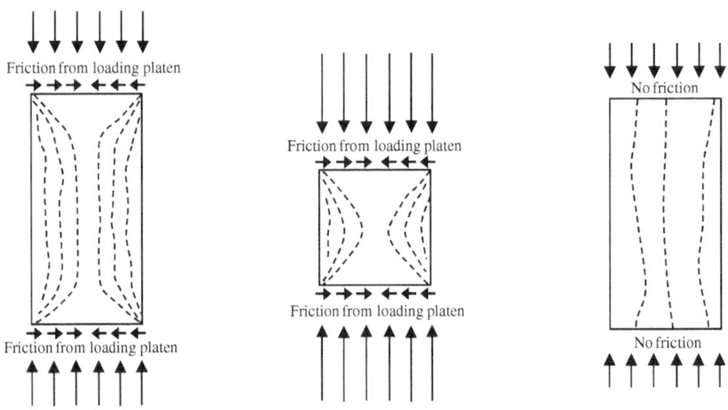

(A) Cylinder with friction (B) Cube with friction (C) Cylinder without friction

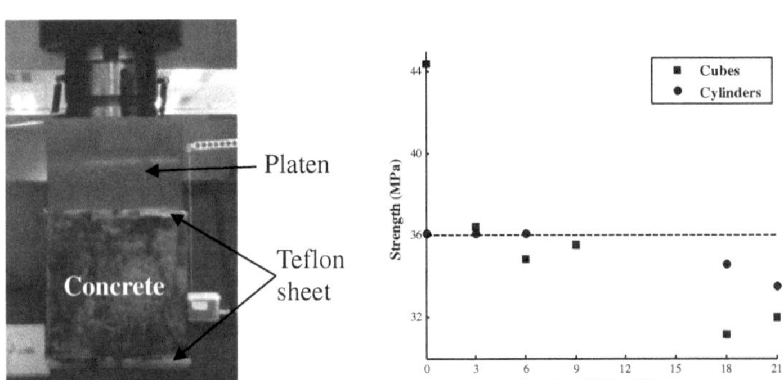

(D) Compression test setup (E) Effect of friction reduction pad

Fig. 8.2 Effect of friction from load platen on concrete strength.

smaller when more layers of Teflon were used. It can be seen from Fig. 8.2E that when there is no friction reduction sheet (Teflon layer = 0), the cylinder strength (36 MPa) is about 81% of the cube strength (44.5 MPa), as expected. When the number of Teflon layers increases to three, the cube and cylinder strength becomes similar. A further increase in the number of Teflon sheets will reduce the strength of both cubes and cylinders, with more reduction in the cubes. This was caused by the lateral expansion of the thick Teflon layer under compression that caused lateral tensile strain of the concrete specimens on the top and bottom faces. From these test results, it can be clearly seen that the restraining/confinement effect at the loading platens against lateral expansion of the concrete can significantly increase the strength of concrete, while the outward expansion of the loading platen, which helps the lateral expansion of the concrete, significantly reduces the concrete strength.

The mechanism of the lateral confinement in increasing concrete strength can be analogized by the compression strut in Fig. 8.3. Concrete fails when its internal bond strength f_t is reached if it is unconfined. The confinement pressure helps to stabilize the strut so that a higher compression f_{cc} can be applied before it fails. The effect of confinement in increasing the strength of the confined concrete can be modeled using an appropriate strut angle α as shown in Fig. 8.3B.

From the previous discussion, it can be seen that concrete under compression can experience a higher stress and deformation when it is restrained in its lateral dilation. As a result, it becomes a stronger and more ductile material when confined. A typical stress-strain curve of the concrete in a column confined by transverse steel reinforcement (stirrups/spirals) is shown in Fig. 8.4, in comparison with that of unconfined concrete. It can be seen from Fig. 8.4 that confined concrete material can be much stronger and more ductile than unconfined concrete. When confined, its failure depends on the failure of its lateral confining system rather than the concrete material itself. This is an important characteristic of concrete material that can be utilized to significantly improve the performance of concrete structures.

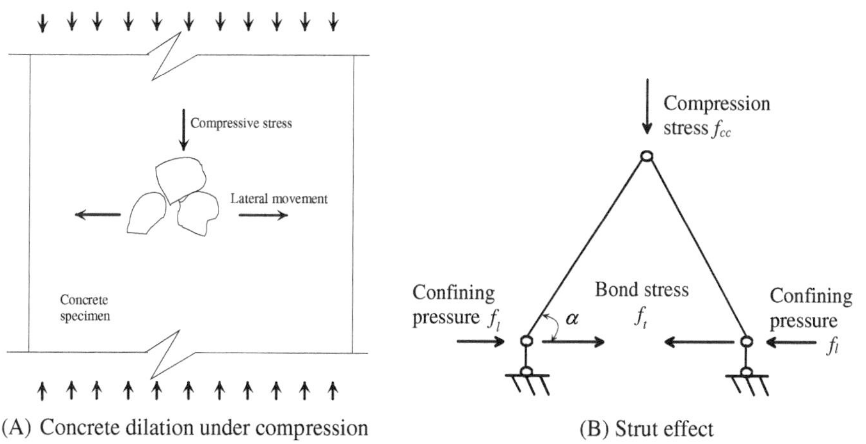

Fig. 8.3 Effect of confinement on concrete strength.

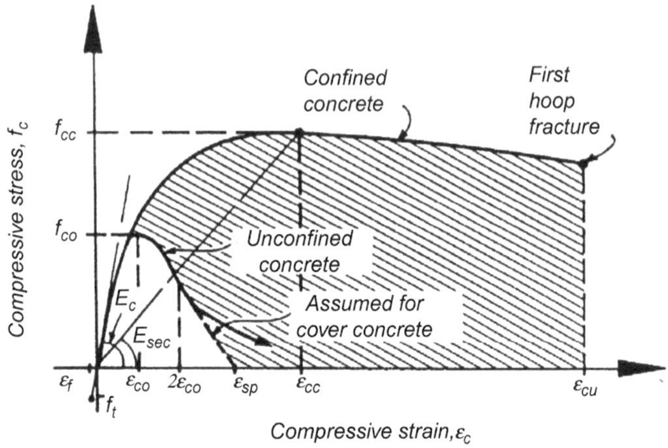

Fig. 8.4 Typical stress-strain model of concrete confined by steel hoops.
From Mander, J. B., M. J. N. Priestley, and R. Park. 1988. "Theoretical stress-strain model for confined concrete." *J. Struct. Eng.* 114 (8), 1804–1826.

8.3 Strength modeling of confined concrete cylinders/ circular columns

It is mentioned in the previous section that concrete can be seen as another stronger and more ductile material when confined. When the stress-strain relationship of a confined concrete is known, the structural behavior can be calculated using the reinforced concrete theories introduced in the previous chapters. For example, the flexural strength of a confined RC member can be calculated using the flexural theorems introduced in Chapters 3 and 4 when the stress-strain relationship of the confined concrete is given.

Extensive research on the stress-strain relationship of confined concrete has been reported in the literature since the early 20th century. A milestone study was reported by Richart et al. (1928, 1929) that forms the cornerstone and basis of current stress-strain models of confined concrete. An extensive experimental program was carried out by Richart et al. (1928) on compression tests of concrete cylinders under hydrostatic lateral pressure using a hydraulic pressure chamber (Hoek cell). The test results are shown in Fig. 8.5. It can be seen from the figure that the additional concrete strength due to lateral confinement is approximately proportional to the lateral pressure. Based on the test results, they proposed the following model:

$$f_{cc} = f_{co} + k_0 f_l \tag{8.1}$$

where f_{cc} and f_{co} are the confined and unconfined concrete compressive strengths, respectively; f_l is the lateral pressure; and k_0 is a confinement effectiveness coefficient. Their test results showed that k_0 slightly decreases as the lateral confining pressure increases, tending to an average value of 4.1, as shown in Fig. 8.5.

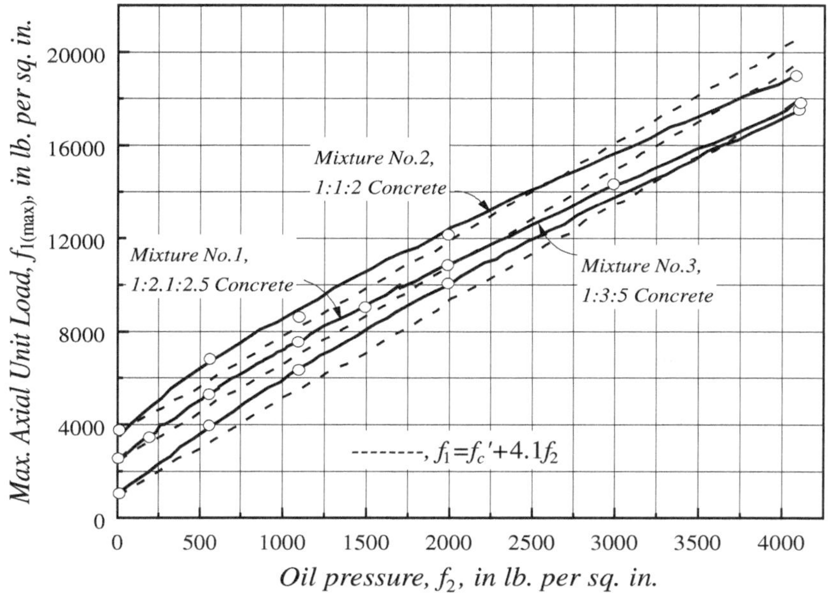

Fig. 8.5 Compression tests of concrete cylinders under lateral liquid pressure.
Based on Richart, F. E., A. Brandtzaeg, and R. L. Brown. 1928. *A study of the failure of concrete under combined compressive stresses*, Bulletin no. 185. Urbana, IL: University of Illinois, Engineering Experimental Station.

Eq. (8.1) can be normalized to a form:

$$\frac{f_{cc}}{f_{co}} = 1 + k_0 \frac{f_l}{f_{co}} \tag{8.2}$$

The model given by Eqs. (8.1), (8.2) is consistent with the model in Fig. 8.3B. Based on Fig. 8.3B, the confined concrete strength is given by

$$f_{cc} = f_t \tan(\beta) + f_l \tan(\alpha) \tag{8.3}$$

where f_t is the internal tensile strength between concrete particles resisting lateral splitting; α and β are the strut angles of the external confinement and internal bond, respectively. Comparing Eq. (8.3) with Eq. (8.1), one has $k_0 = \tan(\alpha)$ and $f_{co} = f_t \tan(\beta)$. For $k_0 = 4.1$, $\alpha = 76.3°$.

Apart from the previous empirical and mechanical models, the confinement effect on concrete strength can also be explained using the simple maximum tensile strain failure criterion and elasticity theory. Concrete material fails under a certain condition that is described by a failure criterion. Numerous failure criteria have been developed for concrete and the oldest and simplest one is the maximum tensile strain criterion proposed by Mariotto in 1682 (Guo 2014), which states that material failure occurs when its maximum tensile strain reaches a certain limit value. This failure criterion has also been considered as a relatively more suitable one for concrete compared with other failure criteria (Carino and Slate 1976; Eyre Nasreddin 2013). "There are strong indications that it is not a limiting stress but a limiting tensile strain that determines the strength of concrete under static loading" (Neville 1981).

For concrete subjected to a three-dimensional stress state with longitudinal stress σ_1 and lateral stresses $\sigma_2 = \sigma_3 = \sigma_r$, assuming it is an elastic and isotropic material, the maximum tensile strain criterion can be expressed as

$$\mu\frac{\sigma_1}{E_c} - \frac{\sigma_2}{E_c} + \mu\frac{\sigma_3}{E_c} = \mu\frac{f_{co}}{E_c} \tag{8.4}$$

where E_c and μ are Young's modulus and Poisson's ratio for concrete, respectively. The left side of Eq. (8.4) is the maximum lateral strain of the concrete in the 2-axis direction under the compressive stress σ_1 in the 1-axis direction and the passive confining stresses σ_2 and σ_3 in the 2- and 3-axis directions, respectively. The right side is the lateral tensile strain of a concrete cylinder without lateral confinement under a uniaxial compression stress f_{co} at onset of its compression strength. Eq. (8.4) can be simplified to

$$\sigma_1 = f_{co} + \frac{(1-\mu)}{\mu}\sigma_r \tag{8.5}$$

It can be seen from Eq. (8.5) that the longitudinal resistance is increased when a lateral pressure is applied to a concrete cylinder. The effectiveness of confinement pressure in increasing longitudinal resistance is magnified by a factor of $k_0 = \frac{1-\mu}{\mu}$, which means even a small confinement pressure may have a significant effect on increasing concrete strength. When $\mu = 0.2$, $k_0 = \frac{1-\mu}{\mu} = 4.0$, which generally agrees with the experimental results of $k_0 = 4.1$ in Eq. (8.1) by Richart et al. (1928). Although concrete material is not linear and isotropic, the general mechanism of the confinement effect can still be well explained by the analytical study in this section.

8.3.1 Active and passive confinement

Confinement pressure can be applied to concrete actively or passively. Active confinement pressure is constant, such as hydrostatic pressure. Passive confinement is due to the passive resistance of confining material such as stirrups and external jackets that resist the lateral dilation of concrete caused by longitudinal compression. The confinement pressure in a passive system depends on the lateral strain of concrete, and therefore is not constant but increases when dilation increases. But at a particular moment with a certain dilation of concrete, the passive confining pressure is constant across the cross section of a circular column. This is because the dilation of a circular column under concentric loading is axisymmetric, which causes a uniform tensioning of the confining material and hence a uniform reactive pressure to the circular column. This uniform passive pressure is the same as the active hydrostatic pressure in nature, and hence causes a uniform internal pressure inside the concrete material (see Fig. 8.9H).

The models of Eqs. (8.1), (8.2) were based on actively confined concrete. Richart et al. (1929) further studied concrete columns passively confined by spiral steel reinforcement. They found that the strength of concrete confined by lateral hydrostatic pressure was approximately the same as that of concrete confined by closely spaced circular steel spirals causing an equivalent lateral pressure. The same relationship as Eq. (8.1) was proposed for steel spiral confined concrete columns, and the confinement effectiveness coefficient k_0 is also equal to 4.1.

The confinement pressure f_l from the spirals at yielding of steel is given by Eq. (8.6). For a unit length of a circular column shown in Fig. 8.6, the uniform confinement pressure is given by

$$f_l = \frac{2F_l}{D} = 2\frac{nA_s f_y}{D} = \frac{\rho_s f_y}{2} \tag{8.6}$$

where D is the diameter of the column; f_y is the yield strength of the steel spiral; A_s is the area of one spiral bar; n is the number of spirals within the unit length; and ρ_s is the spiral reinforcement ratio given by $\rho_s = \frac{nA_s\pi D}{\left(\frac{\pi D^2}{4}\right)\times 1}$.

For a fiber-reinforced polymer (FRP) jacketed concrete cylinder/circular column, the external confinement pressure can similarly be derived from Fig. 8.6:

$$f_l = \frac{2F_l}{D} = 2\frac{f_j \cdot t}{D} = 2\frac{E_f \cdot \varepsilon_j t}{D} \tag{8.7}$$

where f_j and t are the rupture strength and nominal thickness of an FRP jacket, respectively; E_f is the Young's modulus of the FRP; and ε_j is the ultimate strain of the FRP jacket at rupture.

It is noted that concrete confined by transverse steel reinforcement such as steel spiral is a passive confinement strictly speaking, because the confinement is induced by the dilation of concrete under compression. However, steel spiral confinement is often considered as active confinement, because the confinement pressure remains constant after the transverse reinforcement yields. The behavior of concrete under

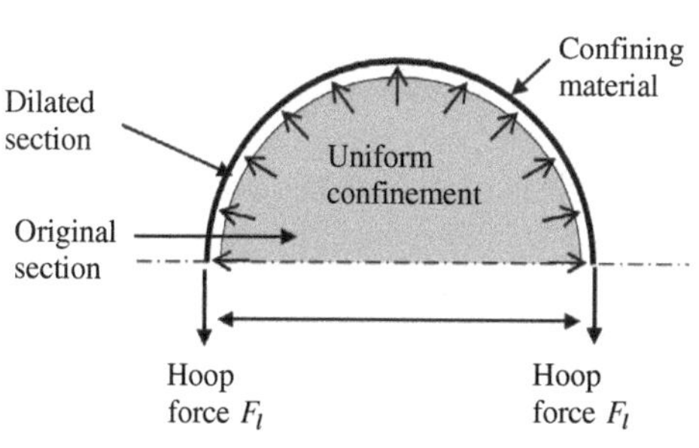

Fig. 8.6 Confining pressure of circular column.

a constant passive confinement pressure due to transverse steel reinforcement is similar to that under a constant active pressure. The behavior of concrete passively confined by FRP is different from that confined by constant lateral pressure, because FRP material is elastic and the confinement pressure keeps increasing until it ruptures.

Active confinement and passive confinement may be mixed in a column. Passive confinement from steel confining materials, such as steel stirrups/spiral and steel plate jackets, can be considered as a mixed case where the confinement is passive before steel yielding and becomes active after steel yielding. FRP-jacketed RC columns with transverse steel stirrups have active confinement from internal steel stirrups (spiral) after steel yielding and passive confinement from external FRP jacket. Initial confinement pressure can be applied to a column during jacketing through prestressing (see Section 8.5.5). In columns with a prestressed jacket, the confinement is also mixed, with an initial active confinement and further confinement due to dilation of concrete in a passive way.

8.3.2 *Typical strength models*

Eq. (8.2) has become the general form, and has been adopted by the majority, of the existing strength models of confined concrete, not only for actively/spiral steel confined concrete but also for concrete passively confined by external FRP jacket. Most of the existing models for the strength of confined concrete were developed by fitting experimental data to the general form of Eq. (8.2) to determine the confinement effectiveness coefficient k_0, which is derived either as a constant, a function of the effective lateral confining pressure f_l, or a function of $\frac{f_l}{f_{co}}$. Typical models are summarized in Table 8.1.

8.4 Strength modeling of noncircular columns

Mathematically speaking, the circular shape encloses the maximum cross-sectional area for a given perimeter length. As a result, any dilation/expansion of concrete within a circular jacket/confining material must result in an increase in the perimeter length of the jacket/confining material and, hence, mobilize the hoop resistance of the jacket, which in turn provides passive confinement to the concrete inside the jacket. Hence, the circular shape is most effective in providing confinement and, subsequently, in increasing the strength and ductility of concrete columns. This is not the case for noncircular-shaped concrete columns. The confinement effectiveness of rectilinear columns is much reduced and less reliable than the case of circular columns. This is because the dilation of concrete on the straight sides may only cause a slight bulging of the jacket outward without significantly mobilizing its membrane resistance.

Table 8.1 Strength models of confined concrete for circular columns.

Model	$\frac{f_{cc}}{f_{co}}$	Model	$\frac{f_{cc}}{f_{co}}$
Fardis and Khalili (1982)	$1+4.1\left(\frac{f_l}{f_{co}}\right)$	Miyauchi et al. (1997)	$1+3.485\left(\frac{f_l}{f_{co}}\right)$
Mirmiran and Shahawy (1997)	$1+4.269\left(\frac{f_l^{0.587}}{f_{co}}\right)$	Samaan et al. (1998)	$1+6.0\left(\frac{f_l^{0.7}}{f_{co}}\right)$
Kono et al. (1998)	$1+0.0572f_l$	Razvi and Saatcioglu (1999)	$1+6.7\left(\frac{f_l^{0.83}}{f_{co}}\right)$
Saafi et al. (1999)	$1+2.2\left(\frac{f_l}{f_{co}}\right)^{0.84}$	Spoelstra and Monti (1999)	$0.2+3.0\left(\frac{f_l}{f_{co}}\right)^{0.5}$
Toutanji (1999)	$1+3.5\left(\frac{f_l}{f_{co}}\right)^{0.85}$	Miyauchi et al. (1999)	$1+2.98\left(\frac{f_l}{f_{co}}\right)$
Xiao and Wu (2000)	$1.1+\left[4.1-0.75\left(\frac{f_{co}^2}{C_j}\right)\right]\left(\frac{f_l}{f_{co}}\right)$	Lam and Teng (2001)	$1+2.0\left(\frac{f_l}{f_{co}}\right)$
Karabinis and Rousakis (2001)	$1+2.1\left(\frac{f_l}{f_{co}}\right)^{0.87}$	Shehata et al. (2002)	$1+1.25\left(\frac{f_l}{f_{co}}\right)$
Xiao and Wu (2003a)	$1.1+\left[4.1-0.45\left(\frac{f_{co}^2}{C_j}\right)^{1.4}\right]\left(\frac{f_l}{f_{co}}\right)$	Campione and Miraglia (2003)	$1+2.0\left(\frac{f_l}{f_{co}}\right)$
Lam and Teng (2003a)	$1+3.3\left(\frac{f_l}{f_{co}}\right)$	Matthys et al. (2005)	$1+2.3\left(\frac{f_l}{f_{co}}\right)^{0.85}$
Youssef et al. (2007)	For ascending curves $1+2.25\left(\frac{f_l}{f_{co}}\right)^{1.25}$, For descending curves $1+3.0\left[\frac{\left(\frac{4E_f\varepsilon_j t}{d}\right)}{f_{co}}\right]^{1.25}$	Kumutha et al. (2007)	$1+0.93\left(\frac{f_l}{f_{co}}\right)$

8.4.1 Confinement effectiveness of square jacket

The confinement pressure produced by a square jacket on a concrete column is not uniform across the cross section. The nonuniformity of the confinement pressure not only greatly complicates the problem but also significantly reduces the confinement effect. Based on engineering common sense, people expect that the concrete near the straight sides of a column is less effectively confined and that around corners and in the center of the column is more effectively confined, as illustrated in Fig. 8.7. Models of effective confinement area have been proposed by researchers. One of these models is the 45 degrees model (Fig. 8.7A) where a parabola that intersects the two ends of a straight side at 45 degrees is used to separate the effectively (shaded zone) and ineffectively (nonshaded zones) confined zones (Mander et al. 1988a). Another one is the model by Lam and Teng (2003b) (Fig. 8.7B), where the parabola ends are parallel to the diagonals of the rectangular cross section. In such a model, the concrete in the effectively confined area is assumed to be as effectively confined as the concrete inside a circular jacket while the concrete in the ineffectively confined area has no confinement at all. Strength models for such rectangular columns have been developed based on the effective confinement area concept, as given in Section 8.4.3.1. Although rough, this approach of modeling has been widely adopted in both the literature and design codes/guidelines due to its simplicity.

A more analytical-based approach has been developed by Wu and coworkers (Wu et al. 2006; Wang and Wu 2008; Wu and Wang 2009; Wu and Wei 2010). To gain an insight into the confinement provided by a square jacket, consider the thought experiment in Fig. 8.8 where a square column with a side length b and corner radius r is confined by an external jacket.

Assume that the confining jacket is made of a membrane material that can only provide tensional resistance without flexural resistance, and that the material of the concrete is linear elastic. For a column cross section without a jacket and subjected to a uniform axial strain ε, the transverse concrete strain in both transverse orthogonal directions will have a constant value of $\mu\varepsilon$, where μ is the Poisson's ratio. Because of the uniform transverse strain across the cross section, the cross-sectional shape of the

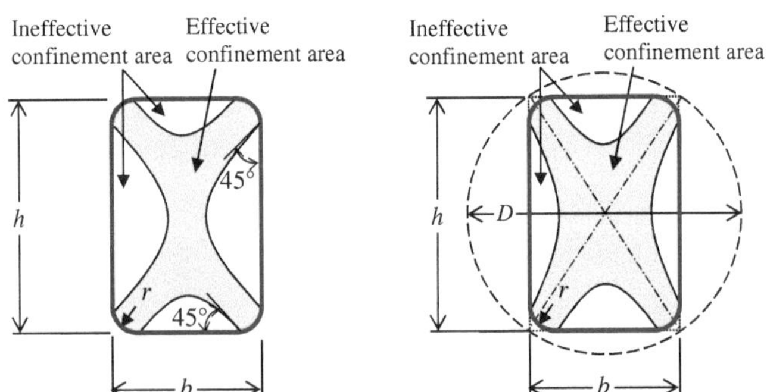

(A) 45° model (B) Lam & Teng's model

Fig. 8.7 Effective confinement area.

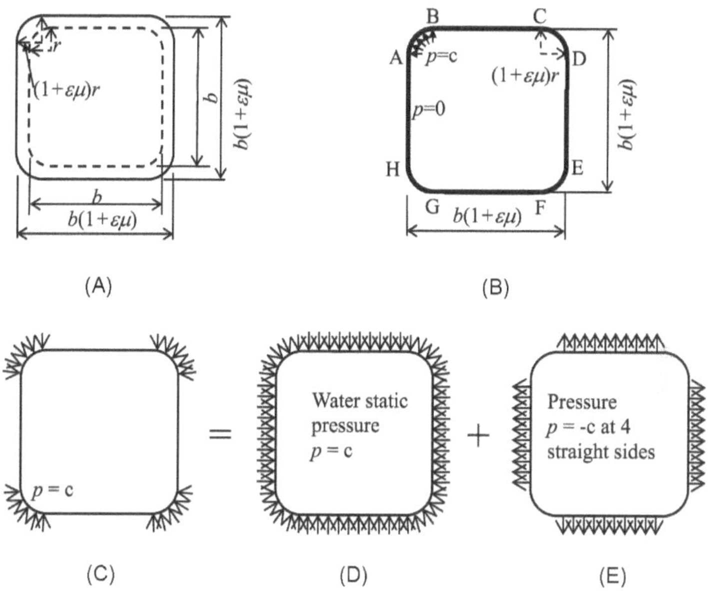

Fig. 8.8 Confinement pressure on square columns.
From Wu, Y. F., T. Liu, and D. J. Oehlers. 2006. "Fundamental principles that govern retrofitting of reinforced concrete columns by steel and FRP jacketing." *Adv. Struct. Eng.* 9 (4), 507–533.

dilated column must be exactly the same as the original one but with an increased side length of $(1 + \mu\varepsilon)b$, as shown in Fig. 8.8A. The membrane jacket must have the same shape as that of the concrete section (Fig. 8.8B). To have a jacket shape such as that in Fig. 8.8B, the only possible pressure on the membrane jacket is a uniform radial pressure p in the four corners and a zero pressure at the four straight sides as shown. Based on the action and reaction law, the confinement pressure exerted on the concrete column must be equal and opposite to that in Fig. 8.8B, as shown in Fig. 8.8C. It can be seen from Fig. 8.8C that there is no lateral pressure or confinement on the concrete on the column surface at the straight sides. Obviously, the internal lateral stress of the concrete in the vicinity of the four straight sides is small, if any.

The confinement pressure in Fig. 8.8C can be visualized as the superposition of a water static pressure on the whole surface of the column as in Fig. 8.8D, and a uniform negative pressure $p = -c$ at the four straight sides as shown in Fig. 8.8E. The water static pressure in Fig. 8.8D provides a uniform confinement that can only produce a uniform transverse stress field across the whole cross section. Therefore, the variation of transverse stress field induced by the external pressure in Fig. 8.8E represents the variation of the superposed transverse stress field of Fig. 8.8D and E, and hence the equivalent case of Fig. 8.8C. In other words, the variation of transverse stress distribution due to Fig. 8.8E portrays the confinement stress field distribution caused by the square jacket. The constant negative pressure at the four straight sides in Fig. 8.8E produces the confining pressure distributions as shown by the contours of transverse stress in Fig. 8.9.

The distributions in Fig. 8.9 are obtained from linear finite element analysis. The magnitude of the external confining pressure will not change the distribution of internal stresses (shape of the contours) but will change the magnitudes of them. It can be seen that the distribution approaches a uniform field of zero lateral stress when the corner radius approaches zero. This can be mathematically explained: when the corner radius approaches zero, the water static pressure of Fig. 8.8D and the negative pressure of Fig. 8.8E cancel out each other, resulting in a zero external pressure and zero confinement stress in the cross section as in Fig. 8.9A. When the straight sides approach zero, a circular section is obtained and all concrete inside the jacket is effectively confined with a uniform internal stress field, as shown in Fig. 8.9H.

For the same confining material, the confinement effectiveness is highest when $2r/b = 1$ in Fig. 8.9H and zero when $2r/b = 0$ in Fig. 8.9A. The confinement pressure for the circular case given by a particular confining material is calculated by Eqs. (8.6) or (8.7). The proportion of well-confined zone relative to the whole cross-sectional area decreases when the corner radius ratio decreases (Fig. 8.7). The corner radius ratio is defined as

$$\rho = \frac{r}{0.5b} = 2\frac{r}{b} \tag{8.8}$$

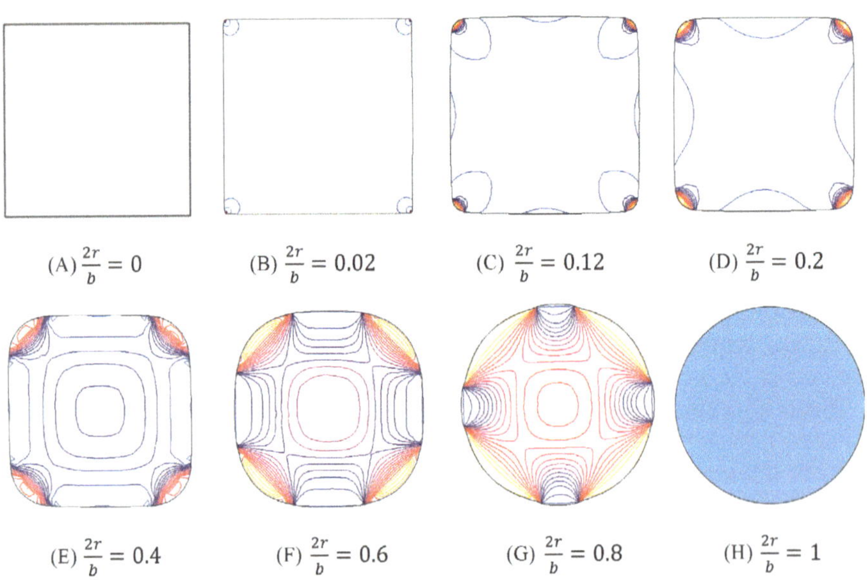

Fig. 8.9 Contours of confining pressure for columns with different corner radius ratios. From Wu, Y. F., T. Liu, and D. J. Oehlers. 2006. "Fundamental principles that govern retrofitting of reinforced concrete columns by steel and FRP jacketing." *Adv. Struct. Eng.* 9 (4), 507–533.

when $\rho = 0$ a sharp corner is obtained and no effective confinement is provided, and when $\rho = 1$ a circular section is obtained, which gives the most effective confinement.

The reduction of confinement effect due to the noncircular shape of the cross section is called the shape effect and is accounted for by multiplying the shape factor k_s by the confinement effectiveness coefficient in Eq. (8.2), i.e.,

$$\frac{f_{cc}}{f_{co}} = 1 + k_s k_0 \frac{f_l}{f_{co}} \tag{8.9}$$

Based on the analytical study earlier, $k_s = 1$ when $\rho = 1$ and $k_s = 0$ when $\rho = 0$. In between these two extreme cases, the shape factor was studied by Wang and Wu (2008) systematically through experimental tests. Fig. 8.10 illustrates a typical relationship between ρ and $\frac{f_{cc}}{f_{co}}$ obtained from their experimental tests. The term $k_0 \frac{f_l}{f_{co}}$ in Eq. (8.9) is for circular columns and can be treated as a constant in Fig. 8.10 for columns confined by a particular FRP jacket. Therefore, the trend of variation of $\frac{f_{cc}}{f_{co}}$ in Fig. 8.10 shows the variation of k_s when ρ changes. Clearly, k_s has approximately a linear relationship with ρ. In other words, one has

$$k_s = 2\frac{r}{b} \tag{8.10}$$

Eq. (8.10) satisfies the two boundary conditions of $k_s = 1$ when $\rho = 1$ and $k_s = 0$ when $\rho = 0$, as discussed earlier.

By collecting a large test database of FRP-confined circular columns ($k_s = 1$) and through regression of the model, Wu and Wang (2009) modified Eq. (8.9) to

$$\frac{f_{cc}}{f_{co}} = 1 + 2.2\left(\frac{f_l}{f_{co}}\right)^{0.94} \tag{8.11}$$

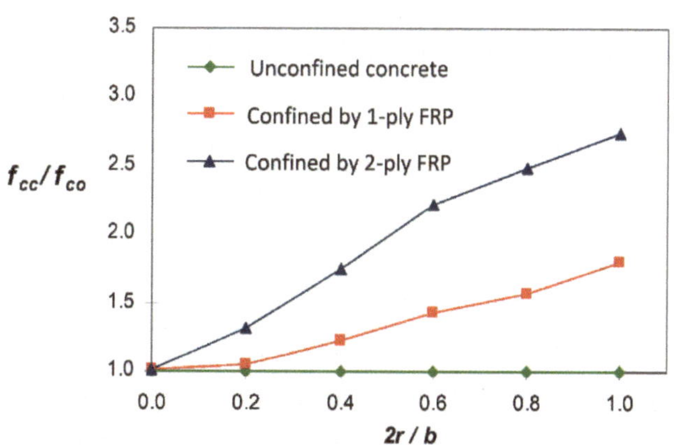

Fig. 8.10 Effect of corner radius ratio on strength of confined concrete.

in which $k_0 = 2.2$ and the exponent of $\frac{f_l}{f_{co}}$ is changed from 1.0 to 0.94 based on the test results. For square columns, the second term of Eq. (8.11) should be multiplied by a shape factor k_s. Using their test database for square columns and through regression analysis of test results, Wu and Wang (2009) further revised the shape factor given by Eq. (8.10) to $k_s = \left(\frac{2r}{b}\right)^{0.72}$ to give the strength of FRP-confined concrete for square columns:

$$\frac{f_{cc}}{f_{co}} = 1 + 2.2\left(\frac{2r}{b}\right)^{0.72}\left(\frac{f_l}{f_{co}}\right)^{0.94} \tag{8.12}$$

It is noted that the modification of the two exponents from 1 in Eqs. (8.9), (8.10) to 0.94 and 0.72 in Eq. (8.12) is purely a best fit to test data. When different test data are used, regression analysis will give different coefficients and exponents of the equation. This is the main reason for the differences in the models summarized in Table 8.1. Therefore, a sufficiently extensive, accurate, and impartial database is the key for derivation of a reliable model from regression analysis.

It is also noted that, due to the nonuniform distribution of lateral stresses, the longitudinal stress distribution is not uniform across the cross section of a column even under concentric (pure axial) loading except for the case of circular columns. Therefore, the longitudinal stress or strength of a column is on an average sense of a cross section.

8.4.2 Effective strain of FRP jacket

The strength of FRP materials is usually obtained from standard flat coupon tests of FRP laminate/plate. Due to the stress concentration caused by concrete cracking and change of curvature (particularly at corners) of an FRP jacket as well as the difference between the workmanship in making FRP jacket and FRP flat coupon, the material strength of FRP from FRP flat coupon tests usually cannot be fully achieved at failure of FRP-confined concrete columns. In other words, the FRP jackets rupture earlier than the FRP flat coupon. An efficiency factor is applicable in calculating the effective confinement pressure f_l using Eq. (8.7) (Wu and Jiang 2013b):

$$\varepsilon_j = k_\varepsilon \varepsilon_{fu} \tag{8.13}$$

where ε_{fu} is the ultimate FRP strain at the rupture of FRP flat coupon and k_ε is the efficiency factor. k_ε is usually less than 1.0 due to the premature failure of FRP in a real column jacket.

Large discrepancies in the efficiency factor have been reported in the literature, with values ranging from 0.3 to 1.1 (Wu and Jiang 2013b). ACI 440.2R-17 (ACI Committee 440 2017) recommended a value of $k_\varepsilon = 0.55$. Since the efficiency factor is significantly smaller than 1, it cannot be ignored in calculating the confinement pressure and jacket design. The large discrepancy in k_ε causes controversy in the development of design guidelines. Wu and Jiang (2013b) studied the problem and concluded that the discrepancy in k_ε is related to (1) inaccuracy of measurements in tests and (2) inconsistency in fiber alignment between the manufacturing of the FRP jacket and the making of the FRP flat coupon.

Since it is difficult to reach consensus on a suitable value of k_ε, Wu and Wang (2009) proposed an approach where the use of k_ε is avoided. The approach directly uses the FRP material strength from flat coupon tests and calculates the confinement pressure by Eq. (8.14) rather than using Eqs. (8.7) together with (8.13), i.e.,

$$f_l = 2\frac{f_f \cdot t}{b} = 2\frac{E_f \cdot \varepsilon_{fu} \cdot t}{b} \tag{8.14}$$

where f_f is the rupture strength of FRP material obtained from the flat coupon test of FRP material. For circular columns, $b = D$ (since $\frac{2r}{b} = 1$, $b = 2r = D$). The confinement pressure calculated from Eq. (8.14) is directly used in Eq. (8.12). The efficiency reduction factor k_ε is not omitted in Eq. (8.12) but combined into the coefficients and indexes of the model. Detailed derivation of the model without using the efficiency reduction factor is given in the following text.

For an FRP material with the ultimate strain ε_{fu}, columns with different shapes may have different ultimate FRP strains ε_j at the onset of peak column strength f_{cc}. For columns with a descending type of stress-strain curve after the peak strength, the columns' strength does not even occur at the onset of the FRP rupture that occurs at the last point of the curve. In this case, there is no direct relation between FRP rupture strength f_f and f_{cc}. Let us consider the general form of the strength gain of an FRP-confined concrete column. The confinement pressure at the onset of f_{cc} occurs at the FRP strain ε_j that is given by Eq. (8.13) in which k_ε is a factor related to the corner radius ratio ρ, i.e., $k_\varepsilon = k_\varepsilon(\rho)$. If $k_s(\rho)$ reflects the shape effect due to the nonuniform pressure distribution (see Eq. 8.9), then based on Eq. (8.9), the strength gain can be expressed in the following general form:

$$\begin{aligned}\frac{f_{cc}}{f_{co}} &= 1 + \mathrm{C}_1 k_s(\rho)\left(\frac{k_\varepsilon(\rho) \cdot f_l}{f_{co}}\right)^{\mathrm{C}_2} \\ &= 1 + \mathrm{C}_1 k_s(\rho)(k_\varepsilon(\rho))^{\mathrm{C}_2}\left(\frac{f_l}{f_{co}}\right)^{\mathrm{C}_2} \\ &= 1 + \mathrm{C}_1 f(\rho)\left(\frac{f_l}{f_{co}}\right)^{\mathrm{C}_2}\end{aligned} \tag{8.15}$$

where C_1 and C_2 are constants to be determined and f_l is calculated from Eq. (8.14), i.e., the ultimate strength of the FRP material f_f. Eq. (8.15) shows that the shape factor and the efficiency factor can be combined to give an overall factor related to the shape of the column, i.e., $f(\rho) = k_s(\rho)(k_\varepsilon(\rho))^{\mathrm{C}_2}$.

Through the same regression analysis as that for the derivation of Eq. (8.12), an appropriate form of $f(\rho)$ is identified as the exponential function of $\frac{2r}{b}$, i.e., $f(\rho) = \left(\frac{2r}{b}\right)^{\mathrm{C}_3}$. Therefore, Eq. (8.15) becomes

$$\frac{f_{cc}}{f_{co}} = 1 + \mathrm{C}_1\left(\frac{2r}{b}\right)^{\mathrm{C}_3}\left(\frac{f_l}{f_{co}}\right)^{\mathrm{C}_2} \tag{8.16}$$

Using the same database for the derivation of Eq. (8.12) and a similar regression analysis, the coefficients C_1, C_2, and C_3 can be identified to give the model:

$$\frac{f_{cc}}{f_{co}} = 1 + 2.23\left(\frac{2r}{b}\right)^{0.73}\left(\frac{f_l}{f_{co}}\right)^{0.96} \tag{8.17}$$

Eq. (8.17) is almost identical to Eq. (8.12). The small difference comes from the different procedures used in the regression analyses for the derivations of the two equations and numerical errors involved in the regressions. More details of the derivation can be found in Wu and Wang (2009).

The derivation of Eq. (8.17) shows that FRP material strength, i.e., f_f or $E_f \cdot \varepsilon_{fu}$, can be used directly in the strength model of FRP-confined concrete columns. The efficiency factor k_ε can be integrated directly into the coefficients of the strength model. One advantage of this approach is that users can save the trouble of selecting an appropriate value of k_ε, which is confusing and controversial. Another advantage is that it avoids an unnecessary procedure that causes error and discrepancy: Eq. (8.13) requires measuring ε_{fu} from the FRP flat coupon and ε_j from the FRP column jacket. The efficiency factor must be different between an FRP jacket with a small corner radius and that with a large corner radius. It is impractical to use different values of k_ε for different shapes of columns. The rationale behind Eq. (8.17) is that different values of k_ε for different shapes of column are automatically integrated into the model, because both the efficiency factor and shape factor affect the test results of f_{cc}. The use of column test results for regression of the model automatically includes these effects into the final strength model of Eqs. (8.12) or (8.17).

It is noted that most existing strength models including those in current major design codes of practice adopt an efficiency factor for the calculation of FRP strength at the onset of column strength.

8.4.3 Development of models for FRP-confined square/rectangular columns

Extensive works have been reported in the literature on modeling the strength of FRP-confined square/rectangular columns with rounded corners. Readers can see an overview of the early development of the modeling and typical models reported in the literature in this section. These models are evaluated at the end of the section.

8.4.3.1 Empirical models

(1) Restrepo and De Vino's model (1996)
This model was one of the earliest strength models for square/rectangular columns confined by FRP jacket. The model was based on the work of Mander et al. (1988a) for steel-confined rectangular columns. The compressive strength of FRP-confined rectangular concrete columns under concentric loading was given as

$$\frac{f_{cc}}{f_{co}} = k_c = \alpha_1 \alpha_2 \tag{8.18}$$

where k_c is the enhancement factor of concrete strength, which depends on the biaxial state of stresses induced by the effective lateral confining pressure. The factors α_1 and α_2 were given by

$$\alpha_1 = 1.25\left(1.8\sqrt{1+7.94\frac{f_{l,j1}}{f_{co}}} - 1.6\frac{f_{l,j1}}{f_{co}} - 1\right) \tag{8.19}$$

$$\alpha_2 = \left[1.4\frac{f_{l,j2}}{f_{l,j1}} - 0.6\left(\frac{f_{l,j2}}{f_{l,j1}}\right)^2 - 0.8\right]\sqrt{\frac{f_{l,j1}}{f_{co}}} + 1 \tag{8.20}$$

in which $f_{l,j1}$ and $f_{l,j2}$ are the larger and smaller values of confinement pressure due to the composite jacket in the two orthogonal directions, respectively, given by

$$f_{l,j1} = \frac{2nt}{b} k_s f_j \tag{8.21}$$

$$f_{l,j2} = \frac{2nt}{h} k_s f_j \tag{8.22}$$

where b and h are the breadth (smaller) and depth (larger) of the column section, respectively; n and t are the number of plies and the nominal thickness of one ply of the wrapping material, respectively. The stress of the jacket f_j can be calculated as $f_j = E_f \varepsilon_j$, and ε_j is assumed to be 0.005 when the peak axial compressive load is attained. k_s is the shape factor that accounts for the effect of nonuniform confinement and is defined as

$$k_s = 1 - \frac{(b-2r)^2 + (h-2r)^2}{3bh(1-A_s)} \tag{8.23}$$

where r denotes the corner radius and A_s is the area of longitudinal steel reinforcement.

(2) Mirmiran et al. model (1998)

Mirmiran et al. (1998) proposed the following model:

$$\frac{f_{cc}}{f_{co}} = 1 + 6.0\left(\frac{2r}{D}\right)\left(\frac{f_l^{0.7}}{f_{co}}\right) \tag{8.24}$$

In this model, the standard form of Eq. (8.9) is used. The confinement effectiveness coefficient k_0 is equal to $6.0 f_l^{-0.3}$, and the shape factor is $k_s = 2r/D$, in which D is the diameter of an equivalent circular column and equated to the side length of a square column or the longer side length in the case of a rectangular section. The lateral confining pressure f_l is calculated by replacing b in Eq. (8.14) with D.

(3) ACI 440.2R02 model
The ACI 440.2R02 model (ACI Committee 440 2002) adopted a form that was originally proposed by Mander et al. (1988a) for steel-confined concrete columns:

$$\frac{f_{cc}}{f_{co}} = -1.254 + 2.254\sqrt{1 + \frac{7.94k_s f_l}{f_{co}}} - 2\frac{k_s f_l}{f_{co}} \tag{8.25}$$

The confining pressure f_l was calculated by

$$f_l = \frac{\rho_f \varepsilon_{fe} E_f}{2} \tag{8.26}$$

where ρ_f is the FRP volumetric ratio, given by

$$\rho_f = \frac{4nt}{D} \text{ (for circular sections)} \tag{8.27a}$$

$$\rho_f = \frac{2nt(b+h)}{bh} \text{ (for noncircular sections)} \tag{8.27b}$$

The nominal hoop rupture strain in the FRP jacket is assumed to be equal to an effective value, ε_{fe}, and it is specified that for RC column members that are completely wrapped by the FRP system around the cross section, the maximum strain used for design should be limited to 0.4%, that is,

$$\varepsilon_{fe} = 0.004 \le 0.75\varepsilon_{fu} \tag{8.28}$$

The shape factor for square and rectangular sections is determined by

$$k_s = 1 - \frac{(b-2r)^2 + (h-2r)^2}{3bh(1-\rho_s)}, \tag{8.29}$$

where ρ_s is the ratio of the longitudinal steel area to the cross-sectional area of the compression member. The confining effect of the FRP jacket should be assumed to be negligible for rectangular sections with an aspect ratio $\frac{h}{b}$ (longer side/shorter side) exceeding 1.5 or side dimensions (b or h) exceeding 900 mm, unless testing demonstrates their effectiveness.

(4) Shehata et al. model (2002)
The equation for the confined concrete strength of square columns is simply given as

$$\frac{f_{cc}}{f_{co}} = 1 + 0.85\left(\frac{f_l}{f_{co}}\right) \tag{8.30}$$

The lateral confining pressure f_l in Eq. (8.30) is calculated by Eq. (8.14).

(5) Campione and Miraglia's model (2003)
This model was developed for FRP-confined square columns. In this model, the confinement effectiveness coefficient k_0 is adopted as a constant of 2.0 and the shape factor k_s is defined as

$$k_s = 1 - \frac{2\left(1 - \frac{2r}{b}\right)^2}{3\left[1 - (4 - \pi)\left(\frac{r}{b}\right)^2\right]} \quad (8.31)$$

The lateral confining pressure f_l is given by

$$f_l = \frac{2\sigma_j t}{b} \quad (8.32)$$

with the FRP stress, σ_j, calculated as

$$\sigma_j = E_f \cdot \varepsilon_{fu} \left[\left(1 - \frac{\sqrt{2}}{2}k_i\right)\left(\frac{2r}{b}\right) + \frac{\sqrt{2}}{2}k_i\right] \quad (8.33)$$

where k_i is a reduction factor, taking into account the stress concentration of FRP at the corners, and was assumed as a constant value of 0.2121 from a best fit to their experimental data. Substituting the value of k_i into Eq. (8.33) they gave

$$\sigma_j = E_f \cdot \varepsilon_{fu} \left[0.85\left(\frac{2r}{b}\right) + 0.15\right] \quad (8.34)$$

Therefore, the model was expressed as

$$\frac{f_{cc}}{f_{co}} = 1 + 2.0k_s\left(\frac{f_l}{f_{co}}\right) = 1 + 2.0k_s\left[0.85\left(\frac{2r}{b}\right) + 0.15\right]\left(\frac{2t}{b}\right)\left(\frac{E_f \cdot \varepsilon_{fu}}{f_{co}}\right) \quad (8.35)$$

(6) Lam and Teng's model (2003b)
Lam and Teng (2003b) proposed the following model:

$$\frac{f_{cc}}{f_{co}} = 1 + 3.3\left(\frac{A_e}{A_c}\right)\left(\frac{f_l}{f_{co}}\right) \quad (8.36)$$

where the value of the confinement effectiveness coefficient $k_0 = 3.3$ was obtained by regression analyses of test results. The confining pressure f_l was calculated from Eq. (8.7) with the equivalent diameter of $D = \sqrt{h^2 + b^2}$, and $\varepsilon_j = k_\varepsilon \varepsilon_{fu}$. Based on the evaluation of their experimental data, they suggested a value of $k_\varepsilon = 0.586$ for carbon FRP (CFRP), 0.624 for glass FRP (GFRP), and 0.851 for aramid FRP (AFRP).

The shape factor k_s was given as the ratio of the effectively confined area A_e (shaded area in Fig. 8.7B) to the concrete area A_c:

$$\frac{A_e}{A_c} = 1 - \frac{\left(\frac{b}{h}\right)(h - 2r)^2 + \left(\frac{h}{b}\right)(b - 2r)^2}{3A_g(1 - \rho_s)} \quad (8.37)$$

where ρ_s = reinforcement ratio of longitudinal steel and A_g = gross area of the column cross section, given by

$$A_g = bh - (4-\pi)r^2 \tag{8.38}$$

(7) Ilki et al. model (2004)
Ilki et al. (2004) proposed the following model:

$$\frac{f_{cc}}{f_{co}} = 1 + 2.4\left(\frac{f'_{lmax}}{f_{co}}\right)^{1.2} \tag{8.39}$$

where $f_{lmax}{}'$ = maximum effective lateral confinement stress, given by

$$f'_{lmax} = \frac{\kappa_a \rho_f \varepsilon_j E_f}{2} \tag{8.40}$$

in which ε_j was assumed to be 70% of the ultimate tensile strain of FRP determined from the flat coupon test, i.e., $\varepsilon_j = 0.7\varepsilon_{fu}$. κ_a = efficiency factor determined as the ratio of the effectively confined cross-sectional area to the gross area, which was assumed to be 1.0 for circular columns. For rectangular sections, it was given by

$$\kappa_a = 1 - \frac{(b-2r)^2 + (h-2r)^2}{3bh}\tan\theta - \frac{(4-\pi)r^2}{bh} - \rho_s \tag{8.41}$$

where θ = arching angle assumed to be 45 degrees.

(8) Al-Salloum's model (2007)
In the model proposed by Al-Salloum (2007), the shape factor k_s was expressed as a function of the breadth of the square section b and the corner radius r, as given by

$$k_s = 1 - \frac{2\left(1 - \frac{2r}{b}\right)^2}{3\left[1 - (4-\pi)\left(\frac{r}{b}\right)^2\right]} \tag{8.42}$$

As the shape factor k_s took into account the effectively confined area of the section, it was suggested that another factor of b/D that accounts for the nonuniformity of the confining pressure be included, where D = diagonal length of the square column, calculated by $D = \sqrt{2}b - 2r\left(\sqrt{2} - 1\right)$. The confinement effectiveness coefficient k_0 was taken as 3.14, which is the average value of 2.98 and 3.3 as proposed by Miyauchi et al. (1999) and Lam and Teng (2003a), respectively. Therefore, the model was expressed as

$$\frac{f_{cc}}{f_{co}} = 1 + 3.14k_s\left(\frac{b}{D}\right)\left(\frac{f_l}{f_{co}}\right) \tag{8.43}$$

(9) Kumutha et al. model (2007)
The model was proposed as

$$\frac{f_{cc}}{f_{co}} = 1 + 0.93\left(\frac{f_l}{f_{co}}\right) \tag{8.44}$$

In this equation, the value of $k_0 k_s$ was found to be a constant of 0.93 from a regression analysis. The confining pressure f_l provided by FRP was given by

$$f_l = \frac{\rho_f f_f}{2} \tag{8.45}$$

where the FRP volumetric ratio ρ_f is given by Eq. (8.27b).

(10) Youssef et al. model (2007)

They proposed different forms for the ascending and descending types of stress-strain curves, given by

$$\frac{f_{cc}}{f_{co}} = 0.5 + 1.225\left(\frac{k_s f_l}{f_{co}}\right)^{0.6} \text{(for ascending curves)} \tag{8.46a}$$

$$\frac{f_{cc}}{f_{co}} = 1 + 1.135\left(\frac{\rho_f E_f \varepsilon_{jt}}{f_{co}}\right)^{1.25} \text{(for descending curves)} \tag{8.46b}$$

In the preceding equations, the shape factor k_s was given by Eq. (8.29); and ε_{jt} = the FRP jacket strain at the transition point from the ascending branch to the descending branch of the stress-strain curves, and was taken as a constant of 0.002. The lateral confining pressure f_l was calculated with Eq. (8.45) and the FRP volumetric ratio ρ_f was given by

$$\rho_f = \frac{2nt[(b+h) - (4-\pi)r]}{bh - (4-\pi)r^2} \tag{8.47}$$

8.4.3.2 Unified models

The empirical models introduced in the previous section cannot accommodate both circular and square/rectangular columns, except for the ones by Mirmiran et al. (1998) and Eq. (8.12). Most researchers developed two separate models, one for circular columns and another for square/rectangular ones. The two separate models have either different formulations or different coefficients for different column shapes, which does not allow a smooth transition from a square/rectangular section to a circular one when the corner radius approaches the case of a circular section. Some models even adopt a constant shape factor regardless of corner radius. Although square/rectangular columns usually have a small corner radius ratio $\rho = 2r/b$ and columns with a large corner radius ratio ρ (close to 1) are rare in practice, unification of models for circular columns and square/rectangular columns into one model is preferable.

The unification of the models for circular, square, and rectangular columns can be achieved by introducing two parameters: (1) the cross-sectional aspect ratio, h/b, and (2) the corner radius ratio, $2r/b$, where h and b are the lengths of the longer and shorter sides, respectively, and r the corner radius, as shown in Fig. 8.11. The aspect ratio, h/b, unifies square and rectangular columns, since square columns are simply the special case of $h/b = 1$. The corner radius ratio, $2r/b$, unifies circular and square shapes, because circular columns are simply the special case of $2r/b = 1$ and sharp-cornered square columns the special case of $2r/b = 0$.

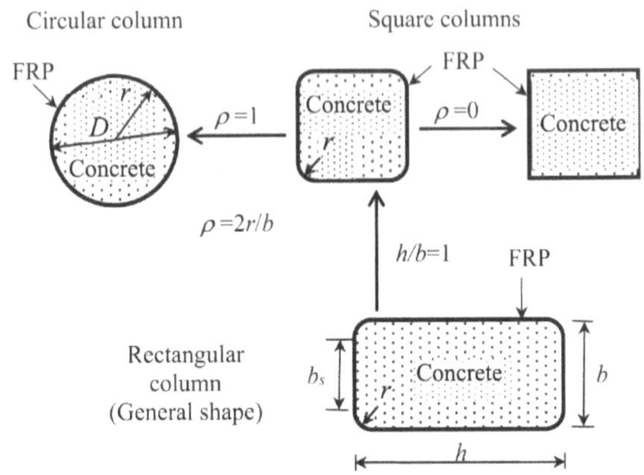

Fig. 8.11 Unification of column shapes.
From Wei, Y. Y., and Y. F. Wu. 2012. "Unified stress-strain model of concrete for FRP-confined columns." *Construct. Build Mater*. 26 (1), 381–392.

A unified model has the following advantages:

(1) It provides a uniform mathematical form and one single model for different shapes of columns, and hence brings convenience to users.
(2) It avoids discontinuity or inconsistency in the results calculated from separate models. When two models are provided, one for circular and another for square/rectangular columns, the result calculated from the square/rectangular model with $h/b=1$ and $2r/b=1$ is usually different from that calculated from the circular model (see Fig. 8.16).
(3) As different columns are evaluated under a single uniform platform, the range of the parameter space is extended. The trend of variation can be captured more comprehensively and hence more accurately when a larger range of parameter variation is allowed in the modeling process, which helps to improve the accuracy of the model.

Eq. (8.12) unifies the cases of circular columns and square ones. Based on further test results for FRP-confined rectangular columns, Wei and Wu (2012) extended Eq. (8.12) to include rectangular columns by including the cross-sectional aspect ratio h/b into the model as another parameter. This extension takes the following form:

$$\frac{f_{cc}}{f_{co}}=1+2.2\left(\frac{2r}{b}\right)^{0.72}\left(\frac{f_l}{f_{co}}\right)^{0.94}\left(\frac{h}{b}\right)^{-1.9} \tag{8.48}$$

8.4.3.3 Hoek-Brown's model

Almost all of the strength models of FRP-confined concrete reported in the literature were based on the mathematical form proposed by Richart et al. (1928) (Eq. 8.2). Wu and Zhou (2010) developed another different one based on the Hoek and Brown failure criterion. By analogy to the nonlinear failure envelope predicted by the classical Griffith crack theory, Hoek and Brown (1980) developed a failure criterion for rock materials. One advantage of the Hoek-Brown failure criterion is its applicability to rock and concrete with existing cracks or damage. Therefore, it is applicable to the design of repair works or strengthening of damaged concrete columns. Another feature of the Hoek-Brown criterion is that it is applicable to both the compression and tension regions of stress so that the entire failure envelope of FRP-confined concrete is available.

By examining a wide range of experimental data for intact rock, rock discontinuities, and jointed rock masses, Hoek and Brown (1980) developed a failure criterion for rocks:

$$\frac{\sigma_1}{\sigma_c} = \frac{\sigma_3}{\sigma_c} + \sqrt{m\frac{\sigma_3}{\sigma_c} + s} \tag{8.49}$$

where σ_1 is the major principal stress at failure, σ_3 is the minor principal stress, and σ_c is the uniaxial compressive strength of the intact rock material. The value of m, a material parameter, depends on the type of rock, whereas s reflects discontinuities in the rock and ranges from 0 to 1. The values of the parameters m and s are available for different kinds of rock with different degrees of discontinuity.

Concrete can be considered as one type of rock and hence Eq. (8.49) is applicable. The parameter s can be determined by letting $\sigma_3 = 0$ in Eq. (8.49), which gives the uniaxial compressive strength as

$$\sigma_1 = \sqrt{s} \cdot \sigma_c \tag{8.50}$$

For concrete without cracks $\sigma_1 = \sigma_c$, hence $s = 1.0$. For completely crushed concrete, $\sigma_1 = 0$ and $s = 0$. For concrete columns with existing damage or cracks, the conventional compression tests can be utilized to determine the value of s from Eq. (8.50):

$$s = \left(\frac{\sigma_1}{\sigma_c}\right)^2 \tag{8.51}$$

where σ_1 represents the strength of the cracked or defective concrete and σ_c is the strength of the original sound concrete.

To determine the value of m for a specific type of concrete, it is assumed $\sigma_1 = 0$ in Eq. (8.49) and in this case failure occurs when the tensile stress is equal to the tensile strength of the concrete, or $\sigma_3 = \sigma_t$, which gives

$$\frac{\sigma_t}{\sigma_c} = \frac{1}{2}\left(m - \sqrt{m^2 + 4s}\right) \tag{8.52}$$

For intact concrete, $s = 1$. When the tensile strength σ_t and the compressive strength σ_c of the concrete are known, the value of m can be obtained from Eq. (8.52).

For confined concrete columns without defect or cracking, $s = 1$. Based on Eq. (8.49), the failure strength can be evaluated as

$$\frac{f_{cc}}{f_{co}} = \frac{f_l}{f_{co}} + \sqrt{m\frac{f_l}{f_{co}} + 1} \tag{8.53}$$

where $f_{co} = \sigma_c$, the compressive strength of unconfined concrete; $f_l = \sigma_3$, the lateral confinement pressure; and $f_{cc} = \sigma_1$, the strength of confined concrete. Similarly, Eq. (8.52) takes the form

$$\frac{f_t}{f_{co}} = \frac{1}{2}\left(m - \sqrt{m^2 + 4}\right) \tag{8.54}$$

where f_t is the tensile strength of concrete. For concrete, the tensile strength f_t can be expressed as a function of f_{co}. Hence, it is reasonable to assume

$$m - \sqrt{m^2 + 4} = C f_{co}^n \tag{8.55}$$

where C and n are coefficients to be determined. The value of m is obtained by solving Eq. (8.55) to give

$$m = \frac{f_{co}^n}{C} - \frac{C}{f_{co}^n} \tag{8.56}$$

Based on laboratory tests on circular concrete specimens confined by different materials, Girgin et al. (2007) found that the constant m for the Hoek-Brown failure criterion depends on the compressive strength of concrete f_{co}. This conclusion is consistent with Eq. (8.56).

Substituting Eq. (8.56) into Eq. (8.53) yields

$$\frac{f_{cc}}{f_{co}} = \frac{f_l}{f_{co}} + \sqrt{\left(\frac{f_{co}^n}{C} - \frac{C}{f_{co}^n}\right)\frac{f_l}{f_{co}} + 1} \tag{8.57}$$

The coefficients of C and n can be determined through a regression analysis using test results. From the regression analysis, Wu and Zhou (2010) identified the coefficients C and n to be -16.7 and 0.42, respectively. Therefore, the strength model for FRP-confined circular concrete columns is given by

$$\frac{f_{cc}}{f_{co}} = \frac{f_l}{f_{co}} + \sqrt{\left(\frac{16.7}{f_{c0}^{0.42}} - \frac{f_{co}^{0.42}}{16.7}\right)\frac{f_l}{f_{co}} + 1} \tag{8.58}$$

For FRP-confined square columns, more parameters affect the strength gain, $\frac{f_{cc}}{f_{co}}$. However, when all other factors are the same as those for circular columns, the only additional variable is the corner radius ratio ρ. In this case, the effective confinement pressure can be expressed as a function of the corner radius ratio, or $f_l \cdot f(\rho)$, where $f(\rho)$ is the shape factor that reflects the reduction in the confinement effect caused by the corner radius. The shape factor $f(\rho)$ reflects two effects: (1) the shape effect due to the nonuniform pressure distribution $k_s(\rho)$ and (2) the reduction of FRP strength

$k_\varepsilon(\rho)$ (see Eq. 8.13). By replacing f_l in Eq. (8.58) with $f_l \bullet f(\rho)$, the following relationship is obtained:

$$\frac{f_{cc}}{f_{co}} = \frac{f_l}{f_{co}} f(\rho) + \sqrt{\left(\frac{16.7}{f_{c0}^{0.42}} - \frac{f_{co}^{0.42}}{16.7}\right)\frac{f_l}{f_{co}} f(\rho) + 1} \tag{8.59}$$

Based on the analysis in Section 8.4.2, the shape factor can be assumed to have the following form:

$$f(\rho) = \rho^{\alpha} \tag{8.60}$$

where α is a constant to be determined from experimental data. Clearly, $f(\rho) = 0$ and 1 are the cases for sharp-cornered square columns and circular columns, respectively. Therefore, when $\rho = 1$, Eq. (8.59) reduces to the strength model for circular columns, i.e., Eq. (8.58). Substituting Eq. (8.60) into Eq. (8.59) gives

$$\frac{f_{cc}}{f_{co}} = \frac{f_l}{f_{co}} \rho^{\alpha} + \sqrt{\left(\frac{16.7}{f_{c0}^{0.42}} - \frac{f_{co}^{0.42}}{16.7}\right)\frac{f_l}{f_{co}} \rho^{\alpha} + 1} \tag{8.61}$$

By regression analysis using a database, Wu and Zhou (2010) identified α to be 0.85. Eq. (8.61) becomes

$$\frac{f_{cc}}{f_{co}} = \frac{f_l}{f_{co}} \rho^{0.85} + \sqrt{\left(\frac{16.7}{f_{c0}^{0.42}} - \frac{f_{co}^{0.42}}{16.7}\right)\frac{f_l}{f_{co}} \rho^{0.85} + 1} \tag{8.62}$$

For columns with existing damage, simply replace 1 in Eq. (8.62) with s to give

$$\frac{f_{cc}}{f_{co}} - \frac{f_l}{f_{co}} \rho^{0.85} + \sqrt{\left(\frac{16.7}{f_{c0}^{0.42}} \quad \frac{f_{co}^{0.42}}{16.7}\right)\frac{f_l}{f_{co}} \rho^{0.85} + s} \tag{8.63}$$

It is noted that the confinement pressure f_l in the preceding equation is calculated from the ultimate strength of FRP from flat coupon tests, i.e., Eq. (8.14).

8.4.4 Assessment of models

Wu and Zhou (2010) evaluated the existing models described in Sections 8.3.2 and 8.4.3 using a database collected from experimental tests published in the literature. The performance of a model was assessed by the deviation between the predictions of a model and relevant test results. The integral absolute error (*IAE*) as given by Eq. (8.64) was used to calculate the deviation

$$IAE = \sum \frac{\sqrt{\left[Expe.\left(\frac{f_{cc}}{f_{co}}\right) - Theo.\left(\frac{f_{cc}}{f_{co}}\right)\right]^2}}{\sum Expe.\left(\frac{f_{cc}}{f_{co}}\right)}, \tag{8.64}$$

where $Expe.\left(\frac{f_{cc}}{f_{co}}\right)$ and $Theo.\left(\frac{f_{cc}}{f_{co}}\right)$ are the experimental and theoretical values of the strength gain, respectively. *IAE* is a normalized statistical index that is sensitive to the deviation of a model to test results. The models were further assessed by conventional statistical analysis using the ratio v, defined as

$$v = \frac{Expe.\left(\frac{f_{cc}}{f_{co}}\right)}{Theo.\left(\frac{f_{cc}}{f_{co}}\right)} \tag{8.65}$$

The correlations between the model predictions and the test results are shown in Figs. 8.12–8.14. The accuracy of a model was classified into three categories: category I (when $IAE \leq 15\%$), category II (when $15\% < IAE \leq 30\%$), and category III (when $IAE > 30\%$), as shown in Fig. 8.12. Fig. 8.12A and B shows that most of the strength models are within category I and thus have a satisfactory accuracy.

As circular columns are a special case of square columns when the corner radius equals half the breadth of the column, a strength model for square columns can in principle be extended to that for circular columns. When this extension is made, the accuracy of many models significantly decreases, as shown in Fig. 8.12C. This reduction in accuracy is as expected because the original models were derived from test results for square columns with small corner radii, except the model by Wu and Wang (2009) for which the full range of $\rho = 0$–1 was allowed when developing the unified strength model, leading to a higher accuracy.

To provide a better picture of the distribution of error, a typical model in each category is selected for further comparison with the test results in Figs. 8.13 and 8.14. Razvi and Saatcioglu's model, which belongs to category III, significantly overestimates the strength gain of FRP-confined circular columns, while Shehata et al. and Youssef et al. models from category II and III, respectively (Figs. 8.13B and 8.14A and B), underestimate the results. Due to the more analytical-based derivation process, the models by Wu and Wang (2009) and Wu and Zhou (2010) show a relatively better performance, with a smaller *IAE* and a more uniform distribution of errors over the parameter space.

As the strength gain of FRP-confined circular and square concrete columns heavily depends on the confinement pressure f_l provided by the FRP and the corner radius ratio ρ, the trends of variation of different models with respect to these two parameters are shown in Figs. 8.15 and 8.16. Fig. 8.15 depicts the influences of the confinement pressure f_l on the strength gain of FRP-confined circular concrete columns. The assessment shows that the strength gain increases approximately linearly as the

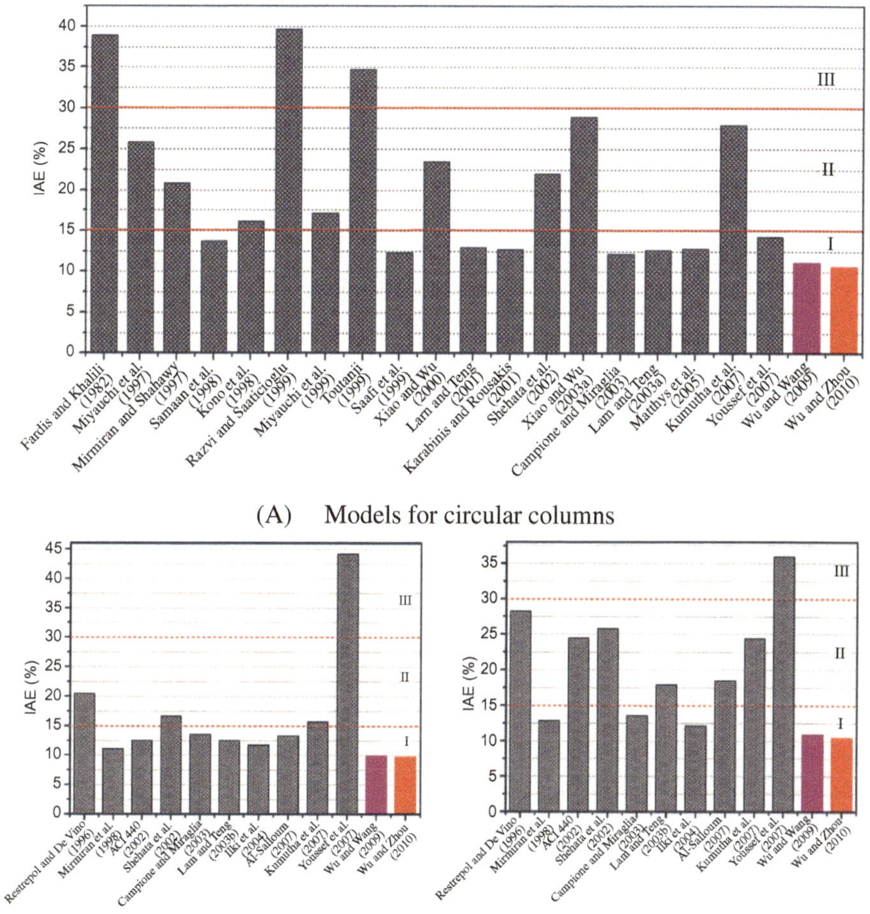

(B) Models for square columns (C) Models for both circular and square columns

Fig. 8.12 *IAE*s of strength models.

From Wu, Y. F., and Y. W. Zhou. 2010. "Unified strength model based on Hoek-Brown failure criterion for circular and square concrete columns confined by FRP." *J. Compos. Constr*. 14 (2), 175–184.

confinement pressure increases. Although these models were derived from two different failure criteria that have different mathematical forms (Eqs. 8.2, 8.49), the trends of the strength gain versus the confinement pressure are very similar. One advantage of the unified form can be seen in Fig. 8.16, because it can avoid the sudden change caused by application of different models for square and circular columns.

Eq. (8.63) can predict the ultimate strength of FRP-repaired defective concrete columns. A test database of predamaged concrete columns that were subsequently repaired by FRP jacketing was collected from the literature for the evaluation of Eq. (8.63), including columns damaged by very low temperature (Green et al. 2006), by compressive crushing (Liu et al. 2004), and by accelerated

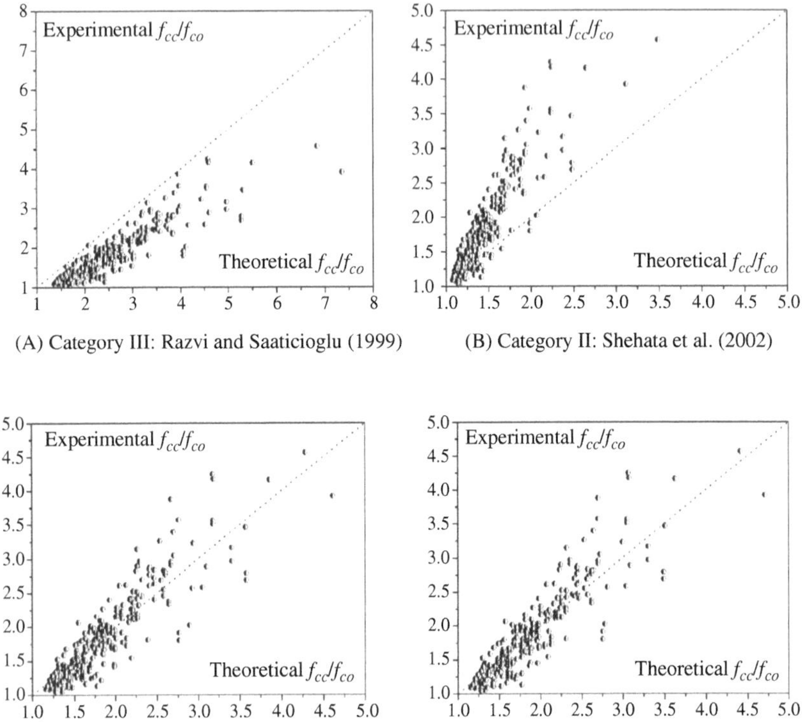

(A) Category III: Razvi and Saaticioglu (1999) (B) Category II: Shehata et al. (2002)

(C) Category I: Wu and Wang (2009) (D) Category I: Wu and Zhou (2010)

Fig. 8.13 Performance of models for circular columns.
From Wu, Y. F., and Y. W. Zhou. 2010. "Unified strength model based on Hoek-Brown failure criterion for circular and square concrete columns confined by FRP." *J. Compos. Constr.* 14 (2), 175–184.

electrochemical corrosion (Tastani and Pantazopoulou 2004). The performance of the equation in predicting the FRP-confined concrete strength for these damaged columns is shown in Figs. 8.17 and 8.18. The damage index s is calculated with Eq. (8.51) using the strength of the plain concrete specimens before and after damage provided in the original references. The *IAE* between the predictions and the test results is 4.1%, and the mean and the variance of v, $\overline{v}$ and v_d, are 1.045 and 0.0044, respectively (Fig. 8.18), which shows a good correlation between the test results and the model prediction by Eq. (8.63). It is noted that damaged and then repaired columns were not used to develop Eq. (8.63) and it was simply obtained by replacing $s = 1$ in Eq. (8.62) by the general case with a damage index s to give Eq. (8.63). The good performance of Eq. (8.63) reflects the rationality of the Hoek-Brown criterion as well as the derivation process for Eq. (8.63). It is also noted that there is a lack of test data in Fig. 8.18 for a very high damage level of $s < 0.49$. Nevertheless, repairing columns with such a high degree of damage is unusual in practice.

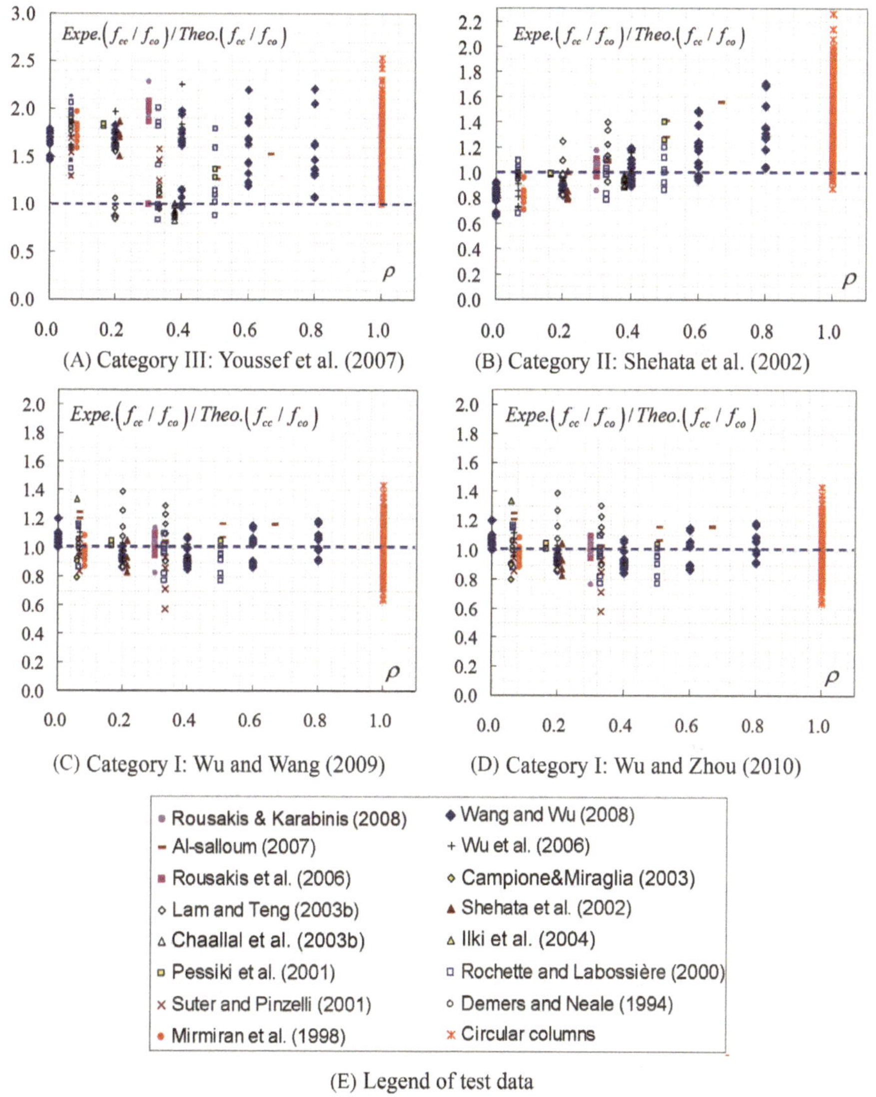

Fig. 8.14 Distribution of error with respect to corner radius ratio.
From Wu, Y. F., and Y. W. Zhou. 2010. "Unified strength model based on Hoek-Brown failure criterion for circular and square concrete columns confined by FRP." *J. Compos. Constr.* 14 (2), 175–184.

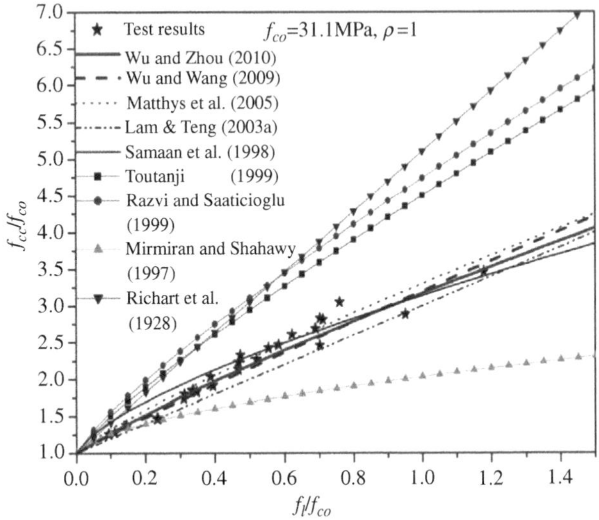

Fig. 8.15 Strength gain vs. confinement pressure.
From Wu, Y. F., and Y. W. Zhou. 2010. "Unified strength model based on Hoek-Brown failure criterion for circular and square concrete columns confined by FRP." *J. Compos. Constr.* 14 (2), 175–184.

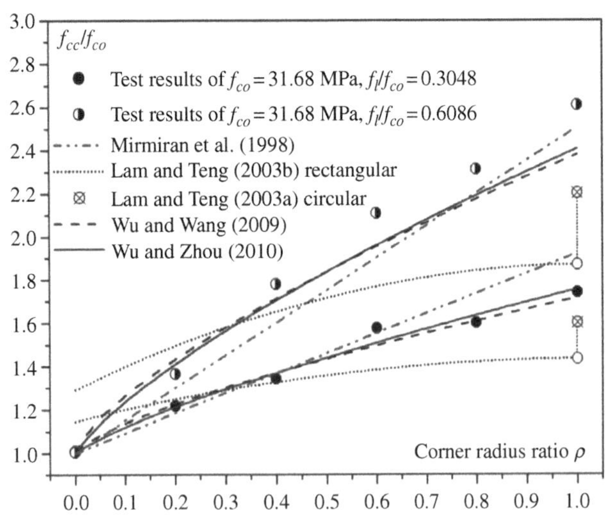

Fig. 8.16 Strength gain vs. corner radius ratio.
From Wu, Y. F., and Y. W. Zhou. 2010. "Unified strength model based on Hoek-Brown failure criterion for circular and square concrete columns confined by FRP." *J. Compos. Constr.* 14 (2), 175–184.

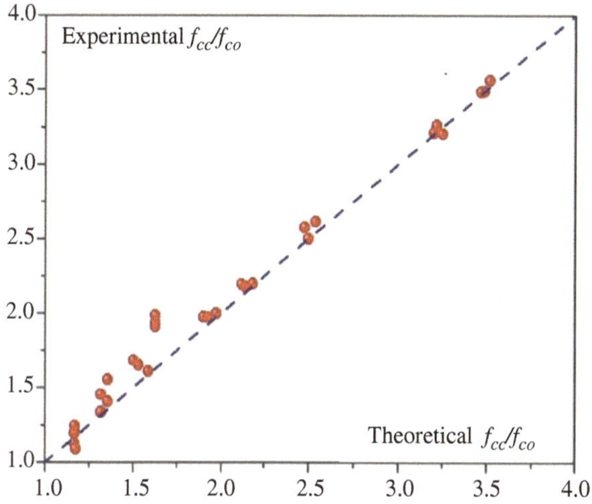

Fig. 8.17 Performance of Eq. (8.63).
From Wu, Y. F., and Y. W. Zhou. 2010. "Unified strength model based on Hoek-Brown failure criterion for circular and square concrete columns confined by FRP." *J. Compos. Constr.* 14 (2), 175–184.

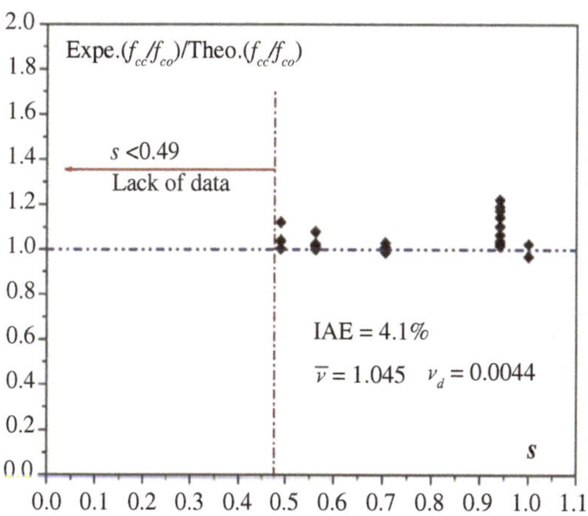

Fig. 8.18 Error distribution of Eq. (8.63).
From Wu, Y. F., and Y. W. Zhou. 2010. "Unified strength model based on Hoek-Brown failure criterion for circular and square concrete columns confined by FRP." *J. Compos. Constr.* 14 (2), 175–184.

8.5 Measures to increase confinement effectiveness for rectilinear columns

Different measures and technologies have been developed to increase the confinement effectiveness of external jacketing. The common approaches reported in the literature are summarized in this section.

8.5.1 Increasing the rigidity of jacketing plate

The analysis in Section 8.4.1 is based on a jacket made from a membrane without flexural rigidity. A thick jacket usually has a certain flexural rigidity that imposes resistance to the out-of-plane bending deformation of the plate, as shown in Fig. 8.19A.

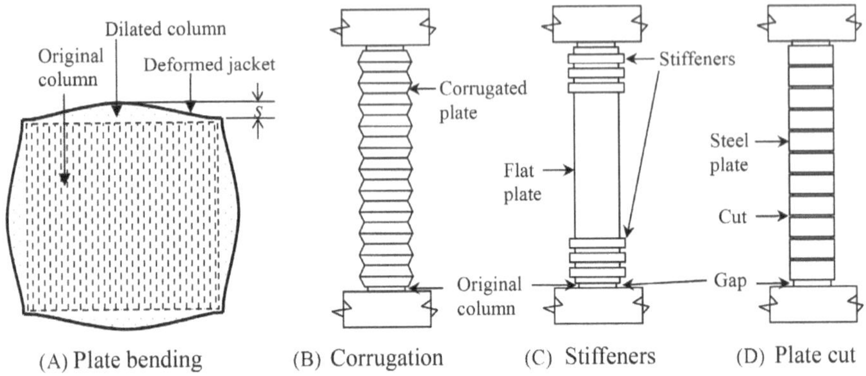

Fig. 8.19 Flexural resistance to dilation of concrete.

The resistance against flexural bending of the jacket also restrains the dilation of the concrete and hence provides certain confinement to the concrete. To increase the confinement effect to concrete due to the flexural resistance of a jacket, different measures have been proposed by researchers, such as corrugation of the plate (Fig. 8.19B) and addition of stiffeners (Fig. 8.19C) (Xiao and Wu 2003b).

8.5.2 Reducing longitudinal stress

Another way to increase the confining action of the jacket is by reducing the longitudinal stress and hence increasing the transverse strength of a jacket, through different means. When a whole piece of plate is wrapped onto a column, the plate is usually subjected to both lateral hoop stress and axial stress, as shown in Fig. 8.20A.

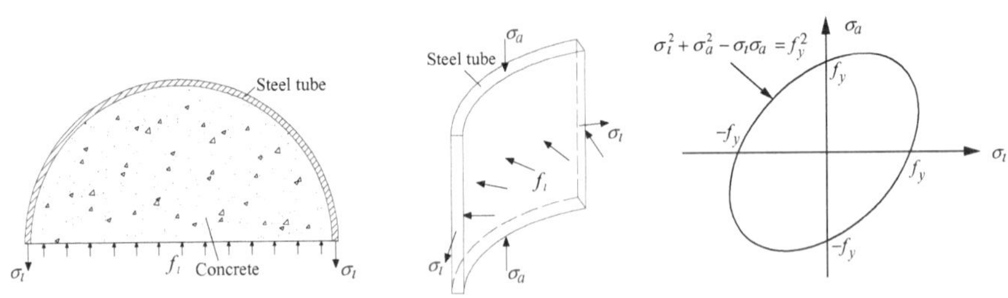

Fig. 8.20 Increasing confinement by reducing axial stress of plate: (A) Stress state of circular steel jacket and (B) von Mises yield plane.
(A) From Wei, Y., C. Jiang, and Y. F. Wu. 2019. "Confinement effectiveness of circular concrete-filled steel tubular columns under axial compression." *J. Constr. Steel Res.* 158, 15–27.

For a steel plate, the von Mises yield criterion can be used to describe its ultimate state:

$$\sigma_l^2 + \sigma_a^2 - \sigma_l \sigma_a = f_y^2 \tag{8.66}$$

where σ_l and σ_a are the lateral hoop stress and axial stress of the steel plate, respectively, at the ultimate state of yielding and f_y is the uniaxial yield strength of the steel plate. Columns are usually under axial compression, and hence the steel jacket is subjected to a compressive axial stress. The dilation of concrete causes tensile stress of the plate in the lateral direction. It can be seen from the von Mises yield plane in Fig. 8.20B that when the plate is under an axial compressive stress, its lateral tensile resistance is reduced.

When the vertical axial stress is reduced or eliminated in some way, such as by cutting it into strips and leaving a gap at the column ends to avoid contact with the support (Fig. 8.19D), the steel plate serves mainly as a hoop confining material. This maximizes the tensile resistance of the plate in the hoop direction, and hence the confinement effect to the concrete. This principle is general and applicable to both circular and rectilinear steel jackets.

8.5.3 Additional anchoring

The ineffectively confined zones (Fig. 8.7) can be significantly reduced by using anchor bolts, as shown in Fig. 8.21. Longer anchors, particularly a through bolt, are more effective and reliable than shorter ones. A nonthrough anchor may be pushed out by the expansion of concrete and should be properly anchored into the effectively confined core concrete zone when there is no nonthrough anchor (the red shaded area in Fig. 8.21). Clearly, use of more anchors can reduce the area of the ineffective confinement zones and increase confinement effect to a column.

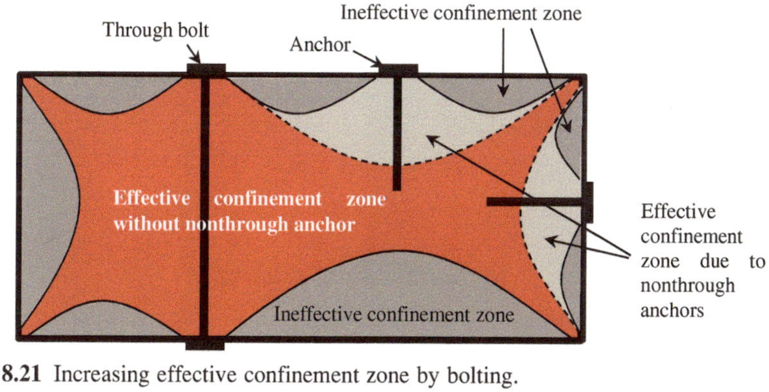

Fig. 8.21 Increasing effective confinement zone by bolting.

The strength of a jacketed rectangular concrete column with anchors may be estimated using an existing model, such as Lam and Teng's model (Section 8.4.3.1(6)) in which A_e is the effectively confined area determined based on the principle illustrated in Fig. 8.21.

8.5.4 Alteration of cross section

The water static pressure in Fig. 8.8D causes a uniform deformation of the cross section, and hence keeps the side straight. The negative side pressure in Fig. 8.8E produces a bulging deformation of the straight sides. Therefore, the straight sides of a rectilinear jacket bulge under concrete dilation, as illustrated in Fig. 8.19A. The dilation of concrete before the onset of the unconfined concrete strength is small. For example, the transverse strain at the peak strength of concrete without confinement can be roughly estimated as $\varepsilon_{co} \times \mu \approx 0.002 \times 0.2 = 0.0004 = 0.04\%$, where μ is the Poisson's ratio. Confinement stress will further reduce this value. A dilation of 0.04% in the dimension cannot be captured by the naked eye, hence no bulging of the side is noticeable in testing before significant concrete softening. However, the lateral dilation and bulging increase quickly after the onset of the peak unconfined concrete strength if it is not firmly restrained by a jacket, as shown in Fig. 8.22.

The side bulging causes little passive confinement from a membrane jacket on a straight side, as a small bulging causes little resistance from the straight membrane. The resistance of the membrane against bulging can be estimated as follows. Assume a straight side bulges to a parabolic-shaped surface (the conclusion is the same with other shapes), it can be derived that a bulging of s at the midpoint of a straight side

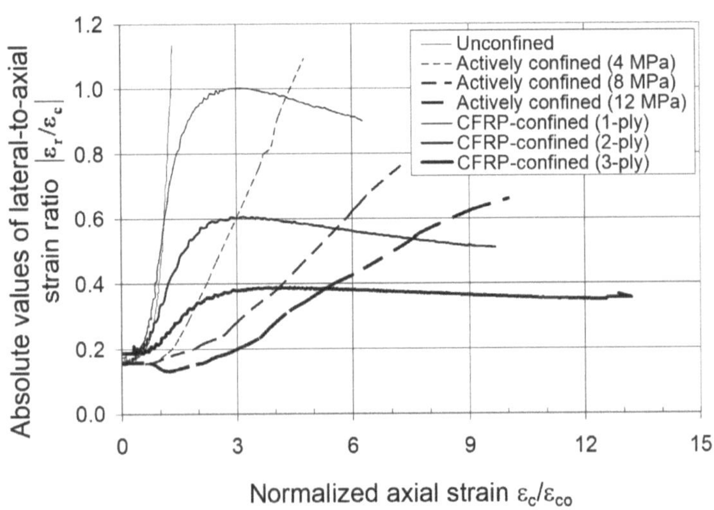

Fig. 8.22 Typical dilation strain of concrete.
From Teng, J. G., and L. Lam. 2004. "Behavior and modeling of fiber reinforced polymer-confined concrete." *J. Struct. Eng.* 130 (11), 1713–1723.

relative to the two ends of the side face (Fig. 8.19A) causes an average lateral strain of the side (=increase of side length/original side length) given by

$$\varepsilon = \frac{8}{3}\left(\frac{s}{b_s}\right)^2 - \frac{32}{5}\left(\frac{s}{b_s}\right)^4 \tag{8.67}$$

where b_s is the length of the original straight side excluding corner radius (Fig. 8.11). For a column with 300 mm straight sides, a 3 mm bulging that corresponds to a dilation strain of 2%, which is a very large value of dilation (Fig. 8.22), causes a strain of $\varepsilon = \frac{8}{3}\left(\frac{3}{300}\right)^2 - \frac{32}{5}\left(\frac{3}{300}\right)^4 = 0.00027$ in the jacket, which is insignificant for the membrane to cause confinement to concrete.

If the straight side is changed to a convex shape, which can be achieved by filling concrete onto the straight side, as shown in Fig. 8.23A, the transverse dilation will be much more effective in mobilizing the hoop strain of the jacket. For the same 300 mm wide column but with a rise of $d = 30$ mm at the middle of its original straight side (Fig. 8.23A), a 3 mm or 2% dilation strain will cause a hoop strain of 0.0051 (0.51%), which is very significant and enough to cause one-third of the ultimate hoop strength of a CFRP jacket. A 2% dilation before the ultimate failure of a column is achievable for steel-confined concrete, as can be seen from the column test results of Mander et al. (1988b), where the ultimate dilation strain at breaking of mild steel stirrups is around 2%–6%. This analysis shows that alterations to the shape of the original straight side by, for example, adding concrete to the side and making it convex is an effective way to increase the confinement (Fig. 8.23A).

The circular-shaped jacket is the most effective in providing confinement as explained in Section 8.4.1, in which the lateral dilation strain of concrete is equal to the tensile strain of the jacket. Using an oval-shaped jacket and filling the gap with additional concrete can also offer very effective and significant confinement to rectangular columns (Fig. 8.23B), which has been experimentally demonstrated (Teng and Lam 2002). The effectiveness comes from the convex curved surface of the oval jacket due to the mechanism discussed earlier. However, it is obvious that the

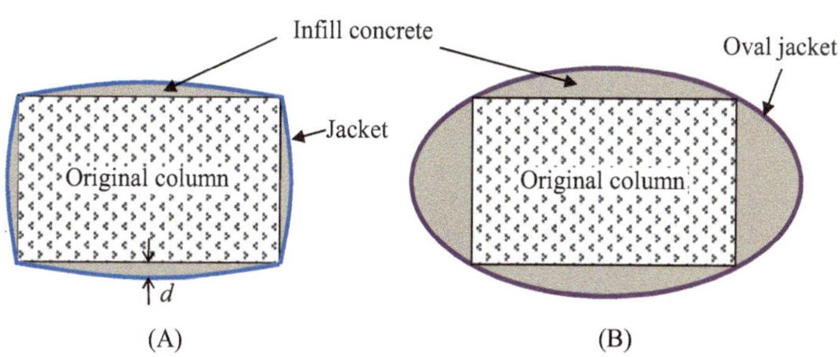

Fig. 8.23 Modification of cross-sectional shape to increase confinement effect.

effectiveness will reduce and diminish when the cross-sectional aspect ratio (depth/breadth) increases. Experimental work by Tan and Yip (1999) showed that the effectiveness of confinement by elliptical hoops diminishes as the side aspect ratio of the major axis to the minor axis of the cross section increases, and becomes insignificant when the aspect ratio is greater than 2.6.

8.5.5 Other means

Apart from the measures discussed in the previous sections, the following means can be used to increase the confinement to a column:

(1) Prestressing the jacket to apply active hoop stress to a column. Various ways can be used to apply active hoop stress, as described in Section 6.1(3).
(2) Rounding off corners. The corner radius ratio $\rho = 2r/b$ is an important factor that affects the confinement. No confinement will be provided by a membrane jacket when the corner radius ratio approaches zero (Fig. 8.10 and Eq. 8.48), no matter how strong the jacket is. A corner radius as large as possible should be made to achieve a better confinement, such as by grinding off concrete at the corners of an existing column.
(3) Avoiding concave surface. A jacket with even a slightly concave surface or recess, such as that caused by poor workmanship in construction, will have a detrimental effect in producing confinement.
(4) A member with a large cross-sectional aspect ratio of depth/breadth, such as a wall, will have a large portion of the longer side being under a small or even no confinement stress, except near the small vicinity of the four corners. In this case, through bolting (Fig. 8.21) should be used for effective confinement.
(5) Increasing jacket stiffness. For the same amount of lateral expansion, a stiffer jacket provides a greater confinement stress. The strength of the jacket limits the maximum confinement that can be provided by the jacket. As higher strength concrete has smaller dilation under compression, a stiffer jacket is usually required to confine higher strength concrete members.
(6) Workmanship. An intimate contact of the jacket with the concrete column is critical. A loose contact will not provide effective confinement to the column. A tight wrapping with even a small prestress (not for active confinement) or use of expansive grout between the jacket and the RC column can significantly improve the contact.

It can be seen from this section that many factors affect the confinement effect on concrete, particularly for rectilinear columns. The preceding factors were consciously or unconsciously varied in construction and experimental studies reported in the literature. The workmanship is especially difficult to specify. This is the reason that large scattering of test results and even contradictory conclusions have been reported in the literature on the effectiveness of confinement on concrete columns, particularly square/rectangular columns.

8.6 Stress-strain relationship of concrete

The stress-strain relationship of concrete is of fundamental importance in studying the behavior of concrete structures. A few popular models of the stress-strain relationship of concrete have been empirically developed by Hognestad (1951), Sargin et al.

(1971), Popovics (1973), and Mander et al. (1988a), among others. These models have only a few parameters and hence are conveniently and widely used in both engineering practice and research works.

With the development of concrete materials such as high-performance concrete, new models for the stress-strain relationship of concrete are required. New construction technologies also require the knowledge of the stress-strain behavior of concrete under complicated stress conditions. For example, FRP-confined concrete has a different stress-strain behavior from that of concrete in reinforced concrete structures because of the passive confinement mechanism. Furthermore, the development of structural theories also requires sophisticated stress-strain models of concrete. One example is the application of the flexural theorems for the derivation of flexural design equations of RC members, which requires more accurate stress-strain relationships of concrete (Chapters 3 and 4).

In the meantime, the stress-strain behavior of concrete under complex stress conditions is complicated. For FRP-confined concrete members with strain gradient in cross section (see Section 8.6.5), many parameters significantly affect the stress-strain behavior. Furthermore, the load path or how loading is applied on a structure affects the stress-strain response of confined concrete (Section 8.6.5.4). Therefore, the stress-strain modeling of concrete is an important but difficult problem and the development of the stress-strain relationships of concrete always plays a pivotal role in structural engineering.

For confined concrete in an axisymmetric system such as in circular columns under uniaxial loading, the stress of concrete across the cross section is uniform. Compression tests can be easily carried out to obtain the load-displacement curve of a structural member. The stress of the confined concrete in such a structural system is simply given by dividing the load by the cross-sectional area and the strain given by dividing the displacement by the gauge length, to give a stress-strain curve. Compared with the uniform confinement and stresses offered by a circular jacket, the nonuniformity of confinement across a cross section, as illustrated in Fig. 8.9, results in a much greater difficulty in the determination of the stress-strain relationship of concrete, as the stresses vary across the cross section significantly under the same axial strain. To overcome this problem, it is a common practice to use an average axial and transverse stress in both design and theoretical modeling in these cases.

Under eccentric loading, i.e., a column with both $N \neq 0$ and $e = M/N \neq 0$, where N, M, and e are the axial force, bending moment, and load eccentricity, respectively, the stress-strain relationship of confined concrete becomes even more complicated. In this case, the load eccentricity as well as other loading factors related to structural design also become factors of the stress-strain relationship. For this reason, there is an argument that the stress-strain relationship of confined concrete is not a constitutive relationship of concrete material but the global behavior of a structural member, and hence should not be treated as a material behavior. This view is apparently correct but one-sided for the following reasons:

(1) In practice, engineers generally use a simple one-dimensional stress-strain relationship $\sigma_z = \varphi(\varepsilon_z)$ for design of reinforced concrete elements, where σ_z and ε_z are the longitudinal stress and strain, respectively. When a three-dimensional constitutive relationship of concrete is used for analysis, the problem becomes too complicated for engineering practice.

(2) Such a relatively simple relationship of $\sigma_z = \varphi(\varepsilon_z)$ does exist and can be developed for engineering uses for the reasons given in Section 8.6.2.

8.6.1 Methods for stress-strain modeling

Currently, a stress-strain relationship for concrete can be developed from (1) constitutive modeling of concrete material and (2) experimental tests. Various constitutive models, such as damage plasticity models (Rousakis et al. 2008; Yu et al. 2010; Jiang and Wu 2012; Mohammadi and Wu 2019), have been developed for research. However, these models are not only complicated and difficult to use by engineers, but also have no reliable approach for determination of the model parameters. In fact, the model parameters are generally determined by matching calculated results to test results. Therefore, it is currently impractical to use such models for daily engineering applications. Current engineering designs are still based on simple empirical models directly derived from experimental tests. These simple empirical models directly give a stress value for concrete for a given strain, and hence are generally one-dimensional models.

Theoretically speaking, stress in one direction is not only determined by the strain in that direction but also related to strains in other directions, and hence is a three-dimensional (3D) problem. However, in a particular structure and under certain conditions, the stress in one direction in a certain part of the structure is actually directly related to the strain in that direction, because strains in other directions are correlated in this case. For example, for FRP-confined concrete columns, the longitudinal stress is directly related to its longitudinal strain.

The majority of the stress-strain models were developed empirically from tests of concrete members under concentric loading. Under concentric loading, the stress distribution is assumed to be uniform across a cross section, and the stress (average stress when stress distribution is not uniform) is calculated by simply dividing the applied force by the cross-sectional area. However, it becomes much more complicated when a member is loaded eccentrically due to nonuniform distribution of longitudinal strain in the cross section. There are three typical approaches to identifying stress from the test results of a member in such cases:

(1) Direct measurement. With this method, small load cells are embedded into a test specimen to measure the local stress directly. However, the presence of a sensor inevitably disturbs the original stress field, thereby leading to unreliability of results. No technique or method is available at present for direct measurement of actual stress in the cross section of a concrete member without disturbing the original stress field.

(2) Regression method. A stress-strain model is assumed, with a few model parameters. The global structural responses can subsequently be calculated using the assumed model. The calculated responses are functions of the parameters of the stress-strain model. By adjusting the values of the parameters, the difference between the calculated responses and the test results can be minimized. The process of error minimization is called regression analysis, which can determine the values of the parameters. This is currently the most popular approach for stress-strain modeling of concrete (Wu and Jiang 2013a; Cao et al. 2018).

(3) Analytical method. Global responses of a member, including the axial force N and bending moment M, are mathematical results of the integration of stress across a cross section. By differentiating the integration equations, the stress is exposed from the integration equations and expressed as derivatives of M and N, and hence can be calculated from the measured global structural responses. This method is introduced in Section 8.6.6.

8.6.2 Existence and form of the one-dimensional stress-strain relationship

Under the framework of conventional continuum mechanics, local stresses are functions of local strains only. The basic relationship that relates longitudinal stress to its strain is given by

$$\sigma_z = \varepsilon_z E + \nu_l \sigma_l + \nu_r \sigma_r \tag{8.68}$$

where σ_l and σ_r are lateral and radial stresses, respectively; E, ν_l, and ν_r are the elastic modulus and Poisson's ratio in lateral and radial directions, respectively. The material properties E, ν_l, and ν_r are constants for linear elastic materials but variables for nonlinear materials. For the case of confined columns, the lateral stresses σ_l and σ_r are confinement pressures.

For a certain column section as in Fig. 8.24, the internal stress fields are fixed for a given deformation of the cross section, i.e., a given longitudinal strain field. Under the

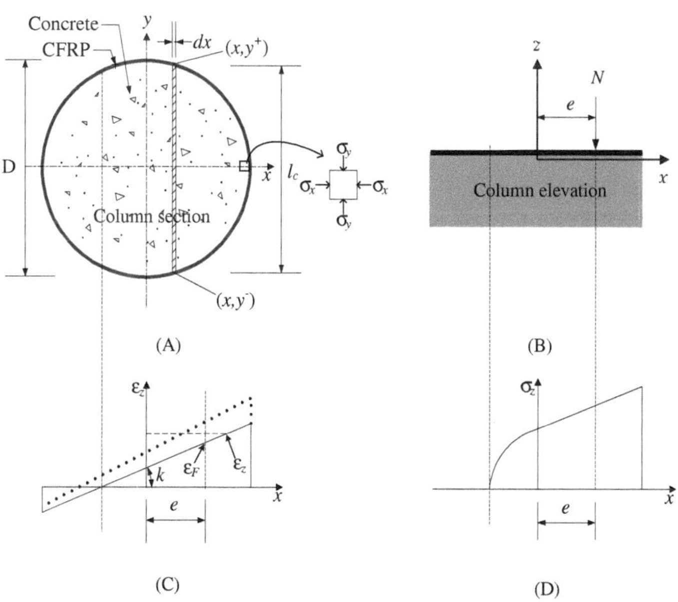

Fig. 8.24 Cross-sectional analysis of a confined column.
From Wu, Y. F., and C. Jiang. 2013. "Effect of load eccentricity on the stress-strain relationship of FRP-confined concrete columns." *Compos. Struct.* 98, 228–241.

assumption of plane sections remain plane, the longitudinal strain field is fixed for the column section under an eccentric axial load when the strain at the loading position ε_F and the curvature k are fixed (Fig. 8.24C). For a given column with a certain jacket rigidity $E_f t_f$, where E_f and t_f are the elastic modulus and nominal thickness of the jacket, respectively, ε_F and curvature k only depend only on the axial force N and its eccentricity e. With a certain strain profile and hence fixed confinement field and longitudinal stress field, the concrete at different locations has different confinement stresses that depend only on the longitudinal strain ε_z as well as its horizontal coordinate y (Fig. 8.24A and C).

If $\overline{\sigma}_l$ and $\overline{\sigma}_r$ are defined as the average stresses of σ_l and σ_r, respectively, for the shaded horizontal strip with a width of dx in Fig. 8.24A, based on the previous analyses, they are functions of $E_f t_f$, N, e, and ε_z, or mathematically,

$$\overline{\sigma}_l = \frac{\int_{y^-}^{y^+} \sigma_l dA}{\int_{y^-}^{y^+} dA} = f\left(E_f t_f, N, e, \varepsilon_z\right) \tag{8.69}$$

$$\overline{\sigma}_r = \frac{\int_{y^-}^{y^+} \sigma_r dA}{\int_{y^-}^{y^+} dA} = g\left(E_f t_f, N, e, \varepsilon_z\right) \tag{8.70}$$

Substituting Eqs. (8.69), (8.70) into Eq. (8.68), one gets

$$\overline{\sigma}_z = \varepsilon_z E + \nu_l \cdot f\left(E_f t_f, N, e, \varepsilon_z\right) + \nu_r \cdot g\left(E_f t_f, N, e, \varepsilon_z\right) \tag{8.71}$$

where $\overline{\sigma}_z$ is the average longitudinal stress of the shaded strip. In engineering design, we do not need to know the stress distribution along the width direction of the shaded strip. Therefore, the average stress $\overline{\sigma}_z$ can simply be treated as the longitudinal stress σ_z. Hence, Eq. (8.71) can be simplified to

$$\sigma_z = \varphi\left(\varepsilon_z, E_f t_f, N, e\right) \tag{8.72}$$

Eq. (8.72) shows that, mathematically speaking, a single stress-strain relationship between σ_z and ε_z does exist. However, it is not only the function of ε_z but also a function of the jacket properties, axial load, and its eccentricity. For concentrically loaded columns, $e = 0$ and hence, ε_z is directly related to N. Therefore, N is not an independent parameter in Eq. (8.72), and hence, $\sigma_z = \varphi(\varepsilon_z, E_f t_f)$, which conforms to the existing stress-strain models for concentrically loaded columns (Sections 8.6.3 and 8.6.4). For eccentrically loaded columns, ε_z, N, and e are all independent and hence must be included in a one-dimensional stress-strain model.

Development of such a stress-strain model generally involves the following three steps:

(1) Selection of a suitable mathematical framework or function for the stress-strain relationship based on experimentally observed stress-strain curves.
(2) Adequate selection/identification of controlling parameters of the stress-strain model through analytical, numerical, or experimental studies.
(3) Derivation of equations for evaluation of the parameters by regression of test results and/or analytical and numerical simulations.

Different mathematical forms for describing the stress-strain curve of confined concrete have been proposed in the literature. The most common forms of the stress-strain curve are (1) the one-expression nonlinear form, (2) two-part form with different nonlinear and linear combinations for different parts of the curve (e.g., Fig. 8.25), and (3) trilinear form, which divides the whole curve into three linear parts. The most frequently used mathematical expressions are Popovics' model (Popovics 1973), Richard and Abbott's four-parameter curve (Richard and Abbott 1975), and Hognestad's parabola (Hognestad 1951).

The main problem with the one-expression form is that it does not suit all types of stress-strain curves. The drawback of Hognestad's parabola as the first part of the stress-strain curve (Fig. 8.25) is that it is unable to connect to the second part smoothly (discontinuous slope at the intersection). To address these issues, Lam and Teng (2003a,b) employed a new parabola, whereas Youssef et al. (2007) used a polynomial curve in the first portion. Most models adopt a straight line for the second part, while a few have used nonlinear ones (Toutanji 1999; Saafi et al. 1999; Harajli et al. 2006).

8.6.3 Stress-strain relationship of concrete confined by steel reinforcement

A stress-strain relationship of concrete for concrete columns confined by steel spiral/stirrups was proposed by Mander et al. (1988a,b) based on their concentric load tests on 31 nearly full-scale RC columns. This model has been widely used in both

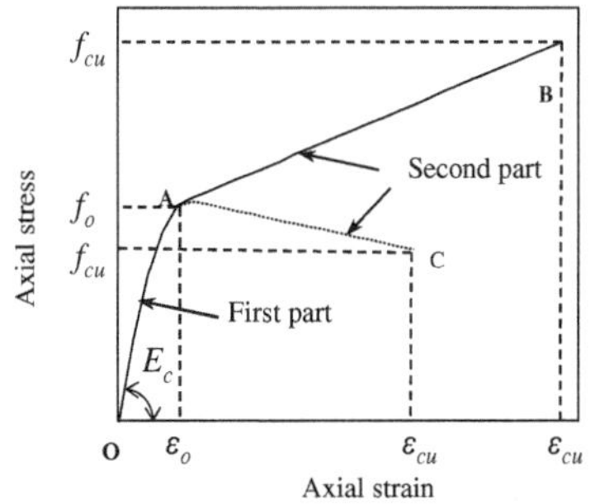

Fig. 8.25 Typical two-part stress-strain model.
From Wei, Y. Y., and Y. F. Wu. 2012. "Unified stress-strain model of concrete for FRP-confined columns." *Construct. Build Mater.* 26 (1), 381–392.

engineering designs and research work. The stress-strain relationship takes the form of Popovics' model (Eq. 3.3):

$$f_c = \frac{f_{cc} \cdot x \cdot \lambda}{\lambda - 1 + x^{\lambda}} \tag{8.73}$$

$$x = \frac{\varepsilon_c}{\varepsilon_{cc}} \tag{8.74}$$

$$\lambda = \frac{E_c}{E_c - E_{sec}} \tag{8.75}$$

where f_c and ε_c are the stress and strain of concrete, respectively, and

$$E_c = 5000\sqrt{f_{co}}\,(\text{in MPa}) \tag{8.76}$$

$$E_{sec} = \frac{f_{cc}}{\varepsilon_{cc}} \tag{8.77}$$

The peak stress f_{cc} and corresponding strain ε_{cc} of the confined concrete are given by

$$f_{cc} = f_{co}\left(-1.254 + 2.254\sqrt{1 + \frac{7.94 f_l}{f_{co}}} - 2\frac{f_l}{f_{co}}\right) \tag{8.78}$$

$$\varepsilon_{cc} = \varepsilon_{co}\left[1 + 5\left(\frac{f_{cc}}{f_{co}} - 1\right)\right] \tag{8.79}$$

Models for calculating the effective confining pressure f_l under various arrangements of lateral reinforcement for circular and rectangular sections were proposed in their paper. The typical stress-strain curve of this model is shown in Fig. 8.4, as well as the dashed lines in Fig. 8.26, which show that a higher confinement produces a higher strength and ductility of concrete. The ultimate strain ε_{cu} is calculated from an energy balance method by equating the maximum energy absorption capacity of the lateral reinforcement, which can be obtained from a rebar pull test, to the additional energy absorbed by the column due to confinement. This additional energy is equal to the difference between the energy absorbed by confined concrete and unconfined concrete as shown by the shaded area in Fig. 8.4, plus the additional energy required to maintain yield in the longitudinal compression steel. However, this energy balance method for the calculation of the ultimate strain was found flawed by Wu and Cao (2017).

Mander et al. (1988b) have shown in their large-scale column tests (circular column $D = 500$ mm, square column 450×450 mm, plus RC walls) that columns designed with reasonable volumetric ratio of confinement reinforcement ($0.6\% \le \rho_{sv} \le 2.5\%$) can develop an ultimate compressive strain of 0.035–0.06, compared with 0.0045 of the unconfined columns. The confined compressive strength of concrete was between

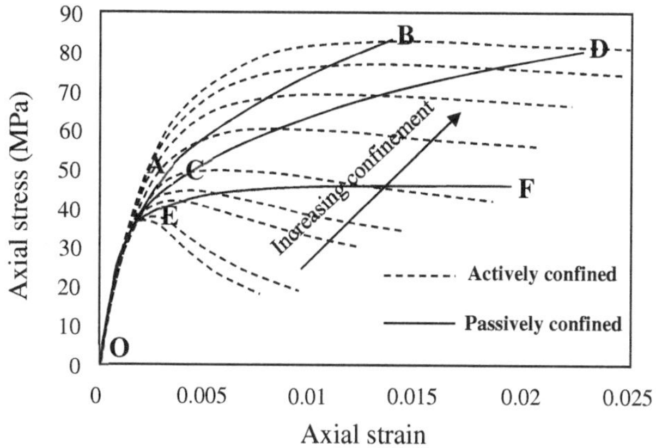

Fig. 8.26 Effect of confinement on stress-strain curve.
Based on: Teng, J. G., and L. Lam. 2004. "Behavior and modeling of fiber reinforced polymer-confined concrete." *J. Struct. Eng.* 130 (11), 1713–1723.

1.3 and 2.0 times that of the unconfined compressive strength of concrete. The effectiveness and efficiency of confinement by transverse steel reinforcement in improving the deformation capacity and strength of RC columns were evident from their test results.

8.6.4 Stress-strain model of concrete for FRP-confined columns

Compared with a constant confinement pressure in actively confined concrete, passive confinement pressure increases when axial strain increases, leading to faster increase in concrete stress after the significant dilation point near the peak of unconfined concrete, as shown in Fig. 8.26. Therefore, the stress-strain curves of actively confined concrete and passively confined concrete are different.

Extensive research on the stress-strain modeling of FRP-confined concrete has been reported in the literature and numerous stress-strain models of FRP-confined concrete for circular and/or rectangular columns have been proposed by researchers. The early stress-strain models for FRP-confined concrete (Fardis and Khalili 1981, 1982; Ahmad et al. 1991; Saadatmanesh et al. 1994) evolved from the models for steel-confined concrete (Richart et al. 1928, 1929; Ahmad and Shah 1982a,b; Mander et al. 1988a). Researchers later realized that FRP confinement differs from steel confinement, and bilinear stress-strain curves were subsequently developed for FRP-confined concrete (Harmon and Slattery 1992; Demers and Neale 1994; Howie and Karbhari 1994; Nanni and Braford 1995; Karbhari and Gao 1997; Xiao and Wu 2000). A few researchers proposed trilinear models to allow greater flexibility and accuracy (Wu et al. 2003; Chaallal et al. 2003a).

Most models adopted nonlinear expression at least for the first portion (the ascending part) of the stress-strain relationship. Samaan et al. (1998) used the four-parameter relationship proposed by Richard and Abbott (1975) to describe the stress-strain response of FRP-confined concrete. The advantage of the four-parameter model is that it allows a nearly bilinear stress-strain curve to be characterized using a single expression. Thereafter, a number of similar FRP-confined concrete models appeared in the literature (Arduini et al. 1999; Yu 2001; Cheng et al. 2002; Moran and Pantelides 2002; Xiao and Wu 2003a; Toutanji and Saafi 2002; Ilki et al. 2004). Miyauchi et al. (1997, 1999) used Hognestad's parabola (Hognestad 1951) to simulate the first portion of the stress-strain curve for FRP-confined concrete, and a straight line to describe the second portion. Toutanji (1999) and Saafi et al. (1999) suggested a stress-strain curve using two segments of parabola with a smooth transition. Similar two-part models have been proposed by Lillistone and Jolly (2000), Li et al. (2003), Jin et al. (2003), Lam and Teng (2003a,b), Harajli et al. (2006), Wu et al. (2007), Youssef et al. (2007), Teng et al. (2009), etc. The difficulty in such models lies in the determination of the ultimate stress and corresponding strain, particularly for stress-strain responses that have a descending curve after the peak.

Another series of stress-strain models adopted an incremental iterative numerical procedure. These models were based on the assumption that the relationship between the incremental axial stress and strain of FRP-confined concrete at a given lateral strain is the same as that of actively confined concrete with a constant pressure calculated from the lateral strain of the FRP jacket. Mirmiran and Shahawy (1997), Spoelstra and Monti (1999), Fam and Rizkalla (2001), Chun and Park (2002), Marques et al. (2004), Jiang and Teng (2007), and Teng et al. (2007) all used the active confinement model by Mander et al. (1988a) as the basis of the incremental procedure. Harmon et al. (1998) proposed a confinement model that was based on the concept of crack slip and separation in concrete, and Becque et al. (2003) modified Gerstle's octahedral stress-strain model (Gerstle 1981a,b) for FRP-confined concrete. These models do not give a simple explicit mathematical expression and are inconvenient for engineering applications.

Most of the models are for FRP-confined circular columns. Models for FRP-confined rectangular columns are more complicated and fewer models have been reported, such as the models by Lam and Teng (2003b), Harajli et al. (2006), Wu et al. (2007), and Youssef et al. (2007), among others. These models are briefly introduced in the following sections.

8.6.4.1 Models developed by Teng's group (Lam and Teng 2003a,b; Teng et al. 2009)

The models developed by Teng's group have been widely applied in both research work and design practice. Lam and Teng (2003b) proposed a stress-strain model for FRP-confined concrete in rectangular columns with an ascending type of stress-strain curve, and subsequently extended it to include circular columns with a descending type of curve for the second portion (Teng et al. 2009).

The model comprises a parabolic first portion whose initial slope is the elastic modulus of unconfined concrete, E_c, and a linear second portion with a reduced slope, E_2, which intersects the stress axis at $f_c = f_{co}$. The model is given by

$$f_c = \begin{cases} E_c\varepsilon_c - \dfrac{(E_c - E_2)^2}{4f_{co}}\varepsilon_c{}^2 & 0 \le \varepsilon_c \le \varepsilon_o \\ f_{co} + E_2\varepsilon_c & \varepsilon_o \le \varepsilon_c \le \varepsilon_{cu} \ \text{ if } \rho_k \ge 0.01 \\ f_{co} - \dfrac{f_{co} - f_{cu}}{\varepsilon_{cu} - \varepsilon_{co}}(\varepsilon_c - \varepsilon_{co}) & \varepsilon_o \le \varepsilon_c \le \varepsilon_{cu} \ \text{ if } \rho_k < 0.01 \end{cases} \tag{8.80}$$

where $\rho_k = 2E_f t_f/[(f_{co}/\varepsilon_{co}) \cdot D]$ is the confinement stiffness ratio and D refers to Section 8.4.3.1(6); ε_o is the axial strain at the intersection between the first and second portions; and f_{cu} and ε_{cu} are the ultimate stress and corresponding axial strain at the ultimate failure point, respectively.

The models are applied to rectangular and circular columns separately and are not suitable for the stress-strain relationship of rectangular columns characterized by strain softening.

8.6.4.2 The model proposed by Harajli et al. (2006)

The model includes two parts. The first part is a second-degree parabola similar to that proposed by Scott et al. (1982), whereas the second part is a complicated nonlinear function. The model is expressed as

$$\begin{cases} f_c = f_o\left[\dfrac{2\varepsilon_c}{\varepsilon_o} - \left(\dfrac{\varepsilon_c}{\varepsilon_o}\right)^2\right] & 0 \le \varepsilon_c \le \varepsilon_o \\ f_c = \sqrt{(k_o^2 - k)} - k_o & \varepsilon_o \le \varepsilon_c \le \varepsilon_{cu} \end{cases} \tag{8.81}$$

where

$$k_o = 0.0031k_1E_f - f_{co} \tag{8.82}$$

$$k_1 = 1.25\left(\frac{f_l}{f_{co}}\right)^{-0.5} \tag{8.83}$$

and

$$k = f_{co}{}^2 - 0.0032k_1E_f f_{co}\left(\frac{\varepsilon_c}{\varepsilon_{co}} + 0.9\right) \tag{8.84}$$

and f_o and ε_o are the stress and strain at the transition point (Fig. 8.25). The model is applicable to the hardening type of stress-strain curves only.

8.6.4.3 The model proposed by Wu et al. (2007)

Wu et al. (2007) proposed three models. Model I used a parabola in the first portion, and the curvilinear function from Xing et al. (1987) for the second portion. The model is applicable only when the confinement effectiveness is relatively weak. Model II originated from Saadatmanesh et al. (1994). Model III is similar to the model by Miyauchi et al. (1999), where the first portion is a parabolic curve connected by a straight line in the second portion without a smooth transition between them. Models II and III can be used for both strain-softening and strain-hardening types of curves, but Model III was recommended. Model III is given as follows:

$$\begin{cases} f_c = f_o\left[\dfrac{2\varepsilon_c}{\varepsilon_o} - \left(\dfrac{\varepsilon_c}{\varepsilon_o}\right)^2\right] & 0 \le \varepsilon_c \le \varepsilon_o \\ f_c = f_o + \dfrac{(f_{cu} - f_o)(\varepsilon_c - \varepsilon_o)}{\varepsilon_{cu} - \varepsilon_o} & \varepsilon_o \le \varepsilon_c \le \varepsilon_{cu} \end{cases} \tag{8.85}$$

8.6.4.4 The model by Youssef et al. (2007)

The two-part model proposed by Youssef et al. (2007) employs a polynomial curve in the first portion and a straight line in the second. It is applicable to both circular and rectangular columns, and for both strain-hardening and strain-softening types of curves for the second portion. The model parameters were determined and evaluated separately for different types of circular and rectangular columns on the basis of regression analysis of the experimental data. This model was expressed as

$$\begin{cases} f_c = E_c\varepsilon_c\left[1 - \dfrac{1}{n}\left(1 - \dfrac{E_2}{E_c}\right)\left(\dfrac{\varepsilon_c}{\varepsilon_o}\right)^{n-1}\right] & 0 \le \varepsilon_c \le \varepsilon_o,\ E_2 > 0 \\ f_c = E_c\varepsilon_c\left[1 - \dfrac{1}{n}\left(\dfrac{\varepsilon_c}{\varepsilon_o}\right)^{n-1}\right] & 0 \le \varepsilon_c \le \varepsilon_o,\ E_2 < 0, \\ f_c = f_o + E_2(\varepsilon_c - \varepsilon_o) & \varepsilon_o \le \varepsilon_c \le \varepsilon_{cu} \end{cases} \tag{8.86}$$

where E_2 is the slope of the second portion, and

$$n = \frac{(E_c - E_2)\varepsilon_o}{E_c\varepsilon_o - f_o} \quad \text{for } E_2 > 0 \tag{8.87}$$

$$n = \frac{E_c\varepsilon_o}{E_c\varepsilon_o - f_o} \quad \text{for } E_2 < 0 \tag{8.88}$$

8.6.4.5 The unified model by Wei and Wu (2012)

The model comprises a parabola in the first portion and a straight line in the second. The model has an initial slope E_c at $\varepsilon_c = 0$ which is the same as the elastic modulus of unconfined concrete, and a smooth transition at the transition point (equal slope) between the two portions. The model is shown in Fig. 8.25 and given by

$$\begin{cases} f_c = E_c\varepsilon_c + \dfrac{f_o - E_c\varepsilon_o}{\varepsilon_o{}^2}\varepsilon_c{}^2 & 0 \le \varepsilon_c \le \varepsilon_o \\ f_c = f_o + E_2(\varepsilon_c - \varepsilon_o) & \varepsilon_o \le \varepsilon_c \le \varepsilon_{cu} \end{cases} \tag{8.89}$$

where the transition strain, ε_o, was calculated from the condition of equal slope at the transition point to give

$$\varepsilon_o = \frac{(f_o + f_{cu} + E_c\varepsilon_{cu}) - \sqrt{(f_o + f_{cu} + E_c\varepsilon_{cu})^2 - 8f_oE_c\varepsilon_{cu}}}{2E_c} \tag{8.90}$$

E_2 is the slope of the second portion, given by

$$E_2 = \frac{f_{cu} - f_o}{\varepsilon_{cu} - \varepsilon_o} \tag{8.91}$$

The elastic modulus of unconfined concrete can be calculated by (ACI Committee 440 2002):

$$E_c = 4730\sqrt{f_{co}} \tag{8.92}$$

The model has three parameters: the ultimate stress f_{cu}, ultimate strain ε_{cu}, and transitional stress f_o, determined from the regression analyses using a test database, given by

$$\frac{f_{cu}}{f_{co}} = 0.5 + 2.7\left(\frac{2r}{b}\right)^{0.4}\left(\frac{f_l}{f_{co}}\right)^{0.73}\left(\frac{h}{b}\right)^{-1} \tag{8.93}$$

$$\frac{\varepsilon_{cu}}{\varepsilon_{co}} = 1.75 + 12\left(\frac{f_l}{f_{co}}\right)^{0.75}\left(\frac{f_{30}}{f_{co}}\right)^{0.62}\left(0.36\frac{2r}{b} + 0.64\right)\left(\frac{h}{b}\right)^{-0.3} \tag{8.94}$$

$$f_o = f_{co} + 0.43\left(\frac{2r}{b}\right)^{0.68}\left(\frac{h}{b}\right)^{-1} f_l \tag{8.95}$$

where ε_{co} can be determined from the model by Popovics (1973), i.e., $\varepsilon_{co} = 0.000937\sqrt[4]{f_{co}}$. The model is applicable to circular, square, and rectangular columns, and is suitable for both ascending and descending types of stress-strain curve.

8.6.5 Stress-strain model of confined concrete under eccentric loading

8.6.5.1 Effect of load eccentricity on the stress-strain relationship of confined concrete

Load eccentricity causes curvature, and hence strain gradient, of the cross section of a member (Fig. 8.24C). Wu and Jiang (2013a) carried out an experimental investigation on FRP-confined concrete columns under eccentric loading. The test scheme is shown

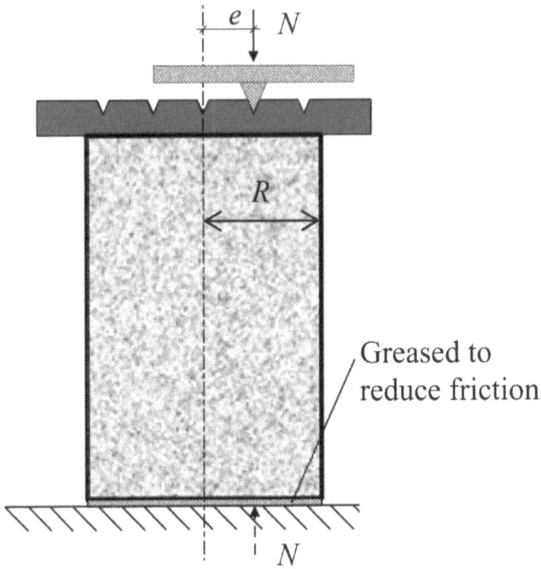

Fig. 8.27 Column test under eccentric loading.

in Fig. 8.27 and the load-displacement response curves at the loading point are shown in Fig. 8.28. For the column under concentric loading (Fig. 8.28A), the stress-strain curve can simply be obtained from the load (load divided by the cross-sectional area) vs. the loading point displacement (displacement divided by the gauge length) curve. Directly using the experimental stress-strain curve obtained from the concentric load test for a particular column as the stress-strain model of concrete for the column under eccentric loading, the load vs displacement curve of the same column under different eccentric loadings can be calculated theoretically.

The theoretical results are compared with the eccentric load test results in Fig. 8.28, where it can be seen that when the stress-strain curve from the concentric loading test is used to calculate the response of columns under eccentric loading, the results are incorrect. The error increases with the increase of load eccentricity, as shown in Fig. 8.29. The error in Fig. 8.29 is given by

$$IAE = \frac{\left[\sum_{n}(Expe. - Theo.)^2\right]^{1/2}}{\sum_{n}|Expe.|}, \tag{8.96}$$

where *Expe.* and *Theo.* are the experimental and theoretical load values at a particular displacement, respectively. The relative error *IAE* is calculated with $n = 20$ points that

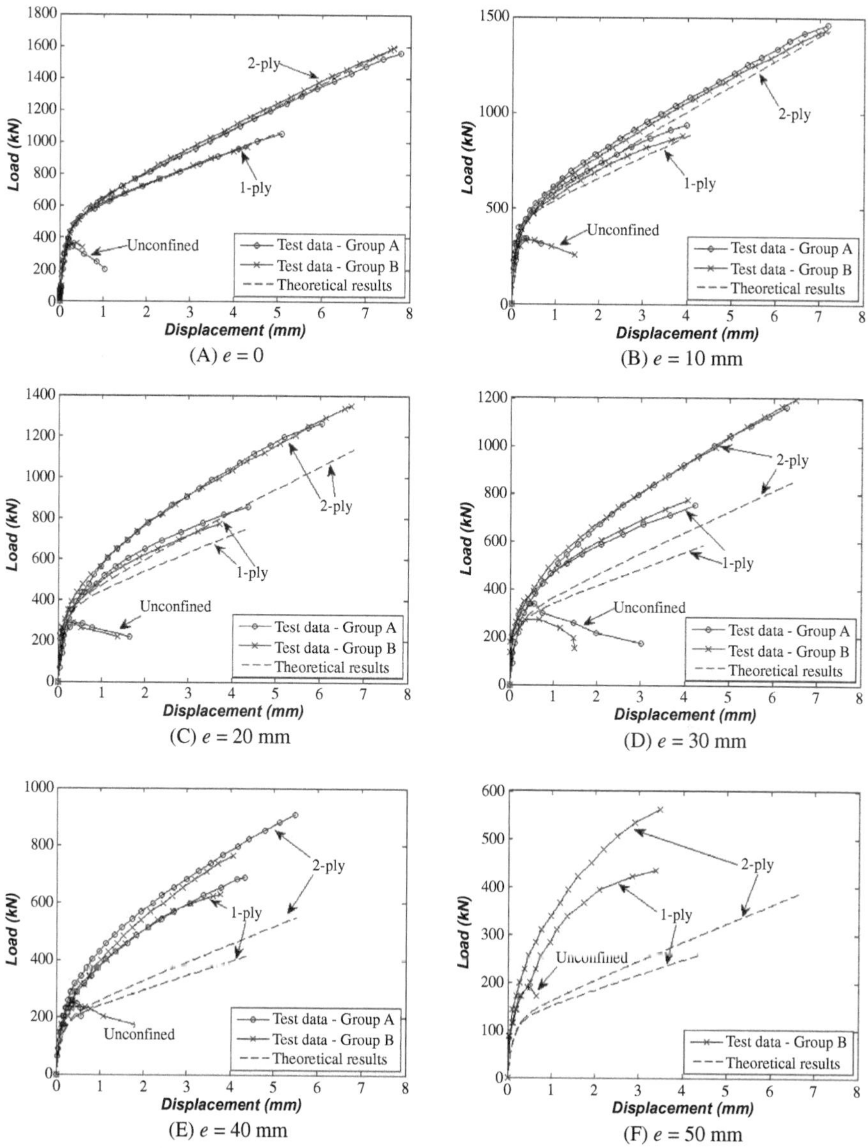

Fig. 8.28 Load vs. load point displacement curves.
From Wu, Y. F., and C. Jiang. 2013. "Effect of load eccentricity on the stress–strain relationship of FRP-confined concrete columns." *Compos. Struct.* 98, 228–241.

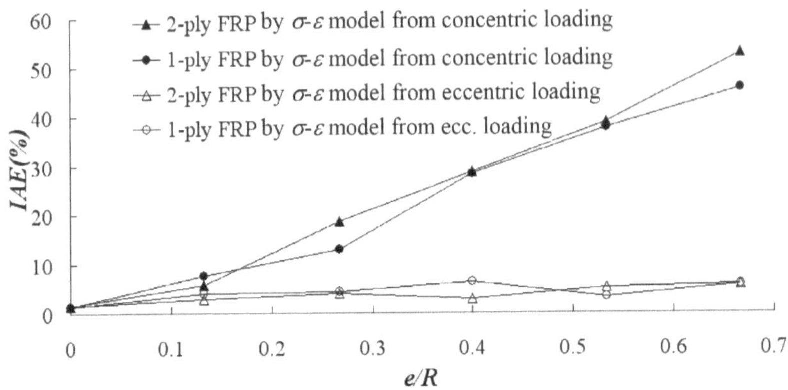

Fig. 8.29 Error caused by using a stress-strain model from concentric loading tests. From Wu, Y. F., and C. Jiang. 2013. "Effect of load eccentricity on the stress–strain relationship of FRP-confined concrete columns." *Compos. Struct.* 98, 228–241.

are evenly distributed on the curves. It can be seen from Fig. 8.29 that, for an error less than 10%, the eccentricity ratio e/R, where R is the radius of the column, should be smaller than approximately 0.18. In other words, when the eccentricity ratio is greater than 0.2, significant error (>10%) occurs if a stress-strain model obtained from concentrically loaded columns is used to predict responses of eccentrically loaded columns.

The poor match with the test results, particularly for cases with larger eccentricity, demonstrates that a stress-strain model developed from concentrically loaded tests cannot be used for columns under eccentric loading. The stress-strain relationship from concentrically loaded columns generally underestimates the stiffness and strength of eccentrically loaded columns. This problem can be overcome when the stress-strain model takes load eccentricity into consideration. Development of the stress-strain model of concrete for eccentrically loaded columns is introduced in the following sections. Fig. 8.29 also shows the significantly reduced error when the stress-strain model developed in Section 8.6.5.2 for eccentrically loaded columns is used to predict the response of the test columns (the hollow circle and triangle points in Fig. 8.29).

8.6.5.2 The stress-strain model by Wu and Jiang (2013a)

Wu and Jiang (2013a) developed a stress-strain model using their test results on FRP-confined circular columns under eccentric loading as shown in Fig. 8.28. Through analytical study and using a regression approach (Approach 2 in Section 8.6.1), the following model is obtained:

$$f_c = \left[(E_1\varepsilon_n - f_o)e^{-\frac{\varepsilon_c}{\varepsilon_n}} + f_o + E_2\varepsilon_c\right]\left(1 - e^{-\frac{\varepsilon_c}{\varepsilon_n}}\right) \tag{8.97}$$

where $\varepsilon_n = n \times \varepsilon_o$, $\varepsilon_o = f_o/E_1$, and n is a curve-shape parameter that controls the curvature of the transition zone satisfying $0 < n \leq 2$, as shown in Fig. 8.30. In the earlier

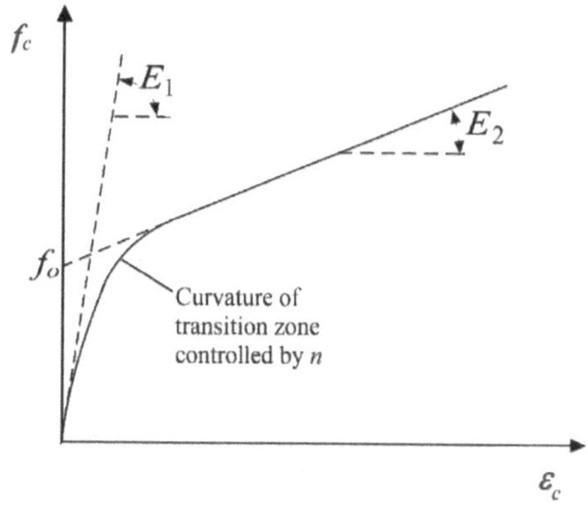

Fig. 8.30 Stress-strain model.
From Wu, Y. F., and C. Jiang. 2013. "Effect of load eccentricity on the stress–strain relationship of FRP-confined concrete columns." *Compos. Struct.* 98, 228–241.

model, there are four parameters, E_1, E_2, f_o, and n. Parameter E_1 is the initial elastic modulus of the stress-strain curve. It is well established in the literature that confinement does not affect the initial elastic modulus of FRP-confined concrete because passive FRP jacketing imposes no confinement pressure on concrete in the initial stages of the stress-strain curves (Wei and Wu 2012). Thus, E_1 is a constant that equals to the initial tangent modulus of concrete (different from Young's modulus of concrete, which is a secant modulus at about one-third of concrete strength). E_2 is the slope of the asymptotic line or the hardening slope (Fig. 8.30), given by

$$\frac{E_2}{E_{2,0}} = 1 + 5.55\left(\frac{e}{R}\right)^{2.49}\left(\frac{E_f t_f}{E_1 R}\right)^{0.11}\left[1.3 - 0.84\left(\frac{e}{R}\right)^{1.74}\left(\frac{N}{f_{co}A_c}\right)^{1.1}\right] \tag{8.98}$$

where A_c is the cross-sectional area of concrete, and f_o is the vertical coordinate of the intersection between the asymptotic line and the y-axis (Fig. 8.30), given by

$$\frac{f_o}{f_{o,0}} = 1 + 7.02\left(\frac{e}{R}\right)^{1.67}\left(\frac{E_f t_f}{E_1 R}\right)^{0.32} \tag{8.99}$$

$E_{2,0}$ and $f_{o,0}$ are values of the corresponding parameters when the load eccentricity is zero, i.e., the respective parameters for concentrically loaded columns. The value of n was found insensitive to the overall stress-strain curve and a constant value of 0.8 was determined by Wu and Jiang (2013a) from concentrically loaded columns.

8.6.5.3 *The stress-strain model by Cao et al. (2018)*

Based on their test results on FRP-confined circular concrete columns that were loaded under a constant axial force and an increasing bending moment, Cao et al. (2018) proposed another stress-strain model. The same Eq. (8.97) was adopted by

Cao et al. (2018) as the basic form of the stress-strain model. However, the equations for evaluation of the model parameters are different, as given by:

$$\frac{E_2}{E_{2,0}}=1-0.58\left(\frac{E_l}{E_c}\right)^{-0.29}\left(\frac{M}{NR}\right)^{0.28}\left(\frac{f_{co}}{f_{30}}\right)^{0.47}\left(1-\left(\frac{N}{N_u}\right)^{1.76}\right) \tag{8.100}$$

$$\frac{f_o}{f_{o,0}}=1+1.27\left(\frac{E_l}{E_c}\right)^{0.32}\left(\frac{M}{NR}\right)^{0.60}\left(\frac{f_{co}}{f_{30}}\right)^{0.32}\left(1-\left(\frac{N}{N_u}\right)^{0.21}\right)^{\alpha} \tag{8.101}$$

where f_{30} is the strength of grade 30 concrete, i.e., $f_{30}=30$ MPa. N_u is the load-carrying capacity of confined concrete under concentric loading, defined as

$$N_u=f_{cc}A_c \tag{8.102}$$

where f_{cc} is given by Eq. (8.11). E_c is the elastic modulus of concrete given by Eq. (8.92). E_l is the confinement stiffness, defined as

$$E_l=\frac{E_f t_f}{R} \tag{8.103}$$

and α is given by

$$\alpha=-10\left(\left(\frac{f_{co}}{f_{30}}\right)^{-3}\left(\frac{E_l}{E_c}\right)^{1.1}+0.027\right) \tag{8.104}$$

When $N=N_u$, or $M=0$, or $E_l=0$, Eqs. (8.100), (8.101) degenerate into concentrically loaded cases.

8.6.5.4 Effect of load path on stress-strain relationship

The assumption that the stress-strain relationship of concrete in an FRP-confined column is load path independent was widely accepted in the literature (Bisby and Ranger 2010; Rocca et al. 2009; Wu and Jiang 2013a). Many researchers used a stress-strain model developed from concentrically loaded tests to evaluate the behavior of columns under eccentric loading. This approach is inappropriate, as explained in Section 8.6.5.1 unless the load eccentricity is small. Few models have been developed for FRP-confined concrete under eccentric loading. Fig. 8.31 compares the typical stress-strain curves from Wu and Jiang (2013a) and Cao et al. (2018). Clearly, these two models predict significantly different stress-strain behavior. The stress-strain curves generally have a softening trend after the turning point based on the Cao et al. model, while those based on Wu and Jiang's model have a hardening trend. Fam et al. (2003) proposed a stress-strain model that also gives a softening trend when load eccentricity increases.

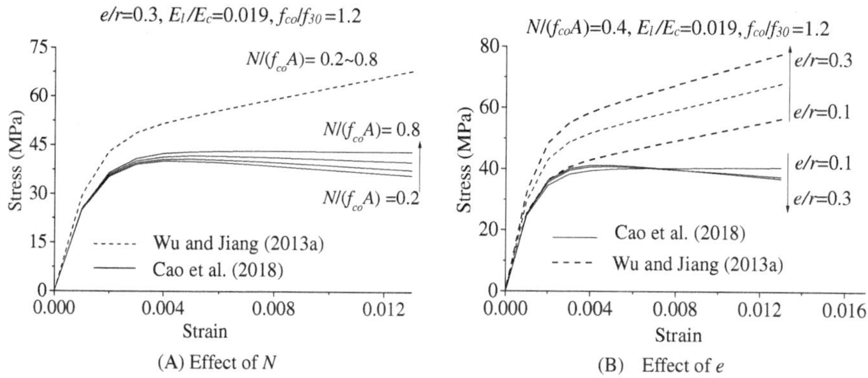

Fig. 8.31 Effect of load path on stress-strain curve.
From Cao, Y. G., Y. F. Wu, and C. Jiang. 2018. "Stress-strain relationship of FRP confined concrete columns under combined axial load and bending moment." *Compos. Part B Eng.* 134, 207–217.

The difference in the stress-strain models is usually caused by the difference in load path. For columns under eccentric loading, there are three typical load paths, as shown in Fig. 8.32:

(I) Increasing axial load N with constant load eccentricity e (Fig. 8.27);
(II) Constant axial force followed by increasing bending moment M or load eccentricity; and
(III) Constant bending moment followed by increasing axial force.

The case with the bending moment or axial force equal to zero can be considered as a special case of the preceding cases. For example, concentric axial loading can be seen as a special case of load path I with $e = 0$. It can also be considered as load path II or load path III with $M = 0$. Beams are a special case of II or III with $N = 0$.

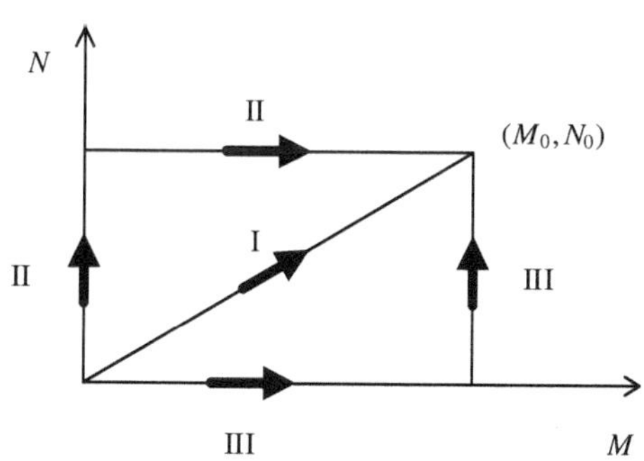

Fig. 8.32 Load paths for columns.
From Cao, Y. G., Y. F. Wu, and C. Jiang. 2018. "Stress-strain relationship of FRP confined concrete columns under combined axial load and bending moment." *Compos. Part B Eng.* 134, 207–217.

Load path I is common for structures under gravitational loading where both axial force and bending moment increase proportionally. Wu and Jiang's model in Section 8.6.5.2 is for this particular load path. Load path II occurs when gravitationally loaded structures are under additional horizontal loading such as wind or seismic load. The model of Cao et al. (2018) in Section 8.6.5.3 covers this load path. Load path III is uncommon in practice. Load path I with $e = 0$ is studied in Sections 8.6.3 and 8.6.4.

The difference in the stress-strain curves under different load paths reflects the fact that dilation of concrete is significantly different when concrete is loaded under different load paths. The hardening trend in the stress-strain model of Wu and Jiang (2013a) is attributed to the increasing axial force under load path I, which causes increasing dilation of concrete and hence increasing confinement, which leads to hardening. The softening trend predicted by Cao et al. (2018) is caused by the softening of concrete material in the postpeak region when total average confinement to concrete remains unchanged with a constant axial force under load path II. The plasticity theory of concrete based on continuum mechanics has also concluded that the dilation property of concrete material is load path dependent (Jiang and Wu 2014), which is supported by triaxial test results (Sfer et al. 2002; Gabet et al. 2008). Therefore, the significant load path dependency reflected by the experimentally based one-dimensional stress-strain model is consistent with the continuum mechanics-based plasticity theory.

The load path dependency of the stress-strain relationship requires different models for different applications. Actually, a difference in cross-sectional shape of columns or the use of different confining materials also results in differences in the form of stress-strain models. Use of many different models is an inconvenience to engineers on selection of an appropriate model. To this end, people have tried to unify different models into one. For example, the unified model from Wu and Wei (2015) unifies the two different stress-strain models, one for steel spiral-confined concrete and the other for FRP-confined concrete, into one. The model from Wu and Wang (2009) (Eq. 8.12) unifies circular and square columns into one, and that from Wei and Wu (2012) (Eq. 8.48) further unifies circular, square, and rectangular columns into one. In principle, the models for different load paths can also be unified under one uniform framework. While unification of models into one avoids selection of models in design, it does further complicate the mathematical form of a model, which also causes extra difficulties. Therefore, a balance between level of unification and model complication should be considered for engineering application in practice. A philosophical discussion of unification of stress-strain models is given in Section 11.3.3.

8.6.6 Analytical method for stress-strain modeling

The analytical method directly calculates the stress-strain curve of concrete from the experimentally measured global structural response of a flexural member. The internal structural responses of N and M can be obtained by the integration of internal stresses (see Eqs. 8.107, 8.108). Prentis (1951) made the first attempt to inverse the integration by differentiating the integration equation and analytically derived the equations to calculate the instantaneous stresses of concrete at the compression and tension faces of a member with a rectangular cross section at respective strains. By applying Prentis's equations to structural member tests, Hognestad et al. (1955) obtained the stress-strain curve of concrete, but with great scattering or strongly amplified experimental errors. The scattering was attributed to the difficulty and inaccuracy in measuring the tensile strain of concrete due to nonuniformity and cracking of concrete. By manually controlling N and M in a flexural test to deliberately keep the strain at the extreme tension face to zero, the tensile strain problem was avoided, and Prentis's equations for concrete stress at a given strain, $f(\varepsilon_c)$, were simplified into the following equations:

$$f(\varepsilon_c) = f_c = \varepsilon_c \frac{dn_o}{d\varepsilon_c} + n_o \tag{8.105}$$

$$f(\varepsilon_c) = f_c = \varepsilon_c \frac{dm_o}{d\varepsilon_c} + 2m_o \tag{8.106}$$

where ε_c and f_c are the concrete strain and corresponding stress, respectively; $n_o = N/(bh)$ and $m_o = M/(bh^2)$ in which b is the width and h the depth of the rectangular section. These two equations are derived and discussed in Section 8.6.6.1. In an experimental test, loading is increased in sufficiently small steps. At each loading step, the strain ε_c can be directly measured from the compression face of the member and the corresponding stress is calculated from Eqs. (8.105) or (8.106). A series of concrete stress vs. strain points, $(\varepsilon_c, f_c)_{i=1\sim n}$, are obtained from the test using this method to form a continuous stress-strain curve when a sufficiently small loading step is adopted.

A few researchers (Soliman and Yu 1967; Kaar et al. 1978; Ibrahim and Macgregor 1996) carried out the test accordingly and produced stress-strain responses of concrete using Eqs. (8.105), (8.106). However, the scatter was still found very large using this method. To reduce scattering, all investigators chose to reduce the number of data points (see Section 8.6.6.1 for the reason), while some further used different smoothing techniques to reduce fluctuation of input data. Kaklauskas (1997, 2004), Gribniak (2009), and Gribniak et al. (2012, 2017) used a numerical process to generate the stress-strain curves at both the extreme tension and compression fibers simultaneously from measured N and M curves, using a particular layered approach. These works directly calculated the tensile stress of concrete while using an assumed model for the compressive stress of concrete. An extensive investigation of the method was reported by Li et al. (2021) that is introduced in the following section.

8.6.6.1 The analytical method by Li et al. (2021)

For an arbitrary cross section (Fig. 8.33) of an RC member, three general assumptions are made in the analytical method:

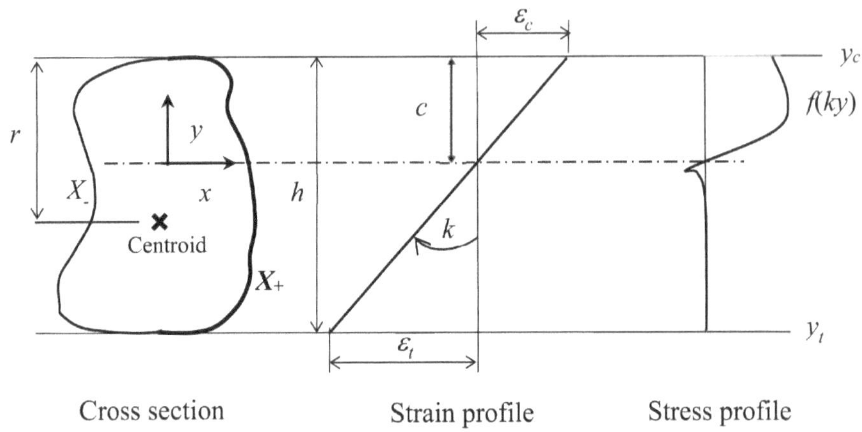

Fig. 8.33 Cross section and stress and strain profiles.
From Li, Z., Y. F. Wu, and P. Li. 2021. "Method to identify stress-strain relationship of FRP-confined concrete under eccentrical load." *J. Compos. Constr.* 25 (1), 04020080.

(1) Plane sections remain plane.
(2) Within a particular monotonic load path, the stress-strain relationship is unique, i.e., for each strain value regardless of its location and the moment of occurrence, there is only one corresponding value of stress.
(3) Only one-dimensional, i.e., the longitudinal stress-strain relationship is considered.

The global responses N and M can be calculated by integration of stress over the cross section:

$$N = \int_{y_t}^{y_c} \int_{X_-}^{X_+} f(ky)dxdy \tag{8.107}$$

$$M = \int_{y_t}^{y_c} \int_{X_-}^{X_+} f(ky)ydxdy \tag{8.108}$$

where the x-axis coincides with the neutral axis and the position of the y-axis is arbitrarily defined; y_c and y_t are the y coordinates of the extreme compression and extreme tension fibers, respectively; M is the bending moment with respect to the neutral axis, and $M = N(M^*/N + c - r)$ in which M^* is the bending moment with respect to the centroid of the cross section; c and r are the depth of the compression zone and the distance from the centroid to the extreme compression face, respectively; and k is the curvature of the section. By introducing the following coordinate transformation:

$$\xi = ky \tag{8.109}$$

Eqs. (8.107) and (8.108) become

$$N = \frac{1}{k}\int_{\varepsilon_t}^{\varepsilon_c}\int_{X_-}^{X_+} f(\xi)dxd\xi \tag{8.110}$$

$$M = \frac{1}{k^2}\int_{\varepsilon_t}^{\varepsilon_c}\int_{X_-}^{X_+} f(\xi)\xi dxd\xi \tag{8.111}$$

where ε_c and ε_t are the strains at y_c and y_t, respectively. The curvature is given by $k = (\varepsilon_c - \varepsilon_t)/h$.

For rectangular sections, Eqs. (8.110) and (8.111) are simplified to

$$N = \frac{b}{k}\int_{\varepsilon_t}^{\varepsilon_c} f(\xi)d\xi \tag{8.112}$$

$$M = \frac{b}{k^2}\int_{\varepsilon_t}^{\varepsilon_c} f(\xi)\xi d\xi \tag{8.113}$$

With the experimental approach by Hognestad et al. (1955), the strain at the extreme tension face is constantly kept to zero, i.e., $\varepsilon_t = 0$, hence $k = \varepsilon_c/h$. Substituting these terms into Eqs. (8.112), (8.113) gives

$$\int_0^{\varepsilon_c} f(\xi)d\xi = \frac{Nk}{b} = \frac{N\varepsilon_c}{bh} = n_o\varepsilon_c \tag{8.114}$$

$$\int_0^{\varepsilon_c} f(\xi)\xi d\xi = \frac{Mk^2}{b} = \frac{M\varepsilon_c^{\ 2}}{bh^2} = m_o\varepsilon_c^{\ 2} \tag{8.115}$$

Differentiating both sides with respect to ε_c and applying the chain rule gives Eqs. (8.105), (8.106).

The measurement of tensile strain is only one source of the scattering error, in fact, not the main source of scattering. In experimental tests, wobbling of measurement results is inevitable due to the limit of accuracy in controlling inputs and measurement of outputs, as well as other uncontrollable factors such as random environmental noises and random cracking in concrete, which causes fluctuation of material resistance. Fig. 8.34A and B shows the typical measured curves of N and M from an experiment by Cao et al. (2018). The overall measured curves show significant fluctuation of the results when it is locally magnified (Fig. 8.34C). The calculated stress from Eqs. (8.105) or (8.106) is not only related to the measured value of N (or M) but also significantly affected by its instantaneous slope $dn_o/d\varepsilon_c$ (or $dm_o/d\varepsilon_c$), which is badly scattered (the slope of the curve in Fig. 8.34C). If one uses the originally measured curves and the immediate adjacent two points in Fig. 8.34A or B to calculate the slopes of the curves, the individual points in Fig. 8.34D are the calculated stresses from Eqs. (8.105), (8.106), which show unacceptable scattering. The large scattering is caused by the fluctuation of the instantaneous slope of N or M, as shown in Fig. 8.34C.

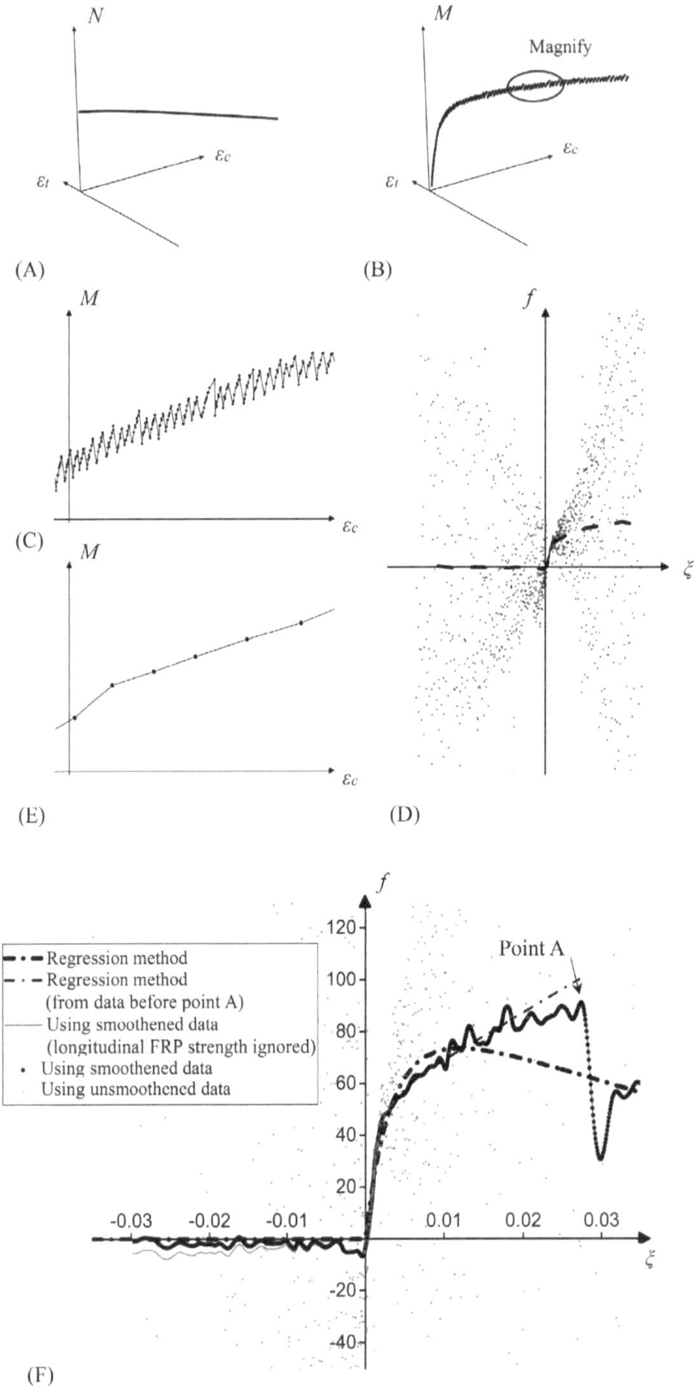

Fig. 8.34 Sample test results: (A) measured curve of *N*; (B) measured curve of *M*; (C) magnified part of (B); (D) stress-strain curve; (E) curve with reduced number of points of (C); (F) typical stress-strain curve produced by Li et al. (2021).
From Li, Z., Y. F. Wu, and P. Li. 2021. "Method to identify stress-strain relationship of FRP-confined concrete under eccentrical load." *J. Compos. Constr.* 25 (1), 04020080.

Researchers have tried to reduce the scattering of stress results by replacing the local slope of curve, $dn_o/d\varepsilon_c$ (or $dm_o/d\varepsilon_c$), with an average slope in a certain range, $\Delta n_o/\Delta \varepsilon_c$ (or $\Delta m_o/\Delta \varepsilon_c$), using a larger step length $\Delta \varepsilon_c$. This approach reduces the scattering due to two mechanisms: (1) local fluctuation can be reduced or even eliminated by averaging many measured data points and (2) for the same error in N or M, a larger step $\Delta \varepsilon_c$ results in a smaller error in its slope of the curve. Fig. 8.34E is an example of applying this approach to Fig. 8.34C and the dashed line in Fig. 8.34D shows the obtained stress-strain curve using this approach, which shows a significant reduction in scattering error.

However, such "averaged" curves for M and N may alter the local features of the original curves, e.g., instantaneous slope is replaced by average slope in a range. Mathematically speaking, the analytical method calculates the stress by differentiating both sides of Eqs. (8.107), (8.108). Therefore, local stress is directly related to local variations of M and N curves. When those local variations are represented by "averaged" variation of the curves, it simply means that instantaneous stress is replaced by an "averaged" stress. While this "averaging" effect may not cause significant error in stress results when the physical process itself is relatively smooth and stable, it could cause significant error when an abrupt physical process is involved, e.g., the concrete cracking process and the stress-strain behavior of high-strength concrete at peak. This problem can be resolved by using adequate curve smoothing technology, as discussed in the paper by Li et al. (2021).

For flexural tests with the strain at the tension face controlled to be zero, Li et al. (2021) derived the following equations:

$$\bar{f}_c = \frac{\Delta n_o}{\Delta \varepsilon_c} \varepsilon_c + n_o + \Delta n_o \tag{8.116}$$

$$\bar{f}_c = \frac{\Delta m_o \, \varepsilon_c{}^2}{\Delta \varepsilon_c \, \varepsilon_{cc}} + 2m_o + \Delta m_o \tag{8.117}$$

where $\Delta n_o = (\Delta N)/(hb)$ and $\Delta m_o = (\Delta M)/(bh^2)$. Eqs. (8.116) and (8.117) become Eqs. (8.105) and (8.106) when the loading step $\Delta \varepsilon_c$ is infinitesimal. Direct use of Eqs. (8.105), (8.106) for practical applications may involve significant error when the load step is not sufficiently small. For too small a loading step, the scattering of the slope is even larger. Therefore, Eqs. (8.116), (8.117) should be used when a finite step size is used.

Either Eqs. (8.116) or (8.117) can be used to calculate the stress-strain curve of concrete. Mathematically speaking, an identical stress-strain curve should be obtained from these two equations. However, it was found that the stress-strain curves obtained from Eqs. (8.105), (8.106) were inconsistent (Hognestad et al. 1955). The reason was discussed in Li et al. (2021), which helps in increasing the accuracy of the method.

For normal flexural tests of rectangular members without controlling the strain of the tension face to zero, Li et al. (2021) derived the equations for calculation of the stresses at the tension face and compression face simultaneously:

$$\overline{f}_t = \frac{\varepsilon_{cc}\Delta(kN) - \Delta(k^2M)}{b(\varepsilon_{tc} - \varepsilon_{cc})\Delta\varepsilon_t} \tag{8.118}$$

$$\overline{f}_c = \frac{\varepsilon_{tc}\Delta(kN) - \Delta(k^2M)}{b(\varepsilon_{tc} - \varepsilon_{cc})\Delta\varepsilon_c} \tag{8.119}$$

where k, N, M, ε_t, and ε_c are directly measured curvature, axial force, bending moment, strain of tension face, and strain of compression face, respectively; Δ gives the increment of their value of a loading step; the calculated results of $\overline{f}_t$ and $\overline{f}_c$ are the tensile stress and compression stress at the tensile strain $\overline{\varepsilon}_t$ and compression strain $\overline{\varepsilon}_c$, respectively; ε_{cc} and ε_{tc} are strain values of a point within the compression strain increment $(\varepsilon_c < \varepsilon_{cc} < \varepsilon_c + \Delta\varepsilon_c)$ and tension strain increment $(\varepsilon_t < \varepsilon_{tc} < \varepsilon_t + \Delta\varepsilon_t)$ of each load step, respectively. $\overline{\varepsilon}_c$, $\overline{\varepsilon}_t$, ε_{cc}, and ε_{tc} can be determined based on local stress distribution and for sufficiently small loading steps linear or constant distribution can be adopted with insignificant error. Eqs. (8.118), (8.119) give a series of points of the tensile stress-strain relationship, $(\overline{\varepsilon}_t, \overline{f}_t)_{i=1\sim n}$, and the compression stress-strain relationship, $(\overline{\varepsilon}_c, \overline{f}_c)_{i=1\sim n}$, from a flexural test. When the loading steps are small enough, the calculated points are close enough to provide two continuous curves simultaneously, one tension stress-strain curve and the other compression stress-strain curve.

Kaklauskas and Ghaboussi (2001) and Gribniak et al. (2017) used an assumed compression stress-strain model and the analytical method to calculate tension stress-strain relationship of concrete. It can be seen from Eqs. (8.118), (8.119) that the two terms $\overline{f}_c$ and $\overline{f}_t$ are interrelated. If an incorrect or inaccurate model of $\overline{f}_c$ is assumed, the result of $\overline{f}_t$ calculated from the analytical method will be incorrect too.

Li et al. (2021) derived another set of equations for calculation of the stress-strain curves from flexural tests of members with a circular cross section. The curve smoothing and numerical procedure were investigated to reduce fluctuation of the calculated results. Other important issues related to the analytical method, including the fundamental assumptions on which this method is based, were also discussed in their paper. A typical stress-strain curve, including both tension and compression parts simultaneously produced from one eccentrically load column test using their method, is shown in Fig. 8.34F.

8.6.7 *Stress-strain model of confined concrete for repair works by Wu et al. (2014)*

A typical stress-strain curve of unconfined concrete without predamage is illustrated by O-P-C-D in Fig. 8.35A. Assume the concrete column is overloaded concentrically up to point C following the path O-P-C, and after the accidental overloading, the column is fully unloaded to point R, and subsequently an FRP jacket is installed to repair the damaged column. The load path of the repaired column will follow R-A-B when it is reloaded. The degree of damage to the column increases when the predamage load moves along the stress-strain curve from point O to P, C, and D.

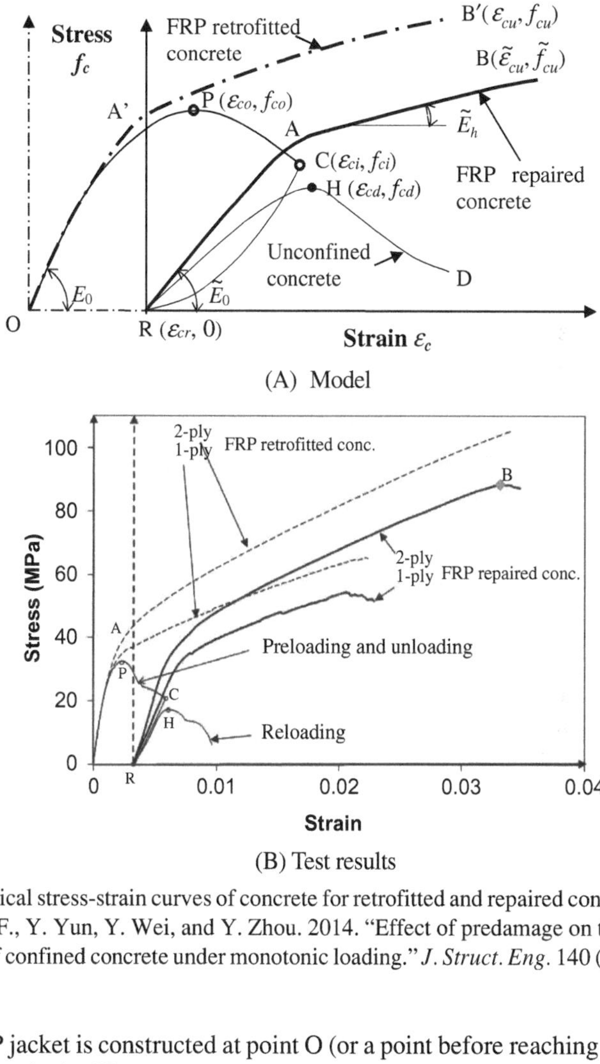

Fig. 8.35 Typical stress-strain curves of concrete for retrofitted and repaired concrete members. From Wu, Y. F., Y. Yun, Y. Wei, and Y. Zhou. 2014. "Effect of predamage on the stress-strain relationship of confined concrete under monotonic loading." *J. Struct. Eng.* 140 (12), 04014093.

If the FRP jacket is constructed at point O (or a point before reaching point P), then no damage (or insignificant damage) has been incurred and thus it is a retrofitting work. Therefore, the stress-strain curve for FRP-retrofitted concrete, as shown by the curve O-A′-B′ in Fig. 8.35A, is essentially a special case of the stress-strain curves of FRP-repaired concrete. This analysis reveals that a stress-strain model developed based on FRP-retrofitted columns is, in principle, not applicable to FRP-repaired columns.

Based on Fig. 8.35, the degree of damage of concrete, δ, can be defined as the strength loss, or

$$\delta = \frac{f_{co} - f_{cd}}{f_{co}} = 1 - \frac{f_{cd}}{f_{co}} \tag{8.120}$$

where f_{co} and f_{cd} are the strengths of undamaged concrete (point P) and damaged concrete (point H), respectively. In continuum damage mechanics, the damage variable D is defined as

$$D = 1 - \frac{\overline{A}}{A} \tag{8.121}$$

where A and $\bar{A}$ are the total and effective cross-sectional areas of a member, respectively, where fibers are assumed to be brittle and break immediately after reaching their strength (Kachanov 1958; Jirasek 2011). Therefore, $f_{cd} = \frac{\overline{A}}{A} \cdot f_{co}$. Substituting this f_{cd} into Eq. (8.120) gives $\delta = D$. Therefore, the damage degree defined by Eq. (8.120) is consistent with the damage variable adopted in classical continuum damage mechanics.

An extensive experimental program was carried out by Wu et al. (2014) on damaged, and subsequently repaired by FRP jacketing, concrete cylinders with various degrees of damage. The predamage on the concrete cylinders was incurred by loading a concrete cylinder to a certain stress level, as shown in Fig. 8.36.

Through theoretical derivation and experimental validation, the following equation was obtained by Wu et al. (2014):

$$\frac{\widetilde{f_{cc}}}{f_{co}} = \frac{f_{cc}}{f_{co}} - \delta \tag{8.122}$$

where $\widetilde{f_{cc}}$ and f_{cc} are strength of confined concrete with and without predamage, respectively. This equation gives a clear effect of damage on confined concrete strength. It is noted that the confined concrete strength is the stress value of the highest point of the stress-strain curve. For the stress-strain curve with a hardening slope, the

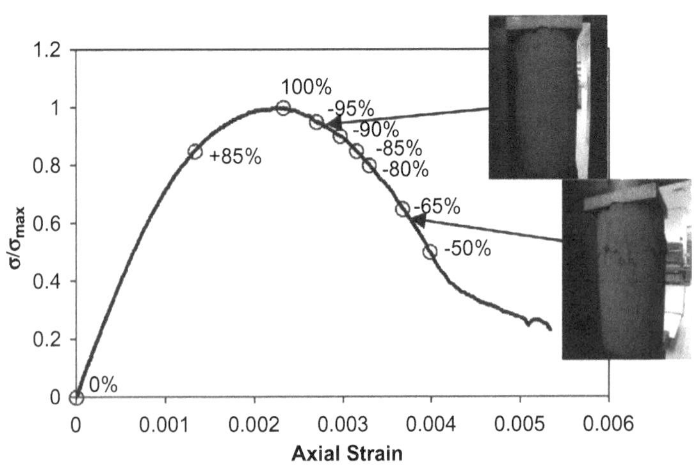

Fig. 8.36 Definition of damage levels.
From Wu, Y. F., Y. Yun, Y. Wei, and Y. Zhou. 2014. "Effect of predamage on the stress-strain relationship of confined concrete under monotonic loading." *J. Struct. Eng.* 140 (12), 04014093.

failure point (B or B′ in Fig. 8.35) gives the confined concrete strength. For a stress-strain curve with post-peak softening, the peak point gives the confined concrete strength. The theoretical derivation and analyses in Wu et al. (2014) also provided substantive remarks on the classical, fundamental but difficult, material failure theory that has not been well developed so far.

Through regression analyses, they provided the following model for the ultimate strain of FRP-repaired concrete columns:

$$\frac{\tilde{\varepsilon}_{cu}}{\varepsilon_{cu}}=1-0.30\left(\frac{f_{co}}{f_{30}}\right)^{2.82}\delta^{0.67}+0.54\delta \tag{8.123}$$

where $\tilde{\varepsilon}_{cu}$ and ε_{cu} are the ultimate strain of confined concrete with and without damage, respectively (Fig. 8.35).

Eq. (8.97) was also adopted as the stress-strain model of FRP-confined concrete with predamage. For damaged columns, the four parameters will have a wave on top, to become $\tilde{E}_1, \tilde{E}_2, \tilde{f}_0$, and $\tilde{n}$. Through regression analyses, the models of the four parameters were obtained as

$$\frac{\tilde{E}_1}{E_1}=e^{\left(3.9\delta^{1.38}-1.65\delta\right)\left(\frac{f_l}{f_{co}}\right)^{-0.016}-2.62\delta^{0.76}} \tag{8.124}$$

$$\frac{\tilde{f}_o}{f_o}=1-0.3\left(\frac{f_{co}}{f_{30}}\delta\right)^2-0.12\left(\frac{f_{co}}{f_{30}}\delta\right) \tag{8.125}$$

$$\frac{\tilde{E}_2}{E_2}=\left(\frac{f_{co}}{f_{30}}\right)^{2.67\delta}-0.46\delta \tag{8.126}$$

$\tilde{n}$ was found insensitive to the entire stress-strain curve and a value of $\tilde{n}=0.56$ was adopted. When the degree of damage δ is zero, the parameters degenerate to those of undamaged ones, i.e., $\tilde{E}_1=E_1, \tilde{f}_o=f_o$, and $\tilde{E}_2=E_2$.

8.6.8 Stress-strain model of confined concrete under cyclic loading

Concrete suffers growing damage under cyclic loading. Confinement to concrete can restrain the damage. Most existing cyclic stress-strain models developed for FRP-confined concrete are applicable only to cases with postpeak strain hardening. Among them, Shao et al. (2006) first developed the cyclic stress-strain model based on experimental test results. Lam and Teng (2009) improved the model by Shao et al. (2006). Wang et al. (2011, 2012) proposed two cyclic stress-strain models for concrete in FRP-confined RC columns, one for a square cross section and the other for a circular one, based on tests of large columns. Hany et al. (2015) conducted experimental tests and proposed a model by revising the coefficients of the monotonic model proposed

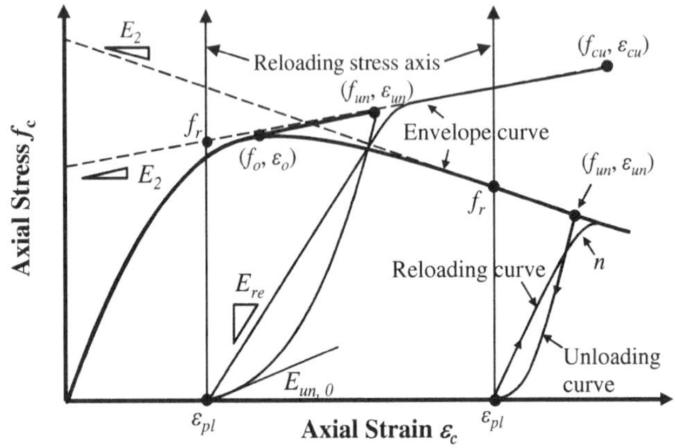

Fig. 8.37 Cyclic stress-strain model.
From Li, P., Y. F. Wu, Y. Zhou, and F. Xing. 2018. "Cyclic stress-strain model for FRP-confined concrete considering post-peak softening." *Compos. Struct.* 201, 902–915.

by Lam and Teng (2003a,b) as the envelope model, and using the unloading model proposed by Wang et al. (2011, 2012). Li and Wu (2015) proposed a cyclic model with a simpler mathematical form, which was later improved to allow for both post-peak hardening and softening (Li et al. 2018).

A stress-strain model of concrete under cyclic compression comprises three main elements: envelope curve, unloading path, and reloading path. The mathematical framework for the cyclic stress-strain model proposed by Li et al. (2018) is shown in Fig. 8.37. The parameters controlling the three elements are also shown in the figure. The factors that control the three elements were studied and the models for evaluation of the parameters were obtained by regression analysis using their test results.

The envelope model is given by Eqs. (8.89)–(8.95). A better model of E_2 than Eq. (8.91) was proposed as

$$E_2 = 1307 - 48842\rho_{eff}^{-2} \tag{8.127}$$

where ρ_{eff} is the effective confinement stiffness given by

$$\rho_{eff} = \frac{2E_f t_f}{bf_{co}}\left(\frac{2r}{b}\right)^{0.68}\left(\frac{h}{b}\right)^{-0.334} \tag{8.128}$$

With a larger database, Li et al. (2018) further revised Eq. (8.93) to

$$\frac{\mathrm{f}_{cu}}{f_{co}} = 0.5 + 2.7\left(\frac{2r}{b}\right)^{0.4}\left(\frac{f_l}{f_{co}}\right)^{0.66}\left(\frac{h}{b}\right)^{-1} \tag{8.129}$$

The unloading model is given by

$$f_c = E_{un,0}\left(\frac{\varepsilon_c}{\varepsilon_{pl}}\right)^m (\varepsilon_c - \varepsilon_{pl}) \tag{8.130}$$

$$m = \log_{(\varepsilon_{un}/\varepsilon_{pl})}\left(\frac{f_{un}}{E_{un,0}(\varepsilon_{un} - \varepsilon_{pl})}\right) \tag{8.131}$$

where f_{un} and ε_{un} are the stress and strain of the unloading point (Fig. 8.37), respectively. The parameter $E_{un,0}$ is equal to the residual tangent modulus when stress decreases to zero, given by

$$\frac{E_{un,0}}{E_c} = 0.21\left(\frac{f_{co}}{f_{30}}\right)^{0.195} \rho_{eff}^{-0.031}\left(\frac{\varepsilon_{un}}{\varepsilon_{co}}\right)^{-1.115} \tag{8.132}$$

The residual plastic strain is given by

$$\varepsilon_{pl} = \begin{cases} 0, & 0 < \varepsilon_{un} < 0.001 \\ 0.353\left(\frac{f_{co}}{f_{30}}\right)^{-0.4}(\varepsilon_{un} - 0.001) + 3.36\rho_{eff}^{-0.178}(\varepsilon_{un} - 0.001)^{1.414}, & \end{cases}$$
$$0.001 < \varepsilon_{un} < \varepsilon_{cu}. \tag{8.133}$$

The reloading model adopts the following mathematical form:

$$f_c = \frac{(E_{re} - E_2)(\varepsilon_c - \varepsilon_{pl})}{\left(1 + \left(\frac{(E_{re} - E_2)(\varepsilon_c - \varepsilon_{pl})}{f_r}\right)^n\right)^{\frac{1}{n}}} + E_2(\varepsilon_c - \varepsilon_{pl}) \tag{8.134}$$

The parameter f_r is the intercept of the asymptotic line of the envelope curve with the reloading stress axis (Fig. 8.37), given by

$$\begin{cases} \frac{f_r}{f_{co}} = 0.693\frac{f_{un}}{f_{co}} + 0.337\rho_{eff}^{-0.053} & E_2 \geq 0 \\ \frac{f_r}{f_{co}} = 0.969\frac{f_{un}}{f_{co}} + 1.981\rho_{eff}^{-2.012} & E_2 < 0 \end{cases} \tag{8.135}$$

The parameter E_{re} is the reloading stiffness, given by

$$\frac{E_{re}}{E_c} = \left(\frac{f_{co}}{f_{30}}\right)^{0.032}(\bar{\varepsilon})^{-0.409} - 0.317\rho_{eff}^{-0.064\bar{\varepsilon}} \tag{8.136}$$

where

$$\bar{\varepsilon} = \begin{cases} \dfrac{\varepsilon_{un}}{\varepsilon_{co}}, \text{when } \varepsilon_{un}/\varepsilon_{co} \le 10 \\ 10, \text{when } \varepsilon_{un}/\varepsilon_{co} > 10 \end{cases} \tag{8.137}$$

The parameter n controls the curvature of the transition part (different from the n for Eq. 8.97), as shown in Fig. 8.37 and given by

$$n = 2.61\left(\frac{\varepsilon_{un}}{\varepsilon_{co}}\right) + 4.88 \tag{8.138}$$

Figs. 8.38 and 8.39 show the predictions by the model in comparison with test results for circular and square/rectangular columns, respectively.

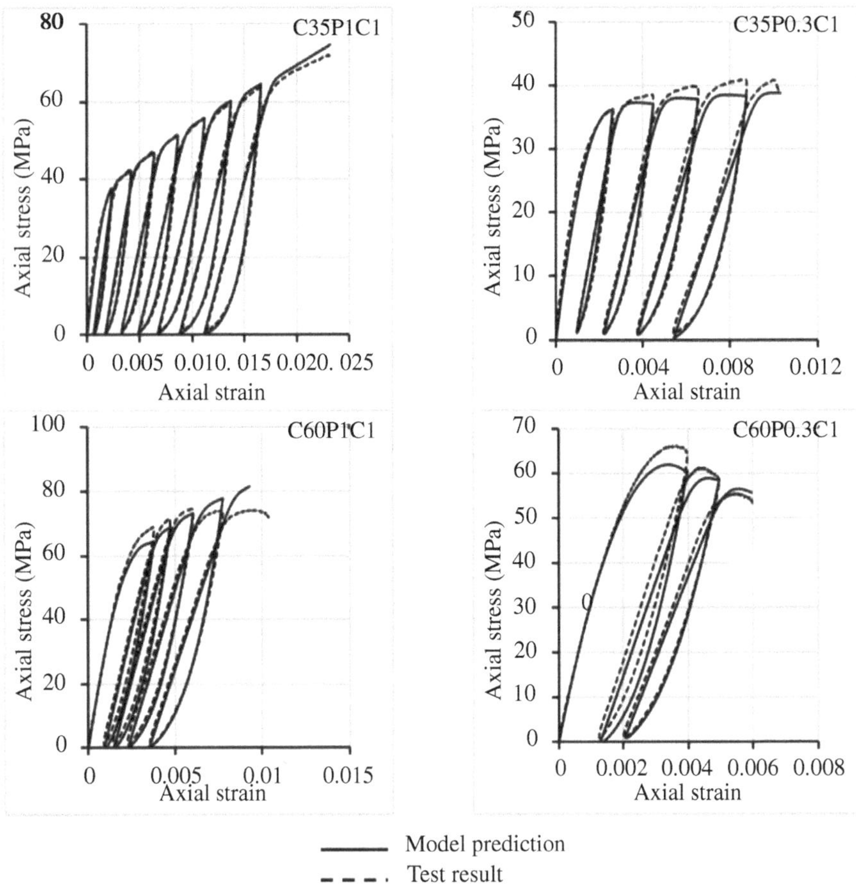

Fig. 8.38 Performance of the cyclic model for circular columns.
From Li, P., Y. F. Wu, Y. Zhou, and F. Xing. 2018. "Cyclic stress-strain model for FRP-confined concrete considering post-peak softening." *Compos. Struct.* 201, 902–915.

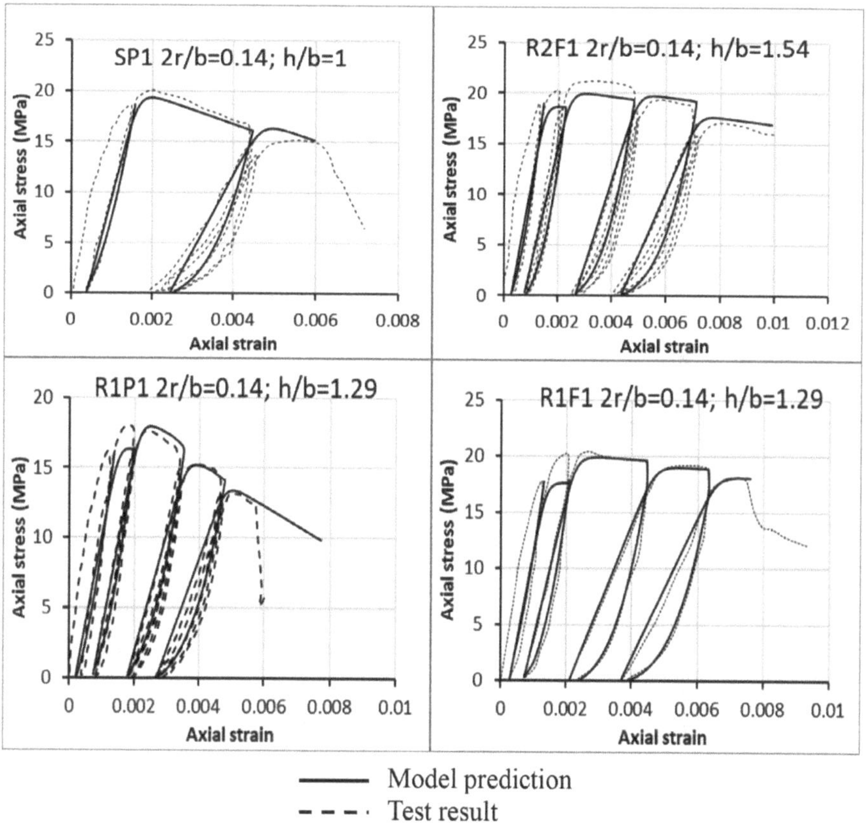

Fig. 8.39 Performance of the cyclic model for square/rectangular columns.

8.6.9 Stress-strain model of concrete for axially loaded members with failure localization

The failure of concrete under tension is concentrated in a narrow fracture zone. This phenomenon is called localization. In fact, localized failure of concrete can also occur under compression, as shown in Fig. 8.40. For a real column with an identical cross-sectional design along the length, the stresses at different cross sections are actually not completely equal under concentric loading due to variations of geometrical and material properties in construction. When the weakest cross section reaches the peak stress of the stress-strain curve of the concrete material and moves toward its descending branch, other parts of the column that have not yet reached the peak stress experience elastic unloading due to equilibrium of force. This is similar to the case of plastic hinge studied in Section 7.3 (see Fig. 7.14A). In this case, the postpeak inelastic strain occurs only in the vicinity of the weakest section (only the weakest link in Fig. 1.9 fails and other links unload), i.e., it is localized. Therefore, the local strains at different cross sections will be very different (Fig. 8.40). Consequently, the

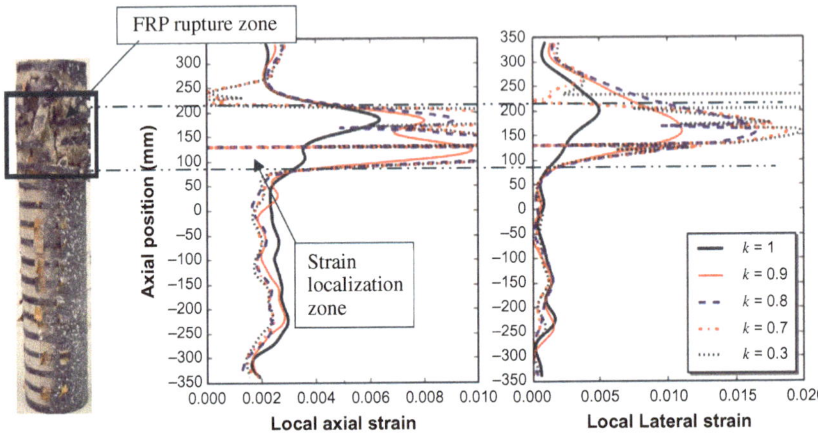

Fig. 8.40 Localization of compression failure.
From Wei, Y. Y., and Y. F. Wu. 2016. "Experimental study of concrete columns with localized failure." *J. Compos. Constr.* 20 (5), 04016032.

global load-deformation behavior of the member will be closely related to the size of the localized failure zone relative to the total length of the member. In other words, the stress-strain behavior of concrete in the failure zone cannot represent the average stress-strain behavior of the whole member, although the whole member is theoretically under a uniform stress. In this case, the average stress-strain curve exhibits size dependency.

In concrete columns under concentric compression, which involves postpeak softening, localization is initiated after it passes the peak stress (Shah and Sankar 1987; Torrenti et al. 1993). Failure of the column only takes place within a zone of limited length, while other parts outside this zone elastically unload in the failure process. This process is similar to the localized tension cracking process. Compared with tension localization, compressive localization is more complex as it is related to many cracks rather than one prominent crack. Generally, the concept of fracture energy is used in the literature for stress-strain modeling involving failure localization.

Before the onset of the peak stress, the stress-strain curve is generally the same in all cross sections of the column if it has an identical cross section throughout its length. During the postpeak softening, the undamaged zone unloads elastically, as shown in Fig. 8.41B. The shaded area enclosed by the loading and unloading path is the energy per unit volume of concrete (g_{fc}) dissipated due to micro cracking during the buildup of the stress to the peak point. The same amount of energy per unit volume of concrete is also dissipated in the failure zone up to the peak stress. During the postpeak softening, additional energy is dissipated only in the failure zone, as shown by the shaded area in Fig. 8.41C. Markeset and Hillerborg (1995) further divided this additional energy into two parts: one related to the formation of longitudinal cracks corresponding to additional lateral strain within the damaged zone and the other due to the shear deformation. In reality, the longitudinal cracks and the inclined shear

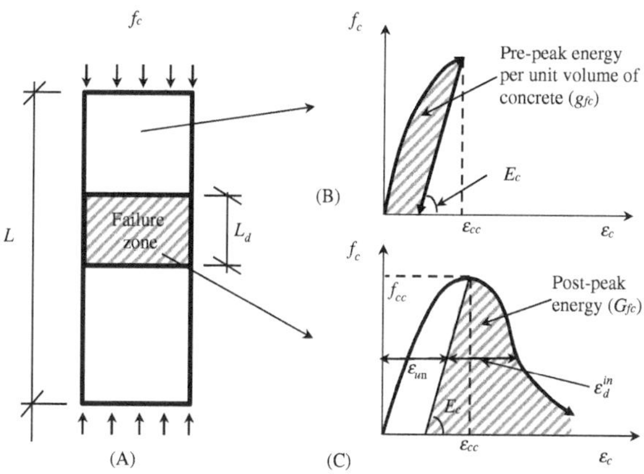

Fig. 8.41 Localization and compressive fracture energy: (A) elevation of member, (B) stress-strain curve of concrete outside failure zone, (C) stress-strain curve of concrete in localized failure zone.
From Wu, Y. F., and Y. Y. Wei. 2016. "Stress-strain modeling of concrete columns with localized failure: an analytical study." *J. Compos. Constr.* 20 (3), 04015071.

cracks are both contained by confinement and it is difficult to quantify them separately. The shaded area in Fig. 8.41C is called the postpeak fracture energy (G_{fc}) that is considered as the key for stress-strain modeling involving failure localization.

Through experimental study (Wei and Wu 2016) and analytical study (Wu and Wei 2016), a stress-strain model for calculating the global stress-strain response of FRP-confined concrete columns involving failure localization was developed. This model is described in this section. It is noted that the theory described in this section is only applicable to structural members that are loaded under a displacement control mode. For a structural member under load control, it can only be loaded to the peak point of its load-deformation curve and fails suddenly. Therefore, failure localization is not applicable to such members.

When a localized failure occurs, the total deformation Δ_c of the compression member in Fig. 8.41A is comprised of two parts, the unloading displacement Δ_{un} outside the failure zone and the displacement occurring in the localized failure zone Δ_d, i.e.,

$$\Delta_c = \Delta_d + \Delta_{un} = \left(\varepsilon_{un} + \varepsilon_d^{in}\right)L_d + \varepsilon_{un}(L - L_d) = \varepsilon_d^{in}L_d + \varepsilon_{un}L = \delta_d^{in} + \varepsilon_{un}L \qquad (8.139)$$

where L is the total length of the specimen; L_d is the length of the localized failure zone; ε_{un} is the unloading strain (Fig. 8.41C) that occurs in the whole specimen; and ε_d^{in} and δ_d^{in} are the strain and displacement of inelastic deformation within the failure zone, respectively. Therefore, the average strain ε_c of the whole column is given by $\frac{\Delta_c}{L}$ or

$$\varepsilon_c = \varepsilon_{un} + \frac{\delta_d^{in}}{L} = \varepsilon_{un} + \varepsilon_d^{in}\frac{L_d}{L} \qquad (8.140)$$

where the unloading strain can be calculated by

$$\varepsilon_{un} = \varepsilon_{cc} - \frac{f_{cc} - f_c}{E_c} = \varepsilon_{cc} - \frac{f_{cc}}{E_c} + \frac{1}{E_c}f_c \qquad (8.141)$$

where f_{cc} and ε_{cc} are the peak stress and the corresponding strain, respectively; f_c and E_c are the stress and elastic modulus of concrete, respectively. It is noted that the slope of the unloading path is assumed the same as the elastic modulus of concrete E_c when unloading occurs before softening.

Eqs. (8.140), (8.141) show that the strain ε_c depends not only on stress f_c but also on δ_d^{in} and L. Substituting Eq. (8.141) into Eq. (8.139) gives

$$\delta_d^{in} = \Delta_c - \left(\varepsilon_{cc} - \frac{f_{cc}}{E_c} + \frac{1}{E_c} f_c \right) L \tag{8.142}$$

It is observed from tests that the postpeak stress vs. inelastic displacement curves for specimens with different values of L are almost the same when L is larger than L_d (Jansen and Shah 1997; Markeset and Hillerborg 1995). This phenomenon indicates that δ_d^{in} is size independent. The postpeak stress-strain curve (Fig. 8.41C) is unified and shown in Fig. 8.42A. The area of the postpeak inelastic deformation (the shaded area in Fig. 8.41C) is separately shown in Fig. 8.42B, which is the postpeak fracture energy G_{fc}.

A majority of the existing models use a nonlinear function to describe the postpeak softening curve (Fig. 8.42B). A summary of the typical postpeak softening models was provided in Wu and Wei (2016). Based on the study of the existing models and the test results, Wu and Wei (2016) proposed the following model for the postpeak inelastic curve:

$$\frac{f_c}{f_{cc}} = \exp\left[-\left(\frac{\delta_d^{in}}{X_0} \right)^m \right] \tag{8.143}$$

where X_0 and m are two parameters that control the shape of the curve. Eq. (8.143) can also be rewritten as

$$\delta_d^{in} = e^{\frac{\ln\left[-\ln\left(\frac{f_c}{f_{cc}}\right)\right]}{m} + \ln(X_0)} \tag{8.144}$$

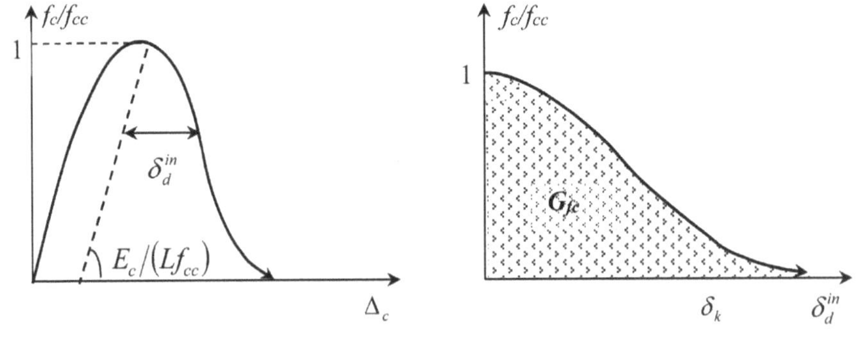

(A) Total deformation (B) Inelastic deformation

Fig. 8.42 Postpeak inelastic displacement curve.

Based on their test results, the two parameters were identified to be

$$X_0 = \frac{G_{fc}}{0.9 f_{cc}} \tag{8.145}$$

$$m = 1.15 + 0.31\left(\frac{f_{30}}{f_{co}}\right)^2 + \frac{1}{3}\rho_{eff}\frac{f_{30}}{f_{co}} \tag{8.146}$$

where ρ_{eff} is the confinement stiffness given by Eq. (8.128). The postpeak fracture energy G_{fc} is given by

$$G_{fc} = \int_0^\infty f_c d\delta_d^{in} = \int_0^\infty f_{cc} \exp\left[-\left(\frac{\delta_d^{in}}{X_0}\right)^m\right] d\delta_d^{in} \tag{8.147}$$

Extensive investigations on the fracture energy G_{fc} have been reported in the literature and a summary of the existing models can be found in Wu and Wei (2016). Based on experimental tests and analytical studies, the following model was proposed by Wu and Wei (2016):

$$\frac{G_{fc}}{G_{fco}} = \frac{f_{cc}}{f_{co}} + 0.25\left(\frac{f_{co}}{\mathrm{f}_{30}}\right)^{2.19} \ln\left(\frac{6.3 + 3.3\rho_{eff}}{6.3 - \rho_{eff}}\right), \ (\rho_{eff} \le 6.3) \tag{8.148}$$

where G_{fco} is the fracture energy of unconfined concrete given by

$$\frac{G_{fco}}{G_{f30}} = \left(\frac{f_{co}}{f_{30}}\right)^{0.15} \tag{8.149}$$

where G_{f30} is the fracture energy of grade 30 concrete and $G_{f30} = 27.5$ N/mm based on test results.

The stress-strain relationship in the postpeak region is obtained by substituting Eqs. (8.144) and (8.141) into Eq. (8.140) to give

$$\varepsilon_c = \varepsilon_{un} + \frac{\delta_d^{in}}{L} = \varepsilon_{cc} - \frac{f_{cc}}{E_c} + \frac{1}{E_c} f_c + \frac{1}{L}\delta_d^{in} \tag{8.150}$$

For the prepeak region, any adequate existing stress-strain model can be adopted. Eqs. (8.73)–(8.77) were adopted by Wu and Wei (2016) for the ascending branch of the stress-strain relationship. The overall stress-strain relationship is therefore given by

$$\begin{cases} \dfrac{f_c}{f_{cc}} = \dfrac{x \cdot \lambda}{\lambda - 1 + x^\lambda}, x = \dfrac{\varepsilon_c}{\varepsilon_{cc}} & (\varepsilon_c \le \varepsilon_{cc}) \\ \dfrac{f_c}{f_{cc}} = 1 + \dfrac{E_c}{f_{cc}}(\varepsilon_c - \varepsilon_{cc}) - \dfrac{E_c}{f_{cc} L}\delta_d^{in} & (\varepsilon_c > \varepsilon_{cc}) \end{cases} \tag{8.151}$$

As f_{cc}, ε_{cc}, and G_{fc} are related not only to concrete material but also to confinement, Eq. (8.151) shows that the postpeak stress-strain relationship of confined columns is affected by concrete material properties, column dimensions, and confinement. The unified strength model, Eq. (8.48), was used for f_{cc}. For the peak strain ε_{cc}, the popular model proposed by Popovics (1973) was used, where the unconfined concrete strength f_{co} was replaced by the confined concrete strength f_{cc}, giving

$$\varepsilon_{cc} = 0.000937 f_{cc}^{\ 1/4} \tag{8.152}$$

Although the stress-strain model developed here is independent of the length of the localized failure zone L_d, the knowledge of the length of the localization zone is desirable in practice. The phenomenon of localization only occurs when column length is greater than the length of the localization zone. Based on the literature, both experimental and analytical studies, the following model was developed by Wu and Wei (2016) through regression analysis:

$$\frac{L_d}{L} = \frac{L_{d0}}{L} + 0.4\left[1 + \tan h\left(\rho_{eff} - 4\right)\right] \tag{8.153}$$

where L_{d0} is the length of the damage zone for unconfined columns, given by

$$L_{d0} = k_f L_{d,Ler} \tag{8.154}$$

and k_f and $L_{d,\ Ler}$ are given by

$$k_f = \begin{cases} 1 & f_{co} \leq 40 \\ 1.15\sqrt{\dfrac{f_{30}}{f_{co}}} & f_{co} > 40 \end{cases} \tag{8.155}$$

$$\frac{L_{d,Ler}}{\sqrt{A_c}} = \begin{cases} 1.36 & \sqrt{A_c} \leq 100 \\ -3.53 \times 10^{-5} A_c + 1.71 & 100 \leq \sqrt{A_c} \leq 180 \text{ (in mm)} \\ 0.57 & \sqrt{A_c} \geq 180 \end{cases} \tag{8.156}$$

where A_c is the cross-sectional area of the concrete member.

It is noted that the theory is based on the assumption that concrete is uniform in a column. In practice, concrete columns are cast vertically, and hence strength variation could exist along the length. If significant strength variation exists in a column, it could affect the location and length of the failure zone.

For FRP-confined columns, the FRP jacket ruptures when the axial strain reaches a certain value that is defined as the ultimate strain. However, the rupture of the FRP jacket is a process rather than a point, particularly for columns with postpeak softening (Wei and Wu 2012). The maximum allowable strain or the ultimate strain of a column should be taken at the point of initial rupture of FRP from a structural safety point of view (Wei and Wu 2012). When the local strain inside the failure zone of a column reaches this ultimate strain, the column should be considered as reaching its ultimate

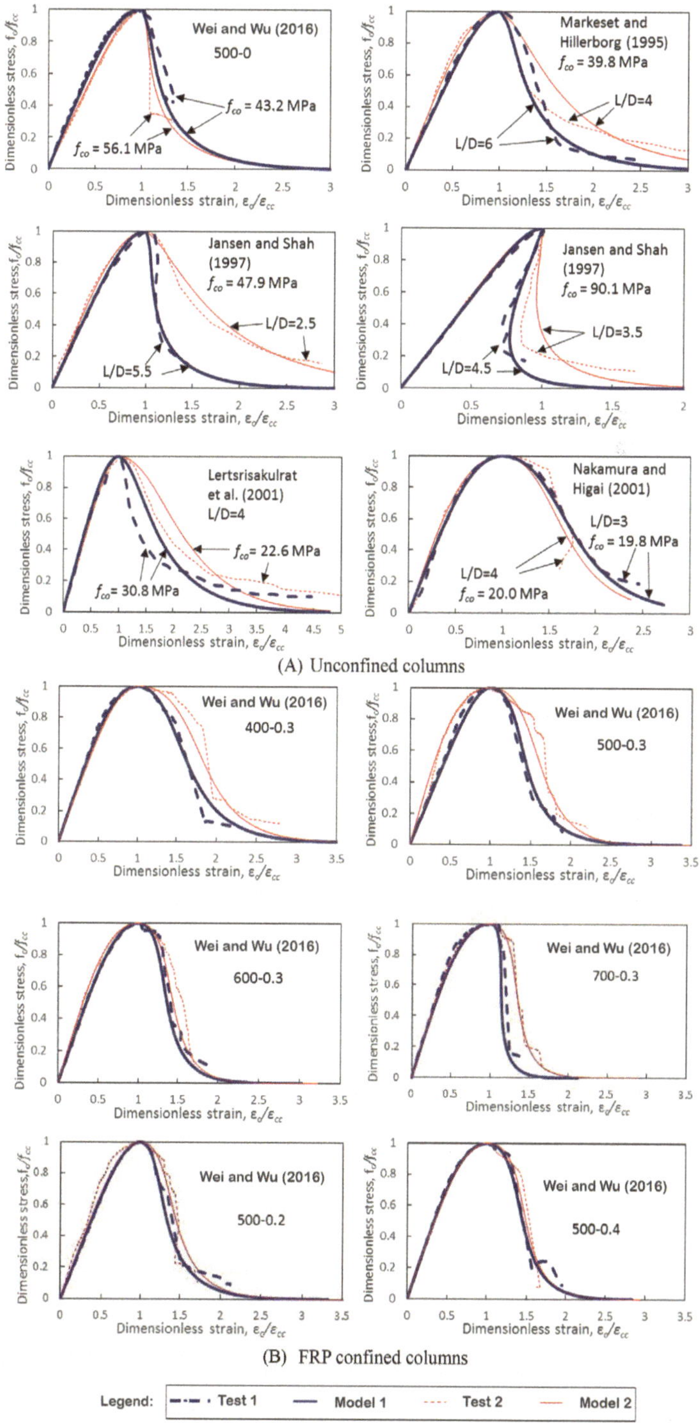

Fig. 8.43 Performance of the stress-strain model considering failure localization. From Wu, Y. F., and Y. Y. Wei. 2016. "Stress-strain modeling of concrete columns with localized failure: an analytical study." *J. Compos. Constr.* 20 (3), 04015071.

state. A model for the ultimate strain of FRP-confined short circular columns without localization was developed by Wu and Wei (2015), as given by

$$\frac{\varepsilon_{cu}}{\varepsilon_{co}} = 1.75 + 140\left(\frac{f_l}{f_{co}}\right)\varepsilon_{fu}^{0.6} \tag{8.157}$$

The ultimate strain models obtained from short column tests without allowing for localization are also applicable to the local ultimate strain of long columns within the local failure zone.

The average strain of a whole column at the onset of the local ultimate strain can be calculated by Eq. (8.140), in which $\varepsilon_d^{in} = \varepsilon_{cu} - \varepsilon_{un}$ (Fig. 8.41C) and ε_{cu} is given by Eq. (8.157). In this case, the value of ε_{un} should be calculated by Eq. (8.141) with the stress $f_c = f_{cu}$, where f_{cu} is the ultimate stress corresponding to ε_{cu} for FRP-confined columns. For circular columns, it is given by (Wu and Wei 2015):

$$\frac{f_{cu}}{f_{co}} = 0.75 + 2.7\left(\frac{f_l}{f_{co}}\right)^{0.9} \tag{8.158}$$

Comparisons of the model predictions with test results are shown in Fig. 8.43, which shows a satisfactory performance of the model. The snapback behavior observed by Jansen and Shah (1997) in tests is also well predicted by the stress-strain model.

References

ACI Committee 440. 2002. *Guide for the design and construction of externally bonded FRP systems for strengthening concrete structures, 440.2R-02*. Detroit, MI: American Concrete Institute.

ACI Committee 440. 2017. *Guide for the design and construction of externally bonded FRP systems for strengthening concrete structures, ACI 440.2R17*. Detroit, MI: American Concrete Institute.

Ahmad, S. H., and S. P. Shah. 1982a. "Complete triaxial stress-strain curves for concrete." *J. Struct. Div. ASCE* 108 (4), 728–742.

Ahmad, S. H., and S. P. Shah. 1982b. "Stress–strain curves of concrete confined by spiral reinforcement." *ACI J.* 79, 484–490.

Ahmad, S. H., A. R. Khaloo, and A. Irshaid. 1991. "Behaviour of concrete spirally confined by fibreglass filament." *Mag. Concr. Res.* 43 (156), 143–148.

Al-Salloum, Y. A. 2007. "Influence of edge sharpness on the strength of square concrete columns confined with FRP composite laminates." *Compos. Part B Eng.* 38 (5–6), 640–650.

Arduini, M., A. Di Tommaso, O. Manfroni, S. Ferrari, and M. Romagnolo. 1999. "II: Confinement o passivo di elementi compressi in calcetruzzo con fogli di materiale composito." *Ind. Ital. Cemento* 11, 836–841.

Becque, J., A. K. Patnaik, and S. H. Rizkalla. 2003. "Analytical models for concrete confined with FRP tubes." *J. Compos. Constr.* 7 (1), 31–38.

Bisby, L., and M. Ranger. 2010. "Axial-flexural interaction in circular FRP-confined reinforced concrete columns." *Construct. Build Mater.* 24 (9), 1672–1681.

Campione, G., and N. Miraglia. 2003. "Strength and strain capacities of concrete compression members reinforced with FRP." *Cem. Concr. Compos.* 25 (1), 31–41.

Cao, Y. G., Y. F. Wu, and C. Jiang. 2018. "Stress-strain relationship of FRP confined concrete columns under combined axial load and bending moment." *Compos. Part B Eng.* 134, 207–217.

Carino, N. J., and F. O. Slate. 1976. "Limiting tensile strain criterion for failure of concrete." *J. Proc.* 73 (3), 160–165.

Chaallal, O., M. Hassan, and M. Shahawy. 2003a. "Confinement model for axially loaded short rectangular columns strengthened with fiber reinforced polymer wrapping." *ACI Struct. J.* 100 (2), 215–221.

Chaallal, O., S. Mohsen, and H. Munzer. 2003b. "Performance of axially loaded short rectangular columns strengthened with carbon fiber-reinforced polymer wrapping." *J. Compos. Constr.* 7 (3), 200–208.

Cheng, H. L., E. D. Sotelino, and W. F. Chen. 2002. "Strength estimation for FRP wrapped reinforced concrete columns." *Steel Compos. Struct.* 2 (1), 1–20.

Chun, S. S., and H. C. Park. 2002. "Load carrying capacity and ductility of RC columns confined by carbon fiber reinforced polymer." In *Proc. Third Int. Conf. on Compos. in Infra*, San Francisco, CA: Univ. of Arizona.

Demers, M., and K. W. Neale. 1994. "Strengthening of concrete columns with unidirectional composite sheets." In *Proc. Fourth Int. Conf. Short and Medium Span Bridges*, Montreal, Canada: Canadian Society for Civil Engineering, 895–905.

Eyre, J. R., and H. S. Nasreddin. 2013. "Tension strain failure criterion for concrete." *Mag. Concr. Res.* 65 (21), 1303–1314.

Fam, A. Z., and S. H. Rizkalla. 2001. "Confinement model for axially loaded concrete confined by circular fiber-reinforced polymer tubes." *ACI Struct. J.* 98 (4), 451–461.

Fam, A., B. Flisak, and S. Rizkalla. 2003. "Experimental and analytical modeling of concrete-filled fiber-reinforced polymer tubes subjected to combined bending and axial loads." *ACI Struct. J.* 100, 499–509.

Fardis, M. N., and H. Khalili. 1981. "Concrete encased in fiber glass reinforced plastic." *ACI J.* 78 (6), 440–446.

Fardis, M. N., and H. Khalili. 1982. "FRP-encased concrete as a structural material." *Mag. Concr. Res.* 34 (121), 191–202.

Gabet, T., Y. Malécot, and L. Daudeville. 2008. "Triaxial behaviour of concrete under high stresses: influence of the loading path on compaction and limit states." *Cem. Concr. Res.* 38 (3), 403–412.

Gerstle, K. H. 1981a. "Simple formulation of biaxial concrete behavior." *ACI J.* 78 (1), 62–68.

Gerstle, K. H. 1981b. "Simple formulation of triaxial concrete behavior." *ACI J.* 78 (5), 382–387.

Girgin, C. Z., A. Nihal, and A. Ergin. 2007. "Evaluation of strength criteria for very-high-strength concretes under triaxial compression." *ACI Struct. J.* 104 (3), 278–284.

Green, F. M., L. A. Bisby, A. Z. Fam, and V. K. R. Kodur. 2006. "FRP confined concrete columns: behaviour under extreme conditions." *Cem. Concr. Compos.* 28 (10), 928–937.

Gribniak, V. 2009. "Shrinkage influence on tension-stiffening of concrete structures." Ph.D. thesis. Vilnius: Vilnius Gediminas Technical University.

Gribniak, V., G. Kaklauskas, R. Kacianauskas, and R. Kliukas. 2012. "Improving efficiency of inverse constitutive analysis of reinforced concrete flexural members." *Sci. Res. Essays* 7 (8), 923–938.

Gribniak, V., G. Kaklauskas, A. Juozapaitis, R. Kliukas, and A. Meskenas. 2017. "Efficient technique for constitutive analysis of reinforced concrete flexural members." *Inverse Prob. Sci. Eng.* 25 (1), 27–40.

Guo, Z. 2014. *Principles of reinforced concrete*, New York: Butterworth-Heinemann.

Hany, N. F., E. G. Hantouche, and M. H. Harajli. 2015. "Axial stress-strain model of CFRP-confined concrete under monotonic and cyclic loading." *J. Compos. Constr.* 19 (6), 04015004.

Harajli, M. H., E. Hantouche, and K. Soudki. 2006. "Stress–strain model for fiber-reinforced polymer jacketed concrete columns." *ACI Struct. J.* 103 (5), 672–682.

Harmon, T. G., and K. T. Slattery. 1992. "Advanced composite confinement of concrete." In *Proc. Int. Conf. on Advanced Composite Materials in Bridges and Structures*, Canadian Society for Civil Engineering, 299–316.

Harmon, T. G., S. Ramakrishnan, and E. H. Wang. 1998. "Confined concrete subjected to uniaxial monotonic loading." *J. Eng. Mech.* 124 (12), 1303–1308.

Hoek, E., and E. T. Brown. 1980. "Empirical strength criterion for rock masses." *J. Geotech. Eng. Div.* 106 (GT9), 1013–1035.

Hognestad, E. 1951. "A study of combined bending and axial load in reinforced concrete members." *Bulletin series no. 399*, Urbana, IL: Engineering Experiment Station, Univ. of Illinois.

Hognestad, E., N. W. Hanson, and D. McHenry. 1955. "Concrete stress distribution in ultimate strength design." *ACI J. Proc.* 52 (12), 455–479.

Howie, I., and V. M. Karbhari. 1994. "Effect of materials architecture on strengthening efficiency of composite wraps for deteriorating columns in the North-east." In *Proc. Third Materials Engineering Conf*, New York: Material Engineering Division, ASCE, 199–206.

Ibrahim, H. H., and J. G. Macgregor. 1996. "Flexural behavior of laterally reinforced high-strength concrete sections." *Struct. J.* 93 (6), 674–684.

Ilki, A., N. Kumbasar, and V. Koc. 2004. "Low strength concrete members externally confined with FRP sheets." *Int. J. Struct. Eng. Mech.* 18 (2), 167–194.

Jansen, D. C., and S. P. Shah. 1997. "Effect of length on compressive strain softening of concrete." *J. Eng. Mech.* 123 (1), 25–35.

Jiang, T., and J. G. Teng. 2007. "Analysis-oriented stress-strain models for FRP-confined concrete." *Eng. Struct.* 29, 2968–2986.

Jiang, J. F., and Y. F. Wu. 2012. "Identification of material parameters for Drucker–Prager plasticity model for FRP confined circular concrete columns." *Int. J. Solids Struct.* 49 (3–4), 445–456.

Jiang, J. F., and Y. F. Wu. 2014. "Characterization of yield surfaces for FRP-confined concrete." *J. Eng. Mech.* 140 (12), 04014096.

Jin, X. N., J. L. Pan, G. Y. Liu, and W. H. Lai. 2003. "Research of stress–strain curve of concrete confined by fiber reinforced plastics under axial compression." *J. Build. Struct.* 24 (4), 47–53 (in Chinese).

Jirasek, M. 2011. "Damage and smeared crack models." In *Numerical modeling of concrete cracking*, Vienna: Springer, 1–49.

Kaar, P. H., N. W. Hanson, and H. Capell. 1978. *Stress-strain characteristics of high-strength concrete*, *Vol.* 55 Special Publication, 161–186.

Kachanov, L. M. 1958. "Izvestiya Akademii Nauk." *SSSR. Otd. Tekhn. Nauk.* 8, 26–31.

Kaklauskas, G. 1997. "A method for determination of stress-strain relations for concrete from experimental data of RC bending members." *Statyba* 3 (10), 60–68.

Kaklauskas, G. 2004. "Flexural layered deformational model of reinforced concrete members." *Mag. Concr. Res.* 56 (10), 575–584.

Kaklauskas, G., and J. Ghaboussi. 2001. "Stress-strain relations for cracked tensile concrete from RC beam tests." *J. Struct. Eng.* 127 (1), 64–73.

Karabinis, A. I., and T. C. Rousakis. 2001. "Carbon FRP confined concrete elements under axial load." In *Proc. Int. Conf. on FRP Composites in Civil Engineering, Hong Kong, China*, 309–316.

Karbhari, V. M., and Y. Gao. 1997. "Composite jacketed concrete under uniaxial compression-verification of simple design equations." *J. Mater. Civ. Eng.* 9 (4), 185–193.

Kono, S., M. Inazumi, and T. Kaku. 1998. "Evaluation of confining effects of CFRP sheets on reinforced concrete members." In *Proc. Second Int. Conf. on Composites in Infrastructure (ICCI), Tucson, Ariz*, 343–355.

Kumutha, R., R. Vaidyanathan, and M. S. Palanichamy. 2007. "Behaviour of reinforced concrete rectangular columns strengthened using GFRP." *Cem. Concr. Compos.* 29 (8), 609–615.

Lam, L., and J. G. Teng. 2001. "A new stress-strain model for FRP-confined concrete." In *Proc. Int. Conf. on FRP Composites in Civil Engineering, Hong Kong, China*, 283–292.

Lam, L., and J. G. Teng. 2003a. "Design-oriented stress-strain model for FRP-confined concrete." *Construct. Build Mater.* 17, 471–489.

Lam, L., and J. G. Teng. 2003b. "Design-oriented stress-strain model for FRP-confined concrete in rectangular columns." *J. Reinf. Plast. Compos.* 22 (13), 1149–1186.

Lam, L., and J. G. Teng. 2009. "Stress-strain model for FRP-confined concrete under cyclic axial compression." *Eng. Struct.* 31, 308–321.

Lertsrisakulrat, T., K. Watanabe, M. Matsuo, and J. Niwa. 2001. "Experimental study on parameters in localization of concrete subjected to compression." *J. Mater. Conc. Struct.* 50 (669), 309–321.

Li, P., and Y. F. Wu. 2015. "Stress-strain model of FRP confined concrete under cyclic loading." *Compos. Struct.* 134, 60–71.

Li, Y. F., C. T. Lin, and Y. Y. Sung. 2003. "A constitutive model for concrete confined with carbon fiber reinforced plastics." *Mech. Mater.* 35, 603–619.

Li, P., Y. F. Wu, Y. Zhou, and F. Xing. 2018. "Cyclic stress-strain model for FRP-confined concrete considering post-peak softening." *Compos. Struct.* 201, 902–915.

Li, Z., Y. F. Wu, and P. Li. 2021. "Method to identify stress-strain relationship of FRP-confined concrete under eccentrical load." *J. Compos. Constr.* 25 (1), 04020080.

Lillistone, D., and C. K. Jolly. 2000. "An innovative form of reinforcement for concrete columns using advanced composites." *Struct. Eng.* 78 (23/24), 20–28.

Liu, H. K., W. C. Liao, L. Tseng, W. H. Lee, and Y. Sawada. 2004. "Compression strength of pre-damaged concrete cylinders reinforced by non-adhesive filament wound composites." *Compos. A: Appl. Sci. Manuf.* 35 (2), 281–292.

Mander, J. B., M. J. N. Priestley, and R. Park. 1988a. "Theoretical stress-strain model for confined concrete." *J. Struct. Eng.* 114 (8), 1804–1826.

Mander, J. B., M. J. N. Priestley, and R. Park. 1988b. "Observed stress-strain behavior of confined concrete." *J. Struct. Eng.* 114 (8), 1827–1849.

Markeset, G., and A. Hillerborg. 1995. "Softening of concrete in compression—localization and size effects." *Cem. Concr. Res.* 25 (4), 702–708.

Marques, S. P. C., D. C. D. S. C. Marques, J. L. D. S. Silva, and M. A. A. Cavalcante. 2004. "Model for analysis of short columns of concrete confined by fiber-reinforced polymer." *J. Compos. Constr.* 8 (4), 332–340.

Matthys, S., H. Toutanji, K. Audenaert, and L. Taerwe. 2005. "Axial load behavior of large-scale columns confined with FRP composites." *ACI Struct. J.* 102 (2), 258–267.

Mirmiran, A., and M. Shahawy. 1997. "Behavior of concrete columns confined by fiber composite." *J. Struct. Eng.* 123 (5), 583–590.

Mirmiran, A., M. Shahawy, M. Samaan, H. E. Echary, J. C. Mastrapa, and O. Pico. 1998. "Effect of column parameters on FRP-confined concrete." *J. Compos. Constr.* 2 (4), 175–185.

Miyauchi, K., S. Nishibayashi, and S. Inoue. 1997. "Estimation of strengthening effects with carbon fiber sheet for concrete column." In *Proc. FRPRCS-3, Sapporo, Japan*, Vol. 1, 217–224.

Miyauchi, K., S. Inoue, T. Kuroda, and A. Kobayashi. 1999. "Strengthening effects with carbon fiber sheet for concrete column." *Proc. Jpn. Concr. Inst.* 21 (3), 1453–1458.

Mohammadi, M., and Y. F. Wu. 2017. "Triaxial test for concrete under non-uniform passive confinement." *Construct. Build Mater.* 138, 455–468.

Mohammadi, M., and Y. F. Wu. 2019. "Modified plastic-damage model for passively confined concrete based on triaxial tests." *Compos. Part B Eng.* 159, 211–223.

Moran, D. A., and C. P. Pantelides. 2002. "Stress-strain model for fiber reinforced polymer-confined concrete." *J. Compos. Constr.* 6 (4), 233–240.

Nakamura, H., and T. Higai. 2001. "Compressive fracture energy and fracture zone length of concrete." *In Modeling of inelastic behavior of RC structures under seismic loads.* ASCE (American Society of Civil Engineers), 471–487.

Nanni, A., and N. M. Braford. 1995. "FRP jacketed concrete under uniaxial compression." *Construct. Build Mater.* 9 (2), 115–124.

Neville, A. M. 1981. *Properties of concrete*, London: Pearson.

Pessiki, S., K. A. Harries, J. T. Kestner, R. Sause, and J. M. Ricles. 2001. "Axial behavior of reinforced concrete columns confined with FRP jackets." *J. Compos. Constr.* 5 (4), 237–245.

Popovics, S. 1973. "Numerical approach to the complete stress-strain relation for concrete." *Cem. Concr. Res.* 3 (5), 583–599.

Prentis, J. 1951. "The distribution of concrete stress in reinforced and prestressed concrete beams when tested to destruction by a pure bending moment." *Mag. Concr. Res.* 2 (5), 73–77.

Razvi, S., and M. Saatcioglu. 1999. "Confinement model for high strength concrete." *J. Struct. Eng.* 125 (3), 281–289.

Restrepo, J. I., and B. De Vino. 1996. "Enhancement of the axial load-capacity of reinforced concrete columns by means of fiber glass-epoxy jackets." In *Proc. Second Int. Conf. on Advanced Composite Materials in Bridges and Structures, Montreal, Canada*, 547–553.

Richard, R. M., and B. J. Abbott. 1975. "Versatile elastic-plastic stress strain formula." *J .Eng. Mech. Div.* 101 (4), 511–515.

Richart, F. E., A. Brandtzaeg, and R. L. Brown. 1928. "A study of the failure of concrete under combined compressive stresses." *Bulletin no. 185*, Urbana, IL: University of Illinois, Engineering Experimental Station.

Richart, F. E., A. Brandtzaeg, and R. L. Brown. 1929. "The failure of plain and spirally reinforced concrete in compression." *Bulletin no. 190*, Urbana, IL: University of Illinois Engineering Experimental Station.

Rocca, S., N. Galati, and A. Nanni. 2009. "Interaction diagram methodology for design of FRP-confined reinforced concrete columns." *Construct. Build Mater.* 23 (4), 1508–1520.

Rochette, P., and P. Labossière. 2000. "Axial testing of rectangular column models confined with composites." *J. Compos. Constr.* 4 (3), 129–136.

Rousakis, T. C., and A. I. Karabinis. 2008. "Substandard reinforced concrete members subjected to compression: FRP confining effects." *Mater. Struct.* 41 (9), 1595–1611.
Rousakis, T. C., A. I. Karabinis, and P. D. Kiousis. 2006. "FRP-confined concrete members: axial compression experiments and plasticity modeling." *Eng. Struct.* 29 (7), 1343–1353.
Rousakis, T. C., A. I. Karabinis, P. D. Kiousis, and R. Tepfers. 2008. "Analytical modelling of plastic behavior of uniformly FRP confined concrete members." *Compos. Part B Eng.* 39 (7), 1104–1113.
Saadatmanesh, H., M. R. Ehsani, and M. W. Li. 1994. "Strength and ductility of concrete columns externally reinforced with fiber composite straps." *ACI Struct. J.* 91 (4), 434–447.
Saafi, M., H. A. Toutanji, and Z. Li. 1999. "Behavior of concrete columns confined with fiber reinforced polymer tubes." *ACI Struct. J.* 96 (4), 500–509.
Samaan, M., A. Mirmiran, and M. Shahawy. 1998. "Model of concrete confined fiber composite." *J. Struct. Eng.* 124 (9), 1025–1031.
Sargin, M., S. K. Ghosh, and V. Handa. 1971. "Effects of lateral reinforcement upon the strength and deformation properties of concrete." *Mag. Concr. Res.* 23 (75–76), 99–110.
Scott, B. D., R. Park, and M. J. N. Priestly. 1982. "Stress-strain behavior of concrete confined by overlapping hoops at low and high strain rates." *ACI J.* 79 (1), 13–27.
Sfer, D., I. Carol, R. Gettu, and G. Etse. 2002. "Study of the behavior of concrete under triaxial compression." *J. Eng. Mech.* 128 (2), 156–163.
Shah, S. P., and S. H. Ahmad. 1994. *High performance concrete: properties and applications*, New York: McGraw-Hill, Inc.
Shah, S. P., and R. Sankar. 1987. "Internal cracking and strain softening response of concrete under uniaxial compression." *ACI Mater. J.* 84 (3), 200–212.
Shao, Y., Z. Zhu, and A. Mirmiran. 2006. "Cyclic modeling of FRP-confined concrete with improved ductility." *Cem. Concr. Compos.* 28, 959–968.
Shehata, L. A. E. M., L. A. V. Carneiro, and L. C. D. Shehata. 2002. "Strength of short concrete columns confined with CFRP sheets." *Mater. Struct.* 35 (1), 50–58.
Soliman, M., and C. Yu. 1967. "The flexural stress-strain relationship of concrete confined by rectangular transverse reinforcement." *Mag. Concr. Res.* 19 (61), 223–238.
Spoelstra, M. R., and G. Monti. 1999. "FRP-confined concrete model." *J. Compos. Constr.* 3 (3), 143–150.
Suter, R., and R. Pinzelli. 2001. "Confinement of concrete columns with FRP sheets." *In Proc., Fifth Int. Conf. on Fibre Reinforced Plastics for Reinforced Concrete Structures, Cambridge, U.K.* 793–802.
Tan, T. H., and W. K. Yip. 1999. "Behavior of axially loaded concrete columns confined by elliptical hoops." *ACI Struct. J.* 96 (6), 967–971.
Tastani, S. P., and S. J. Pantazopoulou. 2004. "Experimental evaluation of FRP jackets in upgrading RC corroded columns with substandard detailing." *Eng. Struct.* 26 (6), 817–829.
Teng, J. G., and L. Lam. 2002. "Compressive behavior of carbon fiber reinforced polymer-confined concrete in elliptical columns." *J. Struct. Eng.* 128 (12), 1535–1543.
Teng, J. G., Y. L. Huang, L. Lam, and L. P. Ye. 2007. "Theoretical model for fiber-reinforced polymer-confined concrete." *J. Compos. Constr.* 11 (2), 201–210.
Teng, J. G., T. Jiang, L. Lam, and Y. Z. Luo. 2009. "Refinement of a design-oriented stress-strain model for FRP-confined concrete." *J. Compos. Constr.* 13 (4), 268–278.
Torrenti, J., E. Benaija, and C. Boulay. 1993. "Influence of boundary conditions on strain softening in concrete compression test." *J. Eng. Mech.* 119 (12), 2369–2384.
Toutanji, H. A. 1999. "Stress-strain characteristics of concrete columns externally confined with advanced fiber composite sheets." *ACI Mater. J.* 96 (3), 397–404.

Toutanji, H., and M. Saafi. 2002. "Stress-strain behavior of concrete columns confined with hybrid composite materials." *Mater. Struct.* 35, 338–347.

Wang, L. M., and Y. F. Wu. 2008. "Effect of corner radius on the performance of CFRP-confined square concrete columns: Test." *Eng. Struct.* 30 (2), 493–505.

Wang, Z., D. Wang, S. T. Smith, and D. Lu. 2011. "CFRP-confined square RC columns. II: cyclic axial compression stress-strain model." *J. Compos. Constr.* 16, 161–170.

Wang, Z., D. Wang, S. T. Smith, and D. Lu. 2012. "Experimental testing and analytical modeling of CFRP-confined large circular RC columns subjected to cyclic axial compression." *Eng. Struct.* 40, 64–74.

Wei, Y. Y., and Y. F. Wu. 2012. "Unified stress-strain model of concrete for FRP-confined columns." *Construct. Build Mater.* 26 (1), 381–392.

Wei, Y. Y., and Y. F. Wu. 2016. "Experimental study of concrete columns with localized failure." *J. Compos. Constr.* 20 (5), 04016032.

Wu, Y. F., and Y. Cao. 2017. "Energy balance method for modeling ultimate strain of confined concrete." *ACI Struct. J.* 114 (2), 373–381.

Wu, Y. F., and C. Jiang. 2013a. "Effect of load eccentricity on the stress–strain relationship of FRP-confined concrete columns." *Compos. Struct.* 98, 228–241.

Wu, Y. F., and J. F. Jiang. 2013b. "Effective strain of FRP for confined circular concrete columns." *Compos. Struct.* 95, 479–491.

Wu, Y. F., and L. M. Wang. 2009. "A unified strength model for square and circular concrete columns confined by external jacket." *J. Struct. Eng.* 135 (3), 253–261.

Wu, Y. F., and Y. Y. Wei. 2010. "Effect of cross-sectional aspect ratio on the strength of CFRP-confined rectangular concrete columns." *Eng. Struct.* 32 (1), 32–45.

Wu, Y. F., and Y. Wei. 2015. "General stress-strain model for steel- and FRP-confined concrete." *J. Compos. Constr.* 19 (4), 04014069.

Wu, Y. F., and Y. Y. Wei. 2016. "Stress-strain modeling of concrete columns with localized failure: an analytical study." *J. Compos. Constr.* 20 (3), 04015071.

Wu, Y. F., and Y. W. Zhou. 2010. "Unified strength model based on Hoek-Brown failure criterion for circular and square concrete columns confined by FRP." *J. Compos. Constr.* 14 (2), 175–184.

Wu, G., Z. Lu, and Z. Wu. 2003. "Stress-strain relationship for FRP-confined concrete cylinders." In *Proc. Sixth Int. Symp. on Fibre-Reinforced Polymer (FRP) Reinforcement for Concrete Structures (FRPRCS-6)*, Vol. 1. Singapore: World Scientific, 551–560.

Wu, Y. F., T. Liu, and D. J. Oehlers. 2006. "Fundamental principles that govern retrofitting of reinforced concrete columns by steel and FRP jacketing." *Adv. Struct. Eng.* 9 (4), 507–533.

Wu, G., Z. S. Wu, and Z. T. Lu. 2007. "Design-oriented stress-strain model for concrete prisms confined with FRP composites." *Construct. Build Mater.* 21, 1107–1121.

Wu, Y. F., Y. Yun, Y. Wei, and Y. Zhou. 2014. "Effect of predamage on the stress-strain relationship of confined concrete under monotonic loading." *J. Struct. Eng.* 140 (12), 04014093.

Xiao, Y., and H. Wu. 2000. "Compressive behavior of concrete confined by carbon fiber composite jackets." *J. Mater. Civ. Eng.* 12 (2), 139–146.

Xiao, Y., and H. Wu. 2003a. "Compressive behavior of concrete confined by various types of FRP composite jackets." *J. Reinf. Plast. Compos.* 22 (13), 1187–1201.

Xiao, Y., and H. Wu. 2003b. "Retrofit of reinforced concrete columns using partially stiffened steel jackets." *J. Struct. Eng.* 129 (6), 725–732.

Xing, Q. S., Y. J. Weng, and J. M. Shen. 1987. "Experimental study on the complete stress-strain curve of confined concrete." In *Proc. Nat. Conf. on Application and Theory of Common and Confined Concrete, Yin Tai, China*, 81–87 (in Chinese).

Youssef, M. N., Q. Feng, and A. S. Mosallam. 2007. "Stress–strain model for concrete confined by FRP composites." *Compos. Part B Eng.* 38, 614–628.

Yu, Q. 2001. "A study on the stress–strain relationship of FRP-confined concrete in axial compression." *Ind. Constr.* 31 (4), 5–8 (in Chinese).

Yu, T., J. G. Teng, Y. L. Wong, and S. L. Dong. 2010. "Finite element modeling of confined concrete—II: plastic-damage model." *Eng. Struct.* 32 (3), 680–691.

Ductility modification technologies

9

9.1 Introduction

Concrete is a brittle material. Reinforced concrete (RC) structures rely largely on the deformation and yielding of the tensile reinforcement to satisfy the ductility demand. The application of low ductility high-strength steel reinforcement in concrete structures has a significant drawback from a ductility point of view. The application of fiber-reinforced polymer (FRP) reinforcement in RC structures has a detrimental effect on structural ductility, as FRPs have a low strain capacity and linear elastic stress–strain behavior up to rupture without yielding. The ductility of RC members reinforced with nonductile bars, especially FRP bars, has been a major concern of researchers and engineers in practice. Conventional engineering methods including confining concrete usually cannot provide adequate ductility to such RC structures.

This chapter introduces a new approach to achieving sufficient flexural ductility of RC members through compressive yielding instead of tensile yielding. The compression yielding approach opens up another avenue for the design of concrete structures. Using the compression yielding technology, a special fused structure is developed that not only increases the ductility of RC structures but also reduces the costs of the structures. Other new technologies related to the ductility of RC structures are also introduced in this chapter.

9.2 Ductility deficient RC structures reinforced with nonductile bars

The flexural deformation of structural members comes from the curvature of sections. For RC members, the cross-sectional curvature comes from the tensile straining of the reinforcement and the compressive straining of the concrete. As concrete is a brittle material with little ductility and strain capacity, RC members achieve ductility or adequate deformability mainly through the tensile yielding or straining of the reinforcement. When the tensile straining of the reinforcement is limited, such as in the case of overreinforced RC beams and RC columns with large axial loads, whereby the tensile reinforcement does not yield and the member fails due to concrete crushing, the ductility of the RC members is limited. The application of nonductile tension reinforcement such as high-strength steel or FRP reinforcement significantly reduces the ductility of the tensile reinforcement, which results in a substantial reduction in the ductility of the structural members.

The confinement of concrete introduced in Chapter 8 does increase the strain capacity and ductility of concrete material. But the increase in the compressive strain capacity

Analytical Approaches for Reinforced Concrete. https://doi.org/10.1016/B978-0-12-821164-9.00009-7

and the consequent ductility is not to the same scale as the strain capacity from tension yielding of steel reinforcement, and hence is insufficient to deliver adequate structural ductility to RC members. On the other hand, nonductile tension reinforcement, such as FRP bars, may break, which causes catastrophic collapse of the structure when the compression concrete is confined to have a higher compressive resistance than the tensile capacity of the tension reinforcement. Therefore, "Unless ductility requirements are satisfied, FRP materials cannot be used reliably in structural engineering applications" (Naaman 2003). As a result, FRP bars have not been commonly used in new constructions so far, despite its superior material properties and the long-term cost effectiveness of the FRP-reinforced concrete structures. Furthermore, the use of more brittle high-strength concrete (HSC), which has been increasing in practical applications at a fast rate in recent years, has a similar detrimental side effect on the ductility of RC members, especially for concrete columns. In general, the use of more advanced high strength and more brittle materials in RC structures has a significant side effect on ductility, which has been a major concern in structural engineering in recent years.

Naaman (2003) summarized the following conventional engineering approaches that can be applied to improve the deformation capacity and ductility of FRP-reinforced concrete members:

1. Confining concrete to increase ductility
   This method cannot avoid FRP rupture for underreinforced beams. For overreinforced beams, heavy confinement reinforcement is needed to avoid concrete crushing. Furthermore, it will be seen in Section 9.7 that the ductility achievable by confinement is generally insufficient.
2. Designing for progressive rupture of FRP
   Prestressing each layer of reinforcement to different stresses, so as to provide a step-by-step progressive rupture of the reinforcing material with increasing deformation. In this way, a sudden and complete rupture of the tension reinforcement can be avoided.
3. Combining FRP bars with conventional steel reinforcement
   With the mixed reinforcement, sudden rupture of total tension reinforcement is avoided, and certain ductility is provided by the steel reinforcement.
4. Using unbonded rebars
   When the tension reinforcement is unbonded, more deformation can be achieved on the tension side, because the deformation of the reinforcement over the full unbonded length can be utilized so that the deformation capacity of the member is increased.
5. Making use of debonding mechanism
   Designing the interface between the FRP reinforcement and the concrete so that bond failure is triggered when the stress in the FRP reaches a threshold level. The extent of bond failure develops and extends with further loading, thus providing higher tension deformation capacity and ductility to the members.
6. Maximizing the curvature capacity
   Designing the cross section of a member to proportionate the reinforcement in order to take advantage of the full strain capacity of concrete simultaneously with that of the reinforcement, i.e., making full use of the compressive strain capacity of concrete together with the tensile capacity of the FRP reinforcement to provide a maximum available curvature.

All these approaches provide a certain increase in the ductility and/or deformation capacity of the RC members reinforced with FRP bars. However, they are either too complicated and are difficult to realize in practice and have a significant increase in costs, or they have an insufficient effect on ductility increase. For these reasons, the only recommendation in ACI 440.1R-15 (ACI Committee 440 2015) is to design for higher strength reserve or overdesign the structures to avoid failure of the FRP reinforced concrete structures. However, overdesign is not only uneconomical but often impractical or impossible in many cases, especially for large structures such as large-span bridge beams. More importantly, using a higher safety factor does not really protect a structure, as discussed in Section 1.5. On the contrary, it may cause an even more drastic failure of the structure. Therefore, conventional engineering approaches cannot fundamentally resolve the problem of ductility-deficient RC structures.

9.3 Avenues for ductility

To discuss possible ways to increase the ductility of a flexural member, let's look at the following facts:

1. Ductility comes from plastic deformation
   Based on Eq. (1.1) and Fig. 1.11, the ductility is determined by the length of the yield plateau. Elastic deformation only increases the deformation capacity of a member but not the ductility. Therefore, ductility comes from plastic deformation. It is the plastic rotation of the plastic hinge that produces the ductility of a flexural member.
2. Plastic deformation concentrates in plastic hinge zone
   When large flexural deformation occurs in a structural member, the plastic deformation is concentrated in a small area called the plastic hinge zone that has a limited length (Paulay and Priestley 1992).
3. Plastic hinge deformation depends on plastic hinge length and plastic curvature
   As shown in Eq. (7.42), the plastic deformation and flexural ductility are determined by the plastic curvature ($\kappa_u - \kappa_y$) and the plastic hinge length L_p.
4. Plastic hinge length can be increased by rebar debonding
   The plastic hinge length is determined by the factors discussed in Section 7.3.2. Among those factors, an effective and easy way to increase the plastic hinge length is by debonding the tension reinforcement.
5. Plastic curvature relies on tension and compression yielding of materials
   The plastic curvature is due to plastic tension and plastic compression straining at a cross section.

The listing of these facts is actually a process of logical reasoning that reveals three possible ways to provide ductility for an RC flexural member:

1. By bond breaking down, particularly progressive debonding and slipping of bar inside the RC member. As bond breakdown increases the deformation of the tension bar while maintaining its tensile force, a properly designed debonding process can lead to very ductile flexural deformation of the RC member. However, it is not easy to achieve progressive debonding with the current RC construction practice.
2. By tensile yielding of reinforcement, which produces plastic hinge rotation and flexural ductility, as shown in Fig. 9.1B. This is the conventional way to achieve ductility of RC flexural members.

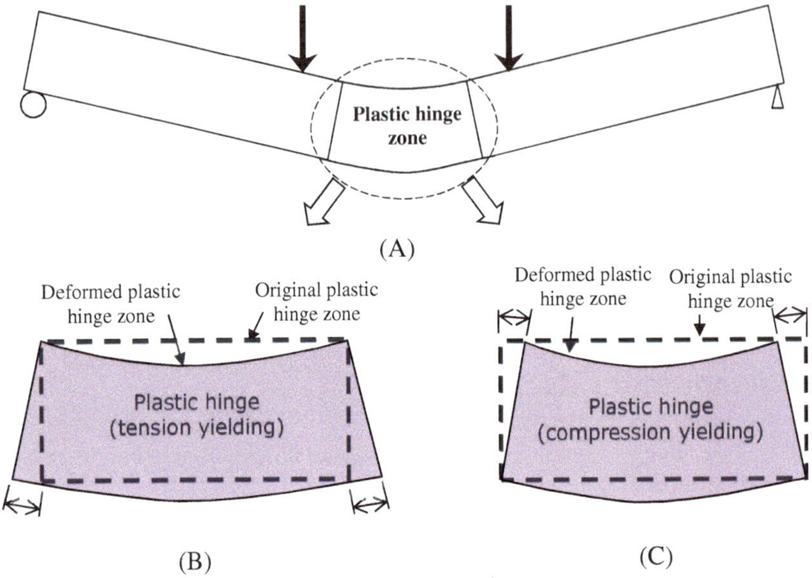

Fig. 9.1 Plastic hinge deformation: (A) plastic hinge of beam, (B) normal RC beam, (C) CY beam.

3. By compression yielding. When large rotation of the plastic hinge cannot be achieved through elongation or yielding of the reinforcement on the tension side, the other avenue to achieve it is by shortening or yielding on the opposite compression side, as illustrated in Fig. 9.1C.

This analysis triggers a new but simple idea: replacing the concrete in the compression zone of the plastic hinge with a more ductile material or mechanism so that ductility can be achieved through yielding of the material in the compression zone. The effectiveness of this idea was demonstrated by the experimental test results shown in Fig. 9.2. By properly designing the compression yielding (CY) instrument, a very ductile response of an RC beam reinforced by nonductile FRP bars can be achieved. Details of the CY structural system are introduced in the following sections.

9.4 Compression yielding structural system

The structural configuration that allows for compression yielding is illustrated with a simply supported beam in Fig. 9.3. The only difference between the CY structural scheme and a conventional RC beam is in the plastic hinge zone where a special ductile CY material or mechanism, namely the CY instrument, is used to replace the concrete on the compression side. Two instruments have been developed to achieve compression yielding in a plastic hinge:

1. a ductile material (Fig. 9.3B) to replace concrete; and
2. a ductile mechanism in the compression zone (Fig. 9.3C and D).

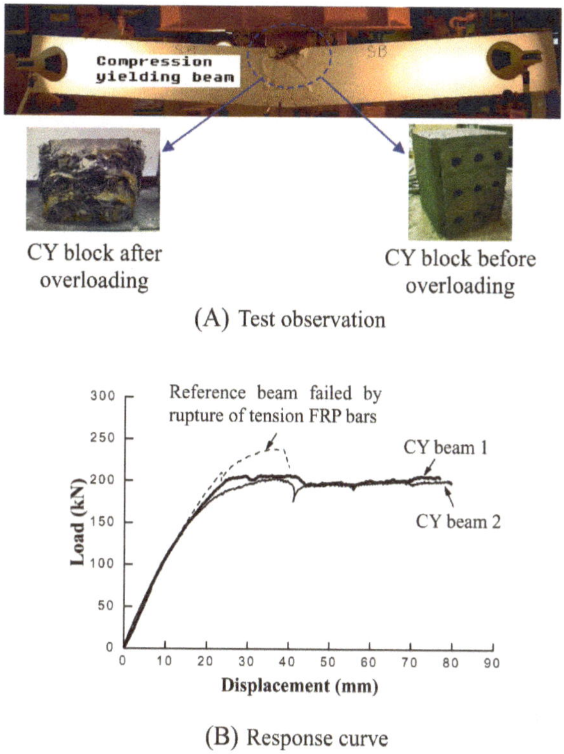

Fig. 9.2 Tests of CY beam.

The CY instruments must satisfy the following general principles:

1. deforming elastically (or almost elastically) under service loads to ensure low creep deformation, sufficient rigidity, and other desirable working requirements;
2. deforming plastically (or almost plastically) in the postpeak response region to ensure sufficient ductility; and

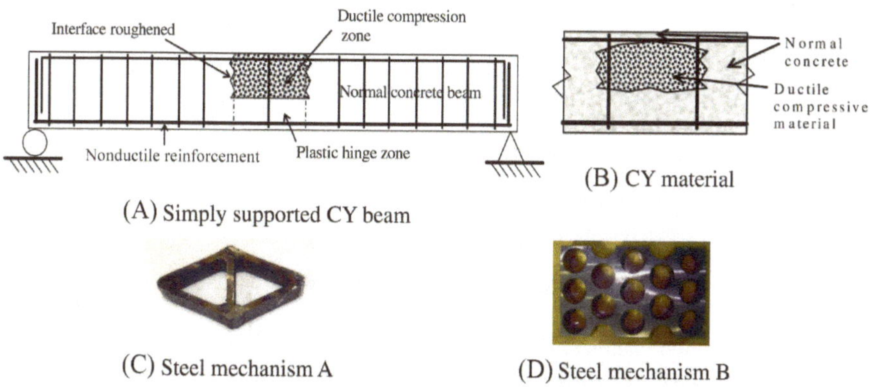

Fig. 9.3 Structural configuration of compression yielding beam.

3. providing a maximum compression resistance, including compression bars if any, that is not greater than the total tensile strength of a section to avoid tensile breaking of the nonductile bars.

It is desirable to place the CY instrument at the location of the maximum bending moment. Nevertheless, it need not coincide exactly with the position of the maximum moment. In fact, the CY instrument acts as a fuse in the structural system, and when an overloading condition occurs, the fuse will be triggered and force the structural system to deform in a (more or less) plastic manner to prevent further increase in loading, and hence avoid abrupt reinforcement rupture, concrete crushing, or other undesired failure modes, including shear failure. If the fuse is not placed at the maximum moment region, the flexural strength of the maximum moment section should be larger than that of the fused section. The CY approach is in principle effective in increasing the ductility and deformability of common ductility-deficient flexural members.

The compression yielding only takes place inside the CY zone. In order to achieve compression yielding in the plastic hinge, the concrete surrounding the CY zone must be stronger than the CY zone. As a result, the plastic deformation is confined within the compression yielding zone. In this case the plastic hinge length is simply the length of the compression yielding zone. This would make the determination of the plastic hinge length simpler than that of the conventional RC members, which are very complicated, as discussed in Section 7.3.

9.5 Compression yielding instruments

Many engineering materials, including metallic, cementitious, plastic, polymeric, elastomeric, rubber concrete, composite materials, etc., have been experimentally tested to find a suitable CY material. However, no single material has been found to satisfy the necessary requirements, i.e., sufficient elasticity and stiffness at service load and a great plasticity, without significant strength variation after yielding. The difficulty is mainly due to the excessive ductility demand on the CY material. The theoretical study in Section 9.7 shows that the ductility demand on the CY instrument is often an order of magnitude higher than that on a beam member. Therefore, finding an existing material that can be directly used to replace the concrete in the CY zone is unlikely.

Inspired by the great deformability of sponge, it was found that such a highly deformable, ductile, yet stiff material can be obtained by creating voids or perforations inside suitable materials (Wu 2008; Wu et al. 2010a). Such perforated CY materials are found to be the most suitable for the CY instrument.

9.5.1 Compression yielding mechanisms

Both steel and nonferrous materials can be used to make a CY mechanism. There is no technical difficulty to design a steel CY mechanism and the requirement is that it should be simple and cheap. One CY mechanism tested in Wu (2006) is the steel strut shown in Fig. 9.3C (CY beam with the steel mechanism; see Fig. 9.12). Its

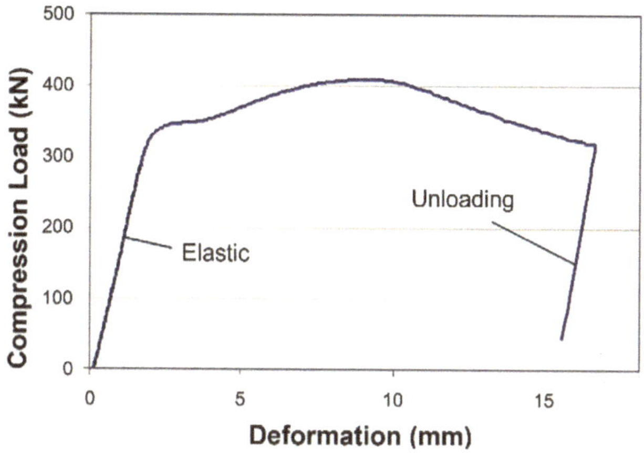

Fig. 9.4 Load vs. deformation curve of the steel strut mechanism.

compression load vs. displace response curve is shown in Fig. 9.4. The mechanism can be encased into a protective material such as lightweight concrete and forms a precast block. This precast block is then cast into a beam to form a ductile compression zone.

A better CY mechanism is the perforated mild steel block, namely P-block (Wu 2008), as shown in Fig. 9.3D. The mild steel P-block allows a nearly elastic-perfectly plastic compressive response to be obtained (Fig. 9.5), which can provide a nearly elastic-perfectly plastic flexural response of the simply supported beam with almost unlimited ductility (see Section 9.6).

The P-block is made by drilling holes in a mild steel or stainless-steel block, as shown in Fig. 9.5. The perfectly elastoplastic compression response is obtained by

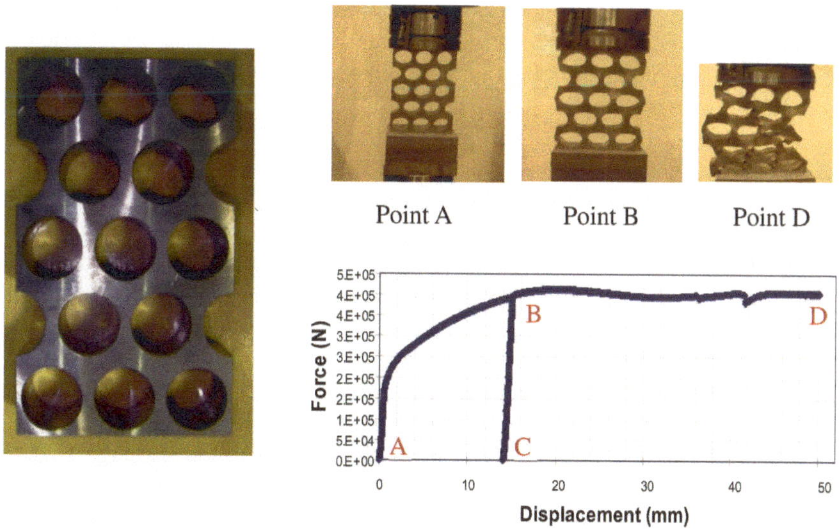

Fig. 9.5 Compression test of perforated mild steel block.

precompressing the block to a certain deformation and then unloading it. Without the precompression, the P-block gives a response of ABD in Fig. 9.5. By precompressing the block to point B and then unloading it, it will exhibit a load path of CBD when it is used as the CY mechanism in a beam. The compression tests of the mild steel P-blocks demonstrated that a ductility ratio of above 50 is easily achievable (Wu 2008).

9.5.2 Compression yielding material

Cement-based composites are considered most suitable as the CY material, since they make the least difference to the materials in normal RC members. Among these composites, slurry infiltrated fiber concrete (SIFCON) is the most promising, as far as ductility is concerned. SIFCON is fiber-reinforced concrete in which the fiber matrix is infiltrated by cement-based slurry, resulting in a material characterized by a very high fiber volume fraction. Due to its large fiber volume, SIFCON exhibits outstanding ductility and compressive strength (Homrich and Naaman 1987), as shown by the typical stress–strain curves in Fig. 9.6B. However, the ductility demand for the CY material is excessively high. If the ductility ratio of 3.5 is required for an RC beam, the required ductility ratio on a CY instrument with a length of 10% of the beam's span is around 20–40 (see Section 9.7). The SIFCON material itself does not provide such a high ductility.

The perforated SIFCON blocks (SIFCON P-block), with the aim of achieving a design that delivers a nearly perfect yield plateau in compression, were developed for this purpose by Wu et al. (2010a). The typical SIFCON P-block and its compression test are shown in Fig. 9.7. Maximum steel fiber content is used in a SIFCON block by compacting the fibers in the mold as much as possible (Fig. 9.6A). The fiber content of the SIFCON P-blocks varied from 14.7% to 19.2% in the tests by Wu et al. (2010a). To enhance the stability of its compression resistance, transverse steel bars were also provided in the SIFCON P-blocks as confining hoops. The test variables included the perforation ratio (volume of hole/volume of block) and the transverse steel bar ratio. The typical response curves of the SIFCON P-blocks are shown in Figs. 9.6 and 9.8.

The up-and-down of the stress–strain curve in Fig. 9.6 was caused by the progressive crushing of the holes, as shown in Fig. 9.9. The holes were crushed row by row in some of the test specimens that exhibited the undulating curve. However, most of the specimens did not show such undulating behavior. The stress–strain response of the SIFCON P-block depends on a number of factors including the arrangement of holes and transverse steel bars. Details of the test specimens and test results can be found in Wu et al. (2010a).

All the response curves exhibited two approximately linear portions (Fig. 9.8). Therefore, a bilinear stress–strain model can be used to fit them. By fitting a bilinear curve model to an actual response curve using least squares regression, the yield point (ε_y, f_y) and ultimate point (ε_{cu}, f_{cu}) (Fig. 9.8) were determined and are summarized in Table 9.1. The test results show that the ductility ratio defined as $\mu = \varepsilon_{cu}/\varepsilon_y$ mainly depends on the perforation ratio v and the transverse steel bar ratio ρ. Their typical relationship is illustrated in Fig. 9.10.

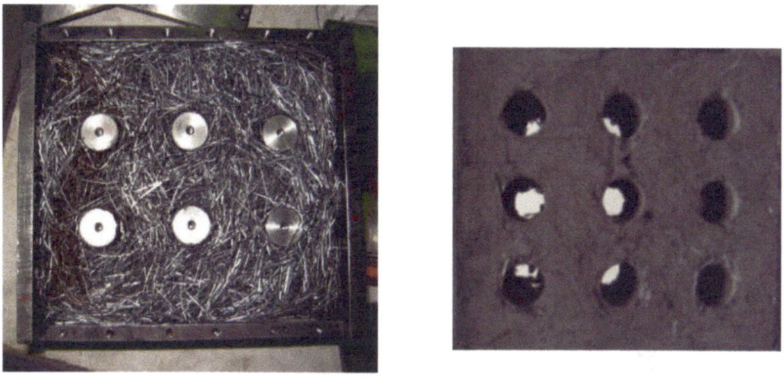

(A) Casting of SIFCON-P block

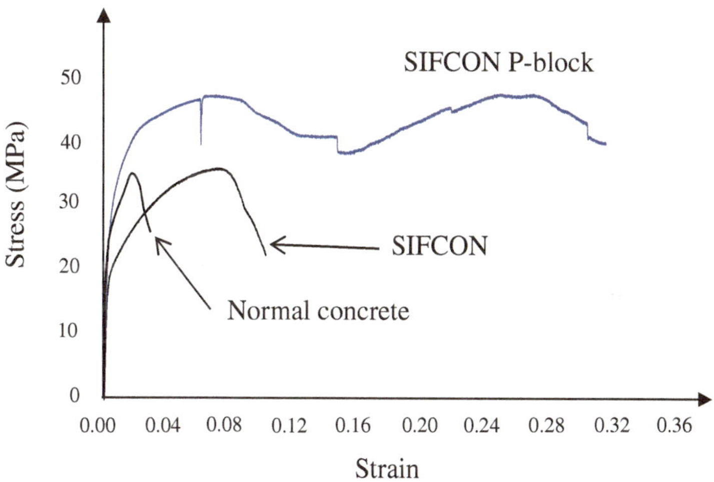

(B) Typical stress-strain curve of cement-based materials

Fig. 9.6 SIFCON materials.

(A) Before test (B) Compression test (C) After test

Fig. 9.7 SIFCON P-block.

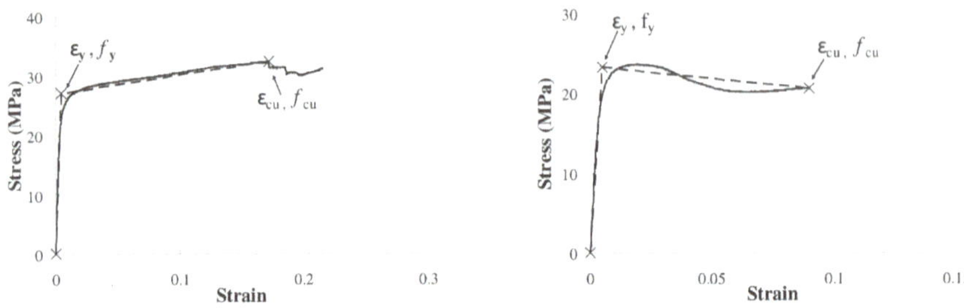

(A) Ascending type of curve (B) Descending type of curve

Fig. 9.8 Typical stress vs. strain curves of SIFCON P-block.
From Wu, Y. F., J. F. Jiang, and K. Liu. 2010a. "Perforated SIFCON blocks—an extraordinarily ductile material ideal for use in compression yielding structural systems." *Construct. Build Mater*. 24 (12), 2454–2465.

9.6 Test of CY beams

Experimental testing was conducted to investigate the structural behavior of the CY structural scheme. The RC test specimens are shown in Fig. 9.11. Glass fiber-reinforced polymer (GFRP) bars were used as the tension reinforcement in all the RC specimens, considering that the ductility problem is of particular interest to FRP-reinforced concrete members. The reference beam, as shown in Fig. 9.11A, was a normal GFRP reinforced concrete beam that was used as a benchmark of the CY beams. In order to avoid the sudden break of the GFRP bars, the beam was designed to be slightly overreinforced, i.e., the tensile resistance of the beam was slightly higher than the compression resistance of the concrete at a cross section. The overall dimensions and tension reinforcement of the CY beams were identical to that of the reference beam, as shown in Fig. 9.11B. Different CY instruments were placed into the void in the plastic hinge region in a beam test. The material properties

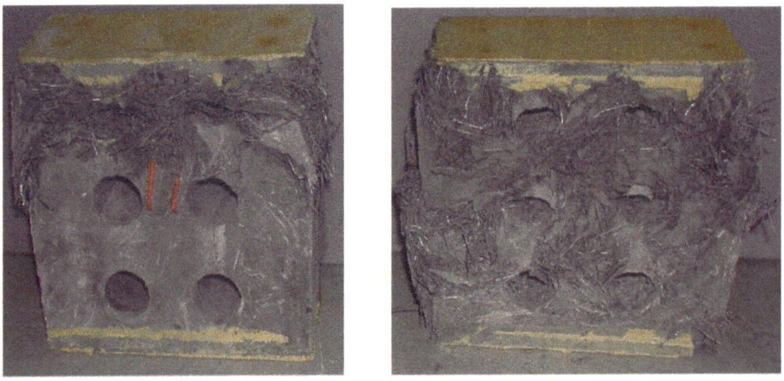

(A) Crushing of one row of holes (B) Crushing of all holes

Fig. 9.9 Failure process of the SIFCON P-block.

Table 9.1 Compression test results of SIFCON P-block.

Group	Spec.	v (%)	ρ (%)	f_y (MPa)	f_y/f_0	f_{cu} (MPa)	f_{cu}/f_0	ε_y (10^{-2})	ε_{cu}	μ
GA	A1	0.00	0.00	22.61	0.64	35.56	1.00	0.34	0.07	20.32
	A2	8.18	0.00	19.03	0.54	37.90	1.07	0.33	0.10	30.09
	A3	8.18	0.94	30.44	0.86	51.80	1.46	0.43	0.14	33.63
	A4	0.00	0.94	34.75	0.98	45.10	1.27	0.37	0.08	22.01
	A5	8.18	0.94	31.68	0.89	28.52	0.80	0.44	0.09	19.91
	A6	8.18	0.94	28.71	0.81	44.80	1.26	0.49	0.14	28.28
	A7	0.00	0.94	42.87	1.21	45.70	1.29	0.53	0.16	29.57
	A8	8.18	0.94	29.00	0.82	29.60	0.83	0.89	0.15	16.77
	A9	0.00	1.41	46.98	1.32	60.70	1.71	0.52	0.13	25.91
	A10	0.00	1.41	41.35	1.16	43.15	1.21	0.80	0.31	38.53
G1	B1			35.75	0.69	49.56	0.96	0.42	0.08	19.05
	B2	0.00	0.00	36.40	0.70	52.73	1.02	0.50	0.11	22.00
	B3			36.05	0.70	53.16	1.03	0.46	0.12	26.09
G2	B4			29.78	0.57	30.00	0.58	0.34	0.10	30.60
	B5	1.88	0.94	38.02	0.73	51.07	0.99	0.47	0.20	42.39
	B6			31.74	0.61	39.30	0.76	0.32	0.39	120.34
G3	B7			29.50	0.57	58.92	1.14	0.46	0.28	60.91
	B8	4.24	0.94	34.40	0.66	52.15	1.01	0.41	0.38	93.74
	B9			28.56	0.55	55.10	1.06	0.42	0.42	100.13
G4	B10			33.36	0.64	34.02	0.64	0.42	0.15	35.65
	B11	8.18	0.94	26.97	0.52	31.02	0.58	0.41	0.22	52.51
	B12			28.32	0.55	32.07	0.62	0.35	0.12	34.29
G5	B13			30.88	0.60	35.56	0.69	0.39	0.15	37.59
	B14	2.51	1.18	31.25	0.60	66.19	1.28	0.53	0.42	80.15
	B15			36.01	0.69	62.17	1.20	0.43	0.39	90.14

Continued

Table 9.1 Continued

Group	Spec.	v (%)	ρ (%)	f_y (MPa)	f_y/f_0	f_{cu} (MPa)	f_{cu}/f_0	ε_y (10^{-2})	ε_{cu}	μ
G6	B16			32.24	0.62	33.97	0.66	0.64	0.13	20.00
	B17	5.65	1.18	27.41	0.53	35.83	0.69	0.47	0.18	38.07
	B18			30.38	0.59	43.63	0.84	0.47	0.24	50.85
G7	B19			31.50	0.61	44.81	0.86	0.45	0.16	35.79
	B20	2.83	1.26	32.93	0.64	64.89	1.25	1.25	0.28	22.69
	B21			31.39	0.61	62.69	1.21	0.55	0.39	71.77
G8	B22			22.10	0.43	34.90	0.67	0.39	0.25	63.15
	B23	6.36	1.26	30.98	0.60	55.27	1.07	0.47	0.40	84.36
	B24			27.04	0.52	49.83	0.96	0.44	0.38	85.95
G9	B25			16.60	0.32	16.72	0.28	0.31	0.19	61.66
	B26	12.27	1.26	23.34	0.45	19.10	0.32	0.49	0.12	24.61
	B27			25.83	0.50	22.44	0.34	0.60	0.14	23.33
G10	B28			–	–	54.09	0.97	0.89	0.01	1.12
	B29	0.00	0.00	–	–	54.56	0.98	0.50	0.02	4.00
	B30			–	–	57.81	1.04	0.61	0.03	4.92

Note: (1) f_0 = average strength of plain SIFCON blocks without holes, GA: 35.56 MPa, G1 to G9: 51.82 MPa, G10: 55.49 MPa. (2) $\mu = \varepsilon_{cu}/\varepsilon_y$, the ductility ratio. (3) v = total volume of holes/volume of block, the void ratio. (4) ρ = stirrup ratio.

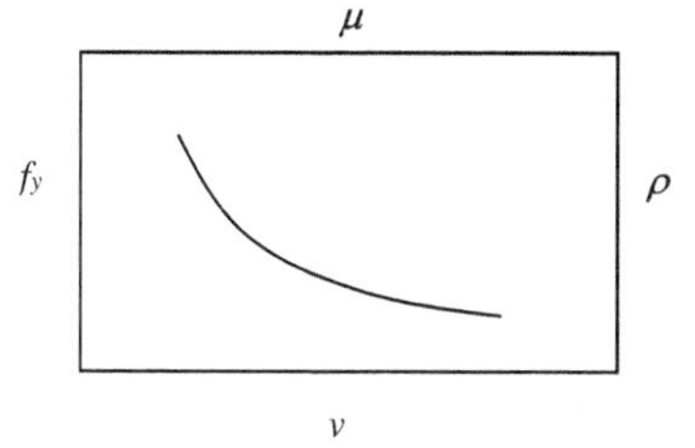

Fig. 9.10 Typical relationship between design variables and strength and ductility of the SIFCON P-blocks.
From Wu, Y. F., J. F. Jiang, and K. Liu. 2010a. "Perforated SIFCON blocks—an extraordinarily ductile material ideal for use in compression yielding structural systems." *Construct. Build Mater.* 24 (12), 2454–2465.

of concrete, steel and GFRP bars, and steel plate are provided in Table 9.2. For more details on the tests, refer to Wu (2006, 2008) and Wu et al. (2010a).

The CY beam test with the steel strut as the CY instrument is shown in Fig. 9.12. The load vs. displacement response of the steel strut was obtained from a compression test of the strut and is shown in Fig. 9.4. The response curves of the test beams are

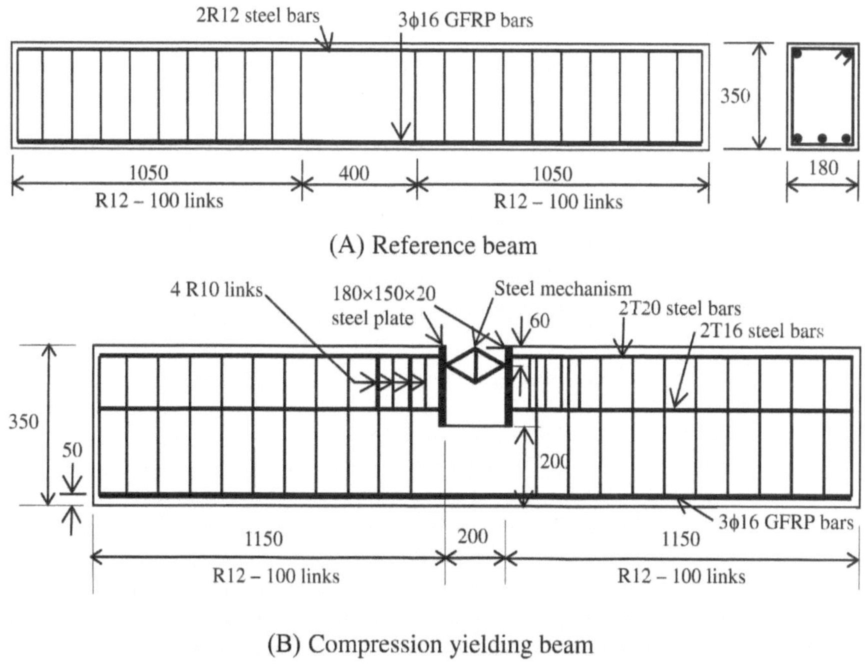

Fig. 9.11 Details of the test beams.
From Wu, Y. F. 2006. "New avenue of achieving ductility for reinforced concrete members." *J. Struct. Eng.* 132 (9), 1502–1506.

Table 9.2 Material properties.

Material		Strength (MPa)	Young's Modulus ($\times 10^3$ MPa)
Concrete	**R1**	f_{cu} = **64.9**	–
	S1	**79.8**	
	S2	**80.4**	
T16 steel bar		$f_y = 548$	–
T20 steel bar		$f_y = 690$	–
Mild steel bar (R)		$f_y = 307$	–
GFRP bar		$f_{fu} = 655$	40.8
Mild steel plate		$f_y = 325$	200.3

Note: R1 = reference beam; S1 = CY beam 1, S2 = CY beam 2; f_{cu} = cube compressive strength at the time of beam testing; f_y = yield strength; and f_{fu} = guaranteed tensile break strength.

shown in Fig. 9.13A. The reference beam failed due to concrete crushing, after which the load dropped quickly. The first CY beam S1 failed catastrophically due to the sudden break of the GFRP bars. This was due to the overstrength of the steel strut, which exceeded the total tensile strength of the GFRP bars. The steel strut was cut smaller (breadth reduced from 70 mm to 59 mm) in the second CY beam test. No final failure occurred in the test of the CY beam S2. The test stopped when the test rig reached its maximum travel. The reference beam failed due to concrete crushing with a ductility ratio of 1.2. The CY beam reached a ductility ratio of 2.75 when the test stopped.

These tests show that the CY instrument can significantly increase the ductility of RC beams. Furthermore, the tensile rupture of nonductile reinforcement can be

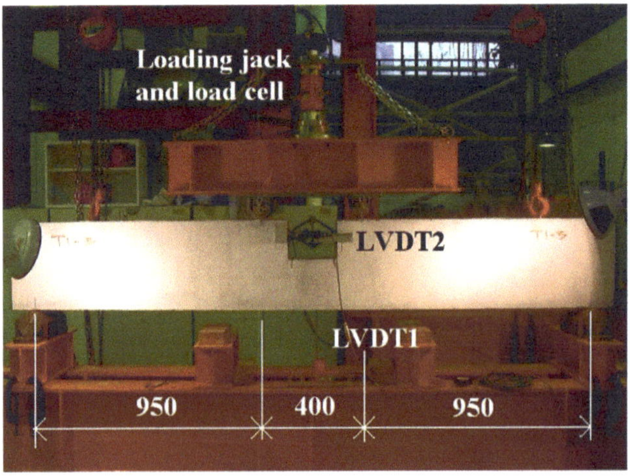

Fig. 9.12 CY beam test with a steel strut.
From Wu, Y. F. 2006. "New avenue of achieving ductility for reinforced concrete members." *J. Struct. Eng.* 132 (9), 1502–1506.

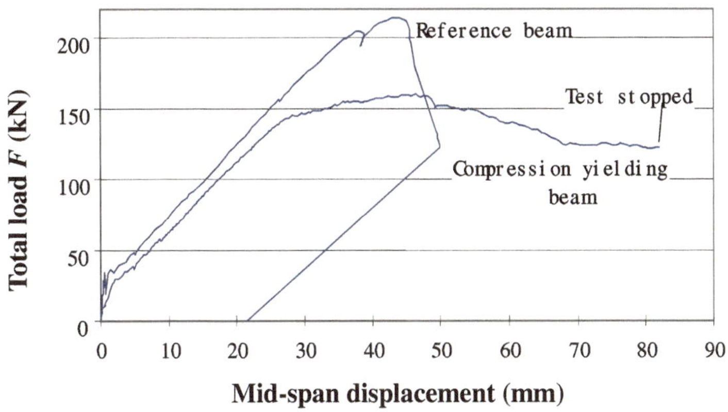

(A) Response curves

(B) CY beam S2 after testing

Fig. 9.13 Beam tests and results.
From Wu, Y. F. 2006. "New avenue of achieving ductility for reinforced concrete members." *J. Struct. Eng.* 132 (9), 1502–1506.

avoided by ensuring that the compression resistance is smaller than the tensile resistance. The rotation of the plastic hinge mainly comes from the plastic shortening of the compression yielding zone. The contribution from the elongation of tension bars is relatively small at large displacement. This can be seen from the measured deformation of the plastic hinge zone of the CY beam S2 shown in Fig. 9.14. The figure shows that the tensile deformation of the GFRP bars was greater than that of the compression mechanism before yielding the CY mechanism. The compression deformation increased quickly and linearly after the yielding point, whereas the tension deformation of the plastic hinge essentially remained unchanged. The compression deformation was almost 10 times that of the tensile deformation at the end of the test. The strain of the GFRP bars in the compression yielding beam reached 0.015 at the maximum displacement. Therefore, the strain capacity of the GFRP bars was almost fully utilized in the test beam. These analyses show that the deformation/ductility contributed by tension straining is very limited and a significant ductility/deformation demand cannot be satisfied without significant compression yielding for beams reinforced with nonductile FRP reinforcement.

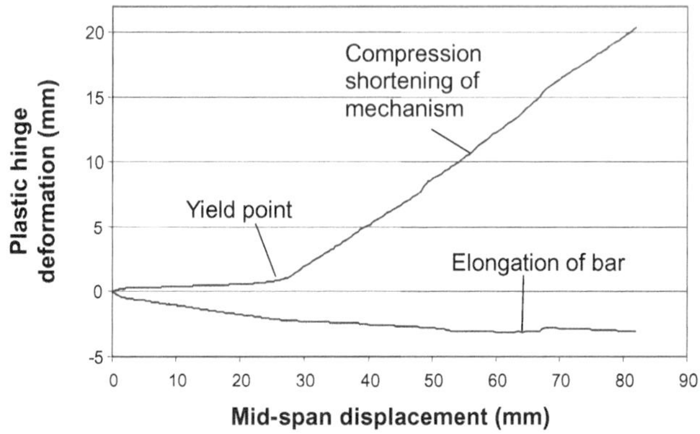

Fig. 9.14 Plastic hinge deformation.
From Wu, Y. F. 2006. "New avenue of achieving ductility for reinforced concrete members." *J. Struct. Eng.* 132 (9), 1502–1506.

The same RC beam design as in Fig. 9.11B was used to test another CY instrument of the steel P-block (Fig. 9.5) by Wu (2008). The details of the mild steel P-blocks are shown in Fig. 9.15. Only one RC beam specimen was cast that was used to do all beam tests. Fig. 9.16 shows the test results. In the first test, one solid concrete block was placed into the void (Fig. 9.11B) in the compression zone of the plastic hinge as the reference beam test. The reference beam test was stopped before concrete crushing as shown in Fig. 9.16. The concrete block in the CY zone was removed and a CY mechanism was placed into the void in the following tests. The second test was unsuccessful due to an inadequate design of another type of mechanism that showed excessive softening after the peak load. The third and fourth tests used the steel P-blocks, BLK 2 and BLK 3 (Fig. 9.15). The response curves of the beam tests in Fig. 9.16 show the nearly ideal elastoplastic load–deformation curves and the high ductility of the CY beams with the steel P-blocks. It was also observed that the CY beams were stiffer than the reference RC beam before yielding (Fig. 9.16) and the tensile cracks of concrete closed up automatically after unloading due to the linear elastic property of the FRP bars.

The CY beams with a SIFCON P-block were tested by Wu et al. (2010a), as shown in Fig. 9.17. The SIFCON blocks for the beams B1a and B2a were perforated (Fig. 9.17A), and that for beam B3a was a solid SIFCON block without perforation. The responses for B1a and B2a in Fig. 9.17B show that a nearly elastic-perfectly plastic and very ductile behavior can be obtained with a SIFCON P-block as the CY instrument. The test of beam B3a resulted in a sudden failure due to rupture of the tensile FRP bars caused by the significant hardening of the solid SIFCON block after yielding. The originally designed compression failure of the solid SIFCON block was not achieved, and hence it could not reflect the true ductility of beams with a solid SIFCON block. This test showed that the hardening of solid SIFCON block is difficult to control. The provision of perforation in the SIFCON block not only is effective in delivering ductility to the beam but also is necessary to avoid the strength hardening of the CY block to avoid abrupt tensile rupture of the FRP bars.

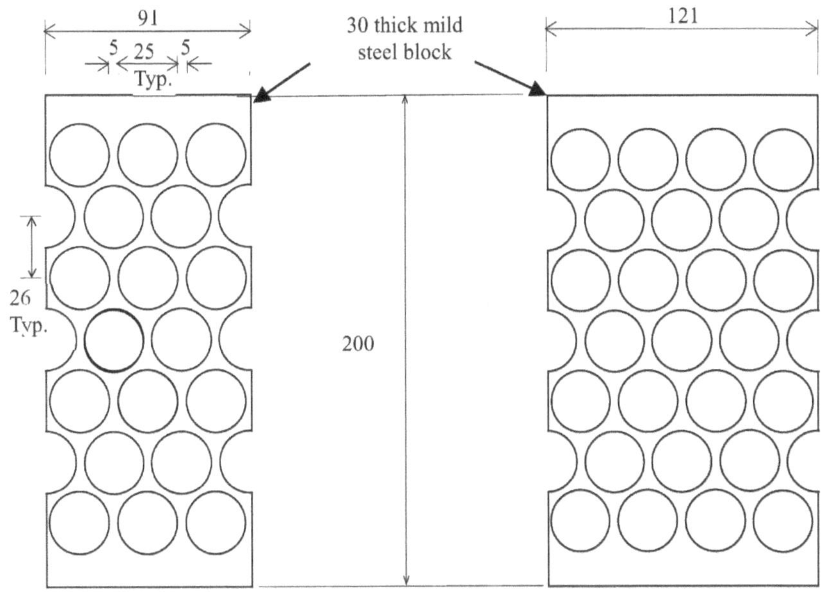

(A) BLK2 (B) BLK3

Fig. 9.15 Details of the test specimens (in mm).
From Wu, Y. F. 2008. "Ductility demand of compression yielding fiber-reinforced polymer-reinforced concrete beams." *ACI Struct. J.* 105 (1), 104–110.

It is noted that the nearly elasto-perfectly plastic response curves in Figs. 9.16 and 9.17B were not only caused by the ideal elastoplastic compressive behavior of the CY instrument (Fig. 9.17C) but were also due to the constant lever arm of the compression force from placing the CY instruments at the top of the beam (Fig. 9.17A). When a CY block is cast into a beam to fulfill the whole compression zone, the depth of compression zone changes when curvature increases that may cause variation of the compression area and its centroid. In this case, care must be exercised to ensure that the total compression resistance does not exceed the total tensile resistance of the beam to avoid rupture of the tension reinforcement. It is also possible by carefully designing the CY zone so that its compressive resultant multiplied by its lever arm (distance from the compression centroid to the tension centroid) keeps constant after yielding of the beam, which will give a constant moment resistance and hence a perfectly flat beam response. Another important observation from the beam tests was that the diameter of FRP bar (or thickness of FRP strip) should be small. Due to the excessive bending inside the plastic hinge zone, an FRP bar with a large diameter will fracture under significant bending as FRP bars are very brittle. Small diameter FRP bar (or thin FRP strip with the thin thickness in the depth direction of beam) is more flexible in bending and will not fracture in the plastic hinge zone.

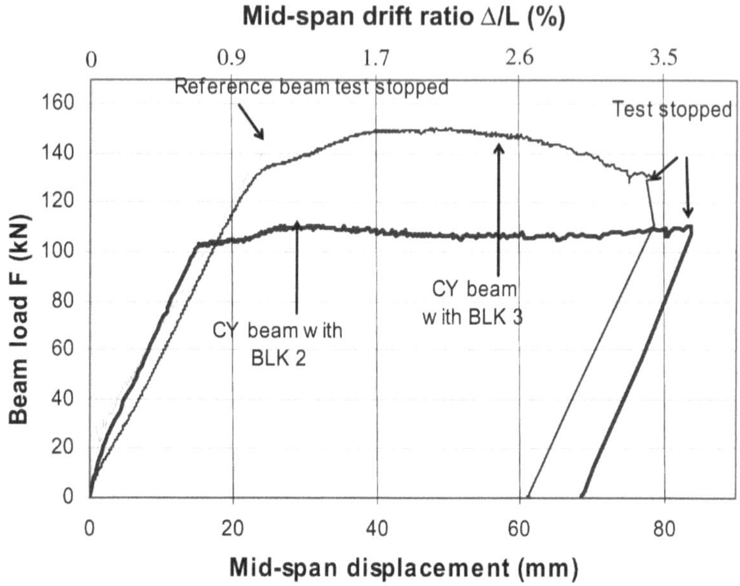

Fig. 9.16 Responses of CY beams with steel P-block.
From Wu, Y. F. 2008. "Ductility demand of compression yielding fiber-reinforced polymer-reinforced concrete beams." *ACI Struct. J.* 105 (1), 104–110.

9.7 Ductility demand on CY zone

The ductility of a beam is directly related to the ductility of the CY instrument. This relation can be derived mathematically. Using Fig. 9.18 that shows half of a simply supported beam, Eq. (9.1) can be derived (Wu 2008).

$$\frac{\mu_c - 1}{\mu_b - 1} = \frac{\Delta_y}{\delta_y} \cdot \frac{4z}{L \cdot \left(1 - \frac{\alpha}{2}\right)} \tag{9.1}$$

where L is the span length of the beam; δ_y and Δ_y are the yield displacement of the CY instrument and the beam, respectively (Fig. 9.18B); z is the lever arm or the distance from the compression centroid to the tension centroid; $\alpha = L_p/L$, the plastic hinge length ratio, where L_p is the length of the plastic hinge zone; $\mu_c = (\delta_y + \delta_p)/\delta_y$, the ductility ratio (demand) of the CY zone, in which δ_p is the plastic deformation of CY zone; and $\mu_b = (\Delta_y + \Delta_p)/\Delta_y$, the ductility ratio of the beam where Δ_p is the plastic deformation of the beam, as shown in Fig. 9.18B.

Eq. (9.1) is further transformed to the following form:

$$\frac{\mu_c - 1}{\mu_b - 1} = \frac{\varepsilon_t}{\varepsilon_y} + 1 + 4\lambda \cdot \frac{\varepsilon_t + \varepsilon_c}{\varepsilon_y} \cdot \frac{(1-\alpha)^2}{\alpha \cdot (1 - 0.5\alpha)} \tag{9.2}$$

(A) SIFCON P-block in the CY zone

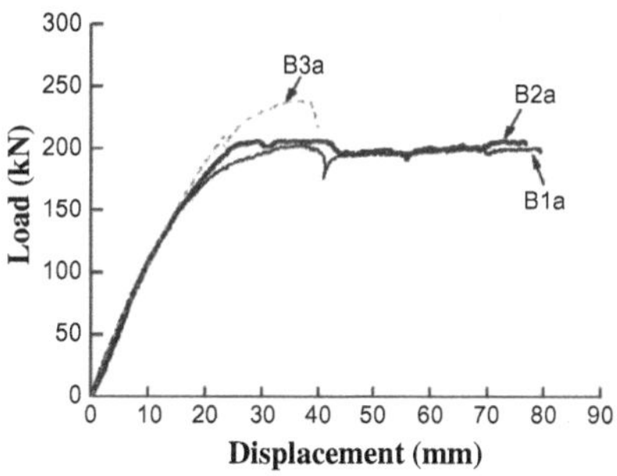

(B) Load vs. mid-span displacement

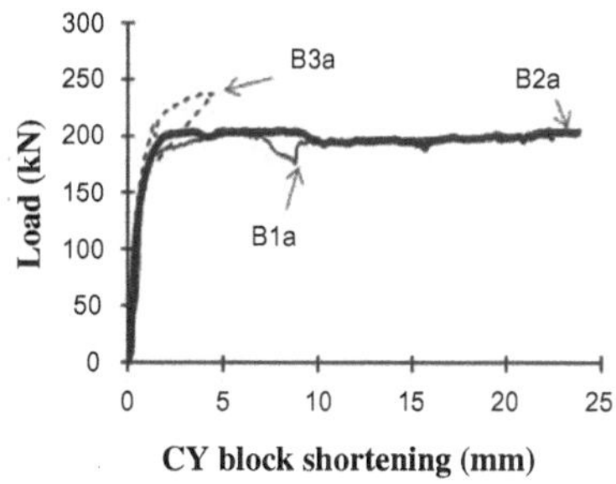

(C) Load vs. CY block shortening

Fig. 9.17 Response of CY beams with SIFCON P-block.
From Wu, Y. F., J. F. Jiang, and K. Liu. 2010a. "Perforated SIFCON blocks—an extraordinarily ductile material ideal for use in compression yielding structural systems." *Construct. Build Mater.* 24 (12), 2454–2465.

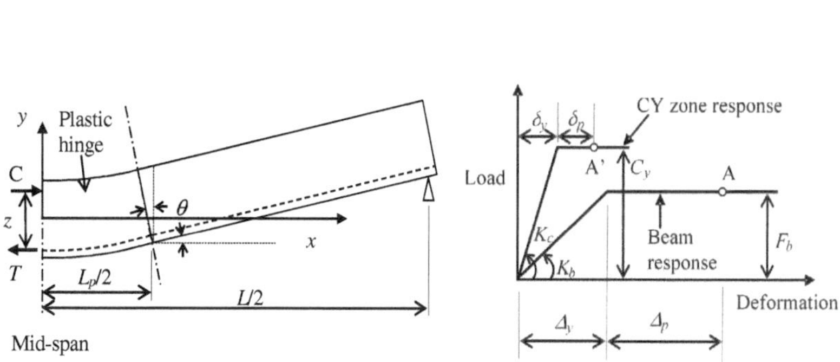

(A) (B)

Fig. 9.18 Plastic deformation of beam.
From Wu, Y. F. 2008. "Ductility demand of compression yielding fiber-reinforced polymer-reinforced concrete beams." *ACI Struct. J.* 105 (1), 104–110.

where ε_t is the strain of the FRP bars and ε_c is the compressive strain of concrete at the compressive centroid of the cross section adjacent to the CY zone; ε_y is the yield strain of the CY instrument; λ is a factor depending on the load distribution and for uniformly distributed load $\lambda = 5/48$ (Table 7.1 in Chapter 7). The ductility ratio of the CY instrument is given by $\mu_c = (\varepsilon_y + \varepsilon_p)/\varepsilon_y = \varepsilon_{cu}/\varepsilon_y$, in terms of strains, where ε_p and ε_{cu} are the plastic and ultimate strains of the CY instrument, respectively.

The results of Eq. (9.2) are evaluated and depicted in Fig. 9.19. The parameters used in the calculations are as follows: $\alpha = 0.025$ or 0.1; $\lambda = 5/48$; $\varepsilon_c = 0.001$. To avoid significant concrete deterioration or crushing outside the CY zone, the concrete strain at the top of the beam for the cross section adjacent to the CY zone should be smaller than 0.002. In this case, the strain at the compressive centroid is approximately less than 0.0013. The variation of ε_c from 0 to 0.0013 causes insignificant change in the results. The three lines in the figure give the results when the FRP strain at the ultimate state equals 75%, 50%, and 25% of the guaranteed rupture strain ε_{fu}^* of FRP (assume $\varepsilon_{fu}^* = 0.015$), respectively, i.e., $\varepsilon_t = 0.01125$, 0.0075, and 0.00375.

Based on Fig. 9.19, typical values of the ductility demand on the CY instrument, μ_c, are listed in Table 9.3. Assuming a required beam ductility ratio of $\mu_b = 3$, it can be seen from the table that the ductility demand of the CY zone is usually very high, especially when the CY zone is short (for small α values).

As discussed in Chapter 8, confining concrete can significantly increase the ductility of the concrete in the plastic hinge zone. When the beam is appropriately confined to have an overreinforced failure mode, it is actually a special case of the CY system. However, the increase in the ductility by confining concrete with steel spiral is limited. The yielding of confined concrete generally starts at the strain of $\varepsilon_y = 0.2\%$–0.3% (Mander et al. 1988). For the case shown in Fig. 9.19A, $(\mu_c - 1)/(\mu_b - 1) = 15.3$–$10.5$ for $\varepsilon_t = 0.75\%$. For a beam ductility ratio of $\mu_b = 3$, this requires a ductility demand of $\mu_c = 31.6$–22. Based on the experimental tests of Mander et al. (1988), the maximum axial strain of confined concrete at the fracture of the mild steel stirrups is about $\varepsilon_{cu} = 3$–5%, which gave a ductility ratio of $\mu_c = 10$–16.6 for $\varepsilon_y = 0.3\%$ ($\mu_c = \varepsilon_{cu}/\varepsilon_y$), or $\mu_c = 15$–25 for $\varepsilon_y = 0.2\%$, where the

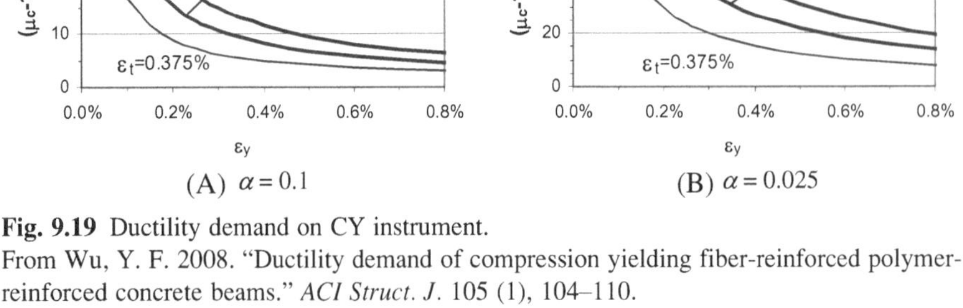

Fig. 9.19 Ductility demand on CY instrument.
From Wu, Y. F. 2008. "Ductility demand of compression yielding fiber-reinforced polymer-reinforced concrete beams." *ACI Struct. J.* 105 (1), 104–110.

Table 9.3 Ductility demand on the CY zone.

Ductility demand	ε_t	$\alpha = 0.1$		$\alpha = 0.025$	
		$\varepsilon_y = 0.3\%$	$\varepsilon_y = 0.7\%$	$\varepsilon_y = 0.3\%$	$\varepsilon_y = 0.7\%$
$(\mu_c-1)/(\mu_b-1)$	0.01125	14.9	7.0	50.6	22.3
	0.0075	10.6	5.1	35.3	15.7
	0.00375	6.2	3.2	20.0	9.2
μ_c	0.01125	30.8	15	102.2	45.6
(for $\mu_b = 3$)	0.0075	22.2	11.2	71.6	32.4
	0.00375	13.4	7.4	41	19.4

ductility calculation is based on the ultimate concrete strain that ignores the degradation of concrete strength. These values are generally insufficient to satisfy the ductility demand of 31.6–22. On the other hand, this calculation is based on Fig. 9.19A that assumes a plastic hinge length of 10% of total span and an FRP utilization level of 50% of its strength. Experience of beam tests has shown that the crushing zone of concrete is quite small even for beams with closely spaced stirrups at the plastic hinge zone. In other words, the plastic hinge length should be smaller than that of the case in Fig. 9.19A, particularly for FRP-reinforced beams due to the elastic nature of the beam. As a result, the ductility demand on the confined concrete would be much larger than the previous estimated values. This example shows that the ductility capacity of confined concrete is insufficient to satisfy the ductility demand on the CY zone, unless a very low level of FRP strain is utilized (i.e., much more FRP than required is provided), which is uneconomical.

External confinement of concrete members with FRP jackets does not provide a higher ductility of concrete than that confined by steel stirrups. In fact, confining concrete with FRP can only provide a maximum ultimate concrete strain of about $\varepsilon_{cu} = 2\%$ when FRP breaks (Lam and Teng 2003). This ultimate strain is smaller than the ultimate concrete strain of steel-confined concrete, and therefore can only result in a smaller ductility of concrete members than that of the steel-confined ones. In conclusion, confining concrete is unlikely to provide sufficient ductility of FRP-reinforced concrete members.

9.8 Design of CY members

For a typical CY beam, as shown in Fig. 9.20, there are a total of four variables or design parameters when the properties of all materials are given. For convenience of the mathematical study, these parameters are converted to dimensionless ones, as shown in Fig. 9.20: the relative height of the CY block η; its relative length α; tensile reinforcement ratio ρ_f; and compression reinforcement ratio ρ_{sc}. As the compression reinforcement is mainly as hanger bars, it is not considered as a variable in the following study. The material properties of the CY block are defined in Fig. 9.21: relative

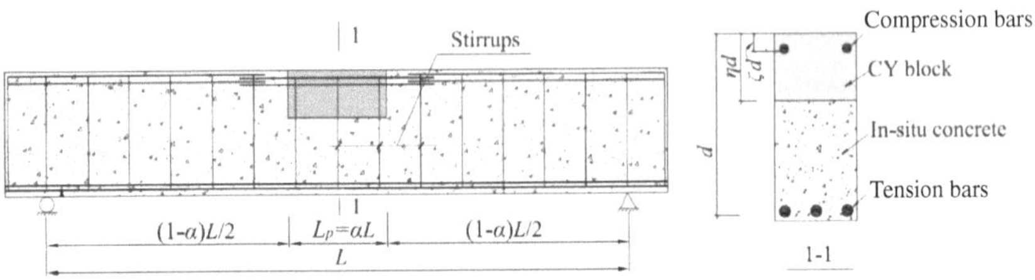

Fig. 9.20 Structural configuration of typical CY beams.
From Wu, Y. F., Y. W. Zhou, and X. Q. He. 2010b. "Performance-based optimal design of compression-yielding FRP-reinforced concrete beams." *Compos. Struct.* 93 (1), 113–123.

yield strength f_b/f_{co}, yield strain ε_{by}, ultimate strain ε_{bu} and strain hardening/softening modulus ratio ξ. Apart from the strength of the CY block f_b, the other three material properties are usually considered as given. Therefore, the design of CY members has a total of four variables: η, α, ρ_f, and f_b. Once these four variables are fixed, the CY beam design is fixed. Mathematically speaking, the design of the CY beam is simply the determination of these variables based on design requirements.

In the CY material model, the CY instrument, regardless of the steel mechanism or SIFCON P-block, is considered as a prefabricated block that fills the whole gap of the plastic hinge zone, and its load–deformation relationship is converted to an equivalent stress–strain relationship, as shown in Fig. 9.21A. The tension reinforcement is considered elastic, which is applicable to FRP bar or steel bar in overreinforced beams.

Greater ductility demand generally requires a larger CY block. At the same time, there are many possible designs (different combinations of the design variables) that satisfy a given set of design requirements. An optimal design would have the lowest construction cost for a given set of design requirements. As the cost of the CY block is relatively higher than that of reinforced concrete material, and furthermore a smaller CY zone is preferred for convenience of construction, the ductility design of CY

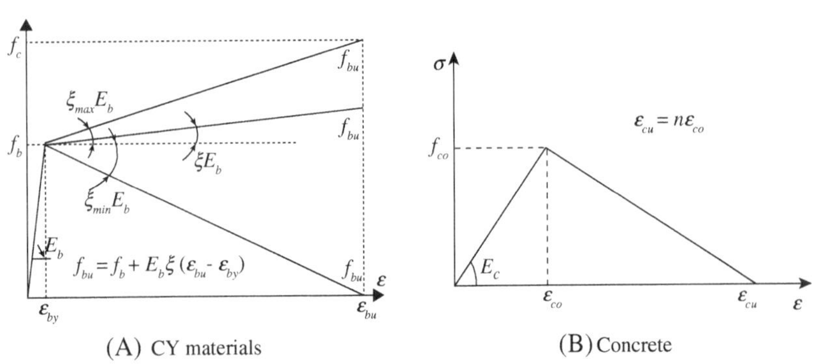

Fig. 9.21 Stress–strain relationships of materials.
From Wu, Y. F., Y. W. Zhou, and X. Q. He. 2010b. "Performance-based optimal design of compression-yielding FRP-reinforced concrete beams." *Compos. Struct.* 93 (1), 113–123.

beams can be considered as an optimization problem whose target is the minimum volume of the CY zone, V_{cy}, or mathematically:

$$\text{Minimize}\, V_{cy} \tag{9.3}$$

s.t.

$$\mu_\Delta \geq \mu_d \tag{9.4}$$

$$f_{bu} < f_{co} \tag{9.5}$$

$$\Delta_e \leq \Delta_d \tag{9.6}$$

$$M_m \geq M_d \tag{9.7}$$

and

$$\frac{M_m - M_{min}}{M_m} \leq \delta_d \tag{9.8}$$

Eq. (9.4) specifies that the beam ductility μ_Δ shall not be less than the required minimum ductility ratio μ_d. Eq. (9.5) limits the strength of the CY block to ensure that the concrete adjacent to the block does not crush. Eq. (9.6) ensures that the deflection of the beam Δ_e is not greater than a limit value Δ_d. Eq. (9.7) is employed to ensure that the flexural strength M_m is not less than the design value M_d. It is noted that M_m is the bending moment at the onset of the maximum allowable FRP strain that is used to avoid rupture of the tension reinforcement. It has been shown by Zhou et al. (2009, 2010) that M_m is very close to the peak moment resistance M_u of a section and can be used as M_u for mathematical convenience. Eq. (9.8) limits the drop of moment after the peak flexural strength, where M_{min} is the minimum moment resistance after the peak. Mathematical expressions of M_m and M_{min} were derived in Wu et al. (2010b).

Conventional nonlinear mathematical programming can be employed to solve this optimization problem. However, it is excessively complicated for design engineers, and a simple criteria-based optimal design approach was developed by Wu et al. (2010b) for engineering designs. The first step of the approach is the cross-sectional design. The typical moment–curvature curves of the RC section with a CY block (section 1–1 in Fig. 9.20) are shown in Fig. 9.22A. The CY section fails under two conditions:

1. Failure mode I—the CY block fails when the ultimate strain of the CY material ε_{bu} (Fig. 9.21A) is reached.
2. Failure mode II—when the allowable rate of the postpeak moment drop (RMD), i.e., $\frac{M_m - M_{min}}{M_m} = \delta_d$ (Eq. (9.8)), is reached.

Under these two failure modes, the relationship between η and the curvature ductility, $\mu_\varphi = \kappa_u/\kappa_y$ where κ_u and κ_y are the ultimate and yield curvature, respectively, can be

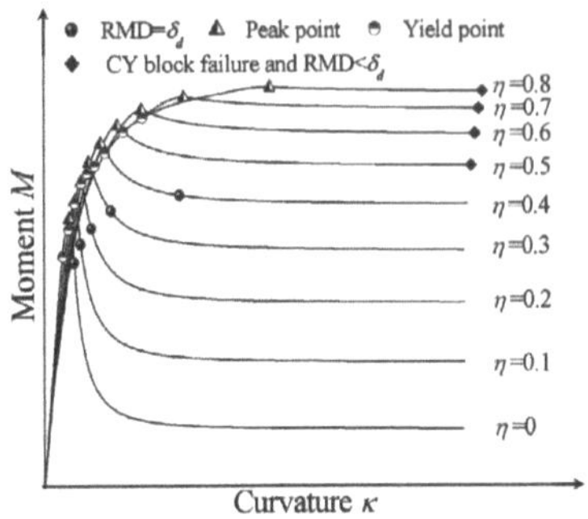

(A) Failure points

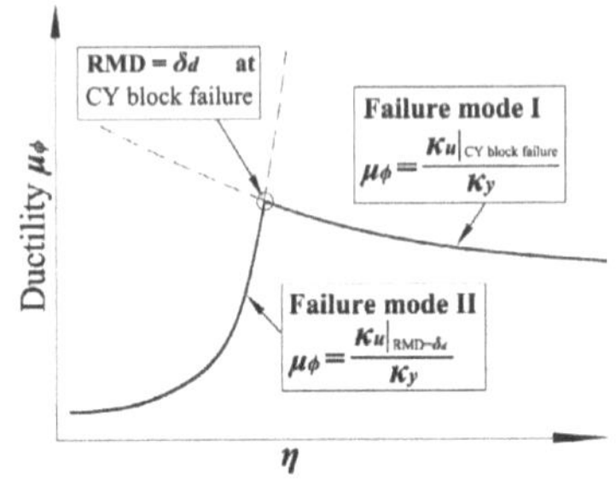

(B) Optimal η

Fig. 9.22 Optimal curvature.
From Wu, Y. F., Y. W. Zhou, and X. Q. He. 2010b. "Performance-based optimal design of compression-yielding FRP-reinforced concrete beams." *Compos. Struct.* 93 (1), 113–123.

derived mathematically (Wu et al. 2010b), as illustrated in Fig. 9.22B. The physical explanations for the two lines in Fig. 9.22B are as follows:

1. It can be seen from Fig. 9.22A that the ductility increases when the depth of the CY block increases, because CY block is more ductile than concrete material. Therefore, for a given value of postpeak moment drop δ_d, the curvature ductility increases when η increases, as shown by the line for failure mode II in Fig. 9.22B.
2. For failure mode I, the yield curvature κ_y increases when η is increased, whereas the ultimate curvature κ_u at CY block failure is barely affected by η (see Fig. 9.22A). Hence, the curvature ductility, $\mu_\varphi = \kappa_u/\kappa_y$, decreases when η increases as shown by the line for failure mode I in Fig. 9.22B.

The solid line in Fig. 9.22B shows the relationship between μ_φ and η. Since the design aim is to find a minimum volume of the CY block for a maximum ductility, clearly the intersecting point of the two curves is an optimal solution. The physical meaning of this intersection is that: the optimal ductility of the CY section is achieved when the failure of the CY block at $\varepsilon = \varepsilon_{bu}$ occurs simultaneously with the onset of the maximum allowable moment drop at RMD $= \frac{M_m - M_{min}}{M_m} = \delta_d$.

Based on the above analyses, the CY beam can be designed through the following procedure:

1. Design for optimal CY block depth η
   The design determines the optimal η based on Fig. 9.22B. It can be calculated from the equation $\frac{M_m - M_{min}}{M_m} = \delta_d$ where M_{min} is evaluated when the strain at the top of the CY block is equal to ε_{bu}. The mathematical expressions of M_m and M_{min} are derived in Wu et al. (2010b).
2. Determine CY block strength f_b
   The strength requirement of the CY block depends on the design moment. Therefore, f_b or ε_{by} (Fig. 9.21A) can be calculated from $M_m = M_d$.
3. Determine tension reinforcement ratio ρ_f
   The maximum force of the tension reinforcement is reached at the onset of M_m. The equation of tension reinforcement ratio at this moment is given in Wu et al. (2010b), which can be used to calculate ρ_f.

4. Determination of α

The length ratio α of the CY block can be determined from the following equation:

$$\mu_\Delta = \frac{\alpha(2-\alpha)\mu_\varphi + 8\lambda(1-\alpha)^2 \frac{(EI)_p}{(EI)_e}}{\alpha(2-\alpha) + 8\lambda(1-\alpha)^2 \frac{(EI)_p}{(EI)_e}} = \mu_d \tag{9.9}$$

where

$$(EI)_p = \frac{M_m}{\kappa_y} \tag{9.10}$$

$$(EI)_e = E_c I_e \tag{9.11}$$

where I_e is the effective moment of inertia given by Eq. (7.12).

5. Check for serviceability limit states

The serviceability limit states, including deflection and crack width, should be checked after the aforementioned design processes. As the CY block behaves elastically at service loading, the design approach at the serviceability limit states can be similar to those for conventional beams. If it is not satisfied, small adjusts can be made to satisfy design requirements in a similar process as that for the design of conventional RC members.

6. Design for shear

The shear strength of the members in a CY zone is usually not a problem for the following reasons: (i) significant amounts of steel material such as steel fibers are usually provided in a CY material to increase its ductility. Therefore, its shear strength is larger than normal concrete material; (ii) the CY zone is usually located around the maximum moment region where shear force is small; and (iii) the interfaces between the CY material and normal concrete are roughened (Fig. 9.3A) to have good interfacial shear resistance. Therefore, a normal shear design procedure for RC members can be used to design for the shear strength of the CY members in the shear dominant regions, e.g., near supports, where no CY instrument is provided. In other words, the shear design of CY beams can be identical to that of the conventional RC flexural members.

9.9 CY columns

The formation of plastic hinges cannot be completely avoided in the columns of bridges and frame structures in unusual events such as earthquakes, especially on the first floor. The large flexural deformation in an RC column may result in not only a large reduction in the moment resistance but also a fast deterioration of the axial load-carrying capacity, causing possible collapse of the structure. As discussed in Sections 3.2.3 and 3.4, the loss of axial load-carrying capacity of a column occurs when the maximum concrete strain of the column reaches the ultimate strain of concrete ε_{cu}. Providing confinement as discussed in Chapter 8 is a good approach in increasing the ductility of RC columns, because confinement can significantly increase ε_{cu}. However, sufficiently strong and effective confinement is not always possible, e.g., in the case of square/rectangular columns. In these cases, CY material can be used in the plastic hinge zone of a column to increase its ductility for the same reason, because the value of ε_{cu} of CY material is much larger than that of normal concrete.

Wu and Zhou (2011) studied the flexural behavior of the RC column with a CY block placed inside the plastic hinge zone, as shown in Fig. 9.23A. The typical response curves from numerical simulations are shown in Fig. 9.23B, which shows that the CY block can significantly increase the ductility of the RC column. For details of the numerical modeling, refer to Wu and Zhou (2011).

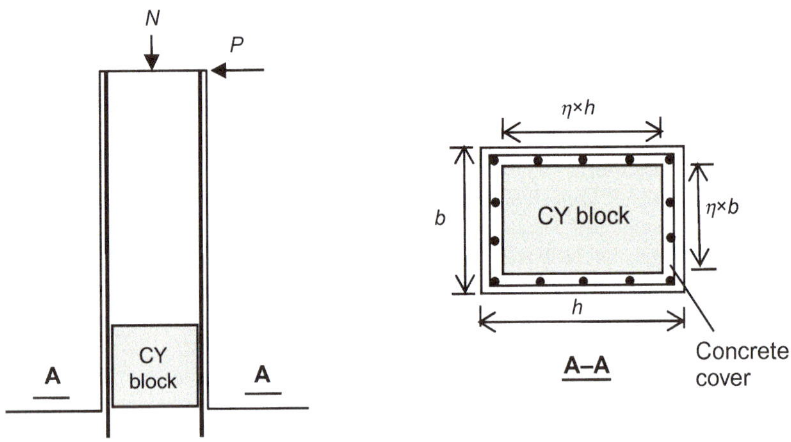

(A) Structural scheme of CY column

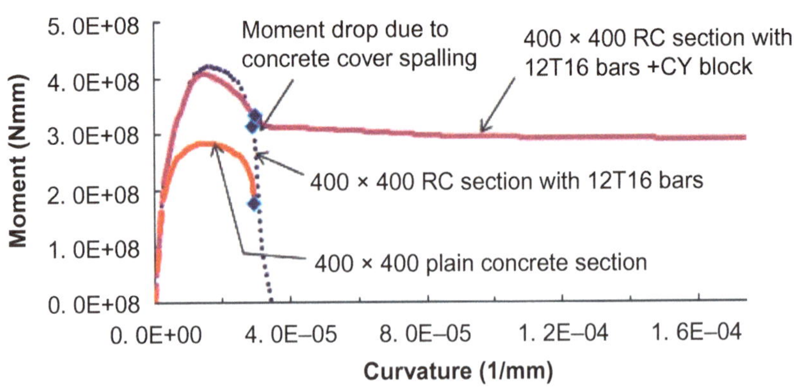

(B) Response of CY column with $N = 0.3\ N_c$

Fig. 9.23 RC column with a CY block.
From Wu, Y. F., and Y. W. Zhou. 2011. "Controlling the damage of concrete columns through compression yielding." *Struct. Control Health Monit.* 18 (8), 890–907.

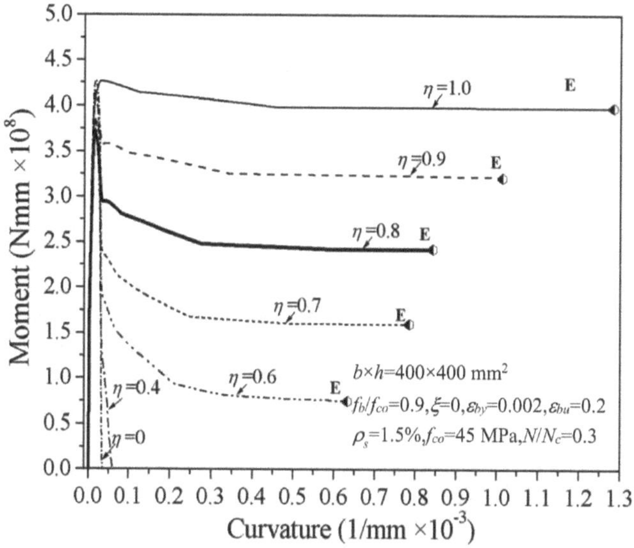

Fig. 9.24 Effect of CY block relative size on moment–curvature curves.
From Wu, Y. F., and Y. W. Zhou. 2011. "Controlling the damage of concrete columns through compression yielding." *Struct. Control Health Monit.* 18 (8), 890–907.

The typical moment–curvature curves of the column cross section with different relative sizes of CY block are shown in Fig. 9.24, where the CY block relative size parameter η is defined in Fig. 9.23A. It can be seen from the figure that the ductility is closely related to the relative size of the CY block. Nevertheless, the existence of the CY block in an RC column may reduce the strength of the column, simply because the strength of the CY block f_b is purposely designed to be less than the strength of concrete f_{co} in order to achieve compression yielding without crushing adjacent concrete in the plastic hinge zone. This reduction in moment resistance at a given axial load level is shown in the *M-N* interaction chart in Fig. 9.25 by the solid lines. The strength drop leads to higher ductility of the member, as shown in Fig. 9.23B. However, the reduction in moment capacity can be minimized or avoided by increasing the size of the CY block relative to the column cross-sectional size, as shown by the dash-dotted lines in Fig. 9.25.

The CY scheme can also be realized by replacing the bottom part of the RC column with a steel stump that yields before the onset of the maximum moment capacity of the RC column, i.e., connecting the RC column to a steel stump that is subsequently connected to foundation or its support. Such a scheme works on the same principle as that of the compression yielding system and is probably more acceptable and feasible in practice for column construction. The discussion in this section shows that the CY scheme can significantly increase the deformation capacity and ductility of the RC columns, hence increasing the survivability of RC columns under extreme events such as an earthquake.

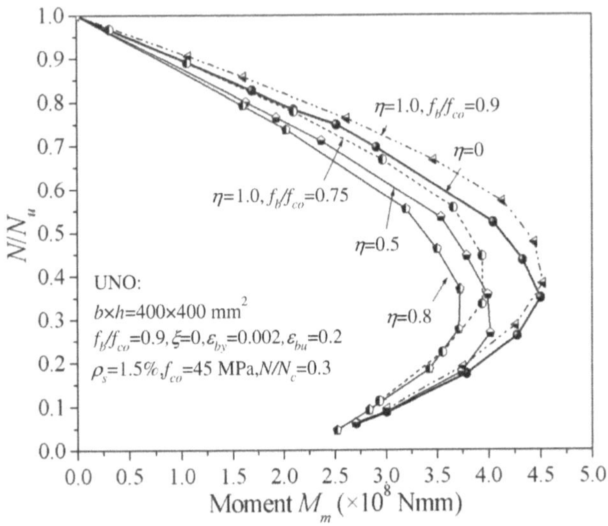

Fig. 9.25 Loading capacity of CY columns.
From Wu, Y. F., and Y. W. Zhou. 2011. "Controlling the damage of concrete columns through compression yielding." *Struct. Control Health Monit.* 18 (8), 890–907.

9.10 Fused structures

A higher safety margin usually requires a heavier structure and consequently higher construction costs. Therefore, safety margin and construction costs are usually two conflicting goals in structural design. In contrast, the safety and costs are generally not considered as conflicting factors in electrical or electronic circuit design, where important electrical components are protected from damage through the use of a fuse that is embedded within an electrical circuit. The triggering of the fuse stops "overloading." Consequently, measures allowing for overloading and its associated additional costs are not required. Therefore, the provision of a fuse in electrical circuits eliminates the conflicting nature between safety and costs in the production of electrical devices. In fact, the use of a fuse in an electrical circuit greatly reduces the consequences of failure. In other words, it changes the circuit to possess a safer and less costly failure mode, thereby, providing a cheaper solution.

The above analysis leads one to ponder the application of the fuse concept to structures. In structural designs, the magnitude of the safety factor for a structure depends on the failure mode. A more dangerous failure mode requires a higher safety factor (or lower failure probability) (see Table 1.1). Therefore, a less dangerous failure mode would require a lighter structure with less structural costs. The provision of a fuse in a structure would lead to a safer and more predictable failure mode, thereby a lighter structure, by changing a more dangerous failure mode to a less dangerous one.

Numerous studies on different structural fuses have been reported in the literature. Fuses have been incorporated into a structure in different structural forms, materials, and loading scenarios. In a majority of cases, the dissipation of energy from overloading

Fig. 9.26 Rubber bearing as a fuse.

due to seismic attack has been the motivator of structural fuses (Vargas and Bruneau 2009; El-Bahey and Bruneau 2011; Cui 2012; Kauffman and Memari 2014; Li et al. 2015; Shoeibi et al. 2017). In the case of overloading, these fuses act as a sacrificial element, thus preventing the other structural components attached to them from overloading. A common fuse is the lead rubber bearing that is used to control (isolate) excessive ground motion from transferring to the upper part of structures, as shown in Fig. 9.26. Generally speaking, any element, instrument, or material that reduces the effect of loading (action) can be considered as a fuse. This section introduces a new type of structural fuse that has been driven by the need to increase the ductility of concrete structures reinforced with FRP bars.

9.10.1 A new type of structural fuse

The tension yielding of steel reinforcing bars provides the ductility of conventional reinforced concrete structures. The yielding of steel bars inhibits further increase in internal forces in an RC structure and prevents fast deterioration of the structure. From this point of view, the yielded steel bar is actually a fuse for RC structures. When the tension reinforcement in an RC member does not yield at failure (e.g., overreinforced beams or columns with high axial load level, and FRP-reinforced members), the fusing action from the rebar is absent and the structure would fail in a nonductile and dangerous failure mode. In this case, the compression yielding instrument discussed in Section 9.5 can be built into an RC structure as a structural fuse.

The use of the CY fuse produces the following effects:

1. It inhibits further increase of loading by triggering the "fuse." As further increase in loading is blocked by the fuse in a safe manner, the characteristics of overloading are changed (the statistical distribution of loading is changed) and the partial safety factor for loading should be reduced (see further discussion in Section 11.8).
2. It transforms a more dangerous failure mode, such as abrupt FRP rupture, concrete crushing, or shear failure, into a ductile and safer one, so a smaller partial safety factor for material (or capacity reduction factor) can be used (see Section 1.6).
3. It localizes the damage to the CY zone. Repair of the member is therefore confined to the CY zone by simply replacing the fuse with a precast CY block rather than reconstruction of the whole member. Therefore, the repair costs are reduced, in terms of both direct and indirect costs particularly the closing down of service. This cost reduction should also be allowed for further reduction of the safety margin.

All these three aspects in the reduction of safety margin would bring a more economical design of the fused structures.

9.10.2 Cost analysis of the fused structures

Neglecting the maintenance, demolition, and associated indirect costs (e.g., losses due to service interruption), the total cost C_t of a structure can be viewed as the product of failure probability P_f and failure cost C_f plus the initial cost C_i (Nowak and Collins 2000), i.e.,

$$C_t = C_i + P_f C_f \tag{9.12}$$

It is noted that the initial cost C_i increases with an increase of reliability index β, because a safer structure normally requires higher initial costs. Generally, P_f is a decreasing function of β since the failure probability decreases with increasing reliability index β. In addition, C_f is believed to be independent of β or P_f.

According to Nowak and Collins (2000), the relationships between C_i and β as well as P_f and β can be approximately expressed as

$$C_i = \alpha(1 + b\beta) \tag{9.13}$$

$$P_f = c \cdot \exp(-\beta/d) \tag{9.14}$$

where a, b, c and d are constants. Substituting Eqs. (9.13) and (9.14) into Eq. (9.12) gives

$$C_t = \alpha(1 + b\beta) + c \cdot \exp(-\beta/d) C_f \tag{9.15}$$

Eq. (9.15) consists of two parts that are both functions of β. The first part is an increasing function of β, whereas the second part is a decreasing one. The qualitative relationship between C_t and β is shown in Fig. 9.27. It is evident that an optimum design can be achieved by adopting a certain value of β that minimizes the total cost C_t while satisfying the minimum safety requirement. This optimal value of β is the target reliability index β_T, as shown in Fig. 9.27.

The use of a CY fuse is not expected to cause a significant increase in the construction costs. Therefore, the first part of Eq. (9.15) can be considered similar for both conventional and the fused structures, that is, the dotted line in Fig. 9.27 does not change. The repair of the structure with a CY fuse can simply be the replacement of the fuse. Hence, the costs of failure, or the second part of Eq. (9.15), are much smaller for the fused structures. The total costs of a conventional RC structure without a fuse and that with a fuse are shown in Fig. 9.27 by the dashed line and the solid line, respectively. It can be seen in the figure that at the target reliability index of a fused structure β_{T2}, the initial cost, failure cost, and the total cost are all smaller than those of the respective conventional structure without the fuse at its target reliability index β_{T1}. In the meantime, the fused structure will have a lower cost if the same reliability index as that used for the conventional structure is adopted.

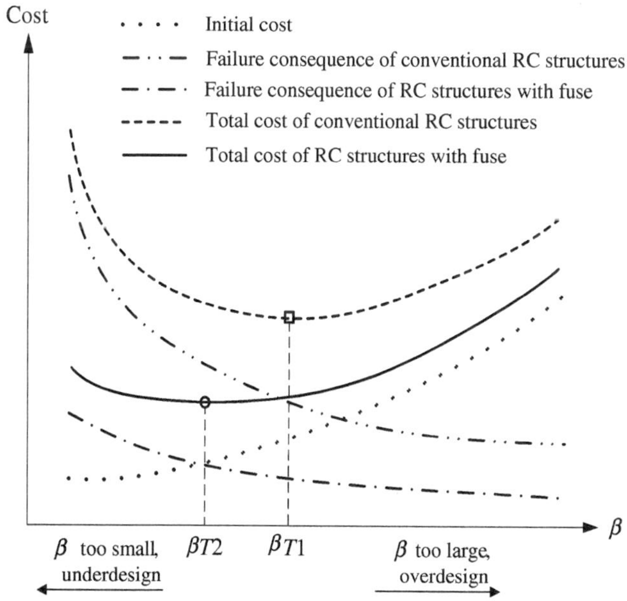

Fig. 9.27 Costs versus reliability index.
From Wu, Y. F., Y. W. Zhou, B. Hu, X. Huang, and S. Smith. 2020. "Fused structures for safer and more economical constructions." *Front. Struct. Civ. Eng.* 14 (1), 1–9.

Table 9.4 Target reliability index β_T at ultimate state of structures under 50-year period.

		Target reliability index β_T	
Structural class	**Consequences of failure**	**Ductile failure**	**Brittle failure**
Class I	High	3.7	4.2
Class II	Medium	3.2	3.7
Class III	Low	2.7	3.2

9.10.3 Determination of safety margin for fused structures

Normally, the target reliability index β_T for a structural component is provided by design codes of practice and such index values may differ among different codes. The target reliability index β_T at the ultimate limit states under a 50-year reference period from the Chinese code GB50068 (MOHURD 2001) is provided in Table 9.4. Since the use of a fuse in RC structures can prevent dangerous failure modes and minimize the losses due to structural failure, RC structures containing a fuse can have a smaller design reliability index β_T. A reliability-based approach can be used to calculate a suitable safety factor or capacity reduction factor of a fused structure.

Considering an arch footbridge as shown in Fig. 9.28 as an example, the main structural components of the footbridge consist of two arches, two longitudinal beams, several trusses and a bridge deck. A CY fuse is inserted into the apex of the arch. As the bridge is symmetrical, only one arch is used to study the costs of the structures with and without a fuse by using a reliability-based approach. For simplicity, only dead and live loads are considered.

The arch axis of the footbridge is determined so that the arch is only under axial compression along its axis under uniformly distributed gravitational loading (Fig. 9.29). This gives the following equation:

$$y = \frac{4f(lx - x^2)}{l^2} \tag{9.16}$$

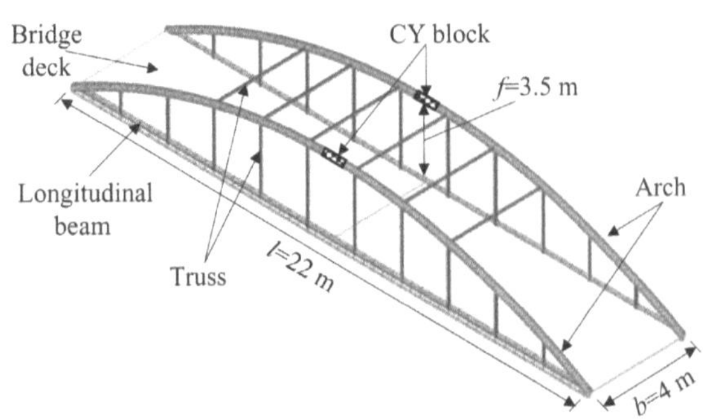

Fig. 9.28 Schematic diagram of footbridge containing a CY fuse. From Wu, Y. F., Y. W. Zhou, B. Hu, X. Huang, and S. Smith. 2020. "Fused structures for safer and more economical constructions." *Front. Struct. Civ. Eng.* 14 (1), 1–9.

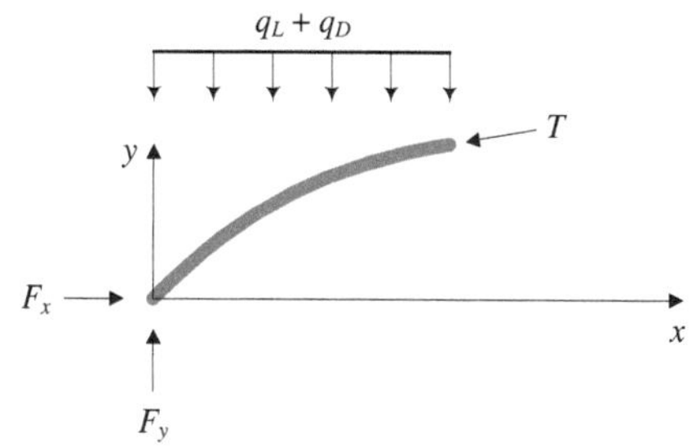

Fig. 9.29 Free body diagram of the arch.
From Wu, Y. F., Y. W. Zhou, B. Hu, X. Huang, and S. Smith. 2020. "Fused structures for safer and more economical constructions." *Front. Struct. Civ. Eng.* 14 (1), 1–9.

where x and y are the coordinates shown in Fig. 9.29; l and f are the span and the rise of the arch (Fig. 9.28), respectively. Both the nominal dead load q_D and nominal live load q_L applied to the arch are assumed to be uniformly distributed, and $q_L = 5$ kN/m^2. The total weight of the footbridge is estimated to be 9000 kg, thus $q_D = 1$ kN/m^2. From equilibrium, one has

$$T(x) = \frac{F_x}{\cos\left(\arctan\left(\frac{dy}{dx}\right)\right)} \tag{9.17}$$

$$F_x = \frac{q_L b l^2}{16f} + \frac{q_D b l^2}{16f} \tag{9.18}$$

where $T(x)$ is the compression force along the arch axis; F_x is the horizontal support force; and b is the width of the bridge (Fig. 9.28).

The common Load and Resistance Factor Design (LRFD) method is adopted to evaluate the safety margin of the structure. Based on ACI 318-14 (ACI Committee 318 2014) and given that the live and dead loads dominate the design, the LRFD expression under the ultimate limit state can be expressed as:

$$\phi R_n = 1.2D_n + 1.6L_n \tag{9.19}$$

where ϕ is the resistance reduction factor; R_n is the nominal resistance of the structure; D_n and L_n are the compression force effect given by Eq. (9.17) due to the nominal dead and live loads, respectively. The coefficients 1.2 and 1.6 are the corresponding partial safety factors for dead and live loads based on GB50068, respectively. The nominal values for D_n and L_n can be obtained from Eq. (9.17), which are

$$D_n = \frac{\frac{q_D b l^2}{16f}}{\cos\left(\arctan\left(\frac{dy}{dx}\right)\right)} \tag{9.20}$$

$$L_n = \frac{\frac{q_L b l^2}{16f}}{\cos\left(\arctan\left(\frac{dy}{dx}\right)\right)} \tag{9.21}$$

The limit state function can be represented by the following equation:

$$Z = R - D - L \tag{9.22}$$

in which R, D and L are the random variables for resistance, dead load effect, and live load effect, respectively. In reliability analysis, failure occurs if $Z < 0$. That is, the combined load effect $(D + L)$ is larger than the resistance (R). Therefore, the failure probability P_f can be expressed as (Rubinstein and Kroese 2016):

$$\begin{aligned} P_f &= \Pr(R - D - L < 0) = \iiint_{R-D-L<0} f_R(R) f_D(D) f_L(L) dR dD dL \\ &= 1 - \Phi(\beta) \end{aligned} \tag{9.23}$$

where $f_R(R)$, $f_D(D)$ and $f_L(L)$ are the probability density functions for R, D and L, respectively; and Φ is the cumulative distribution function of the standard normal distribution; and β is the reliability index.

The resistance (R) is assumed to follow a lognormal distribution, in which the mean value equals the nominal value $R_n = (1.2D_n + 1.6L_n)/\phi$ based on Eq. (9.19). In addition, the coefficient of variation (COV) is equal to 0.1 (GB50068). The dead load effect (D) follows a normal distribution with the mean value equal to 1.05 times the nominal value D_n from Eq. (9.20), while the COV is equal to 0.1 (Galambos et al. 1982). The live load effect (L) follows an extreme value distribution (type I) with the mean value being equal to the nominal value L_n from Eq. (9.21) with a COV of 0.25 (Galambos et al. 1982).

Based on the stochastic information mentioned previously, the probability density functions $f_R(R)$, $f_D(D)$ and $f_L(L)$ can be obtained. In the following reliability analysis, the resistance factors ϕ ranging from 0.7 to 1.0 with a step size of 0.025 are considered. Since the nominal values of D_n and L_n are known, for a given value of ϕ, the nominal value R_n is given by Eq. (9.19). Therefore, the reliability index in Eq. (9.23) can be evaluated by either a Monte Carlo simulation (MCS) or first order reliability method (FORM) (Huang et al. 2016) with the detailed probability density functions $f_R(R)$, $f_D(D)$ and $f_L(L)$. The relationship between the resistance reduction factor ϕ and the reliability index β can be obtained by repeating this process for each ϕ, and is given in Fig. 9.30.

According to Table 9.4, the target reliability index β_T for Class I structures can be 3.7 for a fused structure and 4.2 for a conventional structure with brittle failure modes. Their corresponding resistance reduction factors ϕ can be obtained from Fig. 9.30, which are 0.87 and 0.8, respectively. These results mean that the resistance reduction factor of the footbridge containing a CY fuse is about 9% larger than that of the structure without a CY fuse. Therefore, the initial construction costs of the structure can be reduced by using the fuse.

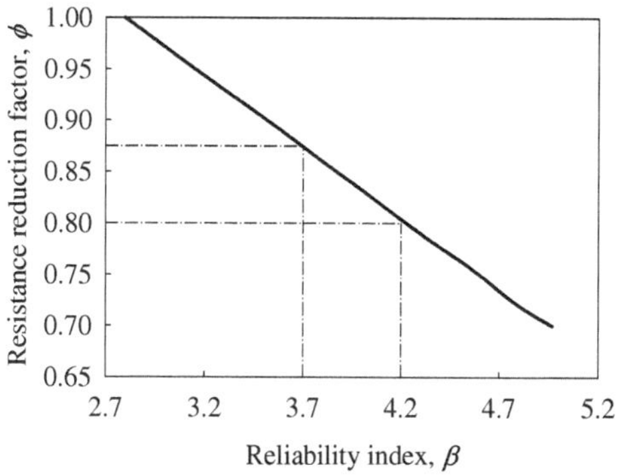

Fig. 9.30 Relationship between β and ϕ. From Wu, Y. F., Y. W. Zhou, B. Hu, X. Huang, and S. Smith. 2020. "Fused structures for safer and more economical constructions." *Front. Struct. Civ. Eng.* 14 (1), 1–9.

As discussed in Section 9.10.1, the provision of the fuse in structures causes three reduction effects in safety margin, i.e., loading, failure damage, and repair costs. In the preceding reliability analysis, only the second effect is considered. The loading effect, i.e., the coefficients 1.2 and 1.6 in Eq. (9.19), is fixed based on current design codes, which are based on the statistic characteristics of normal loading. In fact, the loading is better controlled and its statistics relevant to fused structures would be different from those of ordinary structures. Therefore, the partial safety factors specified in current codes of practice for loading would be inapplicable to fused structures. On the other hand, the third reduction effect, i.e., the reduction in repair costs, is ignored in the current reliability calculation. As discussed earlier, both the direct and indirect costs of the repair for a fused structure would be much lower than those of normal structures. From this point of view, a further reduction in safety margin or increase in the capacity reduction factor is possible by considering this effect.

As shown in Fig. 9.27, the target reliability index for fused structures should be smaller than that for conventional structures. The third reduction effect should be considered by allowing for a smaller target reliability index than those given in Table 9.4. However, no provision is available on target reliability index of fused structures in current codes of practice. More research is required in the development of a reliability-based design approach suitable for different types of fused structures based on first principles. More discussion on the safety margin is given in Section 11.8.

9.11 Increasing concrete strength by reducing perforation

From Section 9.5, it can be seen that the ductility of a material is closely related to its perforation ratio. This is consistent with the behavior of normal concrete. It is well known that the key factor that affects the strength of concrete is the water/cement ratio. A higher water/cement ratio gives a more ductile but weaker concrete. The reason is that a higher water/cement ratio provides more excess water in concrete that becomes voids or perforations when the concrete is hardened. In fact, providing different strength and ductility of concrete by altering its pore/perforation ratio is a basic principle in the design of concrete material. For example, the ultrahigh performance concrete (UHPC) is made by using a very low water/binder (cement, silica fume, etc.) ratio and optimized packing density with ultrafine particles to form an extremely tight mix, which leads to minimized pores between the particles and hence an ultrahigh-strength concrete.

Based on this principle, a compression casting technology was developed to increase the strength of normal concrete or low-strength concrete incorporating recycled materials, by squeezing excess pore water out of the concrete mix during casting of the concrete (Nematzadeh and Naghipour 2012; Nematzadeh et al. 2016; Wu et al. 2020; Kazmi et al. 2021). Wu et al. (2020) applied the compression casting technology to improve the material properties of rubber concrete. The mix design used in their study was the same as that for normal concrete, apart from replacing some of the natural coarse aggregate (NCA) with chipped rubber (Fig. 9.31). Concrete specimens were cast with 0–100% (in volume) of NCA replacement by chipped rubber.

The fresh rubber concrete was then condensed in a specially designed mold by compressing the wet concrete. The manufacturing process for the concrete is shown in Fig. 9.32. The compression force was increased gradually on the wet concrete after casting until the reduced volume of the wet concrete equaled the volume of chipped rubber. The compression pressure applied to the fresh concrete varied from 6.9 MPa to a maximum of 27.7 MPa depending on rubber content. The loading jack was locked to prevent rebound of the compressed concrete for 24 h before demolding. In fact, the pressure on the concrete actually dissipated almost immediately after the loading jack reached the design compression volume, because the compression deformation of concrete was mostly plastic. Further tests showed that the pressure only needs to be maintained for a couple of minutes and can be removed after that.

Fig. 9.31 Chipped rubber.
From Wu, Y. F., S. M. S. Kazmi, M. J. Munir, Y. Zhou, and F. Xing. 2020. "Effect of compression casting method on the compressive strength, elastic modulus and microstructure of rubber concrete." *J. Clean. Prod.* 264, 121746.

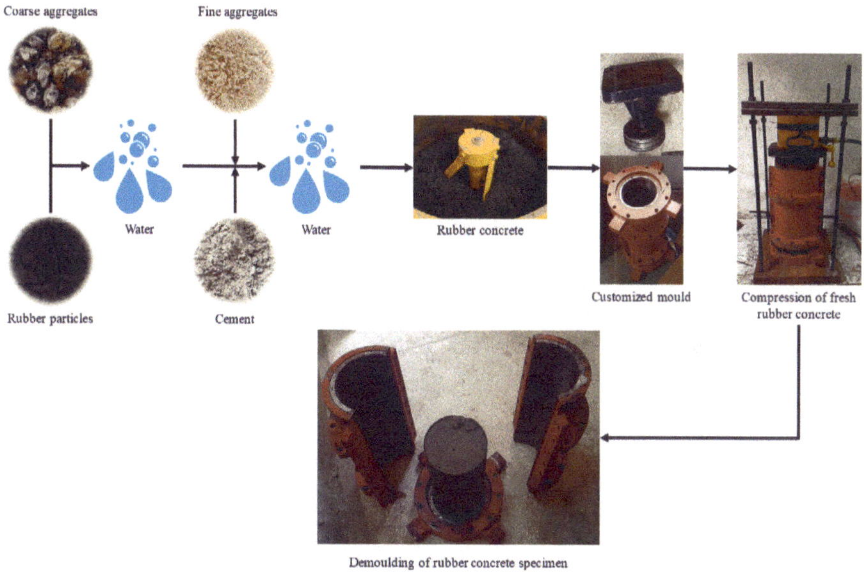

Fig. 9.32 Compression casting technology.
From Wu, Y. F., S. M. S. Kazmi, M. J. Munir, Y. Zhou, and F. Xing. 2020. "Effect of compression casting method on the compressive strength, elastic modulus and microstructure of rubber concrete." *J. Clean. Prod.* 264, 121746.

After hardening of the concrete, the specimens were tested through conventional compression tests of concrete. The typical stress–strain curves of the compressed and uncompressed rubber concrete specimens are shown in Fig. 9.33 together with those of normal aggregate concrete (NAC) without rubber, where R20 represents the rubber concrete specimens with 20% of coarse aggregates replaced by rubber chips without compression and R20-C indicates R20 concrete by compression casting. Fig. 9.34 shows the concrete strength and Young's modulus of the rubber concrete with and without compression casting. These results show that improved stress–strain behavior, i.e., significantly higher compressive strength and elastic modulus, can be achieved from compression casting of the rubber concrete compared with the uncompressed one. The compressed rubber concrete can even have a compressive strength and elastic modulus greater than those of NAC.

Fig. 9.35 depicts the SEM images of the NAC, compressed and uncompressed rubber concrete samples having a 20% replacement ratio of rubber. A smaller number of pores/voids and microcracks and a denser microstructure were observed in the concrete area (without rubber) in the compressed rubber concrete sample than the NAC and uncompressed rubber concrete samples. Denser microstructure and fewer pores/voids and microcracks in the compressed concrete result in enhanced microstructure. The densified microstructure of concrete is the cause of the increase in the elastic modulus and strength of the material. From this point of view, the compression casting technology is generally effective for all concrete materials. In other words, it would also be effective in increasing the strength and stiffness of other concrete materials, such as normal concrete and concrete mixed with other recycled materials. This is evident from Fig. 9.36 in

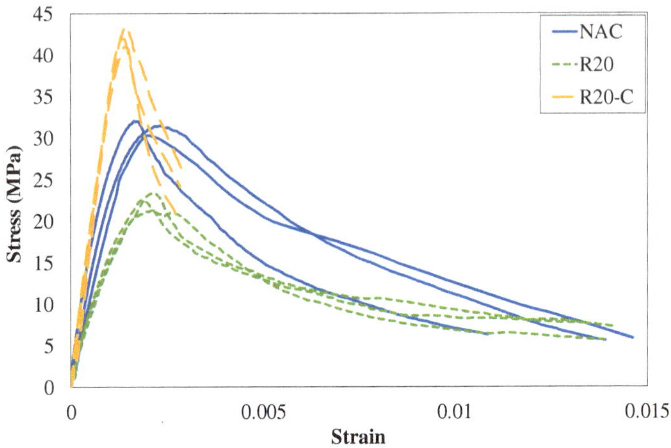

Fig. 9.33 Stress–strain curves of rubber concrete.
From Wu, Y. F., S. M. S. Kazmi, M. J. Munir, Y. Zhou, and F. Xing. 2020. "Effect of compression casting method on the compressive strength, elastic modulus and microstructure of rubber concrete." *J. Clean. Prod.* 264, 121746.

which the recycled aggregate concrete (RAC) after compression casting exhibits a higher strength than the uncompressed RAC. All the RACs in Fig. 9.36 have an identical mix design. The horizontal axis shows the compression pressure applied to the RAC during casting. The RAC in Fig. 9.36 was made by using demolished concrete as coarse aggregates to replace natural coarse aggregates.

The compression casting technology can be readily adopted by the precast concrete industry to manufacture precast concrete products, such as bricks, blocks, panels, and road barriers. The existing manufacturing process and facilities can be retained apart from adding one additional process of compression in the final stage of production. However, the casting mold is required to be sufficiently strong to sustain the compression pressure. The costs of the compressed concrete products made by the compression casting technology would be comparable to the traditional concrete products, because no additional material is required apart from the additional energy cost for compressing the concrete. The preceding discussions are based on normal mix designs. Excess cement paste is squeezed out of the concrete and wasted during compression. Further study is being carried out in the author's group currently by using less cement and water so that little excess cement paste is squeezed out, to save cement material. In this way, higher strength concrete can be achieved with lower costs. Preliminary test results support the prediction and conclusion.

The use of recycled materials in concrete usually leads to inferior material properties that inhibit their widespread application in engineering. Existing technologies for improving the properties of RAC have limited effect apart from the additional costs for the chemical and physical treatment processes. The compression casting technology is relatively simple and cheap yet much more effective than other technologies. The technology can help in facilitating eco-friendly and sustainable constructions while conserving natural resources.

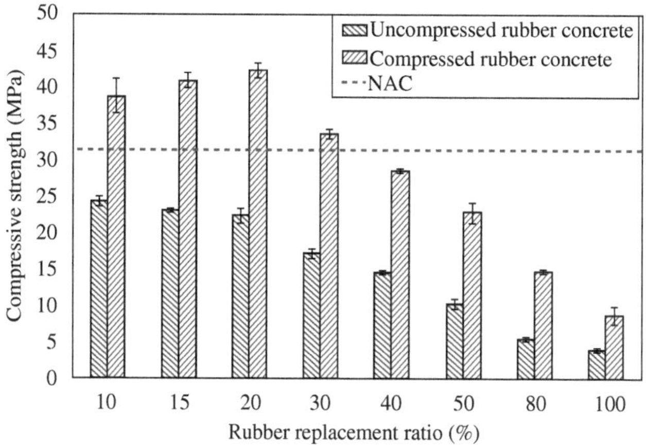

(A) Compressive strength

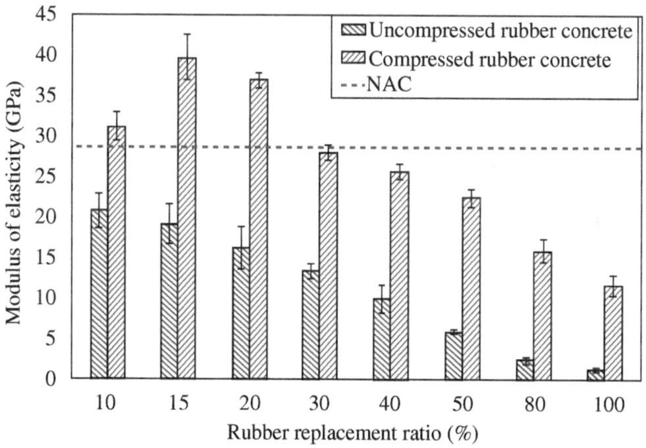

(B) Modulus of elasticity

Fig. 9.34 Concrete strength and Young's modulus of compressed and uncompressed rubber concrete.
From Wu, Y. F., S. M. S. Kazmi, M. J. Munir, Y. Zhou, and F. Xing. 2020. "Effect of compression casting method on the compressive strength, elastic modulus and microstructure of rubber concrete." *J. Clean. Prod.* 264, 121746.

9.12 Failure localization and mitigation

Structures and materials fail locally, even under uniform stress and strain conditions, as illustrated in Fig. 9.37. Localization of failure cannot make full use of the deformability of material outside the local failure zone. Increasing the size of the local failure zone can increase deformability and ductility of materials/structures. The cause and mitigation of the failure localization are studied in this section.

(A) NAC

(B) R20

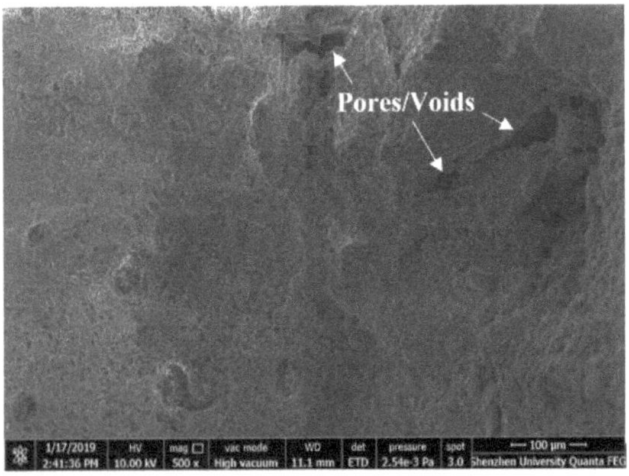

(C) R20-C

Fig. 9.35 Microstructure of concrete.
From Wu, Y. F., S. M. S. Kazmi, M. J. Munir, Y. Zhou, and F. Xing. 2020. "Effect of compression casting method on the compressive strength, elastic modulus and microstructure of rubber concrete." *J. Clean. Prod.* 264, 121746.

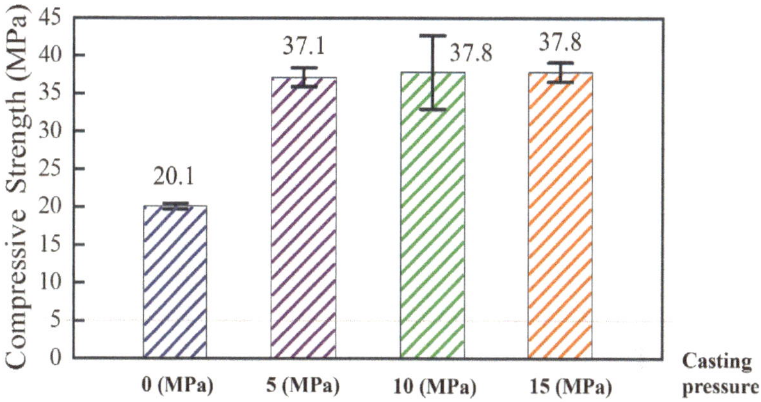

Fig. 9.36 Recycled aggregate concrete.

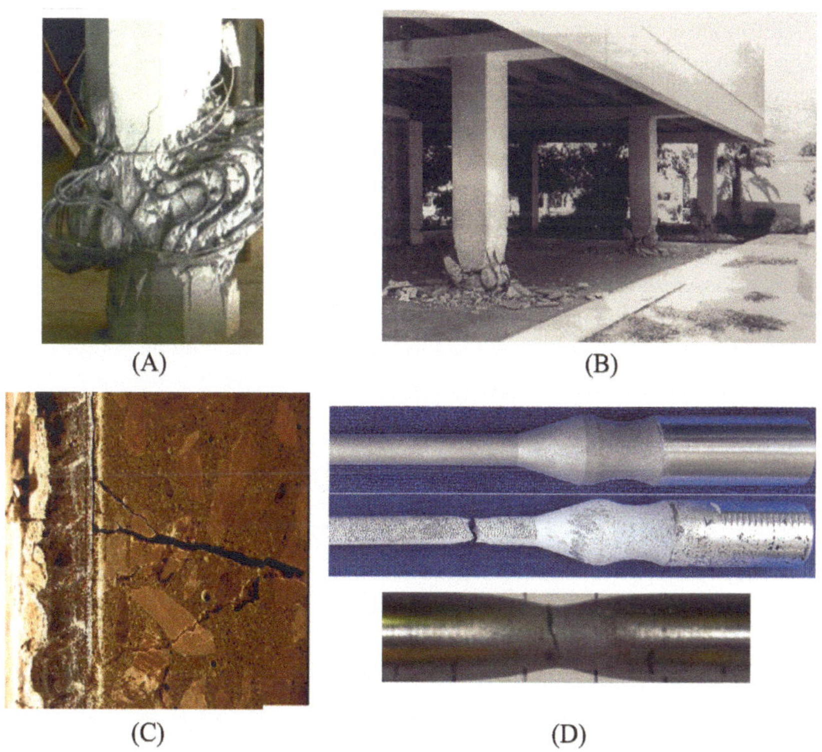

Fig. 9.37 Failure localization: (A) local failure of column under compression, (B) plastic hinge failure, (C) concrete cracking, (D) necking of steel bar.

(A) From: https://web.ics.purdue.edu/~braile/edumod/eqphotos/eqphotos1.htm, (B) From Green, N. B. 1987. *Earthquake resistant building design and construction*. New York: Elsevier Science Publishing Co. Inc. (C) From Li, V. C. 2019. "Chapter 6—Resilience of engineered cementitious composites (ECC) structural members." Engineered Cementitious Composites (ECC). Springer Nature.

9.12.1 Failure of RC column under concentric loading

RC columns fail locally under concentric loading, as shown in Fig. 8.40. The load–deformation behavior of the columns with failure localization is studied in Section 8.6.9. It was experimentally and analytically found that the length of the failure zone (L_d in Fig. 8.41) can be changed by providing confinement to the columns. Fig. 9.38 shows the experimental and analytical results of the length of damage zone of concrete columns confined by FRP jacket.

It can be seen from Fig. 9.38 and Eq. (8.153) that L_d/L approaches 1, i.e., the strain concentration shown in Fig. 8.40 or failure localization disappears, when the confinement stiffness ratio ρ_{eff} is greater than 6.3. Based on Jiang and Wu (2016), the hardening slope of the stress–strain curve of FRP-confined concrete is equal to zero when $\rho_{eff} = 6.3$, as shown in Fig. 9.39. This conclusion is generally consistent with Eq. (8.127), which gives $E_2 = 0$ when $\rho_{eff} = 6.1$. From this study, it can be seen that the failure localization disappears, and a column can deform uniformly along the length when it is confined to have a hardening type of load–deformation behavior.

The mechanism of the localized failure is that, when the axial resistance of the concrete column starts to soften and drop at the weakest cross section, the axial force on other cross sections also drops to maintain the balance of force. Therefore, only the weakest cross section can experience further crushing along the descending branch of the stress–strain curve and all other cross sections unload, for a similar reason to that in Fig. 7.14A. As a result, the failure localizes at the region of the weakest section (only the weakest link fails in Fig. 1.9). When the stress–strain curve has a hardening type of curve, the load resistance of the weakest section increases with further deformation after the turning point P (Fig. 9.39). This will further increase the load applied to the next weakest section and cause the softening (passing the turning point P) of that section. This process continues, which will force all cross sections to pass the turning point P so that failure localization does not occur in columns with a hardening type of load–deformation curve.

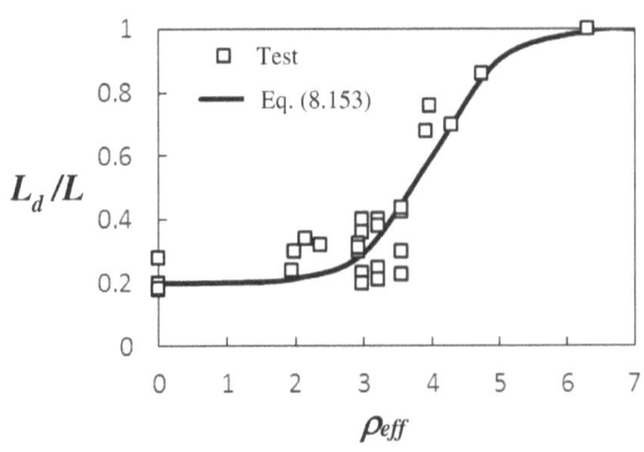

Fig. 9.38 Length of failure zone. From Wu, Y. F., and Y. Wei. 2016. "Stress-strain modeling of concrete columns with localized failure: an analytical study." *J. Compos. Constr.* 20 (3), 04015071.

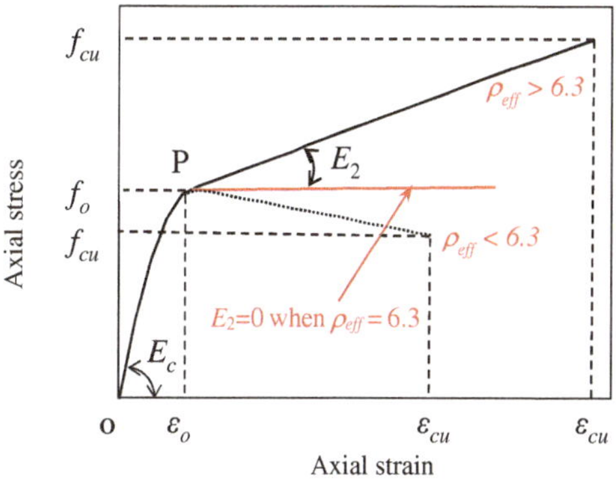

Fig. 9.39 Typical stress–strain curve of FRP-confined concrete.

9.12.2 Concrete cracking

Cracking of concrete is also localized as shown in Fig. 9.40A. In a plain concrete member under tension force, the brittle matrix of concrete loses its tensile load-carrying capacity almost immediately after initial formation of a crack when the stress–strain curve reaches the peak (Fig. 9.41A). Due to the drop of resistance at the cracked section, the stress at other sections experiences unloading, and hence no further cracking occurs in other places. Fiber-reinforced concrete (FRC) has a less steep postpeak descending curve after first cracking (Fig. 9.41A), due to the bridging

(A) Concrete (B) ECC

Fig. 9.40 Cracking of concrete and ECC.
From Li, V. C. 2019. "Chapter 6—Resilience of engineered cementitious composites (ECC) structural members." *Engineered Cementitious Composites (ECC)*, Springer Nature.

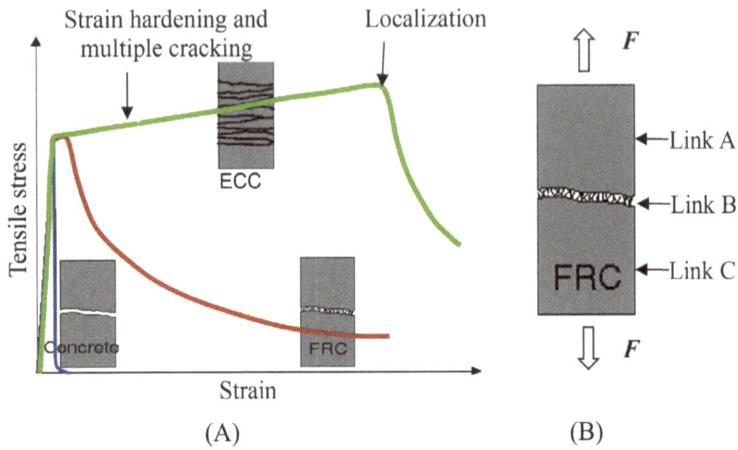

Fig. 9.41 Stress–strain behavior of cementitious materials: (A) typical stress–strain curves, and (B) links of failure.
(A) Based on Fischer, G. 2004. "Characterization of fiber-reinforced cement composites by their tensile stress-strain behavior and quantification of crack formation." *6th RILEM Symposium on Fiber-Reinforced Concretes (FRC)-BEFIB*.

effect of fibers that transfers some stress between the two sides of a crack. As the transferred stress between a crack is less than the cracking stress of the matrix in other sections of FRC, it cannot cause the cracking of the tension member in another section. Therefore, there is still one major crack only in FRC tension members. Although FRC is more ductile than plain concrete, it is still a quasi-brittle material with tension softening behavior (Fig. 9.41A).

Therefore, the crack localization of concrete is caused by the softening of materials, i.e., decaying resistance after the formation of cracking. Can concrete material be reinforced to have a hardening behavior so that failure localization is avoided? The answer is yes. The mechanism can be interpreted through the case of FRC with Fig. 9.41B and Fig. 1.9. When the tension force F increases to a certain value, the member first cracks at the weakest section, i.e., the weakest link in Fig. 1.9 fails first. When a crack occurs in the FRC tension member, the tensile resistance of the matrix diminishes, and the bridging resistance of fibers is mobilized as shown by Link B in Fig. 9.41B. The fiber bridging resistance is usually smaller than the cracking resistance of the matrix in FRC, i.e., the resistance of Link B is smaller than the resistance of Link A and Link C. Therefore, no new crack can be formed in FRC. However, if the fibers are designed in such a way that the resistance at Link B is greater than that at Link A or Link C, then new cracks will be formed in the tension member. As the first crack at Link B is the weakest link, the next crack must occur at a tension force greater than the cracking resistance of the first crack at Link B position. Therefore, the formation of a subsequent crack must require a larger force than that of the previous crack. This means that the load–deformation behavior must have hardening characteristics. In other words, a material characterized by the typical softening load–deformation curve of FRC as shown in Fig. 9.41A can only have one major crack.

Engineered cementitious composite (ECC) can have a hardening stress–strain curve (Fig. 9.41A). Therefore, the concentrated crack can be distributed to many fine cracks in ECC through fiber bridging (Fig. 9.40B). The bridging resistance of fibers in ECC at a crack is greater than the cracking resistance of the matrix, and, consequently, cracking can occur in another place (Fischer 2004). This cracking-and-bridging process continues to form more and more cracks with the continuous increase in the deformation. The increase in force between subsequent cracks reflects the difference in cracking resistance between subsequent cracks. Therefore, the slope of hardening of the stress–strain curve reflects the uniformity of material strength in the member. The fiber bridging resistance must be designed to be greater than the cracking resistance of the matrix in order to form a next crack.

The deformation of ECC is uniform on a macroscale (pseudostrain). No new crack can be formed when the maximum fiber bridging resistance has been activated. In this case, the Link B resistance is less than that at Link A and Link C in Fig. 9.41B. When this happens, crack formation stabilizes and further increase in deformation is predominantly accommodated by opening of existing cracks as opposed to formation of new cracks. The drop of resistance and softening of the material occurs when the bridging resistance of fibers at a particular cracked section starts to decrease. The subsequent further deformation comes from the opening of the softening section so that a localized crack similar to plain concrete and FRC occurs at this particular section, as shown by point F in Fig. 9.42.

The preceding discussion shows that crack localization disappears on two conditions: (1) the peak bridging stress from the fibers at the cracked section exceeds the cracking strength of the matrix, i.e., the applied stress prior to matrix cracking can be carried by the fibers after matrix cracking, and (2) crack propagation occurs at nearly constant ambient stress and constant crack opening in order to achieve a uniform

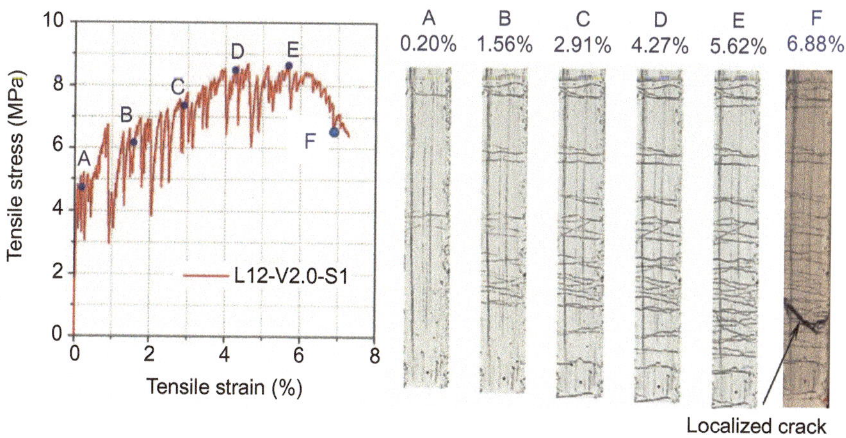

Fig. 9.42 Development of cracking in ECC.
From Huang, B. T., J. Q. Wu, J. Yu, J. G. Dai, C. K. Leung, and V. C. Li. 2021. "Seawater sea-sand engineered/strain-hardening cementitious composites (ECC/SHCC): assessment and modeling of crack characteristics." *Cem. Concr. Res.* 140, 106292.

cross-sectional stress distribution (Li and Leung 1992; Fischer 2004). Therefore, it can be concluded that the condition of crack localization in cementitious materials is that the load–deformation curve softens after cracking. Appropriate use of fibers can change the softening behavior to a hardening one and mitigate the failure localization.

9.12.3 Plastic hinge failure

The plastic hinge failure (Fig. 9.37B and Section 7.3) in RC members is also a localized failure. Based on the same principle as discussed in Sections 9.12.1 and 9.12.2, measures can also be taken to provide a hardening behavior in the plastic hinge region of reinforced concrete members so that the localization of failure is distributed more uniformly to a larger region.

The hardening of the plastic hinge zone can be achieved in different ways. Since softening of the plastic hinge of RC columns is generally caused by concrete crushing, measures can be taken to prevent concrete crushing to avoid excessive degradation of the plastic hinge. Two approaches are shown in Fig. 9.43. The steel plating on the compression face of RC columns (Fig. 9.43A) can hinder concrete crushing and provide a hardening type of the moment–curvature response (load–displacement curve may soften due to the P-Δ effect; see Section 7.3.5). Consequently, the plastic hinge failure can be avoided or delayed by the steel plating, as shown in Fig. 9.43A (Wu et al. 2003). Another effective way to provide hardening in the moment–curvature behavior is by external jacketing of RC columns with FRP, as shown in Fig. 9.43B. Detailed

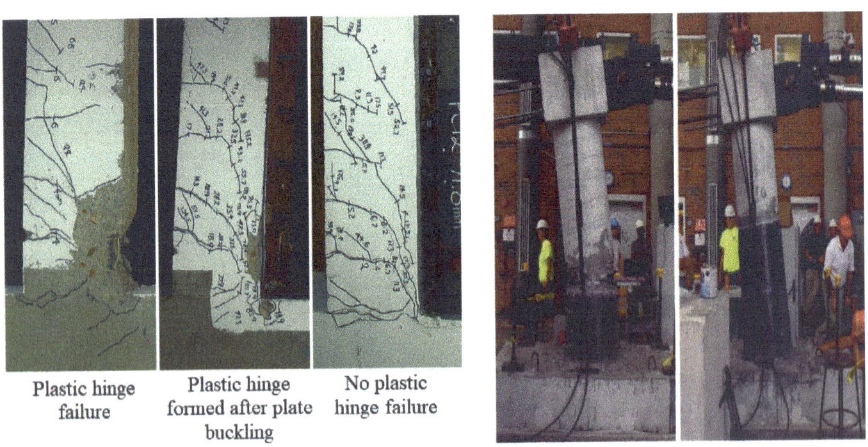

Plastic hinge failure | Plastic hinge formed after plate buckling | No plastic hinge failure

(A) (B)

Fig. 9.43 Methods for mitigation of plastic hinge failure: (A) steel plated RC columns, and (B) FRP jacketed columns.
(A) From Wu, Y. F., M. C. Griffith, and D. J. Oehlers. 2003. "Improving the strength and ductility of rectangular reinforced concrete columns through composite partial interaction: tests." *J. Struct. Eng.* 129 (9), 1183–1190, and (B) from Rutledge, S. T., M. J. Kowalsky, R. Seracino, and J. M. Nau. 2014. "Repair of reinforced concrete bridge columns containing buckled and fractured reinforcement by plastic hinge relocation." *J. Bridg. Eng.* 19 (8), A4013001.

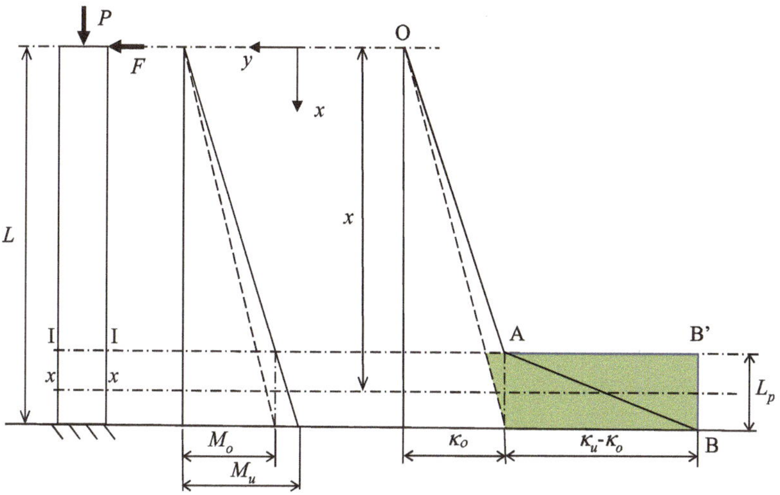

Fig. 9.44 Uniform curvature distribution in plastic hinge of RC column.

study on FRP jacketing to mitigate plastic hinge failure is presented, and a new technique to spread the curvature localization zone to a larger region is developed and introduced in this section.

For the cantilever column in Fig. 9.44, assume the RC cross section is uniform throughout its length. The typical curvature distribution of the original RC columns without FRP confinement is illustrated in Fig. 7.13, which shows significant nonuniformity and localization of curvature in the plastic hinge zone. The significant localization of failure in the plastic hinge zone is due to the softening of the moment–curvature curve as shown in Fig. 7.14A. FRP jacketing in the plastic hinge zone can confine concrete and prevent the concrete in the plastic hinge zone from softening. Consequently, it can significantly increase the deformation capacity of the plastic hinge, and also uniformize the curvature distribution in the plastic hinge region.

The typical moment–curvature curves of FRP-confined RC columns are shown in Fig. 9.45A (Abdallah et al. 2018; Olabi et al. 2020). For simplicity of modeling, the moment–curvature relationship is represented by the bilinear curve in Fig. 9.45B. If uniform confinement is provided to the column along the length, the moment–curvature curve of all cross sections are the same as shown by the line O-A-B in Fig. 9.45B. When the force at the top of the column increases to the ultimate load F_u, the corresponding ultimate moment M_u at the bottom section is given by Eq. (9.24) (the P-Δ effect is ignored for simplicity of discussion):

$$M_u = F_u \cdot L \tag{9.24}$$

The ultimate curvature corresponding to M_u is κ_u based on Fig. 9.45B. The bending moment at section I—I just reaches the yield moment M_o (Fig. 9.44B). The curvature distribution corresponding to this load is shown by the line O-A-B in Fig. 9.44C based on the assumed moment–curvature relationship in Fig. 9.45B. Although this curvature

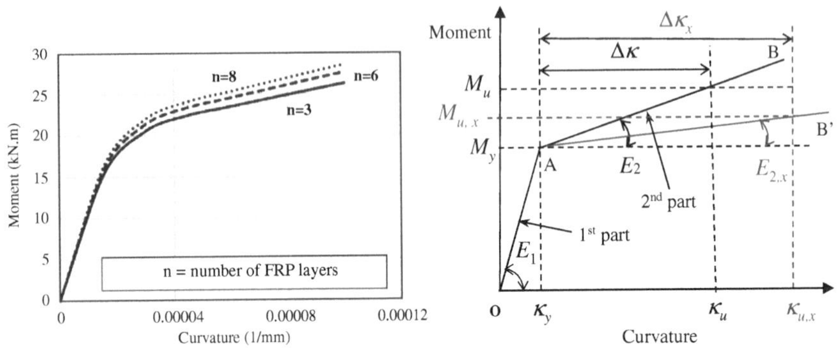

(A) (B)

Fig. 9.45 Moment–curvature relationship of FRP-confined RC columns: (A) analytical results, and (B) Idealized model.
(A) From Abdallah, M. H., H. M. Mohamed, R. Masmoudi, and A. Moussa. 2018. "Analytical modeling of moment-curvature behavior of steel and CFRP RC circular confined columns." *Compos. Struct.* 189, 473–487.

distribution still concentrates in the bottom part of the column, concrete crushing can be prevented in the FRP-confined region of the column (Fig. 9.43B). Therefore, the FRP confinement can prevent the failure localization.

Although the FRP confinement in the plastic hinge zone can prevent local concrete crushing in the plastic hinge zone, strong confinement may reduce the plastic hinge length, as discussed in Section 7.3.4.6 (Fig. 7.32), because the strong confinement restrains the spread of rebar yielding in the plastic hinge zone. This would intensify the curvature concentration in the plastic hinge zone. Such concentration of curvature can be eased by providing nonuniform confinement. If nonuniform confinement is provided so that the FRP rigidity $E_f t_f$ gradually reduces from the bottom toward the column top, the moment–curvature curve will be different at different cross sections. Assume the moment–curvature curve at the section x-x is represented by O-A-B′ in Fig. 9.45B. When the bottom section reaches its ultimate moment M_u, the corresponding bending moment at section x-x is (Fig. 9.44):

$$M_{u,x} = F_u \cdot x = \frac{x \cdot M_u}{L} \tag{9.25}$$

From Fig. 9.45B, it can be seen that the curvature increment from the yield moment at the bottom section and section x-x is $\Delta\kappa$ and $\Delta\kappa_x$, respectively, and $\Delta\kappa_x > \Delta\kappa$.

It is therefore theoretically possible to design the distribution of $E_f t_f$ along the column length to adjust the hardening slope of the 2nd part of the moment–curvature curve, $E_{2,x}$, so that the curvature corresponding to the ultimate load at section x-x is equal to that at the bottom section, i.e.,

$$\kappa_u = \kappa_{u,x} \tag{9.26}$$

In this case, the curvature distribution in the plastic hinge zone will be uniform at the ultimate load, so that the same curvature is achieved in the full plastic hinge length from the bottom of the column to the section I—I, as shown in Fig. 9.44C. The plastic hinge length in this case is

$$L_p = \frac{M_u - M_0}{F_u} = L\frac{M_u - M_0}{M_u} \tag{9.27}$$

A simple way to provide the variable confinement is to reduce $E_{2,x}$ linearly toward the top of the column. In this case, the additional curvature $\Delta\kappa_x$ after yielding due to incremental loading ΔF is equal in all the cross sections in the plastic hinge region. When the load at the top of the column F is increased to $F + \mathrm{d}F$, the incremental moment at a given cross section is

$$dM(x) = x \cdot dF \tag{9.28}$$

As the hardening slope of the moment–curvature curve at location x is $E_{2,x}$, from Fig. 9.45B one has:

$$dM(x) = E_{2,x} \cdot d\kappa \tag{9.29}$$

Substituting Eq. (9.29) into Eq. (9.28) gives

$$\frac{d\kappa(x)}{dF} = \frac{x}{E_{2,x}} \tag{9.30}$$

If the confinement is provided so that the hardening slope increases linearly toward the column base, then

$$E_{2,x} = \frac{x}{L}E_2 \tag{9.31}$$

Substituting Eq. (9.31) into Eq. (9.30) leads to

$$\frac{d\kappa(x)}{dF} = \frac{L}{E_2} \tag{9.32}$$

Eq. (9.32) means that the curvature at different cross sections in the plastic hinge region increases at the same rate when the lateral force increases.

The preceding discussion is based on the assumed moment–curvature relationship in Fig. 9.45B. As pointed out at the beginning of Section 7.3, strictly speaking, the application of one typical moment–curvature curve to all cross sections of a uniform RC member is not suitable in the plastic hinge zone. It is difficult to get a uniform curvature distribution in the full plastic hinge length in reality. Nevertheless, the mechanism discussed here is still valid and a gradual reduction in confinement from column base toward the top would help the spread of yielding upward and reduce the concentration of curvature at the bottom of the column. Therefore, a more uniform curvature distribution can be obtained in the plastic hinge zone through this approach. This is demonstrated by the numerical example in Fig. 9.46.

The RC column in Fig. 9.46A is confined in the plastic hinge zone by the FRP jacket with different variations of FRP thickness, as shown in Fig. 9.46C–F. The load–deformation responses of the column with different confinements are shown in Fig. 9.46B. It is noted that the moment–curvature curves for the bottom section of the columns relevant to Fig. 9.46B have a hardening slope, i.e., $E_2 > 0$, although the load–displacement curves in Fig. 9.46B have a softening part after the peak point due to the P-Δ effect. The curvature distributions in the plastic hinge zone of the columns are shown in Fig. 9.47. It can be seen in Fig. 9.47 that less curvature concentration and a more uniform curvature distribution can be achieved by using the nonuniform confinement scheme in Fig. 9.46F. Due to the more uniform distribution of curvature in Fig. 9.47D, the curvature demand on the column is significantly reduced, because the area enclosed by the curvature distribution curves for different confinement schemes are similar at the same column displacement, as discussed in Section 7.3. The smaller curvature demand reduces the damage to the column and increases the survivability of the structure under extreme events such as earthquakes.

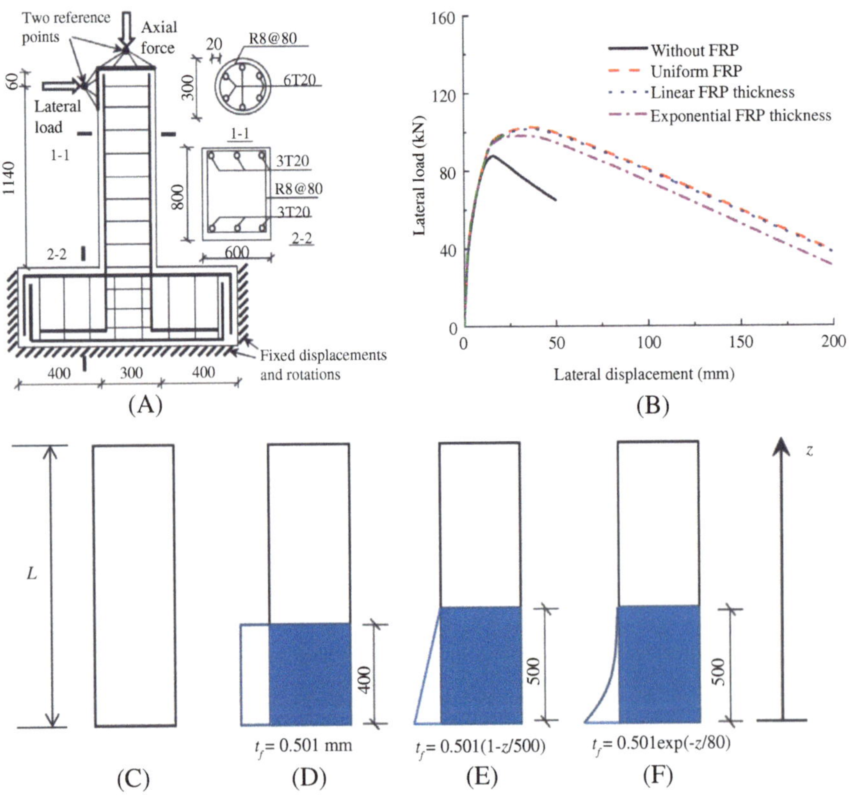

Fig. 9.46 Numerical example of columns with nonuniform confinement: (A) RC column details; (B) load–deformation curves; (C) unconfined column; (D) uniform FRP thickness; (E) linear FRP thickness; and (F) exponential FRP thickness.

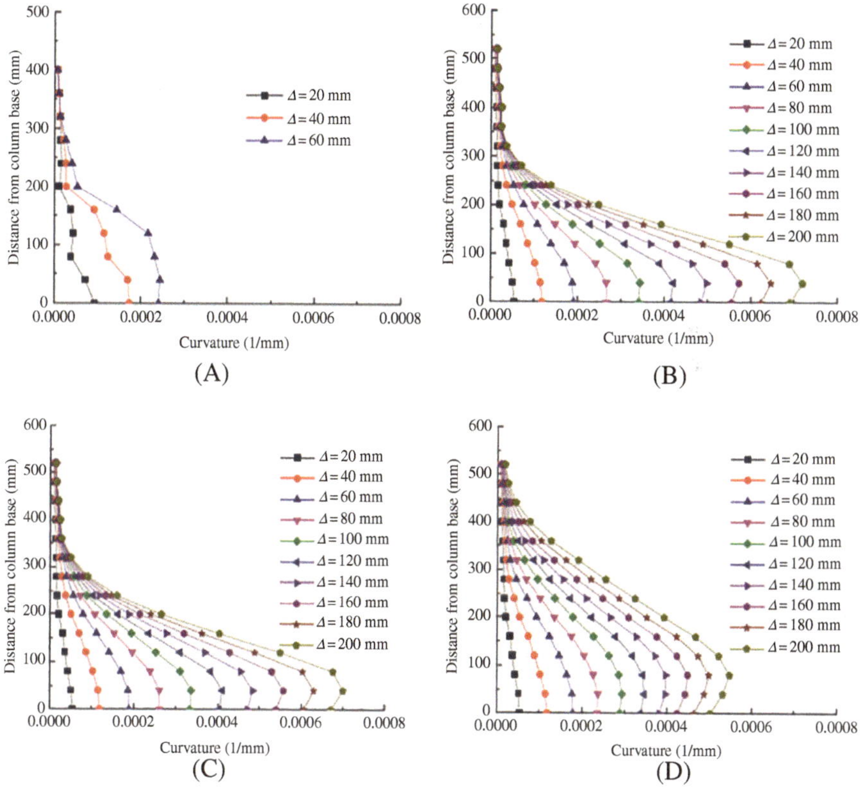

Fig. 9.47 Curvature distribution in plastic hinge zone: (A) unconfined column; (B) uniform FRP thickness; (C) linear FRP thickness; and (D) exponential FRP thickness.

It is noted that, to achieve more uniform curvature distribution in the plastic hinge zone, the FRP strain capacity is required to be significantly larger than that for uniformly confined columns, as shown in Fig. 9.48. This is because the reduced confinement above the column base causes more significant dilation of concrete, which requires the confining material to have a higher strain capacity. For the case of the uniform jacket thickness in Fig. 9.48A, carbon fiber-reinforced polymer (CFRP) is suitable as the maximum CFRP strain is less than 1.5%. However, CFRP is not suitable for the case of the nonuniform FRP thickness in Fig. 9.48B, which requires a confining material with a strain capacity of about 6%. Glass fiber-reinforced polymer (GFRP) and other high strain capacity FRP materials are suitable in this case.

9.12.4 Rebar necking

Necking of steel bars (Fig. 9.37D) is caused by the softening of its tensile load–deformation curve. Therefore, it can also be mitigated if its tensile resistance is augmented to avoid or delay the softening. One approach to augment the tensile resistance of steel bars is developed and introduced in this section. The necking suppression mechanism in Fig. 9.49 makes use of FRP strands that are spirally wound

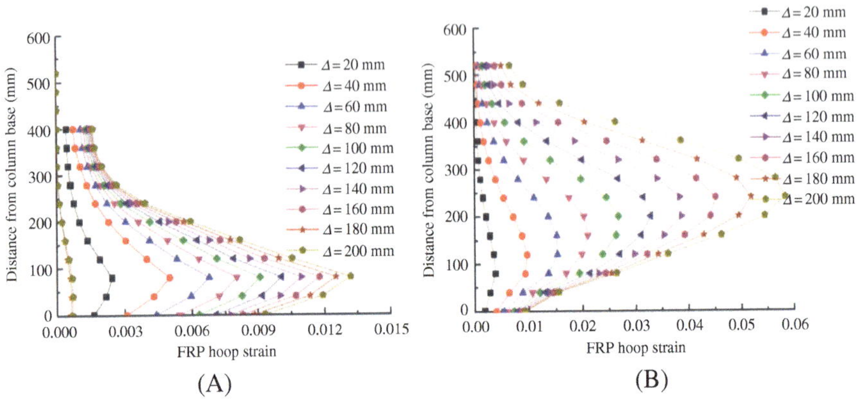

Fig. 9.48 FRP strain distributions in plastic hinge zone: (A) uniform FRP thickness; and (B) exponential FRP thickness.

about and bonded to the metal bar. When necking occurs in a place, the significantly larger local strain in the necking region will mobilize a larger resistance of the FRP strands than in other places. Therefore, it is possible that the combined total resistance of the composite bar in the necking region has a hardening behavior so that it is no longer the weakest section. In this way, the weakest section is shifted to the 2nd weakest section, and later on from the 2nd weakest section to the 3rd one in the same way, and so on and so forth. In this way, multiple necking or no necking would occur in the composite bar.

The behavior of the proposed necking mitigation system is numerically simulated in Fig. 9.50. Comparing Fig. 9.50A with Fig. 9.50B, it can be seen that the localization is significantly reduced by the use of the FRP strands in the composite bar in the

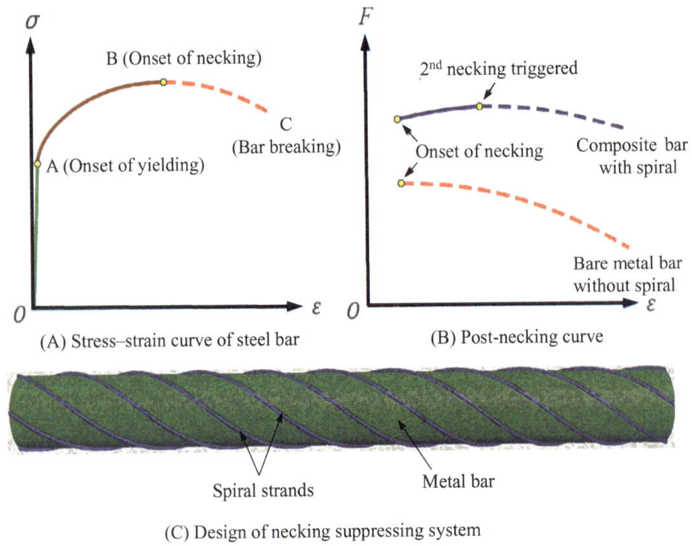

Fig. 9.49 Necking suppression mechanism.

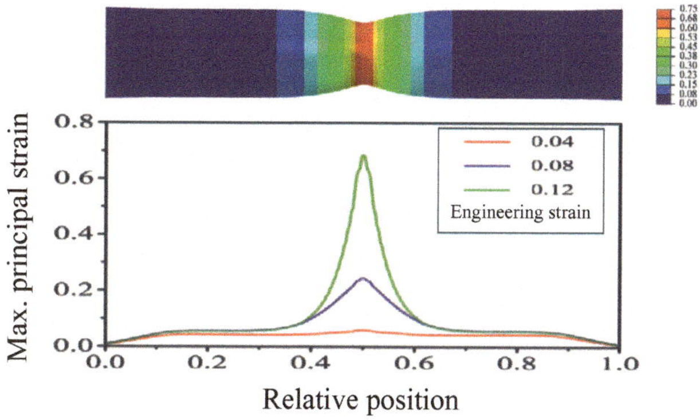

(A) Strain distribution of bare bar

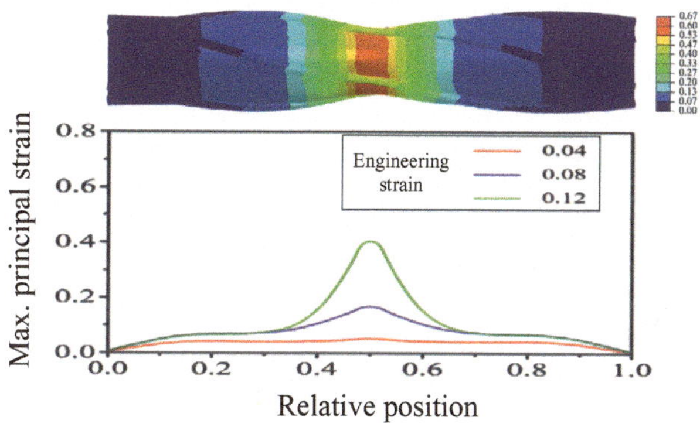

(B) Strain distribution of composite bar

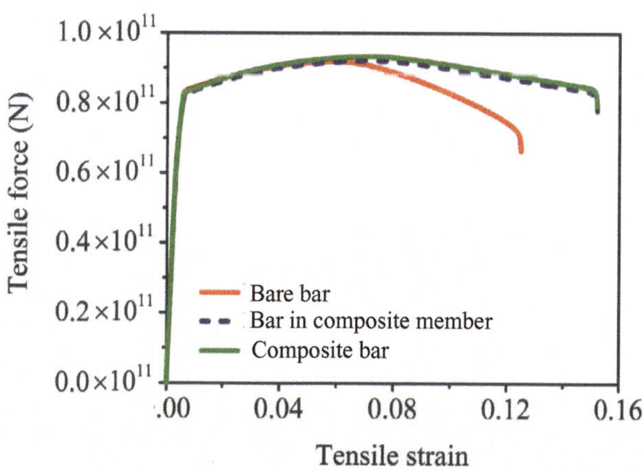

(C) Load-deformation curve

Fig. 9.50 Numerical results of strain concentration of bars.

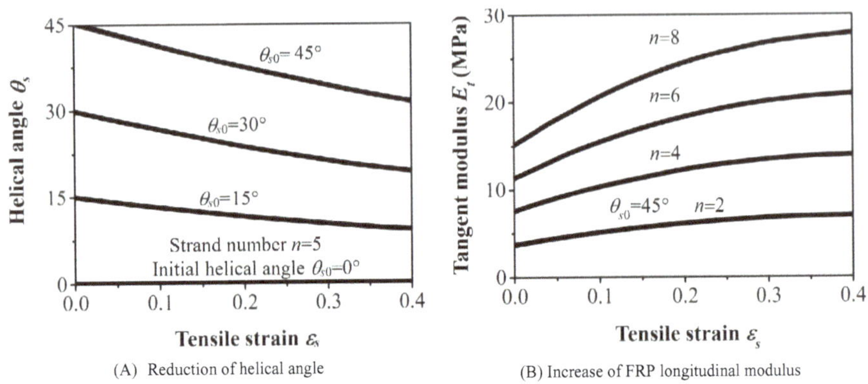

Fig. 9.51 Hardening of FRP strands.

simulated case. More details of the numerical simulation can be found in Wu et al. (2022). The mechanism of the mitigation is shown in Fig. 9.51. The helical angle of the strands, θ_s, reduces when the local tensile strain of the bar increases (Fig. 9.51A). This is because the strands become straighter when the bar is stretched to have a smaller diameter, particularly in the necking zone. A straighter FRP strand results in a greater longitudinal stiffness E_t (Fig. 9.51B). In a necking zone, the diameter of the bar is smaller than that in other places, which produces harder/stiffer FRP strands. It is this hardening resistance of the FRP strands that offsets the softening of the steel bar in the necking zone, and consequently reduces the strain localization. Weak FRP strands are used for the case in Fig. 9.50 to demonstrate that even a small reinforcement by the FRP strands can significantly mitigate the failure localization. When stronger reinforcement is provided, multiple instances of necking can be formed in different places, or even no localization occurs throughout the length.

The preceding strain localization mitigation scheme was experimentally tested. The test specimens are shown in Fig. 9.52 and the test results are shown in Fig. 9.53. Fig. 9.53A compares the tensile load vs. engineering strain curves between the bare bar and the FRP strand reinforced composite bar. As expected, the strand reinforcement

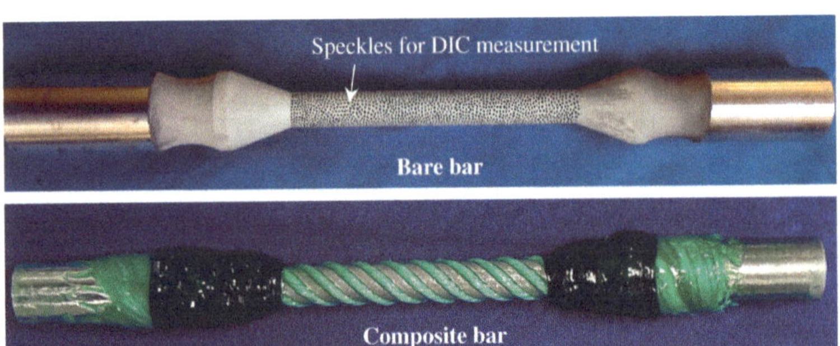

Fig. 9.52 Test specimens.

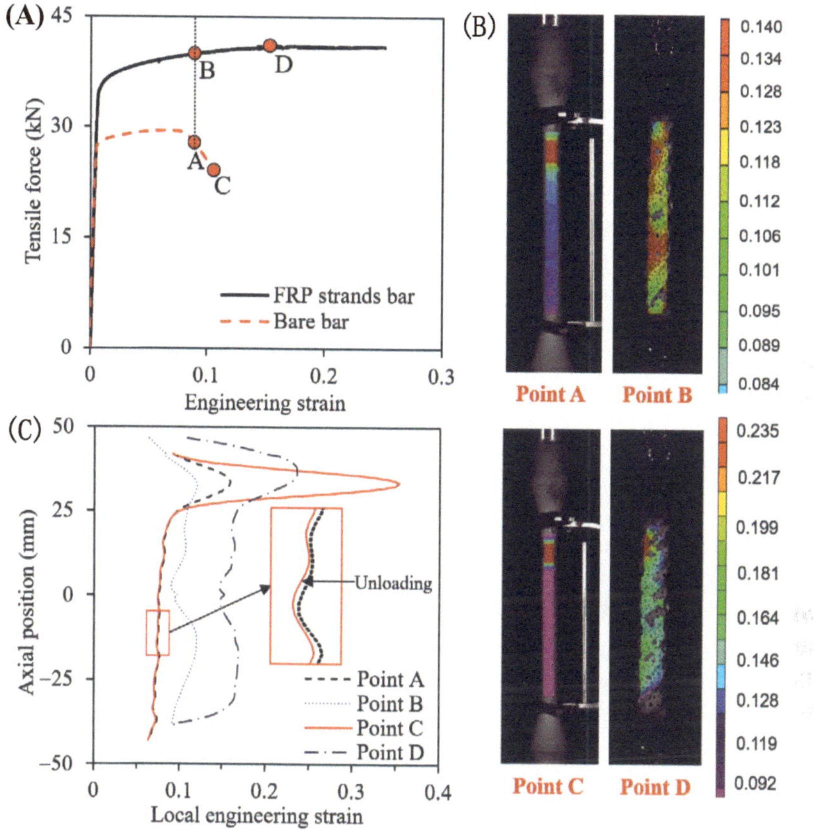

Fig. 9.53 Test results of bars: (A) load–deformation curves; (B) strain distribution patterns; and (C) strain distribution curves.

significantly increased the deformation capacity of the steel bar. The strain fields of the bars were measured by a three-dimensional digital image correlation (3D-DIC) system. Fig. 9.53C shows the longitudinal strain distribution of the bars at different loading points. It can be seen from the figure that the bare bar has one localized strain distribution and failure zone. The composite bar has a more uniform strain distribution and necks at two positions (Fig. 9.53B and C). If a longer bar were used in the test, it is expected that a third, fourth, and more necks would occur one by one with further increase in the deformation of the bar. Although both of the bars fractured at the necking zone, the longitudinal strains of the composite bar outside the necking zone were significantly larger than those of the bare bar. These increased strains contributed to the significant increase in the deformation capacity of the FRP composite bar (Fig. 9.53A). Details of the tests can be found in Wu et al. (2022).

Necking of the bar is caused by the softening of the load–deformation curve. The helical strands can be designed so that the composite bar has a hardening load–deformation behavior. When no softening of the composite bar occurs, no necking occurs theoretically. The total tensile resistance F of the composite bar at a given cross section is contributed by both the FRP strands and the bare bar, or

$$F = \cos\theta_s \times A_f \times E_f \times \varepsilon_f + A_s \times \sigma_s(\varepsilon_s) \tag{9.33}$$

where A_f and E_f are the area and the elastic modulus of the FRP strands, respectively; ε_f is the tensile strain of an FRP strand along its winding direction; A_s is the instantaneous cross-sectional area of the bare bar; and $\sigma_s(\varepsilon_s)$ is the local stress of the bare bar at the local longitudinal strain ε_s. To avoid softening of the load–deformation curve, the following condition has to be satisfied:

$$\frac{dF}{d\varepsilon_s} \geq 0 \tag{9.34}$$

The typical curve of $dF/d\varepsilon_s$ of the composite bar is shown in Fig. 9.54. The curve has a minimum at $\varepsilon_s = \varepsilon_{s,\ min}$ and reaches its posttrough maximum $\varepsilon_s = \varepsilon_{s,\ max}$ where $\varepsilon_{s,\ max}$ is the maximum local strain of the bare bar at its rupture. The values of $\varepsilon_{s,\ min}$ are fixed for a given composite bar design and are given in Wu et al. (2022). Based on Eq. (9.34) and Fig. 9.54, the condition of necking can be categorized as follows:

1. No necking—when the lowest point is a positive value, i.e., $dF/d\varepsilon_s \mid \varepsilon_s = \varepsilon_{s,\ min} \geq 0$, the composite bar has a hardening response curve throughout the whole range of strain. No necking occurs in this case.
2. Multiple necking—when the lowest point is a negative value but it changes to a positive value later, i.e., $dF/d\varepsilon_s \mid \varepsilon_s = \varepsilon_{s,\ min} < 0$ and $dF/d\varepsilon_s \mid \varepsilon_s = \varepsilon_{s,\ max} \geq 0$, necking occurs during the deformation stage when $dF/d\varepsilon_s < 0$. However, further necking stops when $dF/d\varepsilon_s$ changes to a positive value. When $dF/d\varepsilon_s$ increases to a sufficiently large value, necking takes place in another region. This is exactly the case for the test specimen shown in Fig. 9.53.
3. One extended necking—when $dF/d\varepsilon_s \mid \varepsilon_s = \varepsilon_{s,\ min} < 0$ and $dF/d\varepsilon_s \mid \varepsilon_s = \varepsilon_{s,\ max} < 0$. In this case, the helical strands are not strong enough to shift the necking to another place. However,

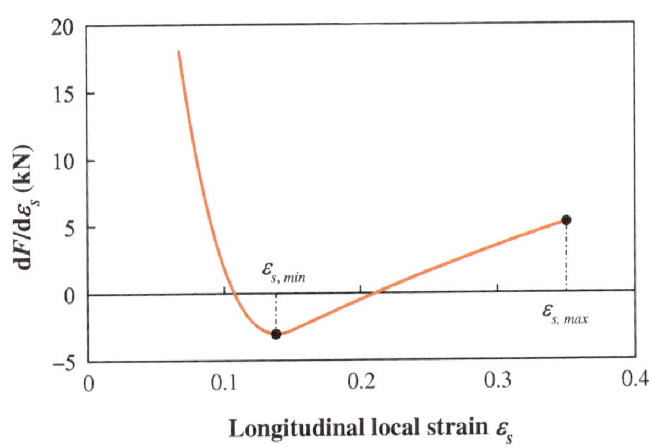

Fig. 9.54 Necking criteria.

the provision of strands still helps to strengthen the necking zone so that the softening of the tensile resistance is slowed down and the necking zone can be extended. This is the case in Fig. 9.50, where the strands extended the necking zone.

9.12.5 General rule for mitigation of failure localization

Strain localization is the failure mode of many materials and structures. Section 9.12 introduces several approaches to suppress failure localization, thereby increasing the deformation capacity and ductility of materials or structures. The mechanism underlying the approaches is the same: to provide a hardening behavior of the member or material so that when the strain at the weakest section tends to localize, its locally larger strain and postyield hardening load–deformation characteristics result in a larger resistance, which causes an increased force applied to the member. When the applied force exceeds the maximum resistance of the next weakest section, failure localization moves to this section. This process continues, resulting in either multiple failure localizations or uniform deformation (nearly uniform in reality) throughout the member, depending on the value of hardening slope (Eq. 9.34). This localization mitigation mechanism mobilizes larger strain throughout the whole length of the member and results in significant increase of the deformation capacity and ductility of the member/material.

For uniform materials or members with a uniform cross section and material property, the rule for mitigation of failure localization is given by Eq. (9.34) and the three conditions following the equation. In reality, a member is not uniform, as illustrated by the tensile resistance of the RC member in Fig. 5.18. In this case, the reinforcing scheme must be such that when local failure occurs at Section A in Fig. 5.18, the hardening reinforcement must be able to provide an additional strength so that the new strength of Section A is greater than the resistance at Section B to cause the local failure at Section B. If the reinforcing system can provide further strengthening of Sections A and B to increase their resistance to a value that is greater than the resistance of Section C, failure localization will occur at Section C. This process continues so that many cross sections fail/soften if the reinforcing system is properly designed.

In fact, the rebar inside the tension member in Fig. 5.18 is another failure localization mitigation scheme. The effect of the rebar is similar to the fibers inside ECC in Section 9.12.2. The criterion of failure transfer from one place to another for all failure localization mitigation schemes is the same, i.e., Eq. (9.34). For the case in Section 9.12.1, F in Eq. (9.34) is the axial compression force. For concrete cracking in Section 9.12.2, F is the tensile force. For plastic hinge failure in Section 9.12.3, F in Eq. (9.34) is the bending moment resistance and ε_s is the curvature. For the case of the RC tension member in Fig. 5.18, F is given by Eq. (9.35)

$$F = R_c + R_s \tag{9.35}$$

where R_c and R_s are the resistance from the concrete and reinforcement, respectively. In cracked sections, $R_c = 0$ and the tensile resistance is only given by the reinforcement. R_s is closely related to the bond between the reinforcement and the substrate concrete. Therefore, the bond controls the cracking patterns in the tension members as discussed in Section 5.8.1; so is ECC.

Therefore, the previous principle for mitigation of failure localization is general. Based on the principle, other failure localization mitigation schemes can be developed and designed.

References

Abdallah, M. H., H. M. Mohamed, R. Masmoudi, and A. Moussa. 2018. "Analytical modeling of moment-curvature behavior of steel and CFRP RC circular confined columns." *Compos. Struct.* 189, 473–487.

ACI Committee 318. 2014. *Building code requirements for structural concrete and commentary*, ACI 318-14. Farmington Hills, MI, USA: American Concrete Institute.

ACI Committee 440. 2015. *Guide for the design and construction of structural concrete reinforced with fiber-reinforced polymer (FRP) bars*, ACI 440.1R15. Farmington Hills, MI, USA: American Concrete Institute.

Cui, S. 2012. "Integrated design methodology for isolated floor systems in single-degree-of-freedom structural fuse systems." PhD thesis, State University of New York at Buffalo.

El-Bahey, S., and M. Bruneau. 2011. "Buckling restrained braces as structural fuses for the seismic retrofit of reinforced concrete bridge bents." *Eng. Struct.* 33, 1052–1061.

Fischer, G. 2004. "Characterization of fiber-reinforced cement composites by their tensile stress-strain behavior and quantification of crack formation." In *6th RILEM Symposium on Fiber-Reinforced Concretes (FRC)-BEFIB*.

Galambos, T. V., B. Ellingwood, J. G. MacGregor, and C. A. Cornell. 1982. "Probability based load criteria: assessment of current design practice." *J. Struct. Div.* 108, 959–977.

Homrich, J. R., and A. E. Naaman. 1987. "Stress-strain properties of SIFCON in compression.". In *Fiber reinforced concrete—properties and applications, ACI SP-105*. Detroit: American Concrete Institute, pp. 283–304.

Huang, X., J. Chen, and H. Zhu. 2016. "Assessing small failure probabilities by AK–SS: an active learning method combining Kriging and subset simulation." *Struct. Saf.* 59, 86–95.

Jiang, J. F., and Y. F. Wu. 2016. "Plasticity-based criterion for confinement design of FRP jacketed concrete columns." *Mater. Struct.* 49 (6), 2035–2051.

Kauffman, A., and A. M. Memari. 2014. "Performance evaluation of different masonry infill walls with structural fuse elements based on in-plane cyclic load testing." *Buildings* 4, 605–634.

Kazmi, S. M. S., M. J. Munir, and Y. F. Wu. 2021. "Application of waste tire rubber and recycled aggregates in concrete products: a new compression casting approach." *Resour. Conserv. Recycl.* 167, 105353.

Lam, L., and J. G. Teng. 2003. "Design-oriented stress-strain model for FRP-confined concrete in rectangular columns." *J. Reinf. Plast. Compos.* 22 (13), 1149–1186.

Li, V. C., and C. K. Y. Leung. 1992. "Steady-state and multiple cracking of short random fiber composites." *J. Eng. Mech.* 118 (11), 2246–2263.

Li, G., Y. Jiang, S. Z. Zhang, Y. Zeng, and Q. Li. 2015. "Seismic design or retrofit of buildings with metallic structural fuses by the damage-reduction spectrum." *Earthq. Eng. Eng. Vib.* 14, 85–96.

Mander, J. B., M. J. N. Priestley, and R. Park. 1988. "Observed stress strain behavior of confined concrete." *J. Struct. Eng.* 114 (8), 1827–1849.

MOHURD (Ministry of Housing and Urban-Rural Development of the People's Republic of China). 2001. *GB50068-2001: unified standard for reliability design of building structures*. Beijing, China: China Building Industry Press (in Chinese).

Naaman, A. E. 2003. "FRP reinforcements in structural concrete: assessment, progress and prospects." In *Proceedings of the Sixth International Symposium on FRP Reinforcement for Concrete Structures*, Singapore: World Scientific, pp. 1–24.

Nematzadeh, M., and M. Naghipour. 2012. "Compressive strength and modulus of elasticity of freshly compressed concrete." *Construct. Build Mater.* 34, 476–485.

Nematzadeh, M., A. Salari, J. Ghadami, and M. Naghipour. 2016. "Stress-strain behavior of freshly compressed concrete under axial compression with a practical equation." *Construct. Build Mater.* 115, 402–423.

Nowak, A. S., and K. R. Collins. 2000. *Reliability of structures*. McGraw-Hill.

Olabi, M. N., N. Caglar, and N. G. Koroglu. 2020. "Numerical investigation for moment-curvature relationship of CFRP jacketed RC columns." *Acad. Platform J. Eng. Sci.* 8 (1), 102–111.

Paulay, T., and M. J. N. Priestley. 1992. *Seismic design of reinforced concrete and masonry buildings*. New York: John Wiley and Sons, Inc.

Rubinstein, R. Y., and D. P. Kroese. 2016. *Simulation and the Monte Carlo method*. John Wiley & Sons.

Shoeibi, S., M. A. Kafi, and M. Gholhaki. 2017. "New performance-based seismic design method for structures with structural fuse system." *Eng. Struct.* 132, 745–760.

Vargas, R., and M. Bruneau. 2009. "Experimental response of buildings designed with metallic structural fuses. II." *J. Struct. Eng.* 135 (4), 394–403.

Wu, Y. F. 2006. "New avenue of achieving ductility for reinforced concrete members." *J. Struct. Eng.* 132 (9), 1502–1506.

Wu, Y. F. 2008. "Ductility demand of compression yielding fiber-reinforced polymer-reinforced concrete beams." *ACI Struct. J.* 105 (1), 104–110.

Wu, Y. F., and Y. W. Zhou. 2011. "Controlling the damage of concrete columns through compression yielding." *Struct. Control Health Monit.* 18 (8), 890–907.

Wu, Y. F., M. C. Griffith, and D. J. Oehlers. 2003. "Improving the strength and ductility of rectangular reinforced concrete columns through composite partial interaction: tests." *J. Struct. Eng.* 129 (9), 1183–1190.

Wu, Y. F., J. F. Jiang, and K. Liu. 2010a. "Perforated SIFCON blocks—an extraordinarily ductile material ideal for use in compression yielding structural systems." *Construct. Build Mater.* 24 (12), 2454–2465.

Wu, Y. F., Y. W. Zhou, and X. Q. He. 2010b. "Performance-based optimal design of compression-yielding FRP-reinforced concrete beams." *Compos. Struct.* 93 (1), 113–123.

Wu, Y. F., S. M. S. Kazmi, M. J. Munir, Y. Zhou, and F. Xing. 2020. "Effect of compression casting method on the compressive strength, elastic modulus and microstructure of rubber concrete." *J. Clean. Prod.* 264, 121746.

Wu, Y. F., P. D. Li, and Z. L. Zhao. 2022. *Ultraductile bar with bioinspired helical strands.* tentatively accepted by J. Struct. Eng.

Zhou, Y. W., Y. F. Wu, J. G. Teng, and A. Y. T. Leung. 2009. "Ductility analysis of compression-yielding FRP-reinforced concrete beams." *Cem. Concr. Compos.* 31 (9), 682–691.

Zhou, Y. W., Y. F. Wu, J. G. Teng, and A. Y. T. Leung. 2010. "Parametric space for the optimal design of compression-yielding FRP-reinforced concrete beams." *Mater. Struct.* 43 (1–2), 81–97.

Shear failure of RC members

10

10.1 Introduction

Failure of an RC member by diagonal cracking is called shear failure. Shear failure is mainly caused by shear force in combination with bending moment. The existence of shear force and diagonal cracking invalidates the plane section condition (Bernoulli's hypothesis) and causes great difficulty in evaluating the stress and strain in the failure zone. Therefore, the shear failure mechanism is much more complicated than that of flexural failure. The shear behavior of RC members is a classical but challenging problem that has attracted enormous efforts and continuous interest of the research community over the past century. Despite significant progress in understanding and modeling the problem, it is still regarded as one of the least understood but most important problems in reinforced concrete. A more rational and accurate, yet simple and understandable, method based on first principles of mechanics, such as that for flexural design of RC members, has been continuously sought. However, it remains a daunting task and major challenge to the research and professional communities to achieve this goal. This chapter introduces the basic mechanism of shear transfer and its failure mode as well as typical modeling approaches for the shear strength of RC members. Evaluation of existing theories and models shows the need for further and more rational development of the shear design approach.

10.2 Shear failure process and failure modes

As discussed in Chapter 1, the shear failure is signified by diagonal shear cracking (Fig. 1.6D and E). Cracking of concrete occurs when the principal tensile stress exceeds the concrete tensile strength. The magnitude and direction of the principal tensile stress in a beam are affected by both the shear stress due to shear force and the normal stress caused by bending moment and axial force, as shown in Fig. 10.1. Cracking develops along a compression trajectory (see Figs. 1.6D & E and 1.2A) because concrete cracks in the direction perpendicular to the principal tensile stress and the compression trajectory is perpendicular to the principal tensile stress direction.

In a flexural dominant region such as the pure bending zone and the shear-bending zone (Fig. 1.3A), cracking occurs first at the bottom of the beam perpendicular to the beam axis, because the maximum principal tensile stress occurs there first. In the shear dominant region (Fig. 1.3A), the maximum principal tensile stress occurs first at the neutral axis position with an angle of 45 degrees to the longitudinal axis, as shown in Fig. 10.1. Based on strength of materials, the shear stress v in an elastic beam is given by

Analytical Approaches for Reinforced Concrete. https://doi.org/10.1016/B978-0-12-821164-9.00010-3

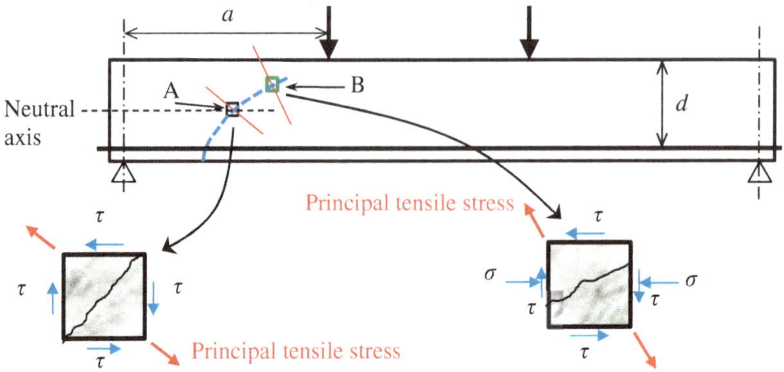

Fig. 10.1 Diagonal tensile cracking.

$$v = \frac{VQ}{Ib} \tag{10.1}$$

where V is the shear force at a given cross section (Fig. 1.1A); b is the breadth of beam cross section; I is the moment of inertia; and Q is the first moment of cross section, which varies along the depth of the beam section and takes the maximum value at the neutral axis. Therefore, the maximum shear stress occurs at the neutral axis where the normal stress is zero (location A in Fig. 10.1). For a shear dominant section where the normal stress due to bending is relatively small compared with the shear stress, cracking does not initiate at the bottom of the beam; rather the maximum principal tensile stress occurs at the neutral axis position and with an angle of 45 degrees, as shown by point A in Fig. 10.1. In the early days, diagonal cracking was considered dangerous and not allowed in design. However, it was later realized that diagonal cracking under service loads cannot be completely avoided and is actually acceptable, provided the crack width is limited in the same way as for flexural cracking.

For RC beams with a flexural strength greater than the shear strength, shear failure will occur when the shear force exceeds the shear strength. The shear strength of an RC member relies heavily on the tensile and compressive strength of concrete material. Therefore, the shear failure process is in general nonductile. For this reason and as discussed in Sections 1.4.2 and 1.5, design codes of practice ensure that the shear failure of RC members does not occur in a structure by providing a higher safety factor for shear failure than that for flexural and other more ductile failure modes.

Shear failure has different modes. Both the failure mode and shear strength are closely related to an important factor, the shear span-to-effective depth ratio a/d, as shown in Fig. 10.1. As a/d is only applicable to members under point loads, this ratio can be rewritten in a more general form as:

$$\frac{a}{d} = \frac{Va}{Vd} = \frac{M}{Vd} \tag{10.2}$$

where M and V are the bending moment and shear force at a particular cross section. Therefore, $\frac{M}{Vd}$ can be used to replace $\frac{a}{d}$ for evaluation of the shear strength of a particular cross section for members under general loading.

Based on the phenomena of shear failure, people divide shear failure into the following categories (ACI-ASCE Committee 326 1962; ASCE-ACI Committee 426 1973; Park and Paulay 1975):

(1) Shear-tension failure without web reinforcement

For slender RC beams without (or with insignificant) web (transverse) reinforcement, diagonal cracking may occur in the web (the red cracks in Fig. 1.3A) after the formation of flexural cracks (the black cracks in Fig. 1.3A) and flexural-shear cracks (the blue cracks in Fig. 1.3A). The flexural cracks usually occur first at the tension side of a flexural member in the direction perpendicular to its longitudinal axis. In a pure bending zone (between the two point-loads in Fig. 1.3A), the cracks keep their direction and extend toward the compression side of the member when load further increases. In a shear span (between the support and the point-load in Fig. 1.3A) where both bending moment and shear force exist, the extension of the flexural cracks at the tension side will change its direction toward the point-load and become inclined gradually due to the change of direction of the principal tensile stress. The inclination of the crack is caused by the presence of shear stress, which increases from the tension side toward the middle of the member. Therefore, these cracks are called flexural-shear cracks. The web cracking initiates near the neutral axis where the shear stress dominates and flexural stress is insignificant.

After formation of diagonal cracking, one of the inclined cracks may propagate quickly toward both the support and the point-load to cause a brittle shear failure (Fig. 1.7). Such a diagonal crack is called the critical diagonal crack (CDC) and the corresponding failure mode is referred to as the diagonal tension failure or shear-tension failure. The CDC may be developed from one of the web cracks or a flexural-shear crack. The shear-tension failure is sudden without warning, splitting a beam into two pieces and causing collapse of the structure (Fig. 1.6D and E). The ultimate failure load is only slightly higher than the load at formation of the CDC. The shear-tension failure usually occurs in a slender RC member with a shear span-to-depth ratio, a/d, larger than about 2.5 without or with insignificant web reinforcement.

(2) Shear-compression failure without web reinforcement

For deep RC beams without significant web reinforcement, the formation of diagonal tension cracking may not cause an imminent brittle shear failure. This is illustrated through the beam tests in Fig. 10.2. As shown by the drops of the load–deflection curves at point A in Fig. 10.2, the diagonal tension cracking occurs at a load of around 80–90 kN for the two beams with $a/d = 2.6$ in Fig. 10.2. This cracking load corresponds to the shear-tension failure load V_c of the RC member (Wu and Hu 2017). These two members should have failed at the load drop if they are shear-tension members. However, these beams did not fail because they are shear-compression members. At the initiation of diagonal cracking, the load drops from Point A (86 kN) to Point B (72 kN) for the beam N10 and from Point C (78 kN) to Point D (77 kN) for the beam N20 (Fig. 10.2). The corresponding cracks developed quickly at the point of load drop as shown in Fig. 10.3 but stabilized without losing their stability.

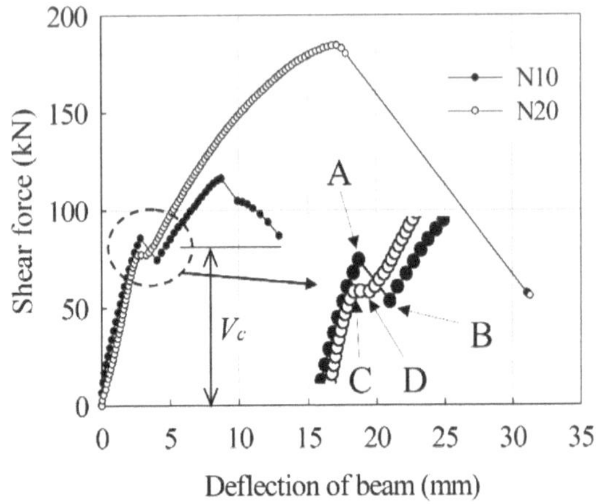

Fig. 10.2 Shear-compression failure after diagonal tension cracking.
From Wu, Y. F., and B. Hu. 2017. "Shear strength components in reinforced concrete members." *J. Struct. Eng.* 143 (9), 04017092.

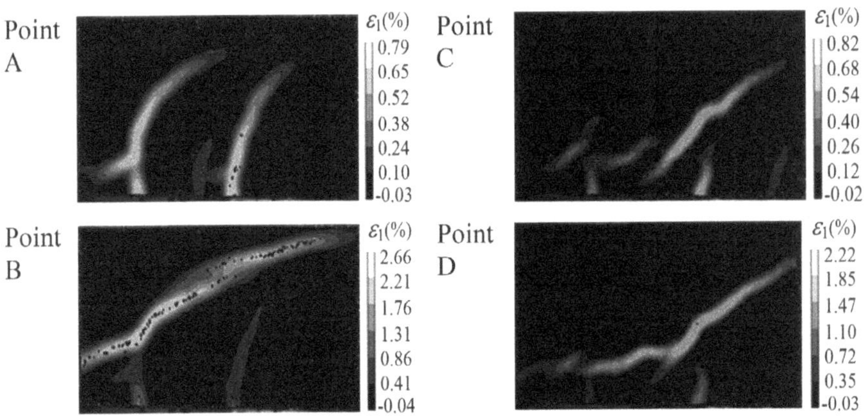

Fig. 10.3 Cracks captured by digital image correlation.
From Wu, Y. F., and B. Hu. 2017. "Shear strength components in reinforced concrete members." *J. Struct. Eng.* 143 (9), 04017092.

As loading continues, the diagonal cracking further develops and propagates at a much slower rate than that of the shear-tension cracking. Several diagonal cracks may be developed at further loading and the beam may sustain a load without collapse that is more than 100% greater than the load at formation of diagonal cracking (Fig. 10.2). Eventually, the concrete in the compression zone of the member above the upper end of the cracks may be destroyed by crushing or shearing off, causing collapse of the beam, as shown in Fig. 10.4. The critical diagonal crack usually does not penetrate through the compression zone at failure. As the failure is signified by concrete failure in the compression zone, the shear failure is referred to as shear-compression failure. The shear-compression failure usually occurs in an RC member with a/d in the range around 2.0–3.0.

(3) Web crushing

For deep RC beams with a/d less than about 2.0 with/without web reinforcement or some slender member with thin web, failure of the beams may be caused by the

Fig. 10.4 Shear-compression failure mode of beam N20.

crushing of web concrete, as shown in Fig. 10.5. It is shown in Figs. 8.1 and 8.2 that crushing of concrete is characterized by longitudinal splitting of concrete. Therefore, web crushing looks similar to web splitting. Under the shear-tension failure and shear-compression failure, the diagonal cracks are caused by the principal tensile stress that exceeds the concrete tensile strength, and are along the principal compressive stress direction. While in a web-crushing failure, the diagonal cracking is due to the splitting of concrete when the principal compressive stress reaches concrete compressive strength, which is also along the principal compressive stress direction. In other words, one is caused by excessive tensile stress and the other by excessive compressive stress. As all cracks are along the principal compressive stress direction, it is sometimes difficult to tell the failure mode just by observing the cracks.

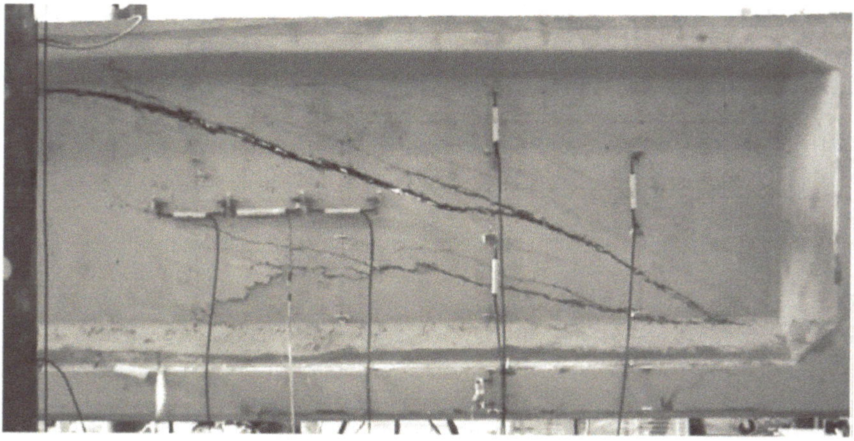

Fig. 10.5 Web crushing failure mode.
From Soetens, T., S. Matthys, S. Hertelé, and W. De Waele. 2017. "Shear behaviour of prestressed precast SFRC girders." *Eng. Struct.* 142, 20–35.

(4) Shear failure with significant web reinforcement

It is shown by Beam 6 in Section 1.3.6 that web reinforcement or stirrups can significantly increase the shear strength of an RC member and prevent the shear failure. The web reinforcement carries little force prior to the formation of diagonal tension cracks (see Section 10.3.2). It has little effect on the cracking load and the behavior of the beam before diagonal cracking. However, the web reinforcing bars intersected by a diagonal crack immediately receive a sudden increase in tensile stress when the diagonal crack occurs (Fig. 10.9), while the web reinforcement not intersected by diagonal cracks remains unstrained. This phenomenon shows that the tensile stress in the original uncracked concrete is immediately redistributed to the web reinforcement after concrete cracking.

For a beam supposed to have shear-tension failure without web reinforcement, the existence of web reinforcement inhibits the rapid propagation of diagonal cracking and stabilizes the cracking process. Further increase in shear force is enabled through two mechanisms: (1) the web reinforcement directly carries a portion of the shear force redistributed from the cracked concrete on formation of the diagonal crack. This portion increases with further increase in loading; and (2) the web reinforcement restrains the development of the diagonal crack and helps in maintaining the integrity of concrete, and hence the load-carrying capacity of the concrete. Further loading may cause yielding of the web reinforcement, which accelerates the opening of diagonal cracking and deteriorates the stress condition of the concrete. Finally, destruction of the compression zone either by concrete crushing or shearing off would occur, causing a shear failure similar to the shear-compression failure, as shown in Fig. 10.6. The dash-dotted lines in the figure show the position of the longitudinal bars inside the beam and the solid lines show the stirrups positions.

This failure process usually occurs gradually following a large increase in diagonal crack width as the web reinforcement yields. It shows significant signs of distressing before final collapse and hence is more ductile than the shear-tension failure mode. For this reason, the ACI Building Codes since 1971 and the current design codes in most countries require a minimum amount of shear reinforcement in most structural members. However, sudden diagonal tension failures similar to shear-tension failure may also occur if the web reinforcement ratio is so low that it cannot carry the internal force redistributed from the cracked concrete at cracking. In this case, the shear failure should be categorized as shear-tension failure without significant web reinforcement. If the web reinforcement ratio is higher than a certain amount or the value of a/d is smaller than a certain value, around 1–2, the concrete crushing in the compression zone may happen before the yielding of the web reinforcement. This failure process is more brittle than the case with yielding of the web reinforcement.

For beams supposed to have shear-compression failure without web reinforcement, the provision of significant web reinforcement can reinforce the stability of the concrete in both the diagonally cracked region and the compression zone of concrete. Consequently, the shear strength of the beams can be increased.

For web crushing failure, as it is caused by web concrete crushing, the web reinforcement can restrain the lateral dilation of concrete under compression and hence increase the compressive strength of the web concrete. Therefore, the effect of the web reinforcement is to confine the concrete as discussed in Section 8.2, which is different from the web reinforcement for shear-tension failure where the web reinforcement directly takes over the excessive tensile stress from cracked concrete. It is shown in Chapter 8 that confinement can significantly increase the strength of concrete. Consequently, it should be able to significantly increase the shear strength of RC members with the web-crushing failure mode. However, experimental observations show that the web reinforcement may not yield at the onset of the shear strength and may not be very effective. This may be caused by two reasons: (1) the local web region may be well confined but disintegration occurs in other parts at further loading, and (2) the web reinforcement only confines the concrete dilation in the on-plane direction of the web (perpendicular to the principal compressive stress) but not the dilation of concrete in the direction perpendicular to the web plane. Concrete is effectively confined only when confinement is provided in all lateral directions. In other words, web reinforcement may not provide effective confinement in both of the lateral directions. From this point of view, the use of lateral links (similar to the through bolts in Fig. 8.21) that connect the web reinforcement on the two sides of the web would be able to provide confinement in the lateral direction perpendicular to the web plane, and consequently increase the relevant shear strength. For the preceding two reasons, the normal detailing of the web reinforcement may not contribute to a significant increase in the shear strength.

Fig. 10.6 Shear failure with web reinforcement.

(5) Other failure modes

Failure of longitudinal bars by debonding or yielding in the region from the CDC to a support may occur after diagonal cracking due to the tension-shift effect. Diagonal cracking causes redistribution of not only the shear stress but also the normal stress. Before diagonal cracking, the normal stresses in the longitudinal reinforcement and the concrete are distributed along the length of the member following the bending moment diagram. The formation of a diagonal crack can change the normal stress distribution, and the tensile stress of the longitudinal bars in the diagonally cracked region cannot be calculated in the normal way described in Chapter 2. When a major diagonal crack forms, a sudden increase in stress also takes place in the longitudinal reinforcement intersected by the diagonal crack, due to the tension shift effect, as discussed in Section 7.3.2(4).

For the beam N20 in Fig. 10.2, a sudden increase in the longitudinal bars was detected from Point C to Point D (Wu and Hu 2017). The strain of the longitudinal bars at section B (closer to the support with a smaller bending moment) in Fig. 10.4 suddenly increased at the load drop of Point D in Fig. 10.2 and was equal to the strain at section A where the bending moment is maximum. In other words, the stress of the longitudinal bars at section B in Figs. 7.18B and 10.4 is not determined by the bending moment of section B but by that at section A, which can be significantly larger than the moment of section B. Therefore, the stress of the longitudinal bars can be significantly increased by the diagonal cracking. In other words, the stress in the tension reinforcement and in the concrete along the beam does not follow the distribution of the bending moment diagram. If the design of the longitudinal bars does not allow for this effect, the beam may fail due to debonding/dowel splitting or yielding of the longitudinal bars.

To allow for this redistribution of internal forces, a longitudinal bar that can be terminated (curtailed) according to the bending moment diagram has to be extended beyond the section where it is no longer needed for bending resistance for a distance equal to the effective depth of the section + anchorage length. There are other possible failure modes related to diagonal cracking of an RC member. For example, for short beams with a/d less than about 1.0, concrete may crush/split at the supports or under a point load before or after formation of diagonal cracking. The two sides of a crack may also relatively slide under the shear force similar to the sliding failure shown in Fig. 5.24A.

Shear failure modes are affected by not only a/d but also many other factors such as detailing of the RC member, loading and boundary conditions, reinforcement types, aggregate size, etc. Therefore, shear failure modes are complicated. New shear failure modes or mechanisms are proposed by researchers from time to time. Shear failure modes may also be grouped or defined in different ways by different people. Nevertheless, all the shear failures are closely related to or dominated by the shear force of RC members. It can also be envisaged from the discussions of the failure modes in this section that the development of a shear failure design model cannot be a simple task.

10.3 Shear resisting mechanisms

It can be seen from the discussions in the previous sections that the shear transfer mechanisms are complicated and affected by many factors. They are difficult to quantify, and what's more, many of them have not been fully understood so far. These are the reasons that a good shear strength design model is difficult to develop, and the current shear design models have a relatively low accuracy compared with those for flexural design.

The free-body diagram of the part of a simply supported beam from the support to a critical diagonal crack is shown in Fig. 10.7, which exposes the components of the shear resistance on a diagonal shear crack. From the equilibrium of the free body in the vertical direction, one has

$$V = V_c + V_s = V_c + \sum_1^n V_{si} \tag{10.3}$$

where V is the total shear resistance; V_s is the shear resistance contributed by the transverse reinforcement, which equals to $\sum_1^n V_{si}$; V_{si} is the force in the ith transverse bar, and n the total number of transverse bars crossing the critical diagonal crack; and V_c is the so-called shear strength of concrete or shear resistance contributed by concrete, which is an aggregate of all other shear-resisting components, including:

1. V_{cc}, the shear resistance from the compression zone of the RC section;
2. S, the interfacial shear transfer from the cracked concrete due to aggregate interlocking, friction, and residual tensile stress. The vertical term, $S{\cdot}sin\theta$, contributes to the shear strength; and
3. V_{cd}, the dowel action from the longitudinal reinforcement.

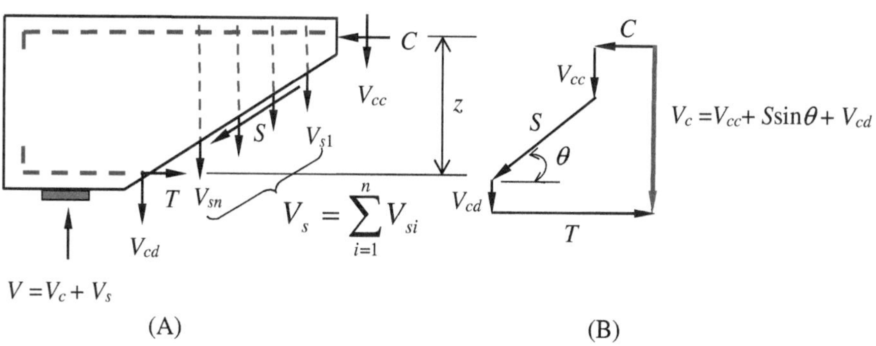

Fig. 10.7 Components of shear resistance.
From Wu, Y. F., and B. Hu. 2017. "Shear strength components in reinforced concrete members." *J. Struct. Eng.* 143 (9), 04017092.

All components of the internal forces excluding the forces from transverse reinforcement are shown in Fig. 10.7B. These internal forces form a resultant vertical force V_c (the red arrow). It can be seen from the figure that V_c is the sum of the vertical components of several shear transfer mechanisms. Research in the 1950s brought about a clear realization that the mechanisms involved in V_c are complex and related to many factors. Extensive efforts have been made by researchers all over the world for decades toward better understanding and modeling of V_c. Although numerous mechanical and analytical models have been proposed in the literature (ACI-ASCE Committee 455 1998), the existing models are far from satisfactory due to their limitations. Until now, only empirical or semiempirical models have been developed based on extensive experimental data. These models have formed the basis of the shear design provisions in current design codes of practice. Generally speaking, the current models of V_c have a large scattering in predicting the shear strength contribution of concrete and are inaccurate when compared with test results.

The diagonal cracking depends on shear stress rather than shear force. Based on the work of E Mörsch, German specifications in 1904 first used the nominal shear stress calculated from $v = V/bz = V/bjd$ to evaluate the shear intensity, where $z = jd$, the internal lever arm (Fig. 10.7). In order to avoid the calculation of the internal lever arm z, the ACI-ASCE Committee 326 (1962) replaced z by d for convenience. The nominal shear stress has been thereafter defined as:

$$v = \frac{V}{bd}, v_C = \frac{V_C}{bd}, v_S = \frac{V_S}{bd} \tag{10.4}$$

With Eq. (10.4), Eq. (10.3) is changed to

$$v = v_C + v_S \tag{10.5}$$

The contribution of shear resistance from the transverse reinforcement or stirrups can be analogized by the so-called truss mechanism, as shown in Fig. 10.8. The cracked RC member in Fig. 10.8A can be modeled by the truss in Fig. 10.8B. The concrete between two diagonal cracks in Fig. 10.8A is visualized as a diagonal compression strut. The tension bars are considered to be the tension chord and the stirrups as the tension links. The compression chord is formed by the uncracked compression zone of the member. One tension link in Fig. 10.8B is formed by a number of stirrups. The area of one tension link A_{link} in Fig. 10.8B is given by

$$A_{link} = \frac{z\cot(\theta)}{s} A_{sv} \tag{10.6}$$

where θ is the angle between the compression strut and the longitudinal direction; A_{sv} is the area of one stirrup (including all legs of a stirrup); z is the internal lever arm from the compression resultant to the centroid of the tension bars; and s is the spacing of stirrups. From the balance of forces in Fig. 10.8C in the vertical direction, one has

$$V_s = A_{link} f_{yv} = \frac{z cot(\theta)}{s} A_{sv} f_{yv} \quad (10.7)$$

or

$$v_s = \frac{V_s}{bz} = \frac{\cot(\theta)}{bs} A_{sv} f_{yv} \quad (10.8)$$

where f_{yv} is the yield strength of transverse steel; and V_s and v_s are the shear strength of the truss system in terms of force and stress, respectively. Considering the 45-degree truss model where $\theta = 45°$, Eq. (10.8) is simplified to

$$v_s = \rho_{sv} f_{yv} \quad (10.9)$$

where $\rho_{sv} = \frac{A_{sv}}{bs}$, the transverse reinforcement ratio. It can be seen from Fig. 10.8C that the shear strength of the truss is determined by the strength of the stirrups (Eq. (10.9)). From the balance of joint A in Fig. 10.8C, the force of the stirrups, $A_{link} f_{yv}$, must be balanced by the compressive resistance of the diagonal strut. If too much transverse reinforcement is provided, the shear resistance V_s is not controlled by $A_{link} f_{yv}$ but by the crushing strength of the strut. Therefore, an upper limit in shear stress is imposed in design codes to prevent the compression failure of the diagonal strut member (diagonal crushing failures of the web concrete).

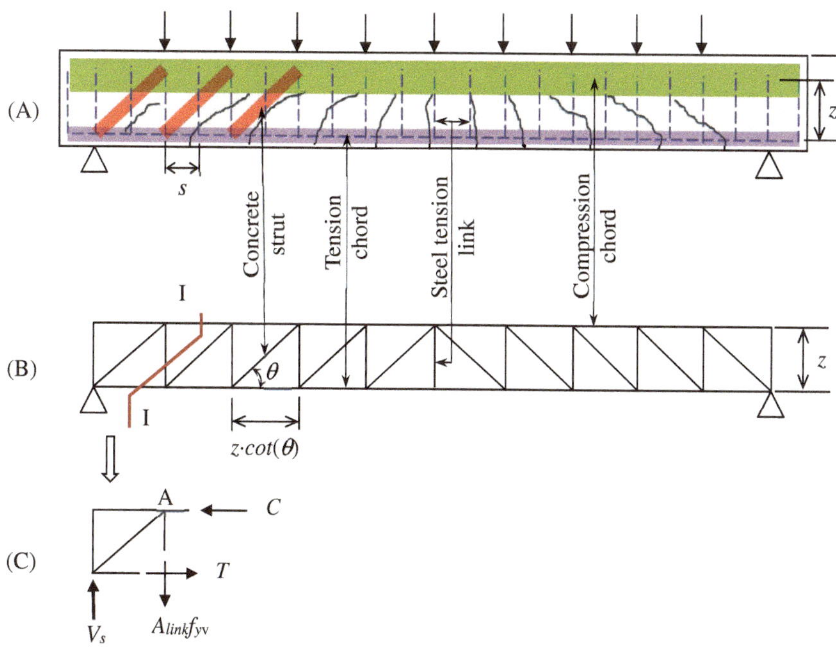

Fig. 10.8 Truss analogy.

10.3.1 Shear transfer mechanisms of RC members without web reinforcement

For RC members without web reinforcement, V_s is absent in Fig. 10.7. When a critical diagonal crack forms, the internal forces along the crack must be redistributed to other places. Before the 1960s, it was generally believed that no force can be transmitted across the crack, and hence the lost internal forces must be redistributed to the remaining shear mechanisms including dowel action and the shear stress in the compression zone of the member (ACI-ASCE Committee 326 1962). However, it was found later that the cracked interface can still transmit a major part of the shear force through the mechanisms of aggregate interlocking, friction, and residual tensile stress. Many factors affect the magnitude of these mechanisms. Although extensive research has been undertaken to quantify these mechanisms, which led to the knowledge of relative proportion of a mechanism with respect to the total shear strength (Park and Paulay 1975), there has been no model that can quantify these mechanisms separately with reasonable accuracy. Redistribution of stress may also occur in longitudinal bars due to the tension-shift, as discussed in Section 10.2(5).

If the preceding redistributions of internal shear and normal stresses in the member cannot reach a new force balance, collapse of the member occurs, resulting in a shear failure. If the collapse occurs abruptly, shortly after the formation of a critical diagonal crack, the failure is a shear-tension failure. If a new balance of internal forces is attained and failure does not occur immediately after the formation of a critical diagonal crack, the member can be further loaded. More diagonal cracks can be formed during further loading. The process continues until the moment that new balance cannot be sustained, when collapse of the member occurs. The final collapse load can be more than double the loads causing the critical diagonal tension crack. This failure process corresponds to the shear-compression failure or web-crushing failure.

Due to the brittle nature of the shear-tension crack and the uncertainty in the additional load from initiation of diagonal crack to collapse of the member, the load causing the formation of a critical diagonal tension crack should be considered as the usable ultimate load-carrying capacity of the RC members with a shear-tension failure mode. For the shear-compression failure and web-crushing failure, the failure processes are generally more stable, and hence the peak resistance of the member can usually be treated as the ultimate load-carrying capacity of the members.

10.3.2 Shear transfer of RC members with web reinforcement

The web reinforcement can usually take over the internal forces redistributed from diagonal cracked concrete and avoid abrupt shear-tension failure unless the web reinforcement is insignificant. The web reinforcement enhances the shear capacity of an RC member through the following mechanisms (Hu 2017):

1. Providing shear resistance V_s directly by the transverse bars crossing the critical diagonal crack (Fig. 10.7);

2. Restraining the opening of diagonal cracks, which helps to enhance friction and aggregate interlock at a crack interface;
3. Restraining the growth of cracks toward the compression zone;
4. Confining the compression zone and hence increasing the shear resistance of the concrete in the compression zone;
5. Supporting the longitudinal bars to enhance the dowel action; and
6. Providing restraint against the splitting of concrete along the longitudinal bars.

It was observed that, under the same transverse reinforcement ratio ($\rho_{sv} f_{yv}$), the effect of transverse reinforcement on the shear capacity is also affected by its detailing such as layout, surface characteristics of reinforcing bar, and anchorage (see Section 10.5).

Extensive investigations have been reported in the early literature on quantification of the shear transfer mechanisms, which have led to a better understanding of them (ACI-ASCE Committee 326 1962; Park and Paulay 1975; ACI-ASCE Committee 455 1998). However, accurate quantification of V_c and V_s from test results was made possible only recently by Wu and Hu (2017), which facilitates better understanding and more accurate evaluation of the shear transfer mechanisms. The typical results obtained from the new experimental approach are illustrated in Fig. 10.9. The beam ID in Fig. 10.9 specifies the type of the transverse reinforcement and the shear span-to-effective depth ratio a/d, where D and R represent deformed bar and plain round bar, respectively, and the number gives the value of a/d. For example, D1.9 represents the beam with $a/d = 1.9$ transversely reinforced by deformed bar, and R3.1 denotes a beam with $a/d = 3.1$ transversely reinforced by plain round bar. All the beams in Fig. 10.9 have the same transverse reinforcement ratio of $\rho_{sv} f_{yv} = 2.07$ MPa. Two critical instants, at the first yield of the transverse bar and at the ultimate shear strength (peak point of the V curve) are marked by the two vertical lines in each figure. The variations of V, V_s, and V_c with respect to a/d are shown in Fig. 10.10. Clearly, V and V_c decrease when a/d increases, but V_s shows an opposite trend.

The following observations are made from the test results in Figs. 10.9 and 10.10:

1. Before the occurrence of the first diagonal crack, all the applied load is taken by concrete (V_c) and there is no significant strain in the transverse bars. The diagonal cracking of concrete is signified by the diverging point of V and V_c curves, as well as the sudden increase of V_s. V_s increases sharply at first diagonal cracking followed by a curve with a gradually reducing slope.
2. For short beams (D1.9 and R1.9), V_c continues to increase after a small drop at diagonal cracking with a relatively smaller slope. The transverse reinforcement also continues to take more load until the first yield of the bars, after which until the peak shear resistance, the increase of loading is solely contributed by concrete shear resistance V_c.
3. For specimens with a median length (D2.5 and R2.5), V_c shows a significant reduction at first diagonal cracking followed by a gradual increase until failure.
4. For longer specimens (D3.1 and R3.1), after the first diagonal cracking, V_c decreases continuously and significantly until yield of the transverse bars, from which point it increases slightly.

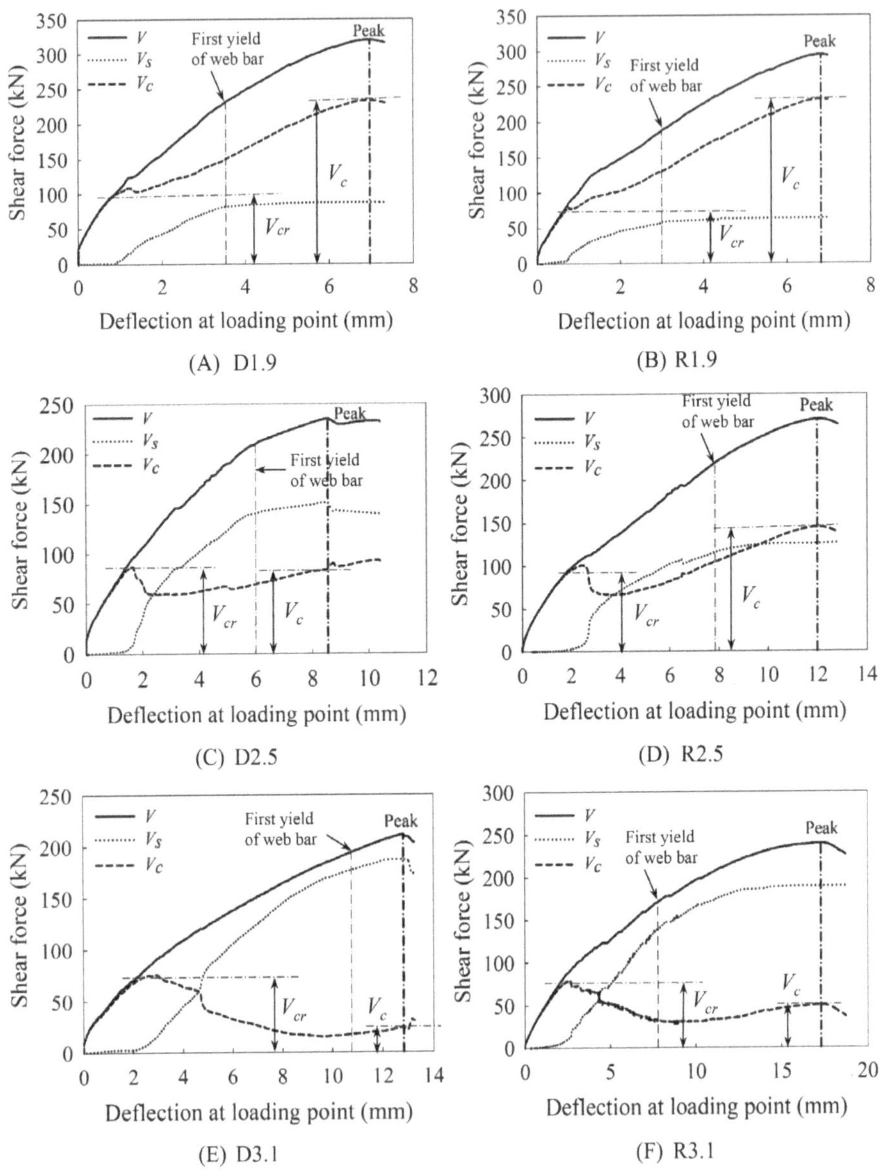

Fig. 10.9 Variation of V_s and V_c.

5. At peak load, the proportion of V_c in the total shear strength V decreases when the slenderness of the beam, a/d, increases. V_c is significantly larger than V_s in short beams (D1.9 and R1.9); and they are similar for a median length beam (D2.5 and R2.5); but V_c is significantly smaller than V_s in longer beams (D3.1 and R3.1), noting that all beams have the same transverse reinforcement ratio.

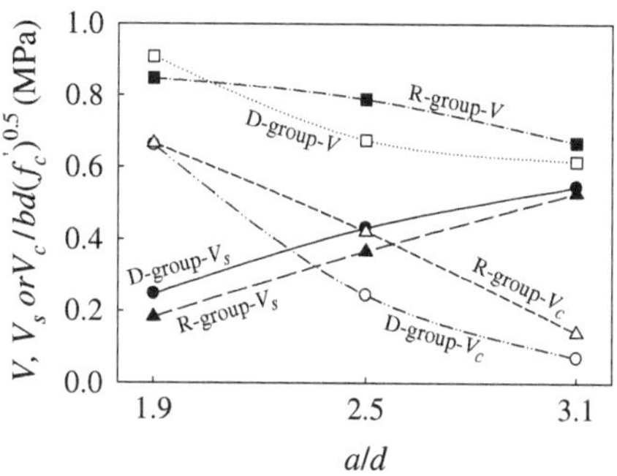

Fig. 10.10 Variations of shear strength components.
From Hu, B., and Y. F. Wu. 2018. "Effect of shear span-to-depth ratio on shear strength components of RC Beams." *Eng. Struct.* 168, 770–783.

6. With an identical transverse reinforcement ratio ($\rho_{sv}f_{yv}$), the values of V_s in the D-groups are greater than those of the R-group, but the difference reduces when the span-effective depth ratio increases (Fig. 10.10).
7. With regard to V_c, both the pre- and postcrack stiffness decline with increasing a/d value. V_c for specimens in the R-group is larger than that in the D-group, which could be attributed to the more even confinement due to closer transverse bars in the R-group. However, since most of the shear in deep beams is transferred via a direct concrete strut, this effect is negligible for specimens with $a/d = 1.9$.

Not all reinforcing bars intersected by the critical diagonal crack are yielded at the peak shear resistance. The efficiency factor k is defined as the ratio of the average stress ($f_{v\text{-}aver}$) of the transverse bars at the point intersecting the CDC at onset of the peak shear resistance to the yield stress (f_{yv}), as follows:

$$k = \frac{f_{v-aver}}{f_{yv}} \tag{10.10}$$

Table 10.1 shows the experimentally determined stresses of the transverse bars along the CDC at the peak shear resistance and the efficiency factor k of these beams. It can be seen from the table that k is not a constant and varies with the shear span-to-depth ratio. All k values are larger than 0.8. In the D-group with high-strength transverse bars, the minimum k value of 0.803 occurs in D2.5 while the maximum is 1.001 in D3.1, which indicates that the transverse bars intersecting CDC fully yield only in D3.1. This result means that the assumption of yielding of all transverse reinforcement for the calculation of V_s (Eq. (10.7)), as adopted in most models and design codes, is un-conservative.

For the R-group with low tensile strength and weak bond reinforcement, however, all k values are larger than 1. This means that all transverse bars intersecting CDC yield in the R-group and enter strain hardening of the steel material, which is the reason that $k > 1.0$. Nevertheless, the stress increase due to the hardening of steel is very

Table 10.1 Stress of transverse bars intersecting CDC.

Spec. ID	$\rho_{sv}f_{yv}$ (MPa)	Rebar stress at CDC (MPa)						Aver. stress (MPa)	Yield stress (MPa)	*k*
		Stirrup 1		Stirrup 2		Stirrup 3				
		Front	Back	Front	Back	Front	Back			
D1.9	2.07	411.44	490.91	–	–	–	–	451.18	492	0.917
D2.5	2.07	496.71	436.99	286.16	360.03	–	–	394.97	492	0.803
D3.1	2.07	494.80	489.64	494.96	490.31	–	–	492.43	492	1.001
R1.9	2.07	333.35	332.86	–	–	–	–	333.11	325	1.025
R2.5	2.07	331.97	330.36	330.50	331.85	–	–	331.17	325	1.019
R3.1	2.07	330.69	330.98	331.27	331.01	332.80	337.28	332.34	325	1.023

limited with a maximum of 2.5% of its yield strength. In this case, the assumption of full yield of transverse bars is valid.

It is noted that stresses in the two legs of a stirrup are significantly different for specimens in the D-group (Table 10.1). However, the difference decreases with the increase of *a/d*. In the R-group, the stresses in the two legs are generally the same. The unequal strains in the front and back legs of a stirrup represent the three-dimensional (3D) nature of cracking. The 3D effect decreases when *a/d* increases, or mild steel is used for the transverse reinforcement.

It can be seen from the preceding test results that the shear transfer mechanisms have complicated trends of variation. These test results are only a few isolated cases that do not give a comprehensive picture of the trends. In fact, many important factors that affect the shear strength of RC members are not included in the previous discussions, such as concrete strength, longitudinal reinforcement ratio, and aggregate size of concrete. In general, more extensive investigations are required before a better understanding and accurate modeling of these mechanisms are possible.

10.4 Development of shear design approaches

Extensive reviews on the development of shear design approaches and models were given in ACI-ASCE Committee 326 (1962), ASCE-ACI Committee 426 (1973), and ACI-ASCE Committee 455 (1998). Early researchers before 1900 developed two schools of thought pertaining to the shear failure mechanism of RC members. One school of thought believed that the horizontal shear between different horizontal layers of concrete (see the bond force T in Fig. 5.4B) is the cause of shear failure. It was believed that concrete alone could only resist low horizontal shear stress, and the vertical stirrups acted as shear-keys to enhance the shear strength. The second school of thought considered the diagonal tension, as shown in Fig. 10.1, as the cause of shear failures, which has been accepted by the majority of the research community today. A literature review could not identify the origin of the concept of diagonal tension, but a clear explanation of diagonal tension was given in the revolutionary work of Ritter (1899) who stated that the function of the stirrups is to resist tension rather than horizontal shear. He further postulated the original 45-degree truss model, as shown in Fig. 10.8 where $\theta = 45°$, and proposed the equation for the design of vertical stirrups based on the truss mechanism, Eq. (10.9). Eq. (10.9) as the model of v_s is still used in most current design codes all over the world.

However, the horizontal shear opinion dominated at that time, and Ritter's viewpoint was not accepted until a decade later, when the pioneering experimental works by Mörsch (1909, 1920, 1922) terminated the horizontal shear theory. From about 1910 until today, Ritter's truss concept has been widely accepted in the literature and applied in practice throughout the world.

The original Ritter's truss model only considered V_s, i.e., the total shear resistance $V = V_s$ without considering V_c. It was found conservative when compared with test results in the early works by Withey (1907) and Talbot (1909). The conservatism of Ritter's truss model can be easily explained: the real beam of Fig. 10.8A is simplified to a truss as in Fig. 10.8B. It is obvious that there are differences between the two structures in Fig. 10.8A and B. These differences are the additional internal actions and forces that are ignored in the truss model. To consider the underestimation of shear strength due to the simplification of shear transfer mechanism by the 45-degree truss model, the total shear strength of an RC member should be supplemented by an additional term, which is referred to as the concrete contribution V_c in most design codes. This V_c term has been included in the ACI codes ever since the truss model was introduced into the American literature by Withey (1907, 1908).

The shear resistant stress due to concrete v_c was a simple constant in early codes. The first report of the National Association of Cement Users (forerunner of the ACI), NACU report of 1908, specified that the shearing strength of concrete corresponding to a compressive strength of 2000 psi shall be 200 psi, and no model was specified for evaluation of the shear resistance due to transverse reinforcement, v_s. The Joint Committee Report of 1909 proposed web reinforcement to be designed to resist two-thirds of the total shear. NACU Standard No. 4 in 1910 specified a v_c value of 40–50 psi depending on concrete strength (from 2000 psi) with the remainder shear stress provided by v_s. ACI 318-51 specified that, if the unit shearing stress at service load v is less than $0.03f'_c$, no web reinforcement is required. If v exceeds this value, the excess shear stress shall be resisted by web reinforcement calculated based on the 45-degree truss analogy.

The collapse of the frame girders in the warehouse at Wilkins Air Force Depot in Shelby, Ohio, United States in 1955, caused by diagonal tension failure, was a milestone event that brought the traditional shear design procedures into question. Careful investigation of the collapse pinpointed the inadequacy of the then design code ACI 318-51. Committee 326 recommended some revisions of the 1956 ACI Building Code to ACI Committee 318, including imposing upper limits on maximum shear stress for members with and without web reinforcement and calling for web reinforcement to resist two-thirds of the total shear force in certain regions of RC members. The intensified research after the collapse of the warehouse indicated that the diagonal shear-tension failure was a complex problem involving many factors. The collapse resulted in a return to forgotten fundamentals and triggered an upsurge in research that had a major impact on the study of shear failure mechanisms and development of shear design guidelines in the following decades.

Based on extensive analytical studies and experimental test results, a more rational semiempirical model of shear strength was proposed in ACI-ASCE Committee 326 (1962) for design of RC members without web reinforcement:

$$v_{cr} = 1.9\sqrt{f'_c} + 2500\frac{\rho_{st}Vd}{M} = 1.9\sqrt{f'_c} + 2500\rho_{st}\frac{d}{a} < 3.5\sqrt{f'_c}\,(\text{in}\,psi) \tag{10.11}$$

where v_{cr} is the shear stress at onset of CDC that was defined as the shear strength of RC members without web reinforcement; ρ_{st} is the tension reinforcement ratio, given by

$$\rho_{st} = \frac{A_{st}}{bd} \tag{10.12}$$

in which A_{st} is the area of the longitudinal bars in tension. It was further specified that the concrete contribution v_c in RC members with web reinforcement in Eq. (10.5) be equal to the shear strength without web reinforcement, i.e.,

$$v_c = v_{cr} \tag{10.13}$$

Eq. (10.11) has remained unchanged since then, including in the most recent ACI 318-14 (2014). However, it can be observed from the latest experimental results in Fig. 10.9 that V_c is not equal to V_{cr} in general.

In contrast to the American approach, where the total shear resistance is given by supplementing a concrete contribution term v_c to the shear resistance from the transverse reinforcement obtained from a 45-degree truss model v_s, another approach is to use a variable angle θ for the truss in Fig. 10.8. It can be seen from Eq. (10.6) that, when θ is less than 45 degrees, the area of the effective tension link A_{link} is increased so that v_s is increased. Therefore, mathematically speaking, the same total shear strength as that from Eq. (10.5) can be obtained by ignoring v_c but properly choosing a θ value for v_s in Eq. (10.8). This approach has been adopted in Europe and is usually referred to as the "standard truss model with no concrete contribution." This was a result of the early European codes/specifications that conservatively ignored v_c and required the web reinforcement to be designed to carry the total shear by the 45-degree truss-analogy approach for RC members with web reinforcement. By allowing for a diagonal angle of less than 45 degrees, the calculated result from the standard truss model with no concrete contribution can be significantly increased. The reduction of the diagonal angle is also explained by the existence of the interfacial shear stress (S in Fig. 10.7), including aggregate interlock and friction, along the crack face, which changes the principal stress direction to an angle less than 45 degrees.

A combination of the variable-angle truss model and a concrete contribution, referred to as the modified truss model approach (CEB 1978a, 1978b; Ramirez and Breen 1991), has also been proposed. As the modified truss model allows for more variables, and hence can be more accurate potentially, it has been used in more recent modeling and design codes. The development of plasticity theories extended the applicability of the model to nonyielding domains (Nielsen 1984; Nielsen and Braestrup 1975). Using elastic theory and the principle of minimum strain energy, Kupfer (1964) analyzed the truss model and provided a solution for determination of θ. Based on a procedure developed by Wagner (1929), Collins and his colleagues (Collins 1978; Collins and Mitchell 1980) extended Kupfer's approach by allowing for nonlinearity and developed the well-known compression field theory (CFT) to determine θ, where the tensile stress of concrete between cracks was omitted. Based on extensive experimental investigation, they later extended the CFT to the modified compression field theory (MCFT) (Vecchio and Collins 1986; Bentz et al. 2006), which included a mechanism to allow for the tensile stress between the cracks that contribute to the shear resistance significantly in the diagonally cracked concrete. MCFT can overcome the problem of CFT that does not work well for RC members with a small amount or no transverse reinforcement. Further discussion on the MCFT is provided in Section 10.6.

Parallel to these developments of the truss model with variable strut inclinations and the CFT, the 1980s also saw the further development of shear friction theory (see Section 5.12) for evaluation of shear transfer along a crack face. In addition, refined truss methods that allow for shear stress transfer across the cracks by friction were introduced (Gambarova 1979; Reineck 1990; Kupfer and Bulicek 1991). The vertical component of the shear stress along the crack face contributes to the shear resistance, which gives rise to the concrete contribution V_c.

Another important development of the shear design approach was the strut-and-tie model (STM), which is a generalization of the truss analogy approach (Collins and Mitchell 1985; Schlaich et al. 1987; Ramirez and Breen 1991). With the strut-and-tie model, engineers can "design" an internal truss structure within an RC member where reinforcement is used as ties and concrete as strut. As different trusses can be designed to suit the same space and loading, multiple designs of the strut-and-tie model are possible. It requires experience to determine a more efficient strut-and-tie model for a given RC member. Schlaich et al. (1987) suggested to choose the angles of the compression diagonals so that they are within ± 15 degrees of the angle of the principal compressive stresses obtained from an elastic analysis. However, redistribution of stresses takes place after cracking, and a strut-and-tie model determined based on elastic stress field may not be most effective at ultimate limit states. Mihaylov et al. (2013) developed kinematic models for RC deep beams while others suggested further improvements of the STM (Hwang and Lee 2002; Russo et al. 2005; Brown and Bayrak 2008; Scott et al. 2012). It is noted that the strut-and-tie structure in B-regions is usually referred to as a truss, while that in a D-region is called an STM (Fig. 2.23). The STM is more suitable for the design of D-regions and deep beams. Further discussions on the STM are given in Sections 10.7.2 and 11.6.2.

Numerical models have also been developed for prediction of shear failure and shear strength, such as fracture mechanics approaches (Bazant and Oh 1983; Reinhardt 1986) and the nonlinear finite-element method (Isenberg 1993; Chen et al. 2012; Feng et al. 2018; Ghahremannejad and Abolmaali 2018). With the numerical approach, analytical models can also be developed by rational analysis in combination with regression analysis using the numerical results.

In summary, due to the great efforts made by the research community over more than a century, a large number of experimental tests have been carried out and a comprehensive test database is available in the literature for the shear strength of RC members. These experimental studies have laid a solid foundation for understanding the mechanisms of shear transfer and the development of theories, models, and design codes of practice. Typical theories and models include the traditional truss model, the compression field theory, the modified compression field theory, plasticity approach, strut-and-tie model, tooth model (Kani 1964), rotating- and fixed-angle softened-truss models (Belarbi and Hsu 1994, 1995; Pang and Hsu 1996), critical shear crack theory (CSCT) (Muttoni and Ruiz 2008; Ruiz and Muttoni 2009), mechanical models (Choi et al. 2007; Cladera et al. 2015; Zhang et al. 2015; Marí et al. 2016), etc.

Most of the existing theories and design codes of practice have adopted the additive model where the shear strength (V) is taken as the sum of the shear strength of concrete

(V_c) and that of transverse reinforcement (V_s), with the exception of the EC2 (CEN 2004) where a variable-angle truss method is recommended. For RC deep beams, in which case a significant portion of shear is transferred directly from the loading point to supports via diagonal compression struts, STM is more appropriate than the sectional approaches (AASHTO 2012; ACI Committee 318 2014).

10.5 Evaluation of existing shear strength models

Extensive experimental tests on the shear strength of various RC members have been carried out by researchers all over the world and a large database is available for development and evaluation of shear design models. To facilitate the development of shear design models, a number of researchers have collected, screened, and processed the shear test data carefully to provide shear test databanks, including the one for RC members without stirrups by Collins et al. (2008) containing 1849 shear tests and with details of the tests, the databank for slender RC beams without and with transverse reinforcement by Reineck et al. (2013, 2014), and the databank for deep RC beams by Mihaylov et al. (2010). These databanks have played an important role in shear strength modeling and have been widely used in the literature.

However, with these databanks the evaluation of shear strength models could only be made on the total shear strength V, because V_c and V_s cannot be measured directly from tests previously. The new experimental approach recently developed by Wu and Hu (2017) that can measure V_c and V_s in a test provides a tool for the first time to assess the models of V_c and V_s separately. A few popular models for the shear strength of RC members are evaluated in this section using the directly measured V_c and V_s from the tests reported in Wu and Hu (2017), Hu and Wu (2017), and Hu and Wu (2018). The models to be evaluated include:

(1) MC-10 model

MC-10 refers to the *fib* CEB-FIP Model Code 2010 (fib 2013). The shear strength models provided in the MC-10 for RC beams with transverse reinforcement are complicated since different mechanical concepts are adopted for different needs of complexity and accuracy. The models allow for four levels of approximation. The Level I Approximation adopts a variable-angle truss approach. The Level II Approximation is based on a generalized stress field approach, allowing the strut inclination θ to be selected within certain limits. The models corresponding to the levels I and II of approximation attribute shear resistance only to transverse reinforcement due to the truss mechanism. The Level III Approximation represents a cross-sectional design approach based on the simplified version of the modified compression field theory, utilizing the additive model $V = V_c + V_s$. The Level IV Approximation requires extensive experience to ensure safe and consistent results by considering the application conditions of equilibrium and compatibility of steel and diagonally cracked concrete. The models for the Levels I and III of approximation are used for beams with and without shear reinforcement, respectively, in the following evaluations.

(2) CSA A23.3-04

The model for the shear strength of RC members in the Canadian design code CSA A23.3-04 (CSA 2004), referred to as CSA-04 hereafter, was developed based on the modified compression field theory. The most recent MCFT in Bentz et al. (2006) is used in the following calculations with the model.

(3) JSCE-07

In the Japanese design code JSCE-07 (JSCE 2010), V_s is determined by adopting the 45-degree truss model, and V_c is empirically modeled considering three factors: member depth, transverse reinforcement ratio, and flexural resistance.

(4) ACI 318-14

The shear strength model in ACI 318-14 (ACI Committee 318 2014) is based on the 45-degree truss model for calculation of V_s supplemented by a semiempirically determined concrete contribution V_c.

(5) Zsutty (1968)

In the well-known Zsutty's model (Zsutty 1968), V_s is modeled by the 45-degree truss model. The model of concrete contribution V_c is a pure empirical model obtained from statistical analysis. The model was referred to as the best-known statistically derived equation (ASCE-ACI Committee 426 1973).

The experimentally measured results for model evaluation include the shear strength (V_{test}), shear strength contribution by concrete $V_{c\text{-}test}$, shear strength contribution by transverse reinforcement $V_{s\text{-}test}$, and the shear force at formation of CDC $V_{cr\text{-}test}$. These test results are compared with the respective theoretical results predicted by the various models. The parameters required in the model predictions of the shear strength of the test beams are summarized in Tables 10.2 and 10.3. The rejointed bars in Table 10.2 refer to the transverse reinforcing bars that were cut into two halves for installation of strain gauges in their center and subsequently rejointed back to form a complete rebar. For more information on the experimental method and the test program, refer to Wu and Hu (2017) and Hu and Wu (2018).

Table 10.2 Mechanical properties of steel reinforcement.

	Longitudinal bars		**Transverse bars**	
Mechanical properties	**T20**	**T14**	**T12 original/ rejoined**	**R12 original/ rejoined**
Yield strength (MPa)	507	561	492.1/491.7	326.0/323.0
Elastic modulus (GPa)	179	228	186.0/184.2	196.6/191.6
Yield strain (%)	0.32	0.20	0.264/0.262	0.165/0.163

Table 10.3 Summary of test specimens.

Beam ID	f'_c (MPa)	ρ_{st} (%)	ρ_{sv} (%)	s (mm)	b (mm)	d (mm)	a/d	V (kN)	Failure mode
N1.9	36.8	3.27	–	–	180	300	1.9	220.7	S
N2.5	43.0	3.27	–	–	180	300	2.5	161.3	S
N3.1	44.6	3.27	–	–	180	300	3.1	96.5	S
D1.9	42.5	3.27	0.42	250	180	300	1.9	319.8	S
D2.5	41.5	3.27	0.42	250	180	300	2.5	234.8	S
D3.1	40.3	3.27	0.42	250	180	300	3.1	211.2	S
R1.9	41.4	3.27	0.64	165	180	300	1.9	294.5	S
R2.5	40.4	3.27	0.64	165	180	300	2.5	270.5	S
R3.1	44.0	3.27	0.64	165	180	300	3.1	239.6	S
D3.1C	35.7	3.27	0.50	250	180	300	3.1	233.2	S
R3.1C	40.4	3.27	0.76	165	180	300	3.1	238.5	S

Notes: ρ_{st} = longitudinal tensile reinforcement ratio; ρ_{sv} = transverse reinforcement ratio; s = spacing of transverse reinforcement; d = effective depth; b = width of beam; V = measured shear capacity; S = shear failure; Beam ID: N = no transverse bar, D = deformed transverse bar, R = plain round transverse bar, the number following bar type = a/d, letter C after 3.1 stands for closed links.

10.5.1 Evaluation of total shear strength

Table 10.4 shows the ratios of the predicted shear strength (V_{theo}) to the experimental result (V_{test}) for the test beams. Generally, all model predictions underestimate the test results by approximately 20%–29%. Among them, ACI 318-14 is the most conservative, with the lowest average value of $V_{theo}/V_{test} = 0.71$ and the highest value of coefficient of variation (CoV) at 13%. The predictions of the model by Zsutty match the experimental results best with an average ratio of 0.8. All models with the exception of JSCE-07 give better predictions for more slender beams (D3.1 and R3.1) than moderately slender beams (D2.5 and R2.5) and deep beams (D1.9 and R1.9). This is because these models are based on the concept of truss-analogy with empirically determined contribution of concrete. They are more suitable for the B-regions than the D-regions (Fig. 2.23).

10.5.2 Evaluation of shear strength components

Summarized in Table 10.5 are ratios of the predicted shear strength contributions due to concrete $V_{c\text{-}theo}$ and those due to transverse reinforcement $V_{s\text{-}theo}$ to the experimental ones $V_{c\text{-}test}$ and $V_{s\text{-}test}$, respectively, for the test specimens. Although the existing models provide conservative predictions of V with a mean ratio of model prediction/test result of approximately 0.75, the accuracy of prediction of the two separate parts is low, ranging from 0.07 to 2.99, which reflects the significant gap between the predicted mechanisms and actual ones.

For slender beams (D3.1 and R3.1), all models are conservative for V_s while they are unconservative for V_c. The ratios of prediction/test result of V_s for D3.1 calculated by MC-10, ACI 318-14, CSA-04, JSCE-07 and Zsutty are 0.89, 0.60 0.74, 0.52 and 0.60, respectively. The case of R3.1 is similar, with the ratios of 0.83, 0.59, 0.73, 0.52 and 0.59, respectively, for the aforementioned five models. The conservatism of the models for V_s can be attributed to the fact that, although all transverse bars crossing the critical diagonal crack yield at the ultimate shear strength (Table 10.1), the actual strut angles are smaller than the predicted ones. The actual numbers of transverse bars

Table 10.4 Comparison of shear strength.

	Test	$V_{theo.}/V_{test}$					
Beam ID	**V_{test} (kN)**	**MC-10**	**ACI 318-14**	**CSA-04**	**JSCE-07**	**Zsutty (1968)**	**Aver.**
D1.9	319.8	0.60	0.58	0.58	0.74	0.69	0.64
D2.5	234.8	0.85	0.77	0.80	0.79	0.90	0.82
D3.1	211.2	0.92	0.84	0.88	0.74	0.88	0.85
R1.9	294.5	0.67	0.63	0.69	0.80	0.75	0.71
R2.5	270.5	0.68	0.67	0.74	0.68	0.78	0.71
R3.1	239.6	0.76	0.75	0.85	0.67	0.78	0.76
Aver.	–	0.75	0.71	0.76	0.74	0.80	0.75
CoV.	–	15%	13%	13%	7%	9%	11.4

Notes: CoV = coefficient of variation; V_{theo} and V_{test} = predicted and measured shear strength, respectively.

Table 10.5 Comparison of shear strength components.

Spec. ID	Test (kN)		MC-10		ACI 318-14		CSA-04		JSCE-07		Zsutty (1968)	
	V_s	V_c	*s*	*c*	*s*	*c*	*s*	*c*	*s*	*c*	*s*	*c*
D1.9	87.2	232.7	1.99	0.07	1.29	0.31	1.59	0.19	1.12	0.59	1.29	0.46
D2.5	150.1	84.7	1.16	0.31	0.75	0.79	0.93	0.54	0.65	1.02	0.75	1.14
D3.1	187.1	24.1	0.89	1.13	0.60	2.65	0.74	1.88	0.52	2.42	0.60	2.99
R1.9	63.3	231.2	2.84	0.08	1.77	0.31	2.20	0.26	1.54	0.59	1.77	0.46
R2.5	125.8	144.7	1.30	0.14	0.89	0.46	1.11	0.42	0.78	0.59	0.89	0.66
R3.1	189.4	50.1	0.83	0.51	0.59	1.33	0.73	1.26	0.52	1.21	0.59	1.48

Notes: $s = V_{s\text{-}theo}/V_{s\text{-}test}$; $c = V_{c\text{-}theo}/V_{c\text{-}test}$.

crossing the CDC are 2 and 3 for D3.1 and R3.1, respectively, larger than the 1.2 and 1.8 predicted by the models. It is noted that MC-10 performs best for V_s since it adopts a variable strut angle, but the predictions are still smaller than the actual ones. The reason for the significant overestimation of V_c is that the models do not allow for the degradation of V_c as shown in Fig. 10.9.

For the moderately slender beams R2.5 and D2.5, almost all models of V_c and V_s give conservative predictions except the V_s models of MC-10 and CSA-04 and the V_c models of JSCE-07 and Zsutty. The models generally perform better for beams with an a/d value of 2.5 than the other two cases of $a/d = 1.9$ and 3.1. An important reason is that, for the two special cases of D2.5 and R2.5 as shown in Fig. 10.9C and D, the values of V_c happen to be close to the shear resistance at the onset of CDC, V_{cr}, that is adopted as V_c in the models.

If the additive models are applied to the deep beams, the performance of the models is of interest. For D1.9, V_s is overestimated by 99%, 29%, 59%, 12%, and 29% by MC-10, ACI 318-14, CSA-04, JSCE-07 and Zsutty (1968), respectively. It is even worse for R1.9, with the overestimation of V_s being 184%, 77%, 120%, 54% and 77%, respectively, for the five models. Careful analysis of the data shows that not all transverse bars intersecting the CDC yield at the ultimate shear strength, particularly for D1.9 with high yield bars. In addition, the number of stirrups intersecting the CDC is not accurately calculated with these models, which involves inaccuracy in predicting the strut angle as well as the calculation of number of stirrups within an actual CDC by using the horizontal length of CDC divided by the spacing of stirrups, L_c/s. Considering the realistic distribution of transverse reinforcement, Loov (2002) suggested the number of stirrups crossing the CDC as $(L_c/s)-1$, which was supported by Kim and Mander (2005) and Laskar et al. (2010). However, this calculation is too conservative for the calculation of V_s based on the test results of the deep beams. In terms of V_c for D1.9 and R1.9, all models substantially underestimate its value. This is because V_c is significantly larger than V_{cr} for the short beams, as shown in Fig. 10.9A and B.

As pointed out previously, the strut-and-tie model is more appropriate to predict the shear strength of deep RC beams. According to Section 23.2.7 of ACI 318-14, for a simple strut-and-tie model, the angle θ of the inclined concrete struts should not be less than 25 degrees. The angle θ can be calculated by

$$\cot\theta = \frac{a}{z} \tag{10.14}$$

For $\theta = 25$ degrees and for the commonly assumed value of $z = 0.9d$, the critical value of a/d is 1.93. Specimens N1.9, D1.9, and R1.9 have the a/d value of 1.9, which is very close to the critical value of 1.93. These experimental results of V_c and V_s provide a test as to whether the additive models are suitable for such beams.

10.5.3 Difference between V_c and V_{cr}

V_{cr}, defined as the shear resistance immediately before the first diagonal cracking (including contributions from both concrete and stirrups), can be identified from the load–deformation response curve and by DIC when the sudden opening of a diagonal crack occurs. It was observed in the tests that the diagonal cracking captured by DIC, the strain jump in transverse reinforcement, and the point where the curves of V and V_c diverge (Fig. 10.9) are essentially the same point. Table 10.6 compares the test results of V_{cr} and V_c and the predicted values of V_c by the different models. ASCE-ACI Committee 426 (1973) recommended (it was extended to ACI 318-14 and adopted by most existing models) that "the shear strength provided by concrete, V_c, is assumed to be the same for members with and without shear reinforcement and is taken as the shear causing inclined cracking" (Clause R22.5.1.1 ACI 318-14).

It can be seen in Table 10.6 that, for deep beams (N1.9, D1.9, and R1.9) and moderately slender beams (N2.5 and R2.5), the test results of V_c at the onset of peak shear resistance (shear strength) are much larger than V_{cr}. For slender beams, however, the test result of V_c is only 33% and 58% of V_{cr} for D3.1 and R3.1, respectively. N3.1 failed shortly after the formation of the first diagonal crack. It can be seen from Table 10.6 that the average ratio of the predicted V_c to the test result of V_{cr} for all specimens are 0.43, 0.81, 0.53, 1.12, and 1.09 for MC-10, ACI 318-14, CSA-04, JSCE-07, and Zsutty (1968), respectively.

The large difference between the experimentally obtained and predicted V_c shows that V_{cr} cannot be considered as V_c except where the beam fails as soon as the formation of the first diagonal crack, such as N3.1. This finding contradicts the conclusion that the shear strength contribution of concrete V_c can be linked to the shear resistance corresponding to the formation of the first diagonal crack, which has been widely adopted in existing codes of practice and supported by some previous experimental results, such as that reported by Bousselham and Chaallal (2008).

By comparing the values of V_{cr} for beams with the same a/d but in different groups (N, D, and R groups), it is found that no clear relation between the transverse reinforcement and the cracking load V_{cr} can be identified. This observation is consistent

Table 10.6 Comparison of V_{cr} and V_c.

	Test			$V_{c\text{-}theo}/V_{cr\text{-}test}$				
	$V_{c\text{-}test}$ (kN)	$V_{cr\text{-}test}$ (kN)	$V_{c\text{-}test}/V_{cr\text{-}test}$	MC-10	ACI 318-14	CSA-04	JSCE-07	Zsutty (1968)
N1.9	220.7	90.3	2.44	0.64	0.75	0.46	1.42	1.13
N2.5	161.3	96.5	1.67	0.65	0.71	0.47	0.91	1.02
N3.1	96.5	82.5	1.17	0.9	0.81	0.56	0.74	0.90
D1.9	232.7	87.5	2.66	0.19	0.82	0.52	1.57	1.22
D2.5	84.7	93.5	0.91	0.28	0.72	0.48	0.92	1.04
D3.1	24.1	72.4	0.33	0.38	0.88	0.61	0.80	0.99
R1.9	231.2	68.2	3.39	0.28	1.04	0.65	1.99	1.56
R2.5	144.7	87.3	1.66	0.24	0.76	0.50	0.98	1.10
R3.1	50.1	86.5	0.58	0.30	0.77	0.53	0.70	0.86
Aver.	–	–	1.65	0.43	0.81	0.53	1.12	1.09
CoV.	–	–	72%	66%	15%	14%	46%	22%

with the conclusion in the Joint ACI-ASCE Committee 326 (1962). Also, no clear relationship between a/d and V_{cr} is found in Table 10.6. The results in Table 10.6 neither support that V_c is the same for beams with and without shear reinforcement nor suggest that V_c can be taken as V_{cr}.

The ACI-ASCE Committee 445 (1998) recommends that the concrete contribution be determined based on the breakdown of the actual shear transfer mechanisms at the critical diagonal crack when the shear failure is reached, rather than directly link the cracking shear resistance V_{cr} to the shear strength contribution of concrete V_c. The previous analysis of test results supports the recommendation. This conclusion points to the fact that all the existing models of V_c were developed based on an inappropriate database. However, the appropriate approach to measuring V_c from tests has been developed only recently and no extensive database of correct V_c is available so far for modeling of V_c.

With regard to the widely adopted additive models where the superposition of V_c and V_s gives the shear strength V, the latest test results in this section show that V_c and V_s vary when a beam deforms (Fig. 10.9). The code-recommended values of V_c and V_s neither remain unchanged after the onset of CDC during member deformation nor occur at the onset of peak shear resistance of a beam.

10.5.4 Other test observations

The following further observations can be made from the two experimental test programs reported by Wu and Hu (2017), Hu and Wu (2017), and Hu and Wu (2018):

1. Stress and strain distribution in transverse reinforcement

At peak shear resistance, the strain in transverse bars is related to a/d. For deep beams with a smaller a/d (1.9), the strains in transverse bars are much less than the yield strain, indicating that the transverse bars are not fully engaged in resisting the shear. Therefore, the shear capacity of deep beams is not sensitive to the transverse reinforcement ratio. For specimen D2.5 with a moderate a/d, two stirrups were intersected by the critical diagonal crack. However, only the stirrup closer to the loading point reached its yield value and the other one did not yield. In other words, not all transverse bars yielded at the peak shear resistance. This contradicts the assumption made in most models and design codes for the calculation of V_s. Similar results were reported by Zhang et al. (2015). For the specimen R2.5, however, all bars intersecting the CDC yielded at the peak shear resistance since the yield strain of the plain round bar is relatively smaller and the two stirrups intersect the CDC at its widest part of the crack. For more slender beams with the a/d of 3.1, all transverse bars intersecting the CDC yielded.

The typical strain distributions along a transverse bar intersecting a CDC are illustrated in Fig. 10.11. The number following the beam ID, i.e., 393, 320, and 427, indicates the distance from the transverse bar to the loading point. Nine strain gauges were distributed evenly inside a transverse bar along the beam height direction, with SG1 closest to the tension side of the beam and SG9 closest to the compression face. The figure reflects that the strains at different points along a transverse bar are significantly different. Only the point of a bar where it intersects the CDC and its surrounding region reaches the yield strain. Due to the higher yield strength and stronger bond of the deformed D bar than the plain R bar, beam D3.1 has a smaller region of yielding around the peak load, while specimen R3.1 shows a larger yielding zone and relatively more uniform strain distribution due to the lower yield strength of the R bar and a weaker bond. For beam D1.9, even the most strained transverse bar did not yield at the peak shear force (Fig. 10.11C).

2. Effect of type of transverse reinforcement on shear strength

Although all the beams were reinforced with the same transverse reinforcement ratio, $\rho_{sv}f_{yv}$, the shear strength of the beams with a moderate to long span length of $a/d = 2.5$ and 3.1 reinforced with plain round stirrups, R2.5 and R3.1, is higher than those reinforced with deformed stirrups, D2.5 and D3.1, by about 15%. Similar test results have also been reported in the literature (Tompos and Frosch 2002; Lee et al. 2015). With the same value of $\rho_{sv}f_{yv}$, plain round stirrups have a higher reinforcement ratio ρ_{sv} due to the smaller yield strength f_{yv}. It is believed that a larger shear reinforcement ratio provides more even confinement in the shear span and can better restrain the opening and propagation of the CDC to increase the resistance at the crack face, as well as better confine the concrete in the compression zone to enhance the shear resistance.

However, this is not the case for the deep beams with $a/d = 1.9$. The shear capacity of R1.9 is 8% lower than that of D1.9. Birrcher et al. (2013) reported that for beams with a/d of 1.2 and 1.85, the shear strength fluctuated in a certain range and no clear trend was observed when the transverse reinforcement ratio increased from 0.2% to 0.31%. This is because for deep beams the majority of shear resistance is transferred via a diagonal compression strut and the transverse reinforcement has a less significant effect on the shear strength. This is also consistent with the previous observation no. 1 that the transverse reinforcement is not effective on the shear resistance of deep beams.

3. Effect of aggregate size (Wu and Hu 2017)

The shear capacity, peak deflection, and postcracking stiffness of the beam increase with aggregate size. The result that the concrete contribution V_c increases with increasing aggregate size is consistent with the models proposed by Vecchio and Collins (1986) and Bazant and Sun (1987). With the same concrete strength, and $a/d = 2.6$ when aggregate size increases from 10 mm to 20 mm, the shear capacity increases by 58.4%, 13.8%, and 9.2% for the N, D, and R groups of the beams, respectively, which indicates that the effect of aggregate size on shear capacity decreases with the increase in the transverse reinforcement.

Comparing beams without transverse reinforcement and with maximum aggregate sizes of 10 mm (beam N10) and 20 mm (beam N20), apart from the rougher crack face of N20 caused by larger aggregate size, the diagonal crack of N20 is also steeper than that of N10 toward the compression side. It has been experimentally observed that the shear capacity of RC beams without transverse reinforcement is sensitive to the inclination of the diagonal cracks. This observation also shows that aggregate size significantly affects shear resistance.

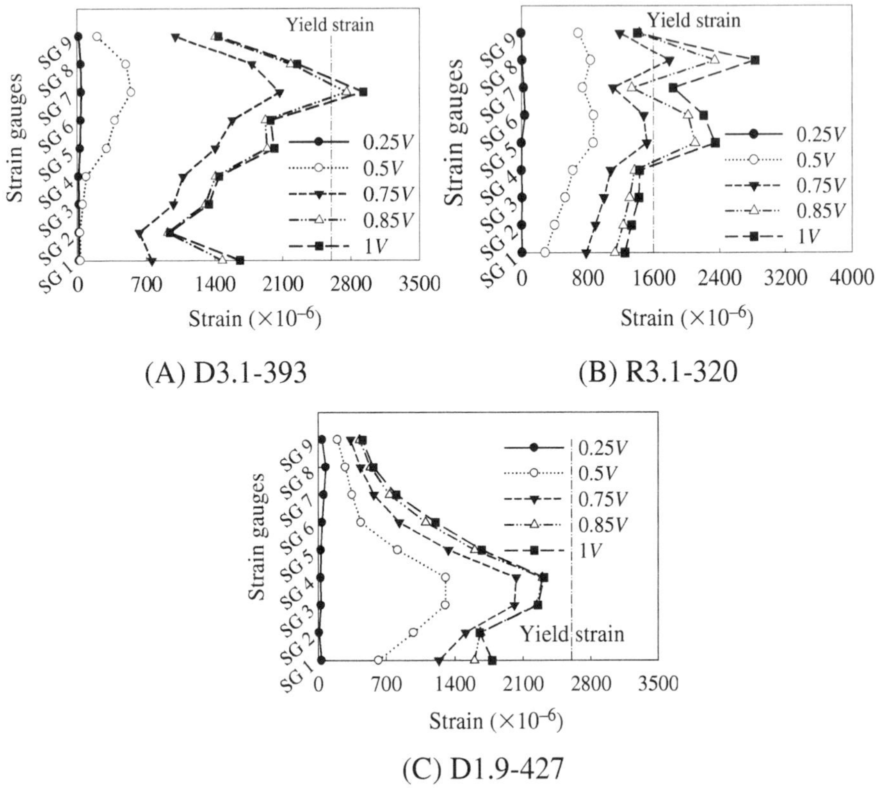

Fig. 10.11 Distribution of strain along transverse bar.
From Hu, B., and Y. F. Wu. 2018. "Effect of shear span-to-depth ratio on shear strength components of RC Beams." *Eng. Struct.* 168, 770–783.

10.5.5 Summary of findings

The following observations are made from the experimental results discussed in Section 10.5, which are related to the shear strength theory and modeling:

1. The initial cracking shear resistance of beam V_{cr} cannot be treated as the concrete contribution to shear strength V_c in general. V_{cr} is significantly larger than V_c when a/d is large, but much smaller than V_c when a/d is small.
2. V_c and V_s significantly vary during the loading process. V_s is essentially zero before diagonal cracking and increases quickly after that. At the peak shear resistance, the transverse reinforcement does not always yield. The peak values of V_c and V_s generally occur at different times. Therefore, if the additive model of $V = V_c + V_s$ is used to calculate the shear strength, V_c and V_s in the equation may not take their maximum values.
3. The values of V_s and V_c are significantly affected by a/d. More specifically, V_c decreases sharply as a/d increases, while V_s has an opposite trend. No clear trend of variation is observed for V_{cr} against a/d.
4. The V_c values for beams with and without shear reinforcement are generally different. This difference can be very large for beams with an a/d value of 2.5 and greater.

5. Although the existing code models usually give a conservative prediction of V, their performance in predicting separate components V_c and V_s is not satisfactory. When a/d is smaller than 2.5, the models are conservative for V_c while overestimating V_s. When a/d is greater than 2.5, the models are conservative for V_s while overestimating V_c. For the specimens tested by Hu and Wu (2018), the error of prediction of V_c and V_s by typical models recommended in popular design codes can be very large.
6. The strain distribution pattern of transverse reinforcement varies significantly when load increases from 25% to 75% of the peak shear resistance, after which the patterns of both the transverse reinforcement strain and crack distribution stabilize, while their magnitude further increases with loading. However, the patterns of transverse strain field and cracks do not change significantly in precracked beams (beams that have been loaded to cracking and then unloaded before the shear tests). This observation shows that cracking significantly changes the distribution of strain in the transverse reinforcement.
7. V_s increases proportionally with critical diagonal crack width before the first yield of transverse reinforcement, after which the crack width increases quickly. The increase in crack width did not cause a significant change in the magnitude of V_c in medium length beams with $a/d = 2.6$ (Hu and Wu 2017). This observation contradicts the existing models such as MCFT and the truss models with crack friction where V_c is related to crack width. A possible explanation for this phenomenon is that, although shear resistance along a CDC reduces when a crack opens, the dowel effect and the shear stress in the compression zone increase on the other hand. For deep and slender beams, the increase in crack width should have a significant effect on V_c as suggested by Fig. 10.9, assuming crack width increases when deflection increases, but with opposite trends for deep beams and slender ones.
8. For medium length beams with $a/d = 2.6$, the inclination of the principal compression strain (θ) matches well with that of principal compression stress (θ_c) before first yielding of shear reinforcement. At the onset of shear strength, θ_c deviates from θ, with a difference that could be greater than 12 degrees. The difference reduces when the transverse reinforcement ratio increases. This result contradicts the assumption made in MCFT.
9. The direction of principal compression strain in shear span varies during loading. For specimens without precrack, the inclination of principal compression strain increases with increasing shear force till the formation of the first diagonal crack. Afterward, it decreases with further increase in shear force. For precracked beams, a monotonic decrease in the inclination was observed.
10. Reinforcement type and aggregate size also affect shear behavior. In particular, aggregate size significantly affects the value of V_c. With the same shear reinforcement ratio $\rho_{sv}f_{yv}$, reinforcement type not only affects crack spacing and width but also shear strength.

These experimental findings indicate that there is room for further improvement in the existing theories and models for the shear design of reinforced concrete members.

10.6 Discussion on shear strength modeling

It can be seen from the previous sections that the prediction of the load-carrying capacity of RC members under shear failure is a highly complex problem due to the many mechanisms and a large number of factors involved in the problem. Significant progress has been made since the 1970s on quantification of shear failure mechanisms using advanced theories/models. Based on extensive experimental investigation on RC panels, Vecchio and Collins (1986) developed the sophisticated modified compression field theory that satisfies equilibrium, compatibility, and constitutive relationships of materials. Other sophisticated models have been proposed continuously, such as the rotating angle softened truss model (RA-STM) (Hsu 1988), fixed angle softened-truss model (FA-STM) (Hsu and Zhang 1997), the disturbed stress field model (DSFT) (Vecchio 2000), critical shear crack theory (CSCT) (Muttoni and Ruiz 2008), etc. These advanced theories are able to predict the shear strength and deformation for particular shear mechanisms with improved performance. However, the calculations with an advanced theory are rather complicated compared with conventional shear design procedures, involving many variables and even iterations.

Although the design theories and models have been significantly and continuously improved, the problem of shear failure modeling of RC members is still far from being resolved satisfactorily. This is reflected by the multifarious shear strength models proposed by researchers continuously, as well as the differences between predicted and measured shear strength values, which are much larger than the scatters in experimental results. This gap indicates room for further improvement (Bazant and Sun 1987). The causes of the larger error in shear strength predicted by existing models are discussed in Section 10.6.2.

After the coordinated efforts of the research community all over the world for decades, the aim of shear strength modeling that "design procedures proposed for regulatory standards should be safe, correct in concept, simple to understand, and should not necessarily add to either design or construction costs. These procedures are most effective if they are based on relatively simple conceptual models rather than on complex empirical equations" (ACI-ASCE Committee 455 1998) has not been achieved so far. The complexity involved in the development of the shear design theory is discussed in this section.

10.6.1 Classic mechanics for derivation of V_{cr}

The flexural strength problem is relatively simple because Bernoulli's plane section hypothesis is applicable, and the only complexity comes from the nonlinear stress-strain relationship of concrete, which has been well established. When the stress-strain relationship of concrete is known, the flexural strength is fixed that can be mathematically derived (see Chapter 3). For the shear problem, both the strain and stress fields are complex and cannot be calculated by classic mechanics in general, particularly after cracking. While it is possible to calculate the stress and strain field of a beam using a high-order beam theory (Section 6.2.7) during its elastic deformation stage, diagonal cracking significantly disturbs and redistributes the strain and stress fields that cannot be predicted further with classic mechanics.

There is one exception. For members with shear-tension failure without web reinforcement, the members fail quickly after the onset of diagonal cracking. The stress condition of the member immediately before cracking can reasonably be evaluated by elastic theory, because an RC member is essentially elastic and its stress and strain distributions can be calculated by the elasticity theory with relatively high accuracy. In this case, it is possible to develop the shear strength model based on elastic analysis. In fact, the V_c model proposed by ACI-ASCE Committee 326 (1962) was developed through elastic analysis in which the key parameters, basic form, and relationship with key parameters were derived from elastic theory. The coefficient of the derived model was then determined by regression analysis using a test database. Therefore, the model is semiempirical and more rational than other pure experimentally based empirical models. The approach is rational for the following reasons: (1) the onset of diagonal cracking is correctly adopted as the failure point, i.e., $V = V_c = V_{cr}$, (2) a theory is available to evaluate the stress at the onset of diagonal cracking, and (3) a sufficiently large database is available for the determination of the model coefficients. As a result, this model is rational and robust and has lasted in ACI-318 from the 1960s until now.

To be conservative in the shear design, the joint ACI-ASCE Committee 326 purposely placed the design equation "near the lower extremes of observed shear stress rather than as average values." This is the reason that its average predicted value of $V_{c\text{-}theo}/V_{cr\text{-}test}$ in Table 10.6 is 0.81, which is lower than 1.0. Clearly, the performance of the ACI-318 model is the best among the models in Table 10.6 in terms of the shear strength for members with a shear-tension failure mode without significant web reinforcement, i.e., $V = V_c = V_{cr}$.

However, this is not the case for other shear failure modes involving significant cracking and nonlinearity before the onset of shear strength, including the common shear-compression failure mode. For the shear-compression failure mode and web crushing failure mode as well as shear-tension failure with significant web reinforcement, the peak shear resistance is significantly higher than the diagonal cracking strength, i.e., $V > V_{cr}$. In this case, the elastic theory is not applicable to the derivation of the shear strength model.

10.6.2 Shear resisting system and individual shear mechanisms

As shown in Fig. 10.7, the shear resistance comes from several mechanisms, including the shear resistance due to transverse reinforcement, the shear force due to the dowel action of the longitudinal bars, the resultant shear force along the cracked interface, and the shear force contributed by the concrete in the compression zone. The relationship between these mechanisms can be analogized by the mechanical system in Fig. 10.12. The total shear force is the sum of the resistances from all the mechanisms, i.e.,

$$V = \sum_{m=1}^{n} V_m(x_m) \tag{10.15}$$

Fig. 10.12 Total shear resistance.

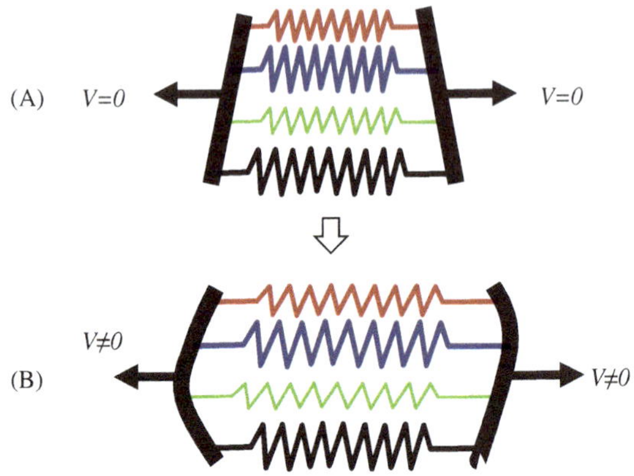

where each mechanism V_m is a nonlinear function of its own deformation x_m.

It is shown in Fig. 10.9 that different mechanisms may reach their own peak resistances at different times. In this case, it is incorrect to calculate the shear strength by adding the peak resistances of different mechanisms together. This is illustrated in Fig. 10.13 for the system with two mechanisms $V1(x)$ and $V2(x)$. For a nonlinear mechanism with certain ductility, the force–displacement relationship is typically featured by an ascending branch, a peak, and a descending branch, such as $V1(x)$ and $V2(x)$ in Fig. 10.13. The peak point of a mechanism is often referred to as an ultimate state of the particular mechanism. For a system with only one mechanism and under load control, the system collapses at the peak point or ultimate state (see Section 2.2.1). However, this is not the case for a system consisting of multiple mechanisms such as that in Fig. 10.13. Under an increasing load (load control), the system does not collapse at the ultimate displacements of the individual mechanisms, xp1 or xp2, but at the displacement xp, the peak point of the whole system. Therefore, the term "ultimate load" or "ultimate displacement" of an individual mechanism in a multiple mechanism system is irrelevant to the failure of the system. The response curve after the peak point of an individual mechanism should be referred to as a softening process of that mechanism and its peak point cannot be treated as a failure point.

The overall shear strength (peak point of the whole system) can be calculated only when the load–deformation curves of all mechanisms are known, as illustrated by the system with two mechanisms in Fig. 10.13. This causes an even greater difficulty than simply finding the peak resistances of individual mechanisms.

The complexity of the shear resisting system as described by Eq. (10.15) comes not only from the difficulty in determining the force–displacement relationship of an individual mechanism, $V_m(x_m)$, but also the relationship among the deformations of different mechanisms, i.e., $x_i \neq x_j$, as illustrated in Fig. 10.12B. The dowel deformation, the web displacement field, and the deformation of the compression zone are evaluated with their own coordinates or strain systems. These displacements must satisfy the compatibility condition, which is also complicated due to the non-Bernoulli strain distribution and discontinuity due to significant cracking.

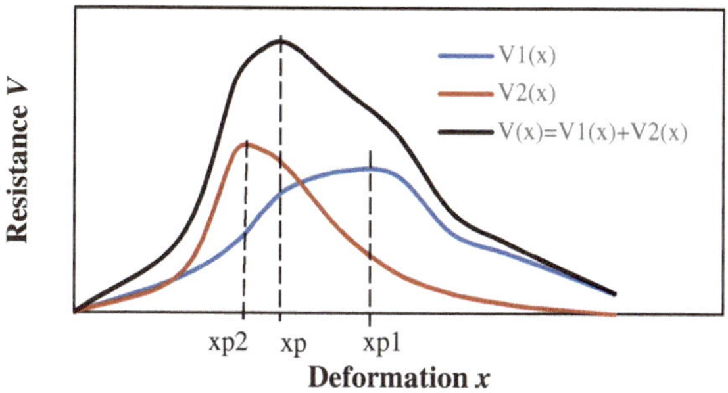

Fig. 10.13 Sum of strength components.

Both the load–deformation relationships, $V_m(x_m)$, for individual mechanisms and the compatibility relationship for the deformations of different mechanisms are not well modeled in extant literature. Therefore, it is difficult to model the shear strength problem accurately by using a rigorous mechanical model such as that in Fig. 10.12 with the current state-of-the-art theories.

10.6.3 Advanced models based on particular shear mechanisms

Numerous advanced models for shear strength modeling have been developed since the 1970s. These models are generally based on, and suitable for, particular mechanisms. A good example is the MCFT. The theory is based on shear tests of RC panels where specimens are subjected to shear in combination with biaxial compression/tension, from which the stress-strain relationships of the RC panels are derived. The advantage of the method over others is that it uses a more rational approach to calculate the shear resistance of the web based on the stress-strain relationship of the reinforced concrete web. Theoretically it can deal with any case of reinforced concrete web to calculate the web shear resistance as long as a reasonable stress-strain relationship of the web is known. The current MCFT model may still have some problems. For example, people believed that the stress-strain relationship obtained from the panel (membrane) tests is not applicable to RC beams where the stress field significantly varies within the web under the combined loading of shear and bending (Lee and Kim 2016). Also, the test result contradicts the assumption made in MCFT on the relationship between the inclination of the principal compression strain (θ) and principal compression stress (θ_c) (Section 10.5.5). However, these are technical problems that can be further improved. The basic idea to make use of the load–deformation characteristics of the RC web for calculation of its shear resistance is, in principle, a general approach that can be used to predict the shear resistance of the web under all stress conditions, regardless of the failure modes. Therefore, it should be able to predict the shear strength of RC web accurately for shear-compression, shear-tension, and web crushing failure modes, in comparison with the classic truss model that can only be applied to RC members with the web reinforcement yielded before failure.

Although these advanced theories have advantages over the classic truss model and other empirical/semiempirical models, they are for particular shear mechanism(s), e.g., the MCFT is only suitable for the web stress condition that is apparently different from that of the uncracked compression zone. Therefore, it is only suitable for calculating the web shear resistance (including both contributions from concrete and web reinforcement) and not suitable for the dowel action and the shear resistance in the compression zone. In other words, these advanced theories are for particular mechanism(s), or for specific spring(s) in Fig. 10.12, rather than for the whole shear resisting system. There are cases where the web shear resistance dominates in the whole shear resisting system, and other mechanisms are insignificant. In such cases, the web softening essentially signifies the failure of the whole shear resisting system and the shear strength can be considered attained when the peak of the dominant shear mechanism is reached. However, there are other cases where one mechanism does not dominate, e.g., in the case where the shear resistance of the compression zone is comparable with, or even larger than, that of the web. In these cases, the shear strength calculation that is only based on the web failure mechanism is inappropriate and cannot be accurate.

Generally, an advanced theory that is based on a particular mechanism is not suitable for calculation of the shear strength of an RC member where at least two mechanisms play significant roles simultaneously. In other words, these advanced shear theories do not provide a general shear design approach.

10.6.4 Criterion-based design approach

People have tried to develop shear strength models by using a failure criterion. For example, a shear strength model based on the failure condition of the compression zone was proposed (Park et al. 2006; Choi et al. 2007, 2014; Choi and Park 2007). Rankine's failure criteria, which defines the failure of material when the principal stresses resulting from the combined stresses reach material strength, were used to determine the failure condition of the concrete in the compression zone under combined normal and shear stresses. This failure criterion is not general because it is not applicable to web-crushing failure where the shear failure occurs when the compression zone is still intact, as shown in Fig. 10.5. It is also inapplicable to shear-tension failure where instability of diagonal cracking occurs without damage in the compression zone.

A failure criterion may match the failure phenomenon correctly, such as crushing or shearing-off of the compression zone at shear failure. However, it may not correctly give the failure load. For example, people usually believe that the flexural strength of an RC section is achieved at concrete crushing. It is mathematically derived and explained in Chapter 3 that the so-called "failure" or onset of peak flexural resistance is a concerted result of concrete stress in every fiber of the compression zone: the concrete stresses in all fibers of concrete in the compression zone change when bending increases, which results in the movement of the resultant compression force in the cross section. For underreinforced members, the peak flexural strength is achieved when the resultant compression force is closest to the extreme compression fiber, at which time the stress at the extreme fiber has far exceeded the compressive strength of concrete. From this point of view, it is generally inappropriate to derive the strength of a system simply by judging the condition of a particular place or based on a particular mechanism, because global failure and peak resistance are a concerted result of all parts in an RC member, as shown by Figs. 10.12 and 10.13.

Nevertheless, the compression zone failure criterion points out the fact that the shear strength of RC members is related to the resistance of the compression zone. Based on the truss model, the resistance of the compression zone should affect the shear strength contribution of concrete, V_c (Fig. 10.7). For this reason, the compression bar should have a significant effect on V_c. However, existing V_c models only consider the tension reinforcement A_{st} without considering A_{sc} (Eq. 10.11). This observation reveals a potential way to increase the shear strength of RC members by using compression reinforcement or providing a stronger compression zone in other ways, such as using a higher strength concrete. In the meantime, the accuracy of the V_c model would be increased by including A_{sc} in the model. Preliminary theoretical and experimental studies by the author's group have recently demonstrated that the compression bars do have a significant effect on V_c of RC beams.

Mathematically speaking, the failure criterion based on a local failure condition or for a particular mechanism does not guarantee the onset of global peak resistance, as illustrated in Fig. 10.13. However, it is possible that a simple criterion exists that can capture the correct state of the failure of a system and be used for calculation of the peak resistance of the system, provided the criterion is rigorously derived from the system involving all major mechanisms. The flexural theorems derived in Chapter 3 are examples of such criteria. Theorems 3.1 and 3.3 describe the status of an RC section at the onsets of flexural strength and the loss of flexural integrity, respectively. They are derived from the stress conditions of all fibers and are true failure criteria for flexural designs. They are also conceptually simple and easy for engineering applications. Therefore, it is also possible for the shear strength of RC members that such a simple criterion exists and may be derived one day.

10.6.5 Lower bound solution

No theory or model reported in the extant literature covers all shear mechanisms or the major shear mechanisms of RC members simultaneously. There are cases where a mechanism dominates, and other mechanisms can be neglected with insignificant error. In this case, the relevant model performs well. But the same model would give a significant error in other cases where more nontrivial mechanisms are involved. From this point of view, no existing model can predict the shear strength of RC members as a general theory. This is the state-of-the-art of the shear design theory.

On the other hand, a theory or model developed from a specific mechanism provides a lower bound solution of the shear strength since some internal resistances or mechanisms are omitted. The accuracy of a lower bound solution varies case by case depending on the relevancy of the predicted mechanism to the actual mechanisms involved in the RC member, as well as the performance/accuracy of the model for the relevant mechanism. A lower bound solution does not always give an underestimation of result. An inaccurate model for one mechanism may still overestimate the shear strength of RC members involving several mechanisms.

10.7 Potential solutions

Based on the studies in this chapter, it is clear that developing a rational, more accurate, yet still simple shear design model is a daunting task. The relatively poor accuracy of shear design models compared with those of flexural design models is due not only to the gap between the actual mechanisms involved in the shear problem and those represented by a model, but also to the physical nature of the shear failure that is more unstable than flexural failure. The instability of shear failure is evident from the greater scatter of shear test results than flexural tests, even in repeated tests of identically designed RC members. Nevertheless, it is certain that the gap between a theoretical model and the actual physical mechanisms involved can be further reduced with further and more rational research. Possible improvements that might be made in the future are discussed in the following.

10.7.1 More sophisticated modeling of the load–deformation process for the major shear mechanisms

This approach is to implement the system in Fig. 10.12 more accurately with Eq. (10.15). Sufficiently accurate models of $V_m(x_m)$ and compatibility relationships among the deformations of different mechanisms are required before the system can be analytically, mathematically, or numerically developed for a general shear strength modeling. The ideal outcome of this approach is a simple criterion similar to that for flexural design, i.e., the onset of ε_{cm} at extreme compression fiber for calculation of flexural strength and the onset of ε_{cu} for calculation of ultimate curvature (Section 3.2). The shear strength of an RC member can then be calculated based on the condition set out by the criterion. Complicated equations and lengthy calculation processes are usually an indication that further research can be done.

10.7.2 Design RC details of a potential shear failure zone to have a simple and desirable load path and failure mode

Engineers can actually design the RC details of an RC structure to have a preferred load path and failure mode, which facilitates more convenient calculation of shear strength. Good examples are the strut-and-tie model and the diagonally reinforced coupling beams between shear walls (Paulay and Binney 1974). As shown in Fig. 10.14, the cantilever RC beam can be designed to have different internal strut-and-tie structures. The load paths in Fig. 10.14A and B are different but both can provide a design to transfer the load to its support safely. The coupling beam between shear walls can either be designed as a normal RC beam (Fig. 10.15A) or the diagonally reinforced coupling beams (Fig. 10.15B).

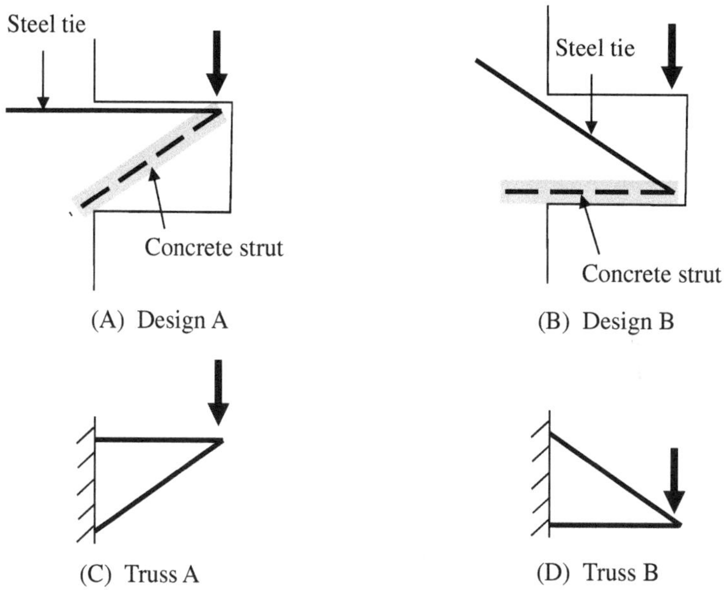

Fig. 10.14 Selection of strut-and-tie structures.

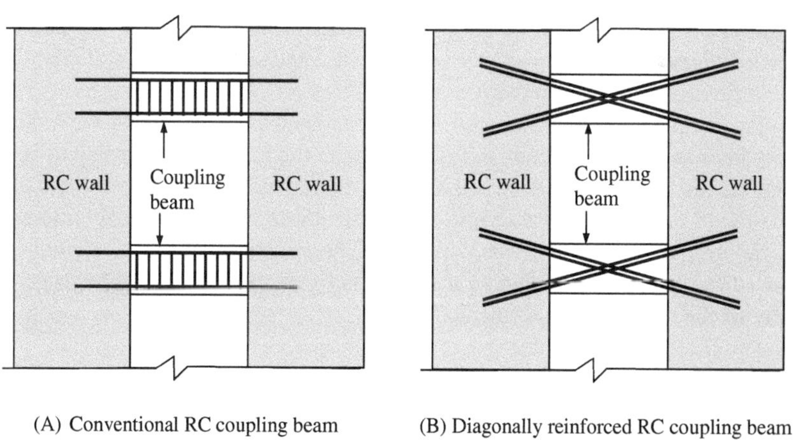

Fig. 10.15 RC coupling beams.

The design of the strut-and-tie structure in Fig. 10.14 and the diagonally reinforced coupling beam in Fig. 10.15B are both simplified to a truss structure. In reality, both beam action and arch (strut action) exist in RC members (Section 5.6). When the real structure is simplified to a pure truss, the beam action or the other mechanism/load path is ignored. Consequently, the calculated load-carrying capacity is usually smaller than the actual one. In general, the structural strength calculated based on a designed load path is a lower bound solution (Section 11.7.1). Therefore, it is possible that the calculated shear strength caused by designing the RC structures with a strut-and-tie model has a poor accuracy when the other neglected mechanisms are significant. Such inaccuracy caused by STM is well noted in the literature (Aguilar et al. 2002; Matamoros and Wong 2003).

Another potential problem with a designed load path is that significant redistribution of internal stress and hence cracking may occur before the load path is changed within the RC member to the designed load path. It can be visualized in Fig. 10.14 that the original stress field of the beam in the elastic stage is clearly different from that of the truss in Fig. 10.14C or D. With cracking and redistribution of the internal stress field under increasing load, the internal stress field will approach that of the truss as designed, but it may require significant cracking to enable the change of load path. As excessive cracking is not allowed under service load, this requires as small stress redistribution as possible under service load. This is the reason that, as a good practice of design, the stress field of the designed strut-and-tie structure should be as close to the elastic stress field of the real structure as possible. To restrain cracking caused by redistribution of internal stress, adequate distribution reinforcement should be provided to limit crack width in structures designed by a strut-and-tie model.

Regardless of the possible inaccuracy in the calculated load-carrying capacity of a structure based on a designed load path such as the strut-and-tie model, such designs are sometimes preferred when conventional beam design theory is unsuitable or incompetent, e.g., for D-regions of an RC member. This is because it has a clear load path and can be designed by a simple structural concept, which gives engineers confidence and assurance, in contrast to a complicated design theory, which often confuses designers and causes a sense of insecurity, and consequently more reinforcement is provided for a "conservative" design.

It is important to note that a structure with a designed load path must have sufficient ductility in order for the structure to finally transfer the load path according to design, i.e., the failure mode should be sufficiently ductile unless the stress field of the designed load path is close enough to the elastic stress field of the real structure. A brittle failure mode may cause the collapse of the structure before its internal stress field is sufficiently redistributed to the designed one. In this case, the design load capacity of the structure cannot be achieved.

10.7.3 Using a structural fuse to convert the shear failure mode to another one

It is discussed in Section 9.10 that a structural fuse can be used to avoid an undesirable failure mode. The concept of structural fuse can also be applied to RC structures to ensure that a desirable failure mode always occurs before shear failure, i.e., designing a suitable structural fuse and ensuring the fuse is triggered before shear failure. In this way, the shear failure would be irrelevant, and hence the shear design might be avoided. One possible way is to use the compression yielding (CY) fuse introduced in Section 9.4 in an appropriate way, as discussed in the following text.

A weaker but more ductile SIFCON material than the surrounding normal concrete in a beam can be used as the CY material. The CY material is cast in the compression zone of the potential shear failure region. For the majority of RC members, particularly RC members with significant web reinforcement, the shear failure occurs in a way (mode) that the concrete in the compression zone above the critical diagonal crack crushes. If that local part of the concrete in the potential crushing zone is replaced by the CY material that yields before concrete crushing while sustaining higher compression and shear strain than normal concrete, the following effects are expected:

(1) The potential concrete crushing zone would deform more than usual before shear failure while sustaining its loading capacity without crushing or being sheared off.
(2) The shear resistance contributed by the compression zone (part of V_c) would be increased, as it does not crush.
(3) The greater compression or shear deformation of the CY material would cause a closing effect of the CDC and consequently increase the aggregate interlocking, friction, and residual tension resistance along the CDC while maintaining its shear resistance at the compression zone, and hence increase the shear resistance of the member (a pilot study by numerical simulations confirmed a higher shear strength of such members). In other words, the shear resistance after compression yielding is greater than that before compression yielding. Consequently, shear failure would not occur during compression yielding.
(4) The failure condition of such beams would be the significant yielding of the ductile CY material above the CDC, which should be easier to calculate than evaluating the shear strength of RC members.

Should these effects be achievable in RC members, the shear design is changed to another simpler problem to calculate the loading condition at which the ductile CY material yields. This approach can be analogized by designing a link in Fig. 1.9 that represents shear failure. The link is designed to open up at a certain load, while its resistance increases in the opening up process (hardening resistance). As the onset of opening up of the link is considered as a failure of the structure and its subsequent maximum resistance is not a concern, the design problem becomes an easier one of calculating the opening up resistance of the link rather than the peak resistance of the link.

10.7.4 Developing shear strength model with AI technology

The modeling of shear strength of RC members is a challenging problem that cannot be well resolved so far. For a problem with many variables, unknown relationships, high complexity, and uncertainty such as that shown in Fig. 10.12, artificial intelligence (AI) technology is usually a good tool that can help to solve complex problems. AI is an approach provided by a machine to mimic intelligent human behavior and use human-inspired algorithms for modeling conventionally defiant problems. AI has been widely and successfully used to solve many structural engineering problems (Salehi and Burgueño 2018). AI technology includes machine learning, pattern recognition, deep learning, expert systems, fuzzy logic, genetic algorithms, artificial neural networks, etc.

AI has also been successfully applied to develop shear strength models (Cladera and Marí 2004; El-Chabib et al. 2005; Jung and Kim 2008; Bashir and Ashour 2012; Lee and Lee 2014; Shahnewaza et al. 2020; Zhang et al. 2020). A model developed from an AI method, including those with a formula derived, is generally empirical. Its performance mainly depends on the extensiveness and accuracy of the shear test database, apart from the suitability of the AI method used for modeling the relevant problem. Very large test databanks are available after the extensive research on the shear strength of RC members by researchers all over the world for over a century (see the beginning of Section 10.5). Therefore, it is possible to derive an empirical model that is more accurate than existing ones using a suitable AI method, provided that the database is well screened and characterized, and covers well all important and necessary aspects of the shear problem of RC members.

References

AASHTO (American Association of State Highway and Transportation Officials). 2012. *AASHTO LRFD bridge design specifications*, seventh Ed. Washington, DC.

ACI Committee 318. 2014. *Building code requirements for structural concrete and commentary*. ACI 318-14, Farmington Hills, MI, USA: American Concrete Institute.

ACI-ASCE Committee 326. 1962. *Shear and diagonal tension: Report of ACI-ASCE Committee 326*. American Society of Civil Engineers.

ACI-ASCE Committee 455. 1998. "Recent approaches to shear design of structural concrete (ACI 445R-99).". *J. Struct. Eng.* 124 (12), 1375–1417.

Aguilar, G., A. B. Matamoros, G. Parra-Montesinos, J. A. Ramirez, and J. K. Wight. 2002. "Experimental evaluation of design procedures for shear strength of deep reinforced concrete beams.". *ACI Struct. J.* 99 (4), 539–548.

ASCE-ACI Committee 426. 1973. "The shear strength of reinforced concrete members.". *J. Struct. Div.* 99 (6), 1091–1187.

Bashir, R., and A. Ashour. 2012. "Neural network modelling for shear strength of concrete members reinforced with FRP bars.". *Compos. Part B Eng.* 43 (8), 3198–3207.

Bazant, Z. P., and B. H. Oh. 1983. "Crack band theory for fracture of concrete.". *Mater. Constr.* 16 (3), 155–177.

Bazant, Z. P., and H. H. Sun. 1987. "Size effect in diagonal shear failure: Influence of aggregate size and stirrups.". *ACI Mater. J.* 84 (4), 259–271.

Belarbi, A., and T. T. C. Hsu. 1994. "Constitutive laws of concrete in tension and reinforcing bars stiffened by concrete.". *ACI Struct. J.* 91 (4), 465–474.

Belarbi, A., and T. T. C. Hsu. 1995. "Constitutive laws of softened concrete in biaxial tension-compression.". *ACI Struct. J.* 92 (5), 562–573.

Bentz, E. C., F. J. Vecchio, and M. P. Collins. 2006. "Simplified modified compression field theory for calculating shear strength of reinforced concrete elements.". *ACI Struct. J.* 103 (4), 614–624.

Birrcher, D. B., R. G. Tuchscherer, M. Huizinga, and O. Bayrak. 2013. "Minimum web reinforcement in deep beams.". *ACI Struct. J.* 110 (2), 297–306.

Bousselham, A., and O. Chaallal. 2008. "Mechanisms of shear resistance of concrete beams strengthened in shear with externally bonded FRP.". *J. Compos. Constr.* 12 (5), 499–512.

Brown, M. D., and O. Bayrak. 2008. "Design of deep beams using strut-and-tie models: Part II—Design recommendations.". *ACI Struct. J.* 105 (4), 405–413.

CEB (Comité Euro-International du Béton). 1978a. *International system of unified standard codes of practice for structures: Volume II, CEB-FIP model code for concrete structures*. Paris: Comite Euro-International du Beton.

CEB (Comité Euro-International du Béton). 1978b. "Shear and torsion: Explanatory and viewpoint papers on model code chapters 11 and 12.". In *CEB bulletin 126*, Paris: CEB Committee V.

CEN (European Committee for Standardization). 2004. *Eurocode 2: Design of concrete structures. Part 1–1: General rules and rules for buildings.* NV 1992-1-1: 2004, Brussels, Belgium: Comité Européen de Normalisation.

Chen, G. M., J. F. Chen, and J. G. Teng. 2012. "On the finite element modelling of RC beams shear-strengthened with FRP.". *Construct. Build Mater.* 32, 13–26.

Choi, K. K., and H. G. Park. 2007. "Unified shear strength model for reinforced concrete beams—Part II: Verification and simplified method.". *ACI Struct. J.* 104 (2), 153.

Choi, K. K., H. G. Park, and J. K. Wight. 2007. "Unified shear strength model for reinforced concrete beams—Part I: Development.". *ACI Struct. J.* 104 (2), 142–152.

Choi, K. K., W. C. Sim, J. C. Kim, and H. G. Park. 2014. "Maximum shear strength of slender RC beams with rectangular cross sections.". *J. Struct. Eng.* 141 (7), 04014184.

Cladera, A., and A. R. Marí. 2004. "Shear design procedure for reinforced normal and high-strength concrete beams using artificial neural networks. Part I: beams without stirrups.". *Eng. Struct.* 26 (7), 917–926.

Cladera, A., A. Marí, C. Ribas, J. Bairán, and E. Oller. 2015. "Predicting the shear-flexural strength of slender reinforced concrete T and I shaped beams.". *Eng. Struct.* 101, 386–398.

Collins, M. P. 1978. "Towards a rational theory for RC members in shear.". *J. Struct. Div.* 104 (4), 649–666.

Collins, M. P., and D. Mitchell. 1980. "Shear and torsion design of prestressed and non-prestressed concrete beams.". *J. Prestress. Concr. Inst.* 25 (5), 32–100.

Collins, M. P., and D. Mitchell. 1985. "Chapter 4—Shear and torsion.". In *CPCA Concrete Design Handbook*, Ottawa, Canada: Canadian Portland Cement Association. 4–1 to 4–51.

Collins, M. P., E. C. Bentz, and E. G. Sherwood. 2008. "Where is shear reinforcement required? Review of research results and design procedures.". *ACI Struct. J.* 105 (5), 590.

CSA (Canadian Standards Association). 2004. *Design of concrete structures.* CSA A23.3-04, Mississauga, Ontario, Canada: CSA Committee.

El-Chabib, H., M. Nehdi, and A. Said. 2005. "Predicting shear capacity of NSC and HSC slender beams without stirrups using artificial intelligence.". *Comput. Concr.* 2 (1), 79–96.

Feng, D. C., X. D. Ren, and J. Li. 2018. "Softened damage-plasticity model for analysis of cracked reinforced concrete structures.". *J. Struct. Eng.* 144 (6), 04018044.

fib (International Federation for Structural Concrete). 2013. *fib Model Code for Concrete Structures 2010.* Lausanne, Switzerland.

Gambarova, P. G. 1979. "Aggregate interlock role in RC thin-webbed beams in shear.". *J. Struct. Div.* 113 (1), 1–19.

Ghahremannejad, M., and A. Abolmaali. 2018. "Prediction of shear strength of reinforced concrete beams using displacement control finite element analysis.". *Eng. Struct.* 169, 226–237.

Hsu, T. T. C. 1988. "Softened truss model theory for shear and torsion.". *ACI Struct. J.* 85 (6), 624–635.

Hsu, T. T. C., and L. X. B. Zhang. 1997. "Nonlinear analysis of membrane elements by fixed-angle softened-truss model.". *ACI Struct. J.* 94 (5), 483–492.

Hu, B. 2017. "New approach to shear resistance of reinforced concrete members." Ph.D. thesis, City University of Hong Kong.

Hu, B., and Y. F. Wu. 2017. "Quantification of shear cracking in reinforced concrete beams.". *Eng. Struct.* 147, 666–678.

Hu, B., and Y. F. Wu. 2018. "Effect of shear span-to-depth ratio on shear strength components of RC Beams.". *Eng. Struct.* 168, 770–783.

Hwang, S. J., and H. J. Lee. 2002. "Strength prediction for discontinuity regions by softened strut-and-tie model.". *J. Struct. Eng.* 128 (12), 1519–1526.

Isenberg, J. 1993. "Finite element analysis of reinforced concrete structures II.". In *Proceedings, International Workshop*, New York: American Society of Civil Engineers. June 2–5, 1991.

JSCE (Japan Society of Civil Engineers). 2010. "Standard Specification for Concrete Structures–2007.". In *JSCE Guidelines for Concrete*, Tokyo, Japan: JSCE No. 15.

Jung, S., and K. S. Kim. 2008. "Knowledge-based prediction of shear strength of concrete beams without shear reinforcement.". *Eng. Struct.* 30 (6), 1515–1525.

Kani, G. N. J. 1964. "The riddle of shear failure and its solution.". *ACI J. Proc.* 61 (4), 441–467.

Kim, J. H., and J. B. Mander. 2005. "Theoretical shear strength of concrete columns due to transverse steel.". *J. Struct. Eng.* 131 (1), 197–199.

Kupfer, H. 1964. "Erweiterung der Mörsch schen Fachwerkanalogie mit Hilfe des Prinzips vom Minimum der Formanderungsarbeit (Generalization of Mörsch's Truss Analogy Using the Principle of Minimum Strain Energy).". In *Comite Euro-International du Beton, Bulletin d'Information*, Paris: CEB, pp. 44–57. No. 40.

Kupfer, H., and H. Bulicek. 1991. "Comparison of fixed and rotating crack models in shear design of slender concrete beams.". In *Progress in Structural Engineering*, D. E. Grierson, et al., eds., Dordrecht, The Netherlands: Kluwer, pp. 129–138.

Laskar, A., T. T. Hsu, and Y. L. Mo. 2010. "Shear strengths of prestressed concrete beams part I: Experiments and shear design equations.". *ACI Struct. J.* 107 (03), 330–339.

Lee, J. Y., and J. Kim. 2016. "Simplified equation based on compatibility-aided truss model for shear strength of reinforced concrete beams.". *ACI Struct. J.* 113 (6), 1301.

Lee, S., and C. Lee. 2014. "Prediction of shear strength of FRP-reinforced concrete flexural members without stirrups using artificial neural networks.". *Eng. Struct.* 61, 99–112.

Lee, J. Y., D. H. Lee, J. E. Lee, and S. H. Choi. 2015. "Shear behavior and diagonal crack width for RC beams with high strength shear reinforcement.". *ACI Struct. J.* 112 (3), 323–333.

Loov, R. E. 2002. "Shear design of uniformly loaded beams.". In *Sixth International Conference on Short and Medium Span Bridges*, Vancouver, BC, Canada.

Marí, A., J. M. Bairán, A. Cladera, and E. Oller. 2016. "Shear design and assessment of reinforced and prestressed concrete beams based on a mechanical model.". *J. Struct. Eng.* 142 (10), 04016064.

Matamoros, A. B., and K. H. Wong. 2003. "Design of simply supported deep beams using strutandtie models.". *ACI Struct. J.* 100 (6), 704–712.

Mihaylov, B. I., E. C. Bentz, and M. P. Collins. 2010. "Behavior of large deep beam subjected to monotonic and reversed cyclic shear.". *ACI Struct. J.* 107 (6), 726–734.

Mihaylov, B. I., E. C. Bentz, and M. P. Collins. 2013. "Two-parameter kinematic theory for shear behavior of deep beams.". *ACI Struct. J.* 110 (3), 447–455.

Mörsch, E. 1909. *Concrete-steel construction*. New York: McGraw-Hill. English translation by E. P. Goodrich.

Mörsch, E. 1920. *Der Eisenbetonbau-Seine Theorie und Anwendung (Reinforced Concrete Construction—Theory and Application)*, fifth Ed. Vol. 1, Part 1 Stuttgart: Wittwer.

Mörsch, E. 1922. *Der Eisenbetonbau-Seine Theorie und Anwendung*, fifth Ed. Vol. 1, Part 2 Stuttgart: Wittwer.

Muttoni, A., and M. F. Ruiz. 2008. "Shear strength of members without transverse reinforcement as function of critical shear crack width.". *ACI Struct. J.* 105 (2), 163–172.

Nielsen, M. P. 1984. *Limit analysis and concrete plasticity*. Englewood Cliffs, NJ: Prentice-Hall.

Nielsen, M. P., and N. W. Braestrup. 1975. *Plastic shear strength of reinforced concrete beams*. Technical Report 3, Bygningsstatiske Meddelelser, V. 46.

Pang, X. B. D., and T. T. C. Hsu. 1996. "Fixed-angle softened-truss model for reinforced concrete.". *ACI Struct. J.* 93 (2), 197–207.

Park, R., and T. Paulay. 1975. *Reinforced concrete structures*. New York: John Wiley & Sons.

Park, H. G., K. K. Choi, and J. K. Wight. 2006. "Strain-based shear strength model for slender beams without web reinforcement.". *ACI Struct. J.* 103 (6), 783–793.

Paulay, T., and J. R. Binney. 1974. "Diagonally reinforced coupling beams of shear walls.". *ACI Spec. Publ.* 42, 579–598.

Ramirez, J. A., and J. E. Breen. 1991. "Evaluation of a modified truss-model approach for beams in shear.". *ACI Struct. J.* 88 (5), 562–571.

Reineck, K. H. 1990. "Mechanical model for the behaviour of reinforced concrete members in shear.". PhD thesis, University of Stuttgart.

Reineck, K. H., E. C. Bentz, B. Fitik, D. A. Kuchma, and O. Bayrak. 2013. "ACI-DAfStb database of shear tests on slender reinforced concrete beams without stirrups.". *ACI Struct. J.* 110 (5), 867–876.

Reineck, K. H., E. C. Bentz, B. Fitik, D. A. Kuchma, and O. Bayrak. 2014. "ACI-DAfStb databases for shear tests on slender reinforced concrete beams with stirrups.". *ACI Struct. J.* 111 (5), 1147–1156.

Reinhardt, H. W. 1986. "The role of fracture mechanics in rational rules for concrete design.". In *IABSE surveys (C-36)*, IABSE: Zurich, pp. 1–15.

Ritter, W. 1899. "Die bauweise Hennebique.". *Schweizer. Bauzei.* 33 (7), 59–61.

Ruiz, M. F., and A. Muttoni. 2009. "Applications of critical shear crack theory to punching of reinforced concrete slabs with transverse reinforcement.". *ACI Struct. J.* 106 (4), 485–494.

Russo, G., R. Venir, and M. Pauletta. 2005. "Reinforced concrete deep beams-shear strength model and design formula.". *ACI Struct. J.* 102 (3), 429–437.

Salehi, H., and R. Burgueño. 2018. "Emerging artificial intelligence methods in structural engineering.". *Eng. Struct.* 171, 170–189.

Schlaich, J., I. Schäfer, and M. Jennewein. 1987. "Towards a consistent design of structural concrete.". *J. Prestress. Concr. Inst.* 32 (3), 74–150.

Scott, R. M., J. B. Mander, and J. M. Bracci. 2012. "Compatibility strut-and-tie modeling: part I—formulation.". *ACI Struct. J.* 109 (5), 635–644.

Shahnewaza, M. D., A. Rteilb, and M. S. Alamb. 2020. "Shear strength of reinforced concrete deep beams—a review with improved model by genetic algorithm and reliability analysis.". *Structures* 23, 494–508.

Talbot, A. N. 1909. "Tests of reinforced concrete beams: Resistance to web stresses series of 1907 and 1908.". In *Bulletin 29*, Urbana, Ill: University of Illinois Engineering Experiment Station.

Tompos, E. J., and R. J. Frosch. 2002. "Influence of beam size, longitudinal reinforcement, and stirrup effectiveness on concrete shear strength.". *ACI Struct. J.* 99 (5), 559–567.

Vecchio, F. J. 2000. "Disturbed stress field model for reinforced concrete: formulation.". *J. Struct. Eng.* 126 (9), 1070–1077.

Vecchio, F. J., and M. P. Collins. 1986. "The modified compression field theory for reinforced concrete elements subjected to shear.". *ACI J., Proc.* 83 (2), 219–231.

Wagner, H. 1929. "Ebene Blechwandträger mit sehr dünnem Stegblech (Metal beams with very thin webs).". *Zeitsch. Flugtech. Motorloftsch.* 20, 8–12. Berlin.

Withey, M. O. 1907. "Tests of plain and reinforced concrete series of 1906.". *Bull. Univ. Wisconsin, Eng. Ser.* 4 (1), 1–66.
Withey, M. O. 1908. "Tests of plain and reinforced concrete series of 1907.". *Bull. Univ. Wisconsin, Eng. Ser.* 4 (2), 1–66.
Wu, Y. F., and B. Hu. 2017. "Shear strength components in reinforced concrete members.". *J. Struct. Eng.* 143 (9), 04017092.
Zhang, T., P. Visintin, and D. J. Oehlers. 2015. "Shear strength of RC beams with steel stirrups.". *J. Struct. Eng.* 142 (2), 04015135.
Zhang, G., Z. H. Ali, M. S. Aldlemy, M. H. Mussa, S. Q. Salih, M. M. Hameed, and Z. M. Yaseen. 2020. "Reinforced concrete deep beam shear strength capacity modelling using an integrative bio-inspired algorithm with an artificial intelligence model.". *Eng. Comput.*. https://doi.org/10.1007/s00366-020-01137-1.
Zsutty, T. C. 1968. "Beam shear strength prediction by analysis of existing data.". *ACI J. Proc.* 65 (11), 943–951.

Modeling

11

11.1 Introduction

Models are of great importance in science and engineering. Theories are almost invariably expressed using mathematical models. All engineering calculations are performed using models. Modeling is the process of generating a model as a conceptual representation of some phenomenon. Important issues related to the development and application of models in reinforced concrete (RC) are discussed in this chapter.

11.2 Types of model

Models are widely used in everyday life and are effective in almost all fields. A map is a model of a location, a watch is a model of time, and a Barbie doll is a model of a woman. The definitions of different models can be found in dictionaries and encyclopedias. Differences exist in the use of terminology and definitions of various models. The definitions of models in this chapter are for engineering and are mainly based on Wikipedia definitions.

A model is an informative representation of an object, person, or system. It can be physical, mathematical, or logical. A proper model is a substitute for direct use of, measurement of, and experimentation with a real object, which is usually expensive, time consuming, and inconvenient. A model can help explain a system and study the effects of different components, and make predictions about behavior. In this book, a model refers to a conceptual or mathematical representation of a material or structural system, e.g., a set of equations attempting to describe a specific behavior of a particular type of structure.

Some of the common terms for models used in structural engineering are given here:

(1) Scientific model

A scientific model is a simplified abstract view of a complex reality. A scientific model represents empirical objects, phenomena, and physical processes in a logical way. The function of the model is to make a particular element or feature of an object easier to understand, define, quantify, visualize, or simulate by referencing it to common sense and knowledge. Different types of models have different aims, e.g., a conceptual model for better understanding, an operational model to operationalize, a mathematical model for quantification, and a graphical model to visualize a subject.

(2) Conceptual model

A conceptual model is the conceptual representation of a subject used to help people know, understand, or simulate the represented subject. The primary objective of a conceptual model is to convey the fundamental principles and basic functionality of the subject. Samples of conceptual models are the flow chart for a process and a composition structure

Analytical Approaches for Reinforced Concrete. https://doi.org/10.1016/B978-0-12-821164-9.00011-5

of a computer program. The truss model in Fig. 10.8B is an example of a conceptual model that represents the truss mechanism of the real structure in Fig. 10.8A.

(3) Mathematical model

A mathematical model is a description of a system using mathematical concepts and language. Mathematical models are usually composed of relationships and variables. Relationships can be described by operators, such as algebraic operators, functions, differential operators, etc. Variables are abstractions of system parameters of interest, which can be quantified. A mathematical model is the set of functions that describe the relations between different variables. Different classifications can be made for mathematical models, such as:

- Linear vs. nonlinear

  Linear models utilize linear mathematical relationship while nonlinear ones adopt nonlinear math functions.
- Static vs. dynamic

  Static models do not involve time and dynamic ones are time-dependent. A quasistatic model is for a process that happens slowly enough for the system to remain in static equilibrium.
- Explicit vs. implicit

  Explicit models give results directly from equations. Implicit models require iterations to get results.
- Discrete vs. continuous

  Discrete models give results at discrete points. Continuous functions are used in continuous models.
- Deterministic vs. probabilistic (statistical, stochastic)

  A deterministic model is for a system in which no randomness is involved. A deterministic model will produce the same output from a given starting condition or initial state. In a probabilistic model, also referred to as a stochastic or statistical model, randomness is involved and variable states are not described by unique values, but rather by probability distributions.
- Deductive vs. inductive

  A deductive model is a result of logical derivation based on a theory. An inductive model arises from empirical findings and generalization from them.
- Quantitative vs. qualitative

  Quantitative models are compact representations where equations are usually used to describe and quantify a specific behavior of system. Engineering design formulas are mostly quantitative models. Qualitative models do not require mathematical formulas, but are used to define, represent, or draw mechanisms, relationships, ideas, hunches, patterns, etc. The strain and stress diagrams for flexural design of RC members (Fig. 2.16) are qualitative models. The truss analogy for shear design is also a qualitative model.
- Black box vs. white box

  A black box model is a system about which there is no a priori information of the system available. A white box model (also called glass box or clear box) is a system where all necessary information of the system is available. When all information is available, one can build a white box model based on first principles. When no system information is available, a black box model can be used. Artificial neural network (ANN) models are typical black box models. A practical system is often somewhere between the black box and white box models, which is called a gray box model.

  In a gray box modeling, although what is going on inside the system is not entirely known, a model based on both insight into the system and experimental data can be constructed. Such models usually have a number of unknown parameters that can be estimated through system identification. Gray box modeling is also known as semi-physical modeling.

Almost all equations and formulas used for design calculations in engineering are mathematical models. Lack of agreement/match between predictions from theoretical mathematical models and experimental measurements often leads to important advances, as better theories are developed.

(4) Phenomenological model

A phenomenological model is a scientific model that describes the empirical relationship of phenomena to each other, in a way that is consistent with fundamental theory, but is not directly derived from theory. In other words, a phenomenological model is not derived from first principles. A phenomenological model forgoes any attempt to explain why the variables interact the way they do, and simply attempts to describe the relationship, with the assumption that the relationship extends beyond the measured values. Regression analysis is sometimes used to create statistical models that serve as phenomenological models. The continuum mechanics based constitutive relationships of materials used in engineering, such as plasticity models and damage plasticity models, are usually phenomenological models.

(5) Nanoscale, microscale, mesoscale, macroscale, and multiscale models

The scale of a model depends on the size of the object to be modeled, as illustrated by the multiscale modeling of concrete structure in Fig. 11.1. Conventional models of structures using the finite element method are at a macroscale level where concrete material is considered as a uniform material and phenomenological models are used as constitutive relationships. Mesoscale models are at a lower level than macroscale models and study the behavior of macroscale material by treating them as heterogeneous materials, e.g., a mesoscale model of concrete considers aggregates, matrix material, and the interfacial transition zone (ITZ) as separate constituents. Microscale models simulate further fine-scale details based on the physical and chemical behavior of fine-scale materials. The results are used as input for mesoscale simulations. Nanoscale models study materials at a molecular level, through quantum mechanics/molecular dynamics. Its results are used for microscale modeling. Multiscale modeling applies models at different scales such as nano-, micro-, meso-, and macroscales simultaneously to solve various physical, chemical, and engineering problems.

(6) Theoretical vs. empirical

A theoretical model is mathematically derived from accepted facts and knowledge. An empirical model refers to any kind of model based on empirical observations rather than on mathematically describable relationships of the system modeled. A semiempirical model is an empirical model involving theoretical and mathematical analysis in the derivation process. The majority of models used in engineering are empirical or semiempirical models.

(7) Analytical vs. numerical model

An analytical model has a mathematical closed form solution or can be expressed as an analytic function. The trend, pattern, or convergency of predicted results from an analytical model can be mathematically studied. Numerical models give purely numerical outputs. To get different predictions under different initial or boundary conditions, one needs to run the model with different inputs. The widely used finite element method is a typical numerical model.

(8) Mechanistic model vs. statistical/empirical model

A mechanistic model is a scientifically based, simplified, but versatile description of how the system responds to inputs. It should be able to provide an explanation for the behavior of the system and a platform for extrapolation beyond the original test database.

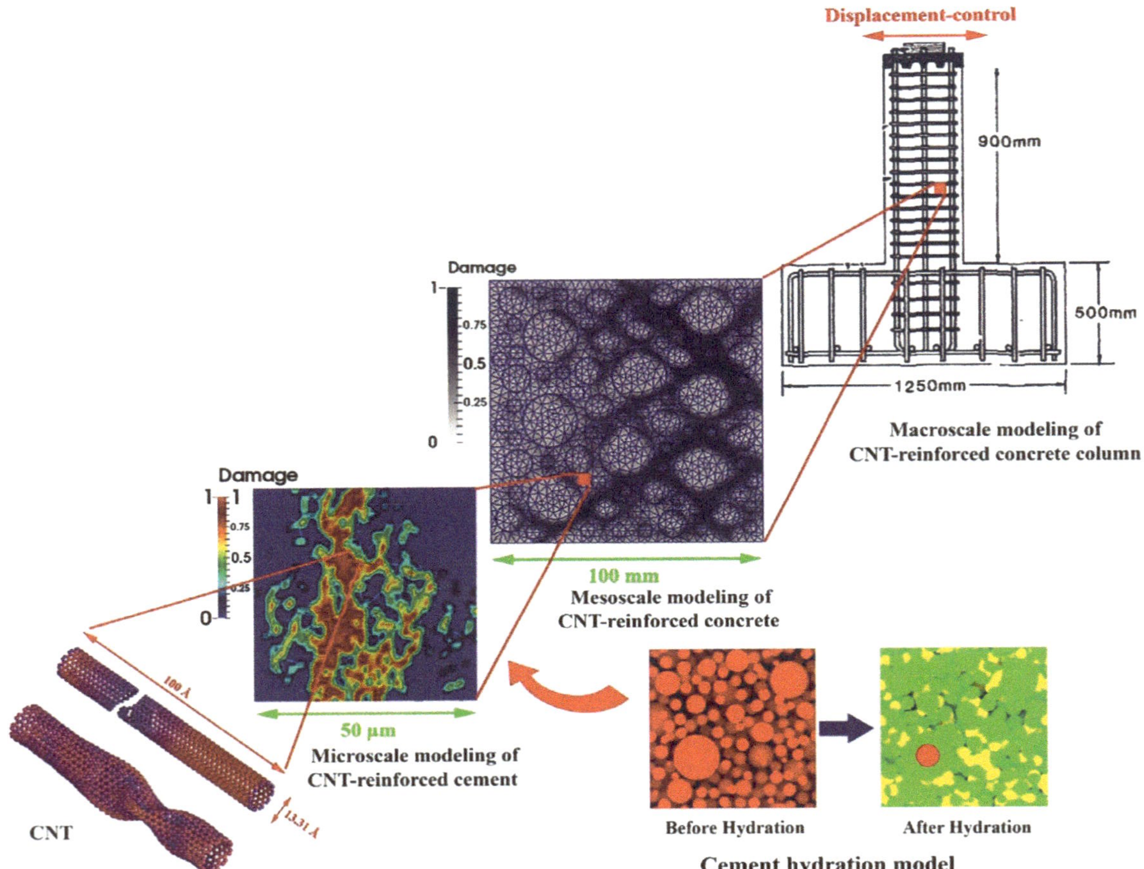

Fig. 11.1 Multiscale modeling of concrete structures.
From Eftekhari, M., S. Mohammadi, and M. Khanmohammadi. 2018. "A hierarchical nano to macro multiscale analysis of monotonic behavior of concrete columns made of CNT-reinforced cement composite." *Construct. Build Mater*. 175, 134–143.

The model parameters of mechanistic models have a physical meaning, which facilitates scientific interpretation of the results.

Statistical models utilize data and statistics to predict trends and patterns. Model predictions are based on data, correlation of data, and the experience learned from data. Since they are based on existing data and not causality, these models are usually bound to their calibration range and cannot be extended beyond the data space they are calibrated from. Statistical models are a black box and do not provide mechanistic understanding. Models developed by conventional data regression and based on big data, machine learning, and artificial intelligence are typical statistical models.

11.3 Principles of modeling

Building a model requires abstraction. A model is a simplified and approximate reflection of reality. The famous British statistician George E.P. Box argued that "all models are wrong, but some are useful," which highlights the limitation of models and the purpose of modeling. It is true that complete and true representation of a real system by a model is impossible. Typically, a model deals with only certain aspect(s) of a system or phenomenon in question. Therefore, it is important to bear in mind that the purpose of modeling is not for correctness but for usefulness.

11.3.1 Accuracy versus complexity of model

Proper trade-off between accuracy and simplicity is important to modeling. A more complex model usually gives a higher level of accuracy but less convenience to use. While added complexity usually improves the realism of a model, it can make the model difficult to understand and analyze, and can also pose computational problems, including numerical instability.

A model with higher accuracy does not necessarily mean a better model. A problem in reality is usually influenced by many factors or variables. Some of the factors have a significant effect on the problem, namely major variables. Others are less important and have a smaller or secondary effect on results. These variables can be treated as minor variables and ignored in modeling. The effects of minor variables can be divided into two categories: (1) noises and (2) secondary effects; both cause fluctuation of results and scattering of a model. The noise-induced fluctuation of results is difficult to control in reality, e.g., the wobbling of results due to ground vibration caused by surrounding vehicle movement. The effects of noise are not related to the major physical nature/mechanism(s) of a problem and should generally not be considered in modeling. The secondary effects do reflect the mechanism(s) related to the problem, but they are of secondary importance. For example, people know that, apart from the major factors, the shear strength of RC members is affected by many other factors including the detailing of reinforcement, loading types, boundary conditions, reinforcement types, aggregate size, etc. The effects of these factors on shear strength are sometimes not negligible. However, these factors are usually excluded in a shear strength model, because it is difficult to control such factors in practice. They also cause extra complication and difficulty not only in developing the model but also in its applications in design. A sensible and wise way to deal with the secondary effects is to treat them as scattering in the model, and allow for it through the use of safety factors. By the use of an appropriate safety factor, the error of a model caused by noises and secondary effects is controlled within its safety margin.

The principle of simplicity in modeling is particularly applicable to structural modeling. When modeling a building for structural design, it is not needed to embed the architectural, mechanical, or electrical components into the model, although it is possible to do so. Those nonstructural or even some secondary structural elements can be omitted in modeling and are treated as loading. If all the nonstructural elements are included in the structural model, the huge number of details significantly increases the efforts to build the model and interpret the excessively large output data, apart from higher computational costs. These practical problems effectively inhibit the usage of such "comprehensive" models. Furthermore, the complexity of the model also greatly increases the chance of making mistakes by both humans and computers because each additional element brings a certain amount of variance and the chance for error into the model.

The accuracy of a model depends on many factors, including the accuracy of test data, breadth of test database, appropriate identification of key variables, rationality of the model, errors involved in modeling, and so on. Complexity, accuracy, and dispersion should be well balanced in modeling to provide a reasonably accurate but relatively simple and convenient application model.

11.3.2 Causality of model

Another more important reason for neglecting secondary effects is to better understand and uncover key mechanism(s). A reduction in the number of variables in a problem can make things clearer, which helps reveal the key underlying physical rules or mechanisms that dominate the problem. Models with many variables show a complex picture that mixes things up and prevents people from visualizing and understanding key features or mechanisms of the problem. The use of key variables only in a model exposes the key characteristics of the problem and helps to capture the dominant mechanism. This modeling approach usually leads to mathematically and physically "beautiful" or "elegant" models.

The best example is Einstein's famous relationship between mass and energy, $E = mc^2$. Another example is the derivation of the flexural theorems in Chapter 3. The assumptions made for the derivations include: (1) plane section remains plane, which is not exactly true for normal flexural members with both bending moment and shear force in a section (see Section 2.8.2); (2) the tensile stress of concrete is ignored, which is incorrect for the tension zone near the neutral axis; and (3) a perfect elastoplastic stress-strain relationship for the reinforcing steel, which is also an approximation. The flexural strength of RC members can certainly be calculated more accurately by including the shear deformation, tensile stress of concrete, and the nonlinearity of the reinforcing bar, through the use of more complicated models such as empirical models and numerical models. However, such extra accuracy covers up the dominant physical rules. Without the omissions/approximations, the flexural theorems in Chapter 3 cannot be derived. The use of simple and conceptually clear models obviously contributes to more efficient, reliable, confident, and enjoyable design works in engineering practice.

It is advisable to try as few variables and as simple a relationship as possible during the modeling process, provided that the model serves the task. A variable number of around three or less and simple relationships such as linear or rectilinear are often adequate for modeling engineering problems. It is also noted that when a factor is a minor variable for many common problems, it may be a major variable for another special problem. For example, the tensile strength is negligible for flexural design, but it is important for the shear design of RC members. Therefore, whether a variable should be included in a model or not can only be determined from its effect on the particular problem.

The principle of simplicity, as an old piece of human wisdom, goes back at least as far as Aristotle (384–322 BCE), who wrote "nature operates in the shortest way possible" (Charlesworth 1956). This wisdom is well reflected in the famous Occam's razor, which is a principle particularly relevant to modeling (Schaffer 2015). Its essential idea is that, among models with roughly equal predictive power, the simplest one is the most desirable. The idea of parsimony or simplicity in deciding between models has been assimilated into common culture as the widespread layman's formulation that "the simplest explanation is usually the correct one." Complexity of a model often reflects its immaturity. Thomas Kuhn argues that as science progresses, explanations tend to become more complex before a paradigm shift offers radical simplification (Kuhn 2004). This is because people learn more and more about a problem over time, eventually finding the most direct interconnection or causality of the problem.

11.3.3 Task driven modeling

Models are task driven. The complexity and accuracy of a model should suit the task. For example, Newton's theory is a special case or simplified version of Einstein's relativity theory when the speed of an object is well below the speed of light. However, Newton's theory is sufficiently accurate, and more importantly, much more convenient for people to study macro-objects for most ordinary-life problems. As all models are task-driven and assumptions apply that specify the domain of application, models always have their boundaries, outside which they are inapplicable. Even a model as general as Newton's laws can only be applied within limits, outside which the theory of relativity and quantum mechanics must be used. It is important for model users to observe the limits and avoid misuse of models.

Models should usually be developed for as specific a task as possible. A model or computer program for a specific calculation is usually more frequently used than a general model or general-purpose computer program. For example, beam or slab design software is much more frequently used by designers than the general purpose software SAP and ABAQUS. A general purpose model also has its own tasks, i.e., for more complicated problems or special applications for which a task-specific model is not available.

Although a task-specific model is preferable in practice, there are cases where too many models are available for slightly different tasks. For example, different models have been developed for confined concrete for different shapes of columns, different confining materials, and different load paths, etc. Too many models bring inconvenience to practical applications. To this end, unification of similar models is often a good idea, as illustrated by the unified model in Section 8.4.3.2. The extent of unification should suit the relevant application. The unification can be regional or global, as illustrated in Fig. 11.2A. A few local models, such as local models j and k in Fig. 11.2A, can be unified to form a regional model. Examples of such regionally unified models can be found in Section 8.4.3.2.

A general mathematical framework has been developed by Wu et al. (2018) to combine local models into a regionally unified model. If there are m local models/functions in the regional space $(0, x_m)$, assume the local function in the ith subdomain between (x_i, x_{i+1}) is given by

$$y_i = y_{0,i} \cdot \overline{f}_i(x) \tag{11.1}$$

where y_i is an arbitrary function defined in the subdomain (x_i, x_{i+1}); $y_{0,i}$ is the y value of the function at $x = x_i$, and $\overline{f}_i(x)$ is a normalized function with its value equal to 1 at x_i. The separate m-segment functions can be transformed into one continuous function by the following equation:

$$f(x) = y_{0,1} \cdot \overline{f}_1(t_1)\overline{f}_2(t_2)\cdots\overline{f}_i(t_i)\cdots\overline{f}_m(t_m) \tag{11.2}$$

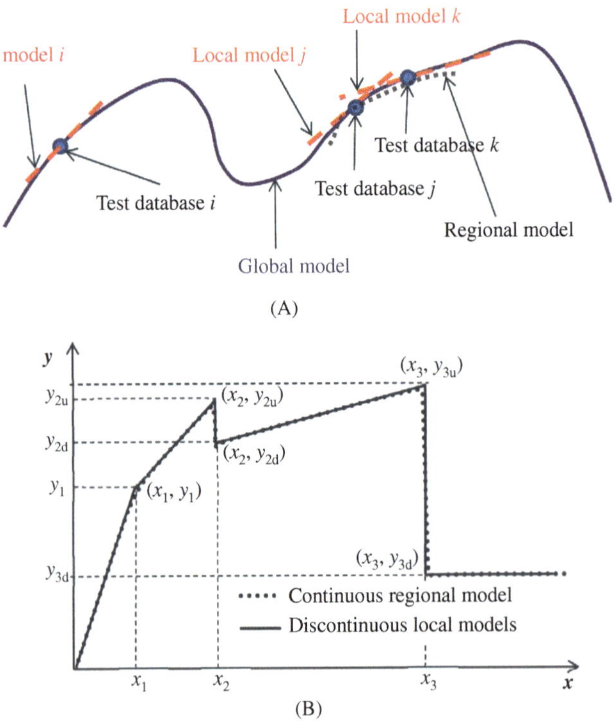

Fig. 11.2 Regional model.

In Eq. (11.2), the regional variable t_i ($i = 1-\text{m}$) is used to replace the global variable x in each individual function $\overline{f}_i(x)$, as given by

$$t_i = \frac{x+\delta}{\left(1+\left(\frac{x}{x_{i+1}}\right)^{n_i}\right)^{1/n_i}}\left(1+\left(\frac{x_i}{x+\delta}\right)^{n_i}\right)^{1/n_i} \tag{11.3}$$

where n_i is a constant that can be selected by users to determine the smoothness of transition at the boundary between two adjacent segments of the function; δ is another small arbitrary constant that is used to avoid singularity of t_i at $x = 0$.

By using the transformation through Eqs. (11.1)–(11.3), any multiple segment local functions with even discontinuity at their boundaries can be transformed into one continuous function. Fig. 11.2B shows an example where the solid lines show individual segments/models at different subdomains with discontinuities at the boundaries. The dotted line shows the continuous smooth curve given by Eq. (11.2), which is a mathematically continuous function. More details of the transformation can be found in Wu et al. (2018).

In principle, Eq. (11.2) can be used to unify any number of local models into one mathematical function to give a global model. However, such unification is simply a mathematical transformation or manipulation that does not involve any unification of mechanisms, theories, or physical principles. A real global model is usually a philosophically sound model (Section 11.5.3). The flexural theorems in Chapter 3 are an example of a global model that is generally applicable to all flexural designs of RC members, unless the assumptions from which the theorems are derived are violated. From this point of view, the pure mathematical unification of models or theories is usually not a real unified theory as it does not uncover the true underlying common mechanism/physics behind the different local models.

11.4 Methods of modeling

Many approaches and techniques are available for modeling. New tools are continuously developed for increasingly complicated and difficult problems. Common modeling approaches can generally be divided into the following categories.

11.4.1 Logic reasoning

A model can be developed through logical reasoning. There are three types of logic reasoning.

(1) Deductive reasoning

Deductive reasoning or deduction allows deriving ***B*** from ***A*** only when ***B*** is a formal logical consequence of ***A***. In other words, deduction derives the consequences of the assumed. Mathematical logic and philosophical logic are commonly associated with this type of reasoning. Given the truth of the assumptions, a valid deduction guarantees the truth of the conclusion. For example, given that "no model is 100% accurate in making predictions" and "the flexural theorem is a model," it follows that "the flexural theorem is not

100% accurate." When its assumptions are valid, a model derived by deductive reasoning makes true predictions. However, its performance is unknown when the assumptions do not hold. Change of assumptions is the point where older theories are succeeded by new ones.

Deduction is the main tool of modeling in physics and mathematics. It is less popular in engineering because engineering problems usually involve many factors, and a solution may not be derived by pure mathematical and philosophical logic. At the same time, solutions must be provided in practice. Therefore, other less stringent approaches such as induction and abduction described in the following sections are widely used to derive models for practical uses. Nevertheless, as a model derived from deduction is more solid and reliable, deductive reasoning should be used whenever possible. The derivation of the flexural theorems in Chapter 3 is an example of deductive reasoning.

(2) Inductive reasoning

Inductive reasoning or induction allows inferring $\boldsymbol{B}$ from $\boldsymbol{A}$, where $\boldsymbol{B}$ does not follow necessarily from $\boldsymbol{A}$. $\boldsymbol{A}$ might give us a good reason to accept $\boldsymbol{B}$, but it does not ensure $\boldsymbol{B}$. It hypothesizes a rule after numerous examples are found to follow such a rule. This type of reasoning is commonly associated with generalization from empirical evidence. While they may be persuasive, these arguments are not deductively valid. For example, if the flexural designs of all concrete structures are carried out without considering the tensile strength of concrete so far, one may induce that the concrete tensile strength can always be neglected for the flexural design of reinforced concrete members. While this induction is true for nearly all flexural designs of RC members, there are cases where the application of this rule causes significant error. For example, when the concrete in an RC member is a new kind of concrete material called engineered cementitious composites (ECC), in which case the special concrete has significant tensile strength that cannot be neglected.

Generally speaking, an induced model can be properly used when it is developed from sufficiently extensive observations (database) and its domain of validity is well defined. For the case of reinforced ECC members discussed previously, the inappropriate use of the design rule can be avoided by specifying the conditions of its application, such as by limiting the scope of the rule to concrete materials with a tensile strength less than a certain value. The majority of models used in engineering are derived by induction. Therefore, care must be exercised in appropriate uses of such models.

(3) Abductive reasoning

Abductive reasoning or abduction allows inferring $\boldsymbol{A}$ as an explanation of $\boldsymbol{B}$. As a result of this inference, abduction allows the precondition $\boldsymbol{A}$ to be abducted from the consequence $\boldsymbol{B}$.

Abductive reasoning is also referred to as abductive inference or retroduction. It is another type of approach of modeling proposed by the American philosopher Charles Sanders Peirce in the 19th century. Based on observations, it seeks or infers the simplest or most likely explanation/model. One may simply understand the abductive reasoning as inference to the best explanation or best guessing. This process is not as strict as deductive reasoning but yields a plausible conclusion without proving it. An abductive conclusion can be described as inference to "best available" or "most likely" explanation. Abduction is the most fertile but least secure mode of inference. A model from abductive reasoning may be subjective because different people may infer different "most likely" explanations from different perspectives or even from the same viewpoint. Coordinated and collaborative research efforts can partially overcome the subjectivity.

Abductive reasoning is widely used in engineering modeling, particularly in the early stage of model development. For example, many different models have been developed for the shear strength of RC members by different people. Even the same person proposed different models at different times. These models have been derived by inductive reasoning but have different interpretations of their mechanisms, and use different mathematical relationships. In other words, these models are from different abductions. While it is necessary to use abductive reasoning for finding engineering solutions before a more sound deductive solution is found, extra care should be taken in using such models, for example by application of a larger safety factor.

Failure of structures has occurred from the use of such models, e.g., the well-known collapse of the frame girders in the warehouse at Wilkins Air Force Depot in Shelby, Ohio, United States in 1955. The structure was designed by the old design model (Section 10.4) that was an abductive one. Incorrect models from abduction are not uncommon in engineering design guidelines and codes of practice. The school of horizontal shear for the shear strength modeling of RC members (Section 10.4) is a typical example of incorrect abduction.

(4) Differences between deduction, induction, and abduction

Deductive reasoning goes in the same direction as that of the conditionals, and links premises to conclusions. A rigorous deduction reaches a conclusion that is necessarily true. Deductive reasoning is a "top-down" logic in contrast to inductive reasoning that is a "bottom-up" logic. In deductive reasoning, a conclusion is reached reductively by applying general rules that hold over the entirety of a closed domain of discourse, narrowing the range under consideration until only the conclusion(s) remains. In deductive reasoning there is no epistemic uncertainty. In inductive reasoning, the conclusion is reached by generalizing or extrapolating from specific cases to general rules, resulting in a conclusion that has epistemic uncertainty. It is noteworthy that the inductive reasoning is not the same as the induction method used in mathematical proofs. Mathematical induction is actually a form of deductive reasoning.

Deductive reasoning differs from abductive reasoning by the direction of the reasoning relative to the conditionals. Deductive reasoning goes in the same direction as that of the conditionals, whereas abductive reasoning goes in a direction that is different from that of the conditionals.

11.4.2 Theoretical vs. empirical modeling

Theoretical modeling derives models from pure deductive reasoning. If all premises are true, the terms are clear, and the rules of deductive logic are followed, then the derived model is necessarily correct. However, the premises or assumptions may not be applicable to practical problems. When a difference exists between the assumed conditions and the practical ones, an error will occur in applications of the model. The error of the model prediction in this case depends on the difference between the assumptions and the real conditions. For example, Bernoulli's hypothesis that plane section remains plane is applied in the derivation of the flexural theorems in Chapter 3. However, Bernoulli's assumption is not applicable in D-regions (Fig. 2.23). Therefore, the application of the flexural theorems will cause an error when they are applied to a D-region of RC members. Nevertheless, the conventional flexural design method, which is consistent with the flexural theorems, is widely used in the flexural design of RC members regardless of B- or D-regions. This is because in most cases the error in the calculated flexural strength caused by the plane section assumption is acceptable from an engineering design point of view.

Phenomenological models are also typical theoretical models that describe a system in terms of laws of nature such as Hooke's law. Such phenomenological models/laws are widely applied in science and engineering. However, all these models/laws have their limitations, including Newton's laws. Significant errors would occur when they are applied outside their domain of validity, e.g., Hooke's law is generally applicable when deformation of the object is small.

Empirical modeling refers to a family of approaches for the development of models based on empirical observations from nature, practice, and experiments, through inductive and/or abductive reasoning rather than by derivations through deductive reasoning. It is based mainly on regression analysis or curve fitting of a suitable experimental database. Different from theoretical models having physical significance, there is not necessarily a physical explanation behind empirically derived models. For this reason, empirical models are usually inapplicable beyond the scope that the database covers.

The main difference between different empirical models lies in the choice of the major variables that are considered in the formulation, and the form of the formula (mathematical equation). A suitable form of formula can be selected based on the observed relationship between the variables and the trend of variation of results. As discussed in Section 11.3, the selection of suitable major variables is critical to an empirical model. The other critical issue of empirical modeling is the collection of test data according to a set of appropriate selection criteria, and screening and processing of data in a uniform manner. The test database should be objective and comprehensive. A biased database leads to a biased model. The performance of a model cannot be evaluated simply by the values of the error indexes, because the error indexes heavily depend on the database that is used for model evaluation. Therefore, models that perform better in one study may underperform in another because different model assessors use different databases.

A purely empirical model is developed solely based on data regression. However, rational analysis and mathematical derivation are sometimes involved in parts of the empirical modeling process. Such models are also referred to as semiempirical models. A semiempirical model is potentially more rational, and hence can perform better than a purely empirical model due to the underlying physical rule/mechanism included in the model. It may also be extended beyond the original scope of the database due to the correct mechanism included in the modeling process. A good example is the semiempirical ACI V_c model for shear design of RC members, which involved analytical study before regressing the model with a suitable database (Section 10.6.1).

Large numbers of empirical models have been developed and used in the designs of RC structures. The empirical procedure may produce a simpler form of model as there is no restriction on the algebraic form of the equation to be used. Such simple models are preferable to those lengthy and complicated equations from a practical application point of view.

11.4.3 Linear vs. nonlinear modeling

Empirical modeling includes linear and nonlinear modeling. Linear models use a linear relationship between the responses of a system and its variables. Linear models are most widely used in science and engineering. Linear regression techniques are applicable for linear modeling (Draper and Smith 1998; Yan 2009). When the system has only one variable, it is called simple linear regression. The process is called multiple linear regression when the variable number is more than one. Linear models are often fitted by minimizing the "lack of fit" in some norm, such as the least squares and least absolute deviations. The techniques for linear modeling are much more efficient than those for most nonlinear modeling. However, linear models may not be applicable to nonlinear problems.

Nonlinear modeling uses a nonlinear relationship between the responses and the variables. Nonlinear models do not satisfy the superposition principle. There are many different types of nonlinear systems. Due to the complexity of the nonlinear relationships, nonlinear modeling has higher flexibility, uncertainty, and difficulty. In conventional nonlinear modeling, explicit mathematical functions are first selected. The parameters or coefficients of the functions are often determined by curve fitting. The selection of the mathematical functions can be based on the observation of the variation trends of the function in relation with the variables based on test results, in combination with common sense, intuition, experience, expert opinion, and convenience of mathematical form.

Many different techniques have been developed for modeling nonlinear systems. The nonlinear system identification technique is useful for nonlinear modeling. System identification is a method of identifying models from the measurements of their inputs and outputs (Torsten and Stoica 1989; Nelles 2003). Numerous system identification methods have been developed that can be broadly categorized into five basic approaches (Nelles 2003; Billings 2013):

- Volterra series models,
- Block-structured models,
- Neural network models,
- Nonlinear AutoRegressive Moving Average model with eXogenous inputs (NARMAX models), and
- State-space models.

A gray box or black box model can be built through system identification. System identification typically involves four steps: data collection, parameter/variable identification, model postulate, and model validation. The NARMAX algorithms can be used to select model terms, determine the model structure, and estimate the unknown parameters in the presence of correlated and nonlinear noise. Compared with the neural network approach that generates opaque models, NARMAX produces models that can be written down and related to the underlying physical process. If the modeling is done by an artificial neural network or other machine learning, the optimization of parameters is called training, while the optimization of model hyperparameters is called tuning and often uses cross-validation (Gurney 1997; Schmidhuber 2015).

11.4.4 Forward vs. inverse modeling

A forward problem starts with the causes and then calculates the effects. An inverse problem starts with effects and then calculates the causes. Deriving models from first principles by deductive reasoning is forward modeling. An inverse modeling determines the causal factors or underlying governing relationship from observed results. Inverse problems are some of the most important mathematical problems in science and mathematics, because they reveal relationships or parameters that we cannot directly observe. A lot of problems in engineering are inverse problems. For example, in Section 8.6.6, the stress-strain relationship of concrete is identified from the measured structural global response of RC members by an inverse modeling technique. In many engineering problems, only the behavior or effects of a system can be observed, but people would like to know the cause or find the underlying mechanisms for modeling. Such a problem is typically an inverse problem. Inverse modeling has wide applications in system identification. Linear inverse problems were well resolved mathematically at the end of the 19th century, but nonlinear ones are much more difficult and only limited types of the problems have been theoretically resolved (Tarantola 1987; Chavent 2010; Aster et al. 2018).

11.4.5 Heuristic modeling

A heuristic is a word from the Greek, meaning "I find out" or "to discover." It is an approach to problem-solving that takes one's personal experience into account. A heuristic method or a heuristic is any approach to problem-solving by self-discovery using a rule of thumb, experience, trial and error, and an educated guess. It employs a practical method that is not guaranteed to be optimal, perfect, or rational, but is nevertheless sufficient for reaching an immediate, short-term goal or approximation. Where finding a more rational/accurate model is impossible or impractical in the short term, heuristic models can be used to solve practical problems.

Most engineering models/solutions are more or less heuristic in the beginning. A good example of such models is the flexural design model for RC members where an "ultimate strain" ε_u is used to calculate the flexural strength, although different values of ε_u are used in different codes (Section 2.7.3). This highly successful flexural design model was also heuristic at first but has since been widely validated by the research community and engineering practice for a century. The method was rigorously proven to be a rational approach by deductive reasoning only very recently (Chapter 3). An example of unsuccessful ones was the longitudinal shear theory (shear between different layers of concrete) for the shear strength of RC members based on the experience and prior knowledge of timber structures and composite structures (Section 10.4). Heuristic models can be biased or even mistaken because of their subjectiveness. Over time, through more holistic efforts, the research community can overcome the subjectivity of the model and eventually develop a rational one.

11.5 Evaluation of model

An important part of the modeling process is the evaluation of performance of a model. The performance of a model can be difficult to evaluate as it may involve many different considerations of modeling. These considerations may include the ability to explain past or existing observations, the ability to predict future or further observations, the costs of using the model, refutability or the degree of confidence in the model, simplicity, or even aesthetic appeal. A comprehensive list of references on model validation can be found in the paper by Hamilton (1991).

A model is usually evaluated first and foremost by its consistency with observations or empirical data. A model inconsistent with observations must be rejected or modified. One common way to modify models is by limiting their scope or domain in which it is credited with validity. However, a good fit to empirical data alone is sometimes insufficient for a good model. Factors to be considered in model evaluation are discussed in the following sections.

11.5.1 Fitting observations

The most important but also the easiest part of model evaluation is to check whether a model well fits experimental measurements or other empirical data. A common approach to test this fit is to split the test data set into two disjoint subsets, a training subset for modeling and another verification subset for model evaluation. The derived model should at least fit the data of the training subset well but may not perfectly fit the data, due to the difference between a model and the real system as well as the inevitable scattering of data. As discussed in Section 11.3, overfit should actually be avoided. A good model should also well match the verification subset even though this data set has not been used for modeling. This practice is referred to as cross-validation. However, many people use one data set only that is used for both modeling and evaluation. While such a method is not prohibited, a model developed in this way is more refutable.

Defining a metric, such as mean, standard deviation, or coefficient of variation, to measure distances or errors between observed and predicted data is a useful tool for assessing model fit. Researchers have the flexibility to define different metrics for different problems and use appropriate weights for different data points. In the decision theory of statistics and some economic models, a loss function plays a similar role.

For models with given mathematical forms, it is rather straightforward to test the fit of model parameters by using a selected metric. Often, researchers also have the liberty to select appropriate mathematical forms. More mathematical tools are available to test the fit of statistical models than models involving differential equations. Tools from nonparametric statistics can sometimes be used to assess the fit between the data and a known distribution or to suggest a model that makes minimal assumptions about the mathematical form of the model (Tedeschi 2006).

11.5.2 Scope of model

For models derived from deductive reasoning, the assumptions applied during derivations define the scope of the models. For a model derived from observations using a data set, it is not straightforward to determine the scope of the model. Although the data set used for modeling is clear, it is usually unknown which situations the data set represents. Therefore, if a model is constructed based on a set of data, one needs to assess the scope of the model.

The predictions of the model for points between those in the observation data set is called interpolation, and the predictions outside the observed data is called extrapolation. It is easy to understand that extrapolated results need to be validated by new test data outside the original data set. Interpolation points may also be invalid. For example, if a linear relationship is adopted for modeling but the actual variation between data points follows a sinusoidal relationship, in this case the predictions may have a significant error in some cases. Therefore, validation of interpolation points is of similar importance. In principle, the validity of interpolated and extrapolated points can only be determined by new observations or facts.

For many engineering problems, people know the relationship or trend of variation of a problem from common sense, practical experience, and observations. The validity of interpolation and extrapolation of a model can usually be judged correctly. Such judgment may be more difficult for a scientific problem due to lack of experience. For example, the Newtonian classical mechanics were developed from ordinary life physics. Its scope of validity was actually unclear when the theory was developed. Many people at that time even thought that the theory was universal and there were no restrictions. It was not until the development of relativity theory and quantum mechanics discovered that Newtonian mechanics does not apply to the fields where objects travel at close to light speed and the object size is on a molecular scale that this understanding changed.

11.5.3 Philosophical considerations

The purpose of modeling is often to enhance the understanding of a problem. In this case, the validity of a model rests not only on its fit to empirical observations within its original scope, but also on its ability to extrapolate to situations or data beyond the original database from which it is developed. In science, some people argue that a model is worthless if it does not provide insight beyond what is already known. A philosophically sound qualitative model is potentially extendable beyond its original domain. In engineering, however, a quantitative model that is valid within a given domain is useful and important for practical designs and calculations. Whether a derived model can make predictions beyond its original scope can only be ascertained by observations.

Nature has certain general rules that are applicable, at least philosophically, to many different problems or areas. Therefore, it is possible for a model developed from a given domain that captures a general rule of nature to be applicable to an extended domain beyond its original base. Newton's laws were developed from limited observations, but it is applicable to a much wider scope beyond its original observation domain. This is because it captures a general rule of nature (here the scope of nature is defined as the domain where Newtonian mechanics is applicable). Another good example is the second law of thermodynamics. This law was first discovered by Nicolas L.S. Carnot in 1824 as an empirical finding concerning the conversion of heat into mechanical work. Statistical mechanics later provided a microscopic explanation of the law. Nowadays, the original scope of the law has been greatly extended to almost all areas of science and engineering. Such great extension is possible only when it correctly captures a general rule of nature.

For engineering modeling, it is also critical for a model to reflect the governing mechanism(s) of the problem. Models constructed based on relevant mechanism(s) are generally more reliable and accurate than those obtained simply by curve fitting to test data. These models are potentially extendable beyond the original domain where it is developed. Therefore, whenever possible, one should try to develop a philosophically sound rational model that correctly reflects its underlying physics or mechanisms.

Although the performance of a model is often related to its rationality, there are exceptions where accuracy of a model is not related to its rationality at all. A model developed from curve fitting and artificial intelligence (AI) can have a high precision in predictions but usually does not reflect its mechanism(s). Phenomenological models may also have a similar problem, as illustrated in the example in Fig. 11.3. The monotonic stress-strain response in the figure can be calculated by three different models:

The damage model:

$$\sigma = \varepsilon \cdot E_{c,damage} \tag{11.4}$$

The plasticity model:

$$\sigma = \left(\varepsilon - \varepsilon_p\right) \cdot E_c \tag{11.5}$$

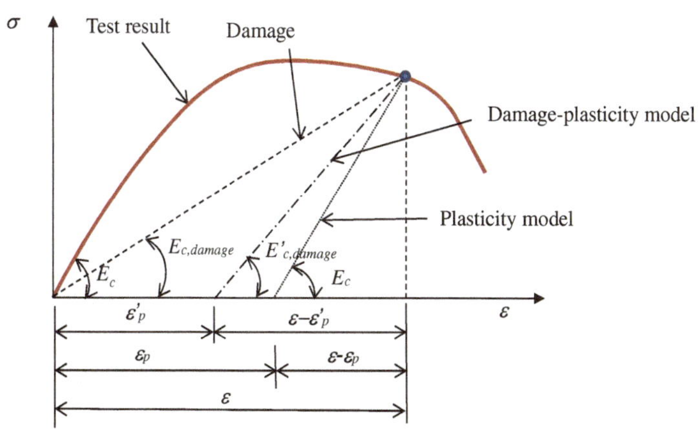

Fig. 11.3 Rationality of models.

The damage plasticity model:

$$\sigma = \left(\varepsilon - \varepsilon_p'\right) \cdot E_{c,damage}' \tag{11.6}$$

where σ and ε are the stress and strain of concrete, respectively; E_c is the elastic modulus of concrete, which is the same as the modulus of concrete for the plasticity model; $E_{c,damage}$ and $E_{c,damage}'$ are the modulus of concrete for the damage model and damage plasticity model, respectively; ε_p and ε_p' are the plastic strain of concrete with the plasticity model and the damage plasticity model, respectively.

All the preceding three models can give precise predictions as long as their model parameters, i.e., ε_p for the plasticity model, $E_{c,damage}$ for the damage model, and ε'_p and $E'_{c,damage}$ for the damage plasticity model, are calibrated to fit the test curve exactly. For a single test curve, there is always a way to make a close match to the test curve. For a test database with many test curves, regression analysis can be used to develop empirical formulas of the parameters, by minimizing the prediction errors.

As all the preceding three models can closely match the test results, this does not mean that they are all correct or rational. It is true that under the condition of monotonic loading, all the three models can make correct predictions, as long as the model parameters are correctly regressed to fit the test results. However, a slight change of the premises of these numerically calibrated models may expose the physical irrationality of them. For example, if the three models are applied to cases with unloading, only the damage plasticity model can correctly unload (along the dash-dotted line) and reload. The other two models will give wrong predictions in this case. This is because the damage plasticity model correctly reflects the characteristics or physics of concrete, particularly that it experiences both plastic straining and damage to its modulus under significant strain.

Therefore, a physically sound and philosophically rational model can often be extended outside the original premises from which it is developed, while a model from simple curve fitting only does not usually have this potential. From this point of view, the extensibility of a model outside its original premises is often a reflection of the rationality of the model.

11.6 Selection of model

When different choices are available, designers sometimes face difficulty in selecting the more appropriate model for a problem. Selection of different models may lead to different outcomes, which confuses junior designers. Therefore, it is important to understand the rationale behind model selection.

11.6.1 Quantitative models

Often, different models are available for the same problem. For example, many different models are recommended for the concrete contribution to shear strength V_c in different codes of practice. Design codes recommend different equations depending on local conditions, as well as the experience, knowledge, and personal preference of the committee members who developed the codes. Such equations are supposed to give similar results. There are cases, however, where significantly different results are given by different models. There are different reasons for the difference, including mistakes and misunderstandings of modelers. Most of the time, the differences are caused by the differences in local conditions or practices. For example, the models for geotechnical designs are often different in different places because the geotechnical conditions are significantly different in different places. Another example is that the model for calculating the Young's modulus of concrete in the Hong Kong RC design code gives smaller values than those calculated from other major design codes (Buildings Department 2013). The reason for this is that the local aggregates in Hong Kong are softer than normal aggregates elsewhere. Therefore, it is sometimes necessary to understand the basis of a model to avoid misuse of model.

11.6.2 Qualitative models

For qualitatively different models, the selection of different models may produce completely different designs. Selection of an appropriate qualitative model requires good understanding and experience of the model users. Two examples, one for strut-and-tie modeling (STM) and the other for selection of structural models, are used to explain the matter. The discussions in this section explain that, as stated in the last sentence of Chapter 1, a good understanding of fundamentals is important for making appropriate choices in practice.

(1) Selection of STM

For design of short RC members, engineers are free to choose between the conventional beam design approach and the strut-and-tie model. Both methods are allowed in design codes, but the results are completely different structural systems. As discussed in Section 5.6, loads on a beam member can be transferred to its supports by both beam action and arch action. The arch action is small and beam action dominates in slender members. It is the opposite in short members. If a short member, such as a pile cap, is designed as a normal beam, which is allowed in design codes, its load transfer relies on the bond of rebars (Eq. 5.19). In the case in which the short member does not have a sufficient length for the rebars to develop sufficient bond strength, the supposed beam action cannot be realized. It is noted that the anchorage of rebars at their ends only provides arch action rather than beam action. In this case, the design as a normal beam is inappropriate and may cause structural problems. The correct choice in this case is the STM. This example shows that there are cases where the selection of an inappropriate model may lead to potential structural problems.

For the strut-and-tie modeling approach, there are also different strut-and-tie models to choose from, as illustrated in Fig. 10.14. Different STMs give different structural designs. The recommendation in design guidelines on the selection of STMs is that the stress field of the selected truss model be as close to the elastic stress field as possible. However, this is easier said than done. The stress fields of an RC beam with the reinforcement as in Fig. 10.14B is shown in Fig. 11.4.

The red color in Fig. 11.4A and B shows tensile stress and others indicate compressive stress, with the maximum compressive stress of concrete shown by the blue color. Fig. 11.4A shows the stress field of the short RC beam just before concrete cracking. Before cracking concrete, the tensile resistance of concrete plays a major role in its internal

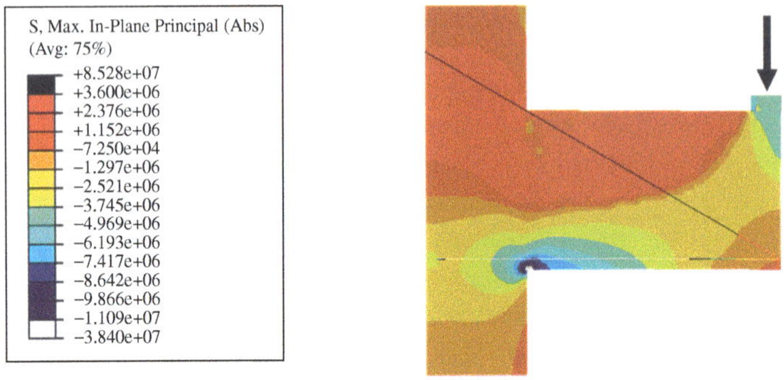

(A) Maximum principal stress contour just before cracking (Unit: Pa)

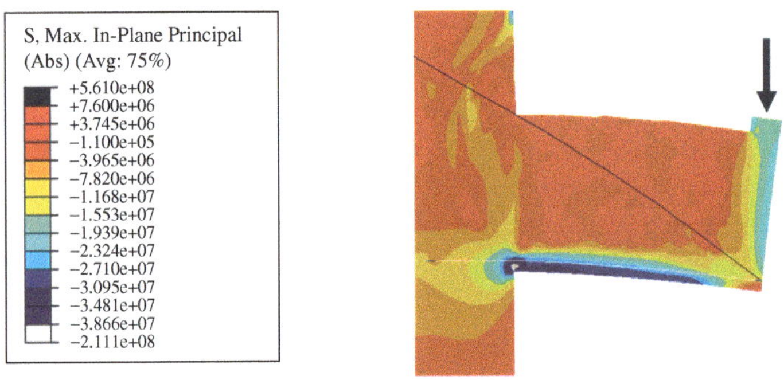

(B) Maximum principal stress contour at peak load (Unit: Pa)

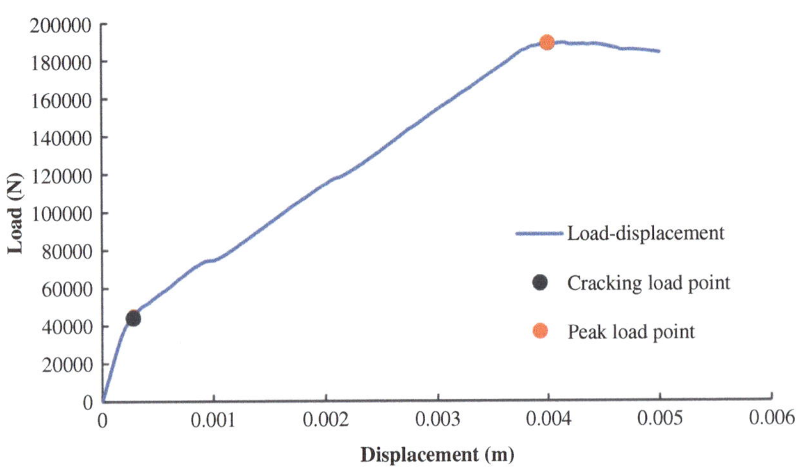

(C) Load versus displacement curve

Fig. 11.4 Stress fields of strut-and-tie model.

resistance and the reinforcement has a minor effect. Therefore, the stress fields of the structures in Fig. 10.14A and B are similar before cracking. It can be seen from Fig. 11.4A that, apart from the vicinity around the point load, the stress field is similar to that of a flexural member, i.e., under tension at the upper part and compression at the lower part, with its neutral axis at about the middle of the member.

The stress field at the peak load is very different, as shown in Fig. 11.4B. The compression zone is concentrated to a small layer at the bottom face of the member, which well matches the concrete strut as indicated in Fig. 10.14B. The red area in Fig. 11.4B shows the residual concrete tensile stress, which is small compared with the compressive stress of concrete and the tensile stress in the diagonal steel bars. When the small concrete tensile stress is neglected, the internal stress field exactly matches the truss in Fig. 10.14D.

These results show that, as the load increases from the cracking load to the peak load (Fig. 11.4C), the internal stress distribution gradually changes from Fig. 11.4A to B. Significant redistribution of internal stresses occurs in the process. In fact, the load path changes completely, from a beam action to a truss action. Such a large redistribution of stresses requires significant deformation, including cracking of concrete and yielding of reinforcement, before the load path can be changed from the initial one corresponding to the elastic deformation stage to the final one corresponding to the designed STM. This drastic redistribution process can only be achieved in a ductile structure, because a brittle structure will fail before the redistribution process is completed. It can be seen from Fig. 11.4C that the load-deformation curve of the structure is somehow ductile for the redistribution process to complete.

The preceding discussion provides the rationale behind the STM approach that engineers have the liberty to select/design different struts and ties (a truss structure) to transfer design loads. Although different STMs can be designed to fulfill the same design requirements, as shown in Fig. 10.14A and B, the performances of different designs can vary significantly. For the structure in Fig. 10.14B, the top face is under tension, but the diagonal reinforcement is placed near the middepth. Significant cracking on the top face is required before the diagonal bars can pick up the tensile stress redistributed from the top face. In other words, massive cracking and excessive deformation of the member are required for the designed STM to be in full action. This is the reason that appropriate distribution reinforcement is important for structures designed by STM to limit crack width at service loads. From this discussion, one will be able to see that the STM in Fig. 10.14A is a better design because the tension reinforcement is placed at the top tension face, which can pick up the tensile stress even in the elastic deformation stage without the need of significant change of load path. Therefore, a good STM should place the ties as close to the original elastic tension zone of the member as possible.

(2) Selection of structural models

With a given space and loads, engineers are free to choose different structural models for design. This is illustrated using the example in Fig. 11.5. Selecting the structural scheme in Fig. 11.5A causes smaller bending moments in the superstructure but a larger bending moment at the supports (substructure). Another structural option in Fig. 11.5B produces larger bending moments in the superstructure but no bending moment in the substructure. The design in Fig. 11.5A gives lower costs for the superstructure but higher costs for the substructure, as opposed to that in Fig. 11.5B. Engineers need to make appropriate choices based on the local conditions of the particular project. For the example in Fig. 11.5, the choice can be made based on the geotechnical condition of the structure. Foundation structures are usually costly. Therefore, the structure in Fig. 11.5B would be a better choice for structures with poor foundation conditions.

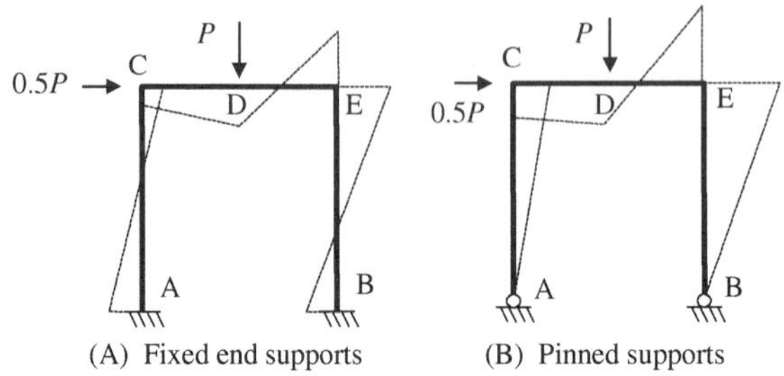

Fig. 11.5 Bending moment diagrams for different structural models.

The example shows that when qualitatively different models are available, it is often a matter of choice by model users. It is, however, noted that whenever a certain model is selected, the following designs should be consistent with the model. For example, for the model in Fig. 11.5B, the reinforcement at the supports should be detailed in such a way that their moment resistance is minimized. Otherwise, a significant moment would be transferred into the foundation, which may cause its failure.

It is also important that the selected structural model match the actual load path of the real structure. This is explained through the structure in Fig. 11.6. In Example 1 of Fig. 11.6, the minor structural member N3 in Fig. 11.6B was not included in the original structural design model of Fig. 11.6A. During renovation, N3 was added as an additional support. The engineer approved the addition, considering it as a strengthening to the structure. However, the actual effect of this addition is an increase of the axial force applied to the member N2, which may cause the failure of N2. The second example is the pad footing in Fig. 11.6C. The original size of the footing was B, but the actual size of the footing was increased to $B1$ due to a construction error. The engineer ignored the error, as he believed that it was safe due to providing a larger one than the design. However, the actual effect of this enlargement of the footing is an increase in its applied moment, because the maximum bending moment of the footing is given by

$$M = \frac{1}{2}P \cdot \frac{1}{4}B1 = \frac{1}{8}P \cdot B1 \tag{11.7}$$

where if $B1 > B$, the bending moment is increased. The preceding examples show that it is important to match the structural model to the actual structure.

Nevertheless, exact match of the design model to a real structure is difficult for complex structures. In reality, structural failures caused by such mismatching are rare. This is due to

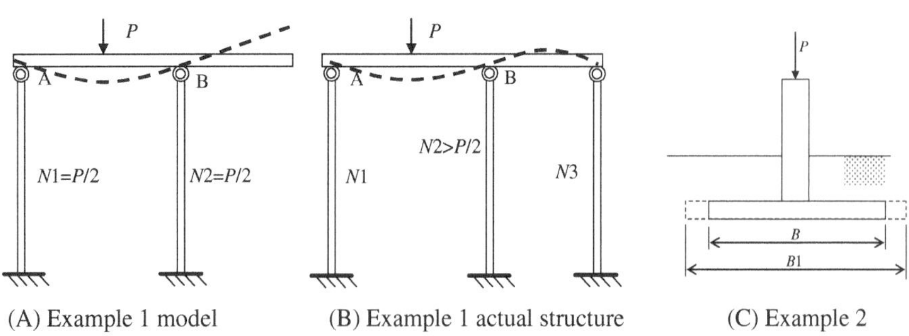

Fig. 11.6 Structural modeling.

the ductility of structures. In Example 1, when the member N2 "fails" due to the overloading caused by N3, the point B will move downward. This movement will quickly release the additional load applied to N2. Therefore, the structure will not fail as long as the member deforms in a ductile manner. For Example 2, the overloading causes flexural yielding of the footing, which would lift up the edge of the footing, and consequently reduce the soil pressure at the edge and increase the soil pressure at its center. This redistribution of the soil pressure will reduce the bending moment. Therefore, the structure would not fail as long as it has sufficient ductility to allow for redistribution.

These discussions show that it is important to select an appropriate qualitative model that matches the real problem as much as possible. Nevertheless, a structure designed based on an assumed model would not fail even there is a gap in the load path between the real structure and its model, provided that sufficient ductility is available and excessive deformation and damage are allowed; and furthermore, the assumed model provides an adequate load path to transfer the applied loads. This is due to the redistribution of internal stresses, as illustrated by the structures in Figs. 11.4 and 11.6. The theoretical basis behind this is the lower bound theorem discussed in the next section.

11.7 Modeling with theorems of plasticity

The theorems of plasticity are useful in deriving models for reinforced concrete structures. The following three theorems are applicable to any elastic-perfectly plastic body or structures (Chen and Scawthorn 1970; Duthinh and Carino 1996).

11.7.1 Theorems of plasticity

(1) Lower Bound Theorem (static theorem)

If a load path (i.e., distribution of internal stress) can be found within a body or structure that balances the applied load and nowhere violates the yield criterion (not exceeding the plastic limit values), the body will not fail, or be just at the failure load. The corresponding load is a lower bound to the actual load that causes the collapse of the body.

(2) Upper Bound Theorem (kinematic theorem)

The load computed on the basis of an assumed failure mechanism compatible with the geometrical constraints of the structure (compatible collapse mechanism) is an upper bound of the strength of the structure. The upper bound load can be calculated from the condition that the rate of work done by the external loads equals the internal plastic dissipation.

(3) Uniqueness Theorem

An external load computed on the basis of an assumed compatible mechanism, in which the forces are bounded by plastic limit values and are in equilibrium, is equal to the true collapse load.

The preceding plasticity theorems come from three different solutions of boundary value problems. With the plasticity theory, the following three aspects need to be considered in solution:

(a) Equilibrium

(b) Internal resistance $R \leq$ yield (plastic) limit values R_p, i.e., internal resistance not exceeding the full plastic capacity

(c) Compatible collapse mechanism, and $R = R_p$ at all plastic hinges

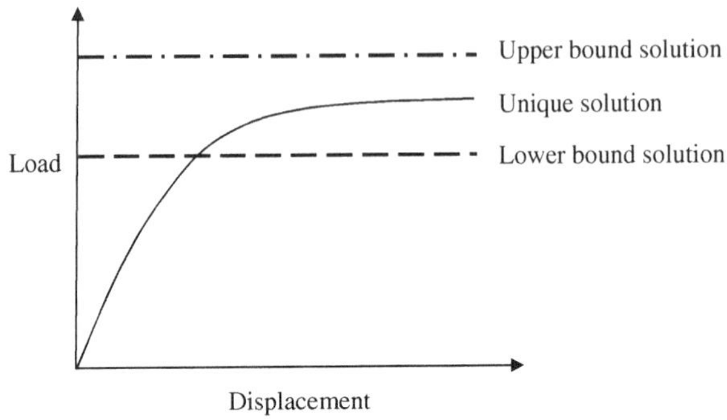

Fig. 11.7 Plasticity solutions.

The solution from the Uniqueness Theorem satisfies (a), (b), and (c). A solution from the Lower Bound Theorem satisfies (a) and (b). A solution from the Upper Bound Theorem satisfies (a) and (c).

The solutions from the three theorems are illustrated in Fig. 11.7. The lower bound solution corresponds to a state of loading condition before a failure mechanism is formed or just formed. The elastic solution of a structure is also a lower bound solution. When loading increases, the structure is subjected to increasing internal stress, which moves the lower bound solution upward. When the load is increased to a limit state where a failure mechanism forms and the structure reaches its loading capacity, the collapse load is reached. Therefore, the unique solution from the Uniqueness Theorem is a special case of the lower bound solution.

For the upper bound solution, any possible failure mechanism can be assumed and used to calculate the collapse load. For example, for the structure in Fig. 11.5A, a failure mechanism is formed when plastic hinges are formed in the following cases: (1) at points C, D, and E, (2) at points A, B, C, and E, (3) at points A, D, E, and B, and (4) at points A, C, D, and B. The condition that the rate of work done by the external loads equals the internal plastic dissipation of the Upper Bound Theorem can be used to calculate an upper bound solution of load for all the possible collapse mechanisms. Each collapse load is an upper bound solution. The smallest value among all the upper bound solutions is the real collapse load. In other words, the smallest upper bound solution is the unique solution from the Uniqueness Theorem.

11.7.2 Application of the plasticity theorems

For structural designs in practice, it is often sufficient to know the ultimate load-carrying capacity of a structure rather than the whole load-deformation process. The corresponding computational method is referred to as the plastic collapse analysis (Vrouwenvelder 2003). The plasticity theorems are applicable in these cases, and hence can be used to develop design models. Numerous design models or theories have been developed using this approach, such as the yield line theory for flexural design of reinforced concrete slabs (Johansen 1998) and various shear strength models of RC members (Duthinh and Carino 1996). Detailed derivations of the collapse load for frames and other structures using the three plasticity theorems can be found in Vrouwenvelder (2003).

Extensive research has been carried out on shear strength modeling of RC members using the plasticity theorems. The frictional shear strength of RC members (Fig. 5.24A) was derived by Nielsen (1984) using the upper and lower bound theorems. Using the truss model, Nielsen (1984) also derived the upper and lower bound solutions for the shear strength of RC members. Based on the truss model, Mitchell and Collins (1974) and Collins and Mitchell (1980) developed the compression field theory (CFT) for the torsion and shear strength of RC members. CFT satisfies equilibrium, strain compatibility, and stress-strain relationships, and hence all the three conditions (a), (b), and (c) mentioned in the previous section. Therefore, it is a unique or "exact" solution to the problem. The exactness here does not mean the accuracy of the method. It simply means that it is a unique solution of the mechanistic model, which is based on the truss analogy, and furthermore, accounts for web shear only without considering other mechanisms. As discussed in Section 10.6, there are other shear mechanisms apart from the web shear mechanism in the shear failure of RC members. Therefore, this "exact" solution may not give accurate shear strength of RC members when other mechanisms are significant.

Therefore, the solution from the plastic theorems depends heavily on the selected mechanistic model. For example, the derived collapse load based on the plastic hinge mechanism for the structure in Fig. 11.5 is applicable to normal frame structures with flexural failure. However, the frame may also fail in shear. In this case the collapse load derived from the plastic hinge failure mechanism is not applicable. Also, in seismic design of RC frames, the plastic hinges may be purposely placed away from the beam/column joints. In this case, the model of collapse load derived from plastic hinging at the joints of frame may also give significant error in its predictions. From this point of view, the selection of a rational mechanistic model is more important than deriving collapse load using a theory for the mechanistic model.

11.8 About safety factor

All models are approximate because no model can consider all factors and quantify the effect of a factor without error. Therefore, all models have a certain degree of uncertainty and inaccuracy. However, engineering practice, in particular structural engineering, cannot afford failure, or more precisely, cannot accept a failure risk that is higher than a certain level. This problem is resolved in engineering by the use of an appropriate safety factor or reliability index. The safety margin is allowed in the structural design process at the following links:

(1) Loading

Different loads have different uncertainty, e.g., live loads have a higher uncertainty than dead loads. The difference in uncertainty of loading requires the use of different safety factors for different loads. The partial safety factors for loading are regulated in loading codes/specifications based on statistics. In current design codes, a load factor for the same type of load is the same for different failure modes. This is questionable because the factors affecting loads can be influenced by the type of construction. For some structures, such as the fused structures (Section 9.10), the statistics of loading will be different from those of normal structures, because overloading is better controlled in a fused structure and hence the partial safety factor can be smaller in principle. Indeed, if the fuse is triggered at a certain load and a load higher than that load is prohibited, e.g., by warning or blocking, then the extra allowance for a higher load is not necessary. The safety factor for loading is determined by the probability of overloading. In situations where overloading is not possible, it is indeed debatable to use a safety factor for extra loading.

(2) Failure mode

The consequences of different failure modes are different: e.g., the shear failure is more dangerous and damaging than the flexural failure. This is covered in design codes by the use of different capacity reduction factors for different failure modes or different partial safety factors for different material failures. Actually, failure is a structural behavior. The same material failure mode may have different structural failure modes. From this point of view, the use of the capacity reduction factor for specific failure mode is more rational than the use of partial safety factors for materials in allowing for proper safety margins for different failure modes. The variability in the quality of construction and consistency of the materials is accounted for in the design values of material properties, e.g., the characteristic values.

The preferred failure mode is achieved in design through the use of a higher safety factor for an undesirable failure mode, e.g., a higher safety factor for shear failure than that for flexural failure. The seismic design principle of strong column, weaker beam, and stronger joint is also realized in the same way.

(3) Repair costs

Repair costs are not directly related to capacity reduction factor in current design codes, but indirectly related through structural class. However, for the same failure mode under the same loading, a structure that requires higher repair costs should be allocated a lower failure probability or a higher safety factor. For example, the failure repair of the CY fused beams (Section 9.10) is simply the replacement of the CY fuse, which costs much less than the repair works of normal RC beams. The indirect costs due to shutdown of service are also lower. Consequently, a larger capacity reduction factor or a smaller partial safety factor can be allocated to such structures.

(4) Accuracy of model

Different models have different accuracies and scatters. A higher safety margin should be assigned to structures designed by a model with larger prediction errors. This safety margin is referred to as the confidence reduction factor of the model. A larger value of safety factor is used when there is not much research on modeling of a specific type of failure mode. The development of a model usually experiences the following stages: simple model with large and unaccountable error, more complex model with smaller error, and finally a radical return to simplicity with accountable and acceptable error. Therefore, as the model matures, a smaller value of the safety factor is used.

For example, in the early days of reinforced concrete construction before 1900, all columns were designed for axial load only, and a safety factor as high as 10 was used (ACI-ASCE Committee 326 1962). The high safety factor was necessary to allow for the effects of flexural stresses induced by eccentric loading. For columns with large eccentricities of load, the high safety factor could usually provide sufficient strength for the members. However, such design models were ultraconservative and uneconomical for members with small load eccentricity. With the improvement of the accuracy of the design model, the safety factor of the model has been gradually reduced.

As both the capacity reduction factor for a failure mode and the confidence reduction factor of a model are specific for a certain failure model, these two margins of safety can be combined into one. In this case, different design models for the same failure mode should have a different capacity reduction factor.

11.9 Detailing of structures

Design is not only a science, but also an art. There are many objectives, requirements, and constraints to be considered in designs, and often they are interrelated. Design usually requires a trial-and-error process to satisfy requirements within constraints and optimize goals. It is a holistic, creative, and satisfying pursuit. From long-term practical experience, people have developed the following effective process for engineering designs: (1) initial design following empirical rules and experience, such as selection of structural system, selection of construction materials, member sizing, etc., (2) structural modeling and calculations, (3) design of structural members as well as cost estimation and evaluation of other objectives, and return to (1) if results are unsatisfactory, and (4) detailing. Each step of this process requires models. The formation of this design process is an optimized result through trial-and-error by the industry over time. It suits the current practice and technologies, and also provides most efficient and effective professional services.

In the design step (1), empirical rules, guidelines, norms, and experience are models of some kind. This step of the design is more or less heuristic. Qualitative and/or quantitative models are used in design steps (2) and (3). Design step (4) is performed in accordance with the detailing requirements/guidelines specified in the design codes of practice, which are actually a special type of model that can be qualitative (e.g., diagrams) or quantitative (e.g., quantity limits).

Many different requirements and conditions need to be considered in engineering designs. Major considerations are usually handled through calculations and simulations using well-developed mathematical models, such as structural analysis and flexural and shear strength calculations. Due to the large number of practical considerations, and sometimes due to unavailability of mathematical models, a lot of design requirements and considerations are satisfied by meeting empirical rules, specifications, and guidelines. These requirements consist of a major part of design codes of practice, which can generally be categorized into design detailing. Therefore, design detailing not only refers to the arrangement of reinforcement (rebar detailing), but also includes the appropriate use of materials, selection of geometric shapes and dimensions of structural details, satisfying limit values of various designs, treatment of surfaces and interfaces, placement of nonstructural elements such as service penetrations and joints, construction specifications, etc.

Therefore, detailing is an important part of modeling in engineering designs. The detailing requirements are provided in design codes of practice, often without explanation. Each and every clause of the detailing requirements has a reason behind it, or was based on a lesson learned from past experience, and hence shall be checked and followed carefully in designs. For example, diagonal cracking causes the tension shift, which increases the stress of the longitudinal bars. This tension shift effect is not considered in the flexural design calculation (Section 10.2(5)). This problem is dealt with in design codes through the rebar detailing requirement that a rebar must be extended for a distance equal to the effective depth plus anchorage length after it is no longer required. Another example is that the spacing of stirrups is limited to half of the effective depth. This is to ensure that every potential diagonal crack is effectively crossed by at least one stirrup.

Detailing should also match the structural design model. The detailing for members designed based on beam action is very different from that designed based on the strut-and-tie model. When a joint is modeled as a hinge while in reality the reinforcement detail at the joint provides a significant moment resistance, it may cause a structural problem for an adjacent structure because no moment is transferred to that structure based on the design model. Therefore, inappropriate detailing may cause the failure of a structural member. Another important reason for good detailing is for construction convenience. A good explanation for various detailing of RC structures can be found in Park and Paulay (1975).

11.10 Size effect

Another important factor to be considered in modeling is the size effect. According to the classic strength of materials, if a large beam and a small beam are made of the same material, both will fail at the same stress. In reality, the larger beam will fail at a smaller stress than the smaller one. In this case, the two beams fail at different strengths of the same material. The difference in material properties due to the difference in size of the member is referred to as the size effect.

The material properties used in engineering designs are based on the classic continuum mechanics where the nominal material properties are independent of structural size. This is not the case in a real structure where people have long realized that the properties of material with the same microstructure, in particular concrete material, significantly change depending on size, shape, boundary conditions, etc., of the structure. This can be seen in Figs. 8.1 and 8.2 where the strengths of different concrete specimens are different depending on sizes and boundary conditions. Therefore, the size effect is not only related to size but also to other conditions, particularly the boundary conditions. The size effect of materials is attributed to two causes (Bazant 2005):

(1) Statistical—the chance of a critical flaw occurring in a failure zone increases with increasing volume of the structure. Consequently, the strength of materials decreases with increasing structural size.
(2) Energetic—due to larger energy release of cracking in a larger member.

Another easier way for engineers to understand the size effect is given in Section 8.2. As shown in Figs. 8.1 and 8.2, the load-carrying capacity of concrete material, i.e., concrete strength, is closely related to the shape, size, and boundary conditions of the test specimens. In fact, the specimen in Fig. 8.1 is a structure. People can only obtain the structural strength of a specimen from testing rather than the "material strength." The so-called size effect is the conversion of the strength of a material in one structure, e.g., standard test specimen, to that in another structure. This understanding is important for engineers, who need to realize that the same material may behave differently in different structural members or even in different parts within the same structural member. The size effect is not only related to sizes of a member but also other conditions, particularly the boundary conditions. The size effect is particularly significant for brittle and quasibrittle material like concrete.

In engineering, a conversion coefficient is used to relate the strength of concrete in one case to another. For example, the strength of concrete from standard cube tests should be multiplied by a factor of about 0.8 to get the strength of standard concrete cylinders. Conversion coefficients are also often used in different structural designs: e.g., the concrete strength for flexural design should be the cylinder compressive strength of concrete multiplied by a factor k_3 (Eq. 2.18). The size effect of concrete for compressive strength is not very significant in ordinary concrete structures, and hence the concrete strength from standard cylinder tests is used as the compressive strength of concrete in normal RC designs regardless of structural size. Although ignoring the size effect in the compressive strength of concrete is not strictly correct, it brings convenience to modeling and engineering designs. The error caused by the approximation is covered by a safety factor in designs.

However, there are cases where the size effect cannot be neglected; e.g., the shear strength of RC members has a significant size effect (Kani 1967; Bazant and Kim 1984). Therefore, the v_c models of the shear strength of RC members usually include a size effect, although different models of the size effect have been proposed (Hu 2017). The size effect is a difficult problem that has attracted continuous and extensive research interests from the research community (Bazant 1999), particularly the renowned research group led by Professor Z.P. Bazant.

The study of the size effect has been a long journey. It was first described in 1638 by Galileo Galilei in the famous book *Dialogues Concerning Two New Sciences* (Galilei 1914). In Proposition IX, Salviati states, "...I once drew the shape of a bone, lengthened three times, and then thickened in such proportion...it would be necessary to either find much harder and more resistant material to form his bones." To paraphrase Galileo, this simply means that weight increases with L^3 and load-carrying capacity (constant strength, variable cross section) increases with L^2. This is the square-cube law, exactly what the size effect law is all about: energy is released from a volume $\propto L^3$ but absorbed by a surface crack $\propto L^2$. According to this law, the load-carrying capacity decreases with increasing size of a body. Intensive search for the answer on the size effect began in the 1980s. But the problem is still far from being resolved, particularly from a practical application point of view (Saouma and Fava 2006).

11.11 Research approach

11.11.1 Scientific method

Aristotle (384–322 BCE) is recognized as the inventor of the scientific method because of his refined analysis of logical implications contained in demonstrative discourse, which goes well beyond natural logic and does not owe anything to those who philosophized before him (Pozzo 2004). The term "scientific method" came into popular use in the 20th century. The scientific method is an empirical method of acquiring knowledge, or the process by which science is carried out. While there have been continuous debates over what scientific methods are and whether there are any universal rules of science (Harrison et al. 2019), the scientific community generally accepts certain rules of method and attempts to justify them with a metamethodology (Nola and Sankey 2014).

Although the scientific method is not a single recipe, the hypothetico-deductive model formulated in the 20th century is a popular scientific method widely used in research work, as shown in Fig. 11.8. It involves (1) careful observation with rigorous skepticism about what is observed, avoiding cognitive assumptions; (2) formulation of a question from curiosity and perception; (3) hypothesis (suggested explanation of a phenomenon) via induction and abduction; (4) predicting consequences of the hypothesis by inductive or deductive reasoning; (5) testing to verify predictions; (6) analysis of the test results for validation of the hypothesis; and (7) conclusion if hypothesis is found true, or repeating the whole process if false.

The hypothetico-deductive model of scientific method is often presented as the applied mathematical approach that is widely used in engineering research. This approach involves the following steps:

1. Experimentally testing → observation and data
2. Mathematical modeling → equation
3. Solving equation → mathematical solution
4. Verifying solution with more data and observations → validated results
5. Drawing general conclusions and rules for practical use → recommendations

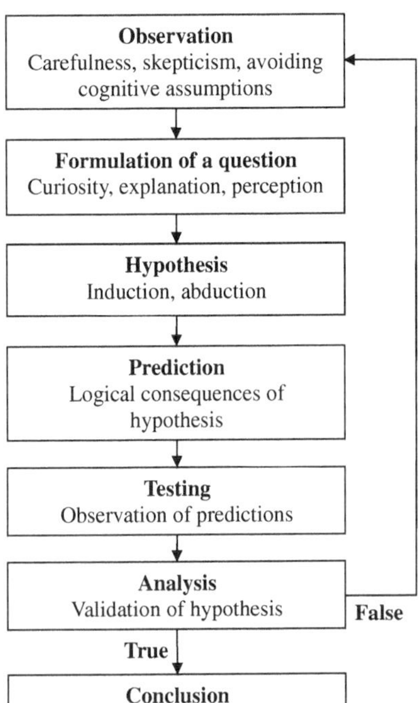

Fig. 11.8 The hypothetico-deductive model of scientific method.

11.11.2 The four dimensions of research approach

New models, or more generally new knowledge, can be acquired from research. Although there have been disputes in the literature on a general approach to more effective knowledge production, there are elements in the knowledge production process that play key roles. The four-dimensional (4D) research approach proposed and promoted by the author in recent years is introduced in this section. The knowledge production process involves four different dimensions (referred to as the 4D research approach hereafter), as shown in Fig. 11.9 and discussed in detail here.

(1) Observation

The starting point of a new family of knowledge usually originates from observations, from which data and facts are gained. For an observation to be a starting point of a new knowledge production process, the observer needs to understand the phenomenon, perceive its meaning, and form an appropriate question. Often, the same phenomenon has been seen by many people but only the right person can capture the phenomenon to form a question.

Penicillin was discovered from an accidental observation. Fleming noticed that the bacteria growth in an open Petri dish was inhibited. Contamination by a mold was apparent. Fleming asked himself: is it the mold that eliminated the bacteria? It was this simple question that led to one of mankind's greatest discoveries. Many people had seen this phenomenon, but only Fleming grasped its significance. Therefore, seeing $\neq$ discovering. Curiosity, insight, depth of understanding, and attention to details are essential traits of observers behind discovery.

(2) Longitudinal thinking

The second dimension of knowledge production is longitudinal thinking. Longitudinal thinking is normally referred to as vertical thinking in the literature. It is a type of approach to problem-solving that involves sequential or analytical logics, such as deductive reasoning. Longitudinal thinking consists of those that use more of a conscious approach via rational assessment and a sequential approach. It produces new knowledge or products that follow the same line of logic. Deduction is a typical longitudinal thinking. Thought experiment is very powerful in finding new knowledge. Good examples are Einstein's thought experiments that led to a number of the greatest discoveries of mankind. Other examples are the thought experiment in Section 2.8.2 that derived the plane section theorem and that in Section 8.4.1 that revealed the distribution of the confinement field and the key parameters governing the confinement effectiveness (Fig. 8.8).

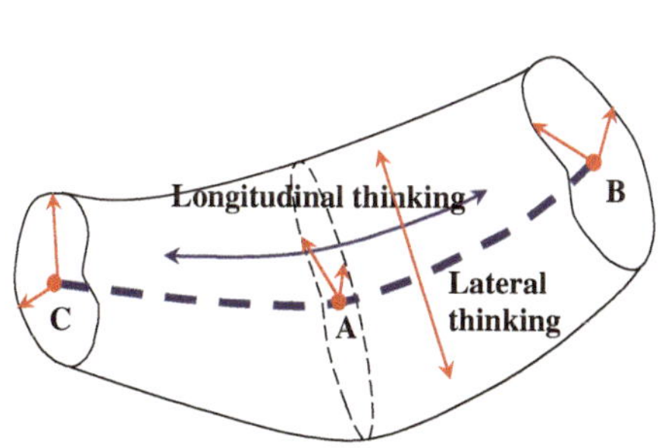

Fig. 11.9 The 4D development of new knowledge.

There are many examples of longitudinal thinking that led to important inventions in reinforced concrete. For example, reinforced concrete was invented based on the following logic: concrete is weak in tension → increasing the tensile resistance by adding reinforcing bars → the invention of RC. Following the same logic, fiber-reinforced concrete (FRC), engineered cementitious concrete (ECC) and many other concrete products with reinforcement have been developed. The logic can also go in the opposite direction through reverse thinking. For example, the problems of concrete structures are mainly caused by concrete cracking → concrete cracking is caused by tensile stress of concrete → axial compression force applied to a member reduces or eliminates the tensile stress → axial compression force can be applied to members by tensioning reinforcement → the invention of prestressed concrete. The invention of the compression yielding system (see Section 9.3) and the discovery of the flexural design theorems were also the results of longitudinal thinking.

These examples show that longitudinal thinking can create new knowledge and products by following a particular logic, as shown by the blue line in Fig. 11.9, starting from the initial observation point A (the first dimension of knowledge production) and moving forward along the line toward B (forward thinking), or in the opposite direction toward point C (reverse thinking). Every point along the line in both directions can be a discovery of new knowledge. Longitudinal thinking is a powerful tool for discovery of new knowledge, particularly for those completely new ideas that cannot be imagined or visualized due to preconceptions and human cognitive limitations.

(3) Lateral thinking

Lateral thinking is a manner of solving problems using an indirect reasoning such as by inspiration and analogy. The term lateral thinking was first used in 1967 by Edward de Bono in his book *The Use of Lateral Thinking* (De Bono 1967). People use lateral thinking to generate a new idea from a known idea that is not directly related. Although lateral thinking is considered pseudoscience by some, as it is not as logical as longitudinal thinking, it has a much greater power and much larger room for new ideas than longitudinal thinking. This is because the movement of ideas in lateral thinking can be in any of the lateral directions in the space, as shown by the red arrows in Fig. 11.9. Longitudinal thinking produces a string of new knowledge along one line following the same logic, while lateral thinking can create a volume of new knowledge as shown by the 3D body in Fig. 11.9.

The great potential of lateral thinking in creating new knowledge can be seen by using the same example of reinforced concrete discussed in longitudinal thinking. Inspired by the reinforcing principle, a whole family of materials unrelated to concrete can be conceived, e.g., fiber-reinforced polymer (FRP), various composite materials, bulletproof glass, multifunctional materials, various materials reinforced by carbon nanotubes (CNTs), and so on. The invention of the CY block was a result of lateral thinking inspired by the great deformability of sponge (Section 9.5). The invention of the HB bonding technology (Section 6.5.1) was also out of the experience of bonding paper onto walls by staples.

Lateral thinking can also be very helpful in disciplines that heavily reply on deductive reasoning (longitudinal thinking). Fermat's Last Theorem held the title of the "most difficult mathematical problem" in the *Guinness Book of World Records*, partly due to the fact that it had the greatest number of unsuccessful proofs. Fermat's Last Theorem belongs to the field of number theory. Andrew Wiles ascribed his insight to lateral thinking from his knowledge of elliptical equations, a different field from number theory, and finally succeeded in proving Fermat's Last Theorem.

(4) Existing knowledge

Existing knowledge or skill is needed for all the three dimensions previously discussed. It is an important factor affecting the new knowledge production process. Therefore, it is the 4th dimension of the knowledge production process. It is the basis and a tool to accomplish the other three dimensions. With better skill and adequate existing knowledge, the process in the previous three dimensions can be accomplished faster and smoother.

However, it is not the decisive factor of new knowledge production, because relevant knowledge can be picked up when needed. For example, Einstein faced difficulty in incorporating his ideas of general relativity into a general physical theory. After a period of hard work to no avail, it was suggested to him to use the mathematics of curvature (Riemann curvature tensor of non-Euclidean geometries) as a means to formulate the new theory. To learn the mathematics, he collaborated with his mathematician friend Marcel Grossmann and they produced the first draft of the general relativity theory together. To pick up relevant existing knowledge when needed is often the case in many research works. In the era of the knowledge explosion, it is impossible to learn too much existing knowledge to do research.

With more relevant existing knowledge and skills, people can expedite the processes of the other three dimensions. Therefore, the fourth dimension is related to the time required for knowledge production, and the other three dimensions determine whether new knowledge can be produced or not and the volume of the new knowledge.

Creativity or insight/perception is the critical trait in the first three dimensions of the research approach. Scientists are free to use whatever means to form hypotheses for a problem under study, including guesswork, imagination, illumination, perception, intuition, instincts, ideas from other fields, analogy, inductive and abductive reasoning, Bayesian inference, and so on. As noted by the famous scientist and philosopher William Whewell (1794–1866), "invention, sagacity, genius" are required at every step of the scientific method. The history of science is full of stories of scientists who made major discoveries from a "flash of inspiration." However, there are difficulties so far in making a formulaic statement of creativity in the literature.

11.11.3 Fostering creativity in research

Although people have diverged dramatically in the precise definition of creativity, a general agreement is that creativity involves the production of novel and useful products. The dominant factors affecting creativity are usually identified as "the four Ps"—process, product, person, and place (Rhodes 1961). Creativity as a subject of proper study began to emerge in the late 19th century, with the increased interest in individual differences.

There are many different theories and models of processes through which creativity occurs and explanations of the sources and methods of creativity (Runco and Pritzker 2020). How to foster creativity in work is an open-ended question. There has been an enormous increase in research on the topic throughout the world in many different disciplines in recent years, such as the arts and humanities, business, education, mental and physical health, neuroscience, psychology, sociology, and economics.

While the studies have found that creativity is closely related to personal traits, there are other factors that can promote better creativity. Creativity is commonly considered to be fostered by a supportive, nurturing, trustworthy environment conducive to self-actualization. Experiments suggest that pressure, anxiety, and depression, whether as a reflection of personality or induced, negatively impact creativity (Chirumbolo et al. 2004). Many investigations have found that good sleep, reading fiction, taking a temporary break, and other leisure activities can result in insight and aid the creativity process (Wagner et al. 2004; Djikic et al. 2013; Runco and Pritzker 2020). Creative work requires coactivation and communication between different regions of the brain that ordinarily are not strongly connected (Mayseless et al. 2015). It involves the forming of associative elements into new combinations in the brain. As a result, creative thinking is slower than other brain activity in which separate parts of the brain work independently. Indeed, quality work takes time and big ships turn slowly.

The fourth dimension, i.e., existing knowledge, is a tricky link in the production of new knowledge. Researchers without a sound training in basic skills and knowledge have difficulty in carrying out far-reaching research. However, learning all (or most) relevant existing knowledge, even in a small area, is impossible due to the vast knowledge available. In fact, learning too much knowledge is not only unnecessary but also harmful to creative thinking. On the one hand, having too much existing knowledge is unnecessary, because it can be acquired when needed, as long as the basic knowledge is well mastered and a solid training in fundamentals has been made by a researcher. On the other hand, possessing too much existing knowledge can constrain creative thinking. In 2018, neuroscience experiments showed that when the brain suppresses obvious or "known" solutions mediated by alpha oscillations in the right temporal lobe, the outcome is solutions that are more creative (Di Bernardi Luft et al. 2018).

Lateral thinking, which relies upon diverse knowledge outside one's own field, is often beneficial to knowledge production, because apparently different but somehow related knowledge is a catalyst for production of new knowledge. Therefore, diverse knowledge in different fields contributes to creative thinking. However, overspecialization can limit one's creativity. According to Damasio's theory of consciousness (Damasio 1999), emotions are associated with learned knowledge and carry what Damasio terms "somatic markers." These markers provide an automatic detection of scenarios they deem more relevant. As emotions filter intuitive thoughts, an individual with "too much" specialized knowledge is likely to have a bias toward knowledge within their own field, which hampers creative thinking. MIT's media lab discovered that its artificial intelligence responded differently to the same images based on different data it was trained on, demonstrating how even artificial intelligence is influenced by exposure to certain knowledge (Wakefield 2018). Hence, "too much" knowledge within one field carries its own emotions, which may limit creative thinking.

As discussed in Section 11.11.2, existing knowledge only affects the rate at which new knowledge is generated. Overwork in the fourth dimension will squeeze the other three dimensions that dominate the generation of new knowledge. Many scientists in modern society lament the fact that people did more and better things in the old days, while modern people cannot even do small things well. The answer to this question may be that modern society pays too much attention, effort, and resources to the fourth dimension, overlooking and squeezing the space of the other three dimensions. Existing knowledge and experience are indeed important to industrial activities and engineering practice, but too much relevant knowledge can be a hindrance to creative thinking in research work. Indeed, young people are often more creative although they are less knowledgeable and less experienced. A smart strategy for dealing with the negative impact of existing knowledge on creativity is to give yourself a chance to think independently before referring to the existing knowledge.

References

ACI-ASCE Committee 326. 1962. *Shear and diagonal tension: report of ACI-ASCE Committee 326*. American Society of Civil Engineers.

Aster, R., B. Borchers, and C. Thurber. 2018. *Parameter estimation and inverse problems*, 3rd Ed. Elsevier, ISBN: 9780128134238.

Bazant, Z. P. 1999. "Size effect on structural strength: a review." *Arch. Appl. Mech.* 69 (9), 703–725.

Bazant, Z. P. 2005. *Scaling of structural strength*, 2nd Ed. London: Elsevier.

Bazant, Z. P., and J. K. Kim. 1984. "Size effect in shear failure of longitudinally reinforced beams." *ACI J. Proc.* 81 (5), 456–468.

Billings, S. A. 2013. *Nonlinear system identification: NARMAX methods in the time, frequency, and spatio-temporal domains*. Wiley.

Buildings Department. 2013. *Code of practice for structural use of concrete*. Hong Kong SAR.

Charlesworth, M. J. 1956. "Aristotle's razor." *Philos. Stud.* 6, 105–112.

Chavent, G. 2010. *Nonlinear least squares for inverse problems*. Springer, ISBN: 978-90-481-2785-6.

Chen, W. F., and C. R. Scawthorn. 1970. "Limit analysis and limit equilibrium solutions in soil mechanics." *Soils Found.* 10 (3), 13–49.

Chirumbolo, A., S. Livi, L. Mannetti, A. Pierro, and A. W. Kruglanski. 2004. "Effects of need for closure on creativity in small group interactions." *Eur. J. Pers.* 18 (4), 265–278.

Collins, M. P., and D. Mitchell. 1980. "Shear and torsion design of prestressed and non-prestressed concrete beams." *J. Prestress. Concr. Inst.* 25 (5), 32–100.

Damasio, A. R. 1999. *The feeling of what happens: Body and emotion in the making of consciousness*. New York: Harcourt Brace.

De Bono, E. 1967. *The use of lateral thinking*. Intl Center for Creative Thinking.

Di Bernardi Luft, C., I. Zioga, N. M. Thompson, M. J. Banissy, and J. Bhattacharya. 2018. "Right temporal alpha oscillations as a neural mechanism for inhibiting obvious associations." *PNAS* 115 (52), E12144–E12152.

Djikic, M., K. Oatley, and M. C. Moldoveanu. 2013. "Opening the closed mind: the effect of exposure to literature on the need for closure." *Creat. Res. J.* 25 (2), 149–154.

Draper, N. R., and H. Smith. 1998. *Applied regression analysis*, 3rd Ed. John Wiley, ISBN: 978-0-471-17082-2.

Duthinh, D., and N. J. Carino. 1996. *Shear design of high-strength concrete beams: A review of the state-of-the-art*. US Department of Commerce, Technology Administration, National Institute of Standards and Technology, Office of Applied Economics, Building and Fire Research Laboratory.

Galilei, G. 1914. *Dialogues concerning two new sciences*. Dover.

Gurney, K. 1997. *An introduction to neural networks*. UCL Press, ISBN: 978-1-85728-673-1.

Hamilton, M. A. 1991. "Model validation: an annotated bibliography." *Commun. Stat.: Theory Methods* 20, 2207–2266.

Harrison, P., R. L. Numbers, and M. H. Shank, eds. 2019. *Wrestling with nature: From omens to science, Chapter 12: Scientific methods*, University of Chicago Press.

Hu, B. 2017. "New approach to shear resistance of reinforced concrete members." Ph.D. thesis, City University of Hong Kong.

Johansen, K. W. 1998. *Yield-line formulae for slabs*. London: CRC Press.

Kani, G. N. J. 1967. "How safe are our large reinforced concrete beams?." *ACI J. Proc.* 64 (3), 128–141.

Kuhn, T. 2004. "Stanford encyclopedia of philosophy." https://plato.stanford.edu/entries/thomas-kuhn/.

Mayseless, N., A. Eran, and S. G. Shamay-Tsoory. 2015. "Generating original ideas: the neural underpinning of originality." *Neuroimage* 116, 232–239.

Mitchell, D., and M. P. Collins. 1974. "Diagonal compression field theory-a rational model for structural concrete in pure torsion." *J. Proc.* 71 (8), 396–408.

Nelles, O. 2003. *Nonlinear system identification*. Berlin: Springer, ISBN: 3-540-67369-5.

Nielsen, M. P. 1984. *Limit analysis and concrete plasticity*. Englewood Cliffs, NJ: Prentice-Hall Inc.

Nola, R., and H. Sankey. 2014. *Theories of scientific method: An introduction*. London: Routledge.

Park, R., and T. Paulay. 1975. *Reinforced concrete structures*. New York: John Wiley & Sons.

Pozzo, R. 2004. *The impact of Aristotelianism on modern philosophy*. CUA Press, ISBN: 0-8132-1347-9.

Rhodes, M. 1961. "An analysis of creativity." *Phi Delta Kappan* 42 (7), 305–310.

Runco, M., and S. Pritzker. 2020. *Encyclopedia of creativity*. Academic Press.

Saouma, V., and G. Fava. 2006. "On fractals and size effects." *Int. J. Fract.* 137, 231–249.

Schaffer, J. 2015. "What not to multiply without necessity." *Australas. J. Philos.* 93 (4), 644–664.

Schmidhuber, J. 2015. "Deep learning in neural networks: an overview." *Neural Netw.* 61, 85–117.

Tarantola, A. 1987. *Inverse problem theory*, 1st Ed. Elsevier, ISBN: 9780444599674.

Tedeschi, L. O. 2006. "Assessment of the adequacy of mathematical models." *Agr. Syst.* 89 (2–3), 225–247.

Torsten, S., and P. Stoica. 1989. *System identification*. New York: Prentice Hall, ISBN: 978-0138812362.

Vrouwenvelder, A. C. W. M. 2003. *Plasticity: the plastic behaviour and the calculation of beams and frames subjected to bending*. Technische Universiteit Delft, Faculty of Civil Engineering and Geosciences, Mechanics, Materials and Constructions, Section of Structural Mechanics.

Wagner, U., S. Gais, H. Haider, R. Verleger, and J. Born. 2004. "Sleep inspires insight." *Nature* 427 (6972), 352–355.

Wakefield, J. 2018. *Are you scared yet? Meet Norman, the psychopathic AI*. BBC News. URL https://www.bbc.com/news/technology-44040008.

Wu, Y. F., L. He, and Z. Li. 2018. "Curve smoothing using a continuous function." *J. Zheijang Univ. Sci. A* 19 (4), 304–314.

Yan, X. 2009. *Linear regression analysis: Theory and computing*. World Scientific, ISBN: 9789812834119.

Index

Note: Page numbers followed by *f* indicate figures and *t* indicate tables.

Kolibri 360 GmbH
Ferdinand-Jühlke-Straße 7
99095 Erfurt, Deutschland
produktsicherheit@kolibri360.de